Design and Analysis of Experiments with SAS

CHAPMAN & HALL/CRC
Texts in Statistical Science Series

Series Editors
Bradley P. Carlin, *University of Minnesota, USA*
Julian J. Faraway, *University of Bath, UK*
Martin Tanner, *Northwestern University, USA*
Jim Zidek, *University of British Columbia, Canada*

Analysis of Failure and Survival Data
P. J. Smith

**The Analysis of Time Series —
An Introduction, Sixth Edition**
C. Chatfield

**Applied Bayesian Forecasting and Time Series
Analysis**
A. Pole, M. West and J. Harrison

**Applied Nonparametric Statistical Methods,
Fourth Edition**
P. Sprent and N.C. Smeeton

**Applied Statistics — Handbook of GENSTAT
Analysis**
E.J. Snell and H. Simpson

Applied Statistics — Principles and Examples
D.R. Cox and E.J. Snell

Applied Stochastic Modelling, Second Edition
B.J.T. Morgan

Bayesian Data Analysis, Second Edition
A. Gelman, J.B. Carlin, H.S. Stern
and D.B. Rubin

**Bayesian Methods for Data Analysis,
Third Edition**
B.P. Carlin and T.A. Louis

Beyond ANOVA — Basics of Applied Statistics
R.G. Miller, Jr.

**Computer-Aided Multivariate Analysis,
Fourth Edition**
A.A. Afifi and V.A. Clark

A Course in Categorical Data Analysis
T. Leonard

A Course in Large Sample Theory
T.S. Ferguson

Data Driven Statistical Methods
P. Sprent

Decision Analysis — A Bayesian Approach
J.Q. Smith

Design and Analysis of Experiment with SAS
J. Lawson

**Elementary Applications of Probability Theory,
Second Edition**
H.C. Tuckwell

Elements of Simulation
B.J.T. Morgan

**Epidemiology — Study Design and
Data Analysis, Second Edition**
M. Woodward

Essential Statistics, Fourth Edition
D.A.G. Rees

**Extending the Linear Model with R — Generalized
Linear, Mixed Effects and Nonparametric Regression
Models**
J.J. Faraway

A First Course in Linear Model Theory
N. Ravishanker and D.K. Dey

**Generalized Additive Models:
An Introduction with R**
S. Wood

**Interpreting Data — A First Course
in Statistics**
A.J.B. Anderson

**An Introduction to Generalized
Linear Models, Third Edition**
A.J. Dobson and A.G. Barnett

Introduction to Multivariate Analysis
C. Chatfield and A.J. Collins

**Introduction to Optimization Methods and Their
Applications in Statistics**
B.S. Everitt

Introduction to Probability with R
K. Baclawski

**Introduction to Randomized Controlled Clinical
Trials, Second Edition**
J.N.S. Matthews

**Introduction to Statistical Inference and Its
Applications with R**
M.W. Trosset

**Introduction to Statistical Methods for
Clinical Trials**
T.D. Cook and D.L. DeMets

Large Sample Methods in Statistics
P.K. Sen and J. da Motta Singer

Linear Models with R
J.J. Faraway

Logistic Regression Models
J.M. Hilbe

**Markov Chain Monte Carlo —
Stochastic Simulation for Bayesian Inference,
Second Edition**
D. Gamerman and H.F. Lopes

Mathematical Statistics
K. Knight

**Modeling and Analysis of Stochastic Systems,
Second Edition**
V.G. Kulkarni

Modelling Binary Data, Second Edition
D. Collett

**Modelling Survival Data in Medical Research,
Second Edition**
D. Collett

**Multivariate Analysis of Variance and Repeated
Measures — A Practical Approach for Behavioural
Scientists**
D.J. Hand and C.C. Taylor

Multivariate Statistics — A Practical Approach
B. Flury and H. Riedwyl

Pólya Urn Models
H. Mahmoud

Practical Data Analysis for Designed Experiments
B.S. Yandell

Practical Longitudinal Data Analysis
D.J. Hand and M. Crowder

Practical Statistics for Medical Research
D.G. Altman

A Primer on Linear Models
J.F. Monahan

Probability — Methods and Measurement
A. O'Hagan

**Problem Solving — A Statistician's Guide,
Second Edition**
C. Chatfield

**Randomization, Bootstrap and Monte Carlo
Methods in Biology, Third Edition**
B.F.J. Manly

Readings in Decision Analysis
S. French

Sampling Methodologies with Applications
P.S.R.S. Rao

Statistical Analysis of Reliability Data
M.J. Crowder, A.C. Kimber,
T.J. Sweeting, and R.L. Smith

Statistical Methods for Spatial Data Analysis
O. Schabenberger and C.A. Gotway

Statistical Methods for SPC and TQM
D. Bissell

**Statistical Methods in Agriculture and Experimental
Biology, Second Edition**
R. Mead, R.N. Curnow, and A.M. Hasted

**Statistical Process Control — Theory and Practice,
Third Edition**
G.B. Wetherill and D.W. Brown

Statistical Theory, Fourth Edition
B.W. Lindgren

Statistics for Accountants
S. Letchford

Statistics for Epidemiology
N.P. Jewell

**Statistics for Technology — A Course in Applied
Statistics, Third Edition**
C. Chatfield

Statistics in Engineering — A Practical Approach
A.V. Metcalfe

**Statistics in Research and Development,
Second Edition**
R. Caulcutt

**Stochastic Processes: An Introduction,
Second Edition**
P.W. Jones and P. Smith

**Survival Analysis Using S — Analysis of
Time-to-Event Data**
M. Tableman and J.S. Kim

The Theory of Linear Models
B. Jørgensen

Time Series Analysis
H. Madsen

Texts in Statistical Science

Design and Analysis of Experiments with SAS

John Lawson

Brigham Young University
Provo, Utah, U.S.A.

CRC Press
Taylor & Francis Group
Boca Raton London New York

CRC Press is an imprint of the
Taylor & Francis Group an **informa** business

A CHAPMAN & HALL BOOK

CRC Press
Taylor & Francis Group
6000 Broken Sound Parkway NW, Suite 300
Boca Raton, FL 33487-2742

© 2010 by Taylor & Francis Group, LLC
CRC Press is an imprint of Taylor & Francis Group, an Informa business

No claim to original U.S. Government works

International Standard Book Number: 978-1-4200-6060-7 (Hardback)

This book contains information obtained from authentic and highly regarded sources. Reasonable efforts have been made to publish reliable data and information, but the author and publisher cannot assume responsibility for the validity of all materials or the consequences of their use. The authors and publishers have attempted to trace the copyright holders of all material reproduced in this publication and apologize to copyright holders if permission to publish in this form has not been obtained. If any copyright material has not been acknowledged please write and let us know so we may rectify in any future reprint.

Except as permitted under U.S. Copyright Law, no part of this book may be reprinted, reproduced, transmitted, or utilized in any form by any electronic, mechanical, or other means, now known or hereafter invented, including photocopying, microfilming, and recording, or in any information storage or retrieval system, without written permission from the publishers.

For permission to photocopy or use material electronically from this work, please access www.copyright.com (http://www.copyright.com/) or contact the Copyright Clearance Center, Inc. (CCC), 222 Rosewood Drive, Danvers, MA 01923, 978-750-8400. CCC is a not-for-profit organization that provides licenses and registration for a variety of users. For organizations that have been granted a photocopy license by the CCC, a separate system of payment has been arranged.

Trademark Notice: Product or corporate names may be trademarks or registered trademarks, and are used only for identification and explanation without intent to infringe.

Library of Congress Cataloging-in-Publication Data

Lawson, John, 1947-
  Design and analysis of experiments with SAS / John Lawson.
    p. cm. -- (Chapman & Hall/CRC texts in statistical science series)
  Includes bibliographical references and index.
  ISBN 978-1-4200-6060-7 (hard back : alk. paper)
  1. Experimental design. 2. SAS (Computer file) I. Title. II. Series.

QA279.L36 2010
519.5'7--dc22   2009052546

Visit the Taylor & Francis Web site at
http://www.taylorandfrancis.com

and the CRC Press Web site at
http://www.crcpress.com

Contents

Preface		xi
1	**Introduction**	**1**
	1.1 Statistics and Data Collection	1
	1.2 Beginnings of Statistically Planned Experiments	2
	1.3 Definitions and Preliminaries	2
	1.4 Purposes of Experimental Design	5
	1.5 Types of Experimental Designs	6
	1.6 Planning Experiments	7
	1.7 Performing the Experiments	9
	1.8 Use of SAS Software	11
	1.9 Review of Important Concepts	12
	1.10 Exercises	14
2	**Completely Randomized Designs with One Factor**	**15**
	2.1 Introduction	15
	2.2 Replication and Randomization	15
	2.3 A Historical Example	18
	2.4 Linear Model for CRD	19
	2.5 Verifying Assumptions of the Linear Model	27
	2.6 Analysis Strategies When Assumptions Are Violated	30
	2.7 Determining the Number of Replicates	37
	2.8 Comparison of Treatments after the F-Test	41
	2.9 Review of Important Concepts	48
	2.10 Exercises	50
3	**Factorial Designs**	**53**
	3.1 Introduction	53
	3.2 Classical One at a Time versus Factorial Plans	53
	3.3 Interpreting Interactions	55
	3.4 Creating a Two-Factor Factorial Plan in SAS	58
	3.5 Analysis of a Two-Factor Factorial in SAS	60
	3.6 Factorial Designs with Multiple Factors - CRFD	80
	3.7 Two-Level Factorials	86
	3.8 Verifying Assumptions of the Model	102
	3.9 Review of Important Concepts	106

CONTENTS

3.10	Exercises	108
3.11	Appendix–SAS Macro for Tukey's Single df Test	112

4 Randomized Block Designs — 115
4.1	Introduction	115
4.2	Creating an RCB in SAS	116
4.3	Model for RCB	119
4.4	An Example of an RCB	121
4.5	Determining the Number of Blocks	124
4.6	Factorial Designs in Blocks	125
4.7	Generalized Complete Block Design	128
4.8	Two Block Factors LSD	131
4.9	Review of Important Concepts	138
4.10	Exercises	140
4.11	Appendix–Data from Golf Experiment	145

5 Designs to Study Variances — 147
5.1	Introduction	147
5.2	Random Factors and Random Sampling Experiments	148
5.3	One-Factor Sampling Designs	150
5.4	Estimating Variance Components	151
5.5	Two-Factor Sampling Designs	161
5.6	Nested Sampling Experiments (NSE)	170
5.7	Staggered Nested Designs	173
5.8	Designs with Fixed and Random Factors	179
5.9	Graphical Methods to Check Model Assumptions	186
5.10	Review of Important Concepts	194
5.11	Exercises	196
5.12	Appendix	198

6 Fractional Factorial Designs — 199
6.1	Introduction	199
6.2	Half-Fractions of 2^k Designs	200
6.3	Quarter and Higher Fractions of 2^k Designs	209
6.4	Criteria for Choosing Generators for 2^{k-p} Designs	211
6.5	Augmenting Fractional Factorials	222
6.6	Plackett-Burman (PB) Screening Designs	232
6.7	Mixed Level Factorials and Orthogonal Arrays (OA)	238
6.8	Review of Important Concepts	246
6.9	Exercises	248

7 Incomplete and Confounded Block Designs — 255
7.1	Introduction	255
7.2	Balanced Incomplete Block (BIB) Designs	256
7.3	Analysis of Incomplete Block Designs	259

7.4	PBIB-BTIB Designs	261
7.5	Youden Square Designs (YSD)	265
7.6	Confounded 2^k and 2^{k-p} Designs	266
7.7	Confounding 3 Level and p Level Factorial Designs	280
7.8	Blocking Mixed-Level Factorials and OAs	283
7.9	Partially Confounded Blocked Factorial (PCBF)	290
7.10	Review of Important Concepts	295
7.11	Exercises	298

8 Split-Plot Designs — 301

8.1	Introduction	301
8.2	Split-Plot Experiments with CRD in Whole Plots CRSP	302
8.3	RCB in Whole Plots RBSP	309
8.4	Analysis Unreplicated 2^k Split-Plot Designs	318
8.5	2^{k-p} Fractional Factorials in Split Plots (FFSP)	324
8.6	Sample Size and Power Issues for Split-Plot Designs	338
8.7	Review of Important Concepts	339
8.8	Exercises	341

9 Crossover and Repeated Measures Designs — 347

9.1	Introduction	347
9.2	Crossover Designs (COD)	347
9.3	Simple AB, BA Crossover Designs for Two Treatments	348
9.4	Crossover Designs for Multiple Treatments	358
9.5	Repeated Measures Designs	364
9.6	Univariate Analysis of Repeated Measures Design	365
9.7	Review of Important Concepts	374
9.8	Exercises	376

10 Response Surface Designs — 381

10.1	Introduction	381
10.2	Fundamentals of Response Surface Methodology	381
10.3	Standard Designs for Second Order Models	385
10.4	Creating Standard Designs in SAS	392
10.5	Non-Standard Response Surface Designs	395
10.6	Fitting the Response Surface Model with SAS	403
10.7	Determining Optimum Operating Conditions	410
10.8	Blocked Response Surface (BRS) Designs	421
10.9	Response Surface Split-Plot (RSSP) Designs	424
10.10	Review of Important Concepts	435
10.11	Exercises	437

11 Mixture Experiments — 443

11.1	Introduction	443
11.2	Models and Designs for Mixture Experiments	445

11.3	Creating Mixture Designs in SAS	452
11.4	Analysis of Mixture Experiment	454
11.5	Constrained Mixture Experiments	461
11.6	Blocking Mixture Experiments	470
11.7	Mixture Experiments with Process Variables	475
11.8	Mixture Experiments in Split-Plot Arrangements	484
11.9	Review of Important Concepts	487
11.10	Exercises	489
11.11	Appendix–Example of Fitting Independent Factors	498

12 Robust Parameter Design Experiments — 501

12.1	Introduction	501
12.2	Noise-Sources of Functional Variation	502
12.3	Product Array Parameter Design Experiments	504
12.4	Analysis of Product Array Experiments	512
12.5	Single Array Parameter Design Experiments	529
12.6	Joint Modeling of Mean and Dispersion Effects	538
12.7	Review of Important Concepts	545
12.8	Exercises	547

13 Experimental Strategies for Increasing Knowledge — 555

13.1	Introduction	555
13.2	Sequential Experimentation	555
13.3	One Step Screening and Optimization	559
13.4	Evolutionary Operation	560
13.5	Concluding Remarks	562

Bibliography — 565

Index — 579

Preface

After studying experimental design a researcher or statistician should be able to: (1) choose an experimental design that is appropriate for the research problem at hand; (2) construct the design (including performing proper randomization and determining the required number of replicates); (3) execute the plan to collect the data (or advise a colleague to do it); (4) determine the model appropriate for the data; (5) fit the model to the data; and (6) interpret the data and present the results in a meaningful way to answer the research question. The purpose of this book is to focus on connecting the objectives of research to the type of experimental design required, describing the actual process of creating the design and collecting the data, showing how to perform the proper analysis of the data, and illustrating the interpretation of results. Exposition on the mechanics of computation is minimized by relying on a statistical software package.

With the availability of modern statistical computing packages, the analysis of data has become much easier and is well covered in statistical methods books. There is no longer a need to show all the computational formulas that were necessary before the advent of modern computing, in a book on the design and analysis of experiments. However, there is a need for careful explanation of how to get the proper analysis from a computer package. The default analysis performed by most statistical software assumes the data have come from a completely randomized design. In practice, this is often a false assumption. This book emphasizes the connection between the experimental units, and the way treatments are randomized to experimental units, and the proper error term for an analysis of the data.

The SAS system for statistical analysis is used throughout the book to illustrate both construction of experimental designs and analysis of data. This software was chosen to be illustrated because it has extensive capabilities in both creating designs and analyzing data, the command language has been stable for over thirty years, and it is widely used in industry. SAS version 9.2 has been used in the text, and all the SAS code for examples in the book is available at http://lawson.mooo.com. The ods graphics used in the book require version 9.2 or later. In earlier versions of SAS similar graphs can be created with the legacy SAS/GRAPH routines and the code to do this is also available on the Web site. Examples of SAS data step programming and IML are presented, and procedures from SAS Stat, SAS QC, and SAS OR are illustrated.

With fewer pages devoted to computational formulas, I have attempted to spend more time discussing the following: (1) how the objectives of a research project lead to the choice of an appropriate design, (2) practical aspects of creating a design, or list of experiments to be performed, (3) practical aspects of performing experiments, and (4) interpretation of the results of a computer analysis of the data. Items (1)-(3) can best be taught by giving many examples of experiments and exercises that actually require readers to perform their own experiments.

This book attempts to give uniform coverage to experimental designs and design concepts that are most commonly used in practice, rather than focusing on specialized areas. The selection of topics is based on my own experience working in the pharmaceutical industry, and in research and development (R&D) and manufacturing in agricultural and industrial chemicals, and machinery industries. At the end of each chapter a diagram is presented to help identify where the various designs should be used. Examples in the book come from a variety of application areas. Emphasis is placed on how the sample size, the assignment of experimental units to combinations of treatment factor levels (error control), and the selection of treatment factor combinations (treatment design) will affect the resulting variance and bias of estimates and the validity of conclusions.

Intended audience This book was written for first-year graduate students in statistics or advanced undergraduates who intend to work in an area where they will use experimental designs. To be fully understood, a student using this book should have had previous courses in calculus, introductory statistics, basic statistical theory, applied linear models such as Kutner *et al.* (2004) and Faraway (2004), and some familiarity with SAS. Matrix notation for analysis of linear models is used throughout the book, and students should be familiar with matrix operations at least to the degree illustrated in chapter 5 of Kutner *et al.* (2004).

However, for students from applied sciences or engineering who do not have all these prerequisites, there is still much to be gained from this book. There are many examples of SAS code to create and analyze experiments as well plentiful examples of (1) diagnosing the experimental environment to choose the correct design, and (2) interpreting and presenting results of analysis. One with a basic understanding of SAS should be able to follow these examples and modify them to complete the exercises in the book and solve problems in their own research, without needing to understand the detailed theoretical justification for each procedure.

For instructors This book can be used for a one-semester or two-quarter course in experimental design. There is too much material for a one-semester course, unless the students have had all the prerequisites mentioned above. The first four chapters in the book cover the classical ideas in experimental design, and should be covered in any course for students without a prior background in designed experiments. Later chapters start with basics, but proceed to the latest research published on particular topics, and they include code to implement all of these ideas. An instructor can pick and choose from

PREFACE

these remaining topics, although if there is time to cover the whole book, I would recommend presenting the topics in order.

Some instructors who do not intend to cover the entire book might consider covering factorial experiments in Chapter 3, fractional factorials in Chapter 6, and response surface methods in Chapter 9, following the pattern established by the DuPont Strategies of Experimentation Short Courses that were developed in the 1970s. I chose the ordering of chapters in the book so that variance component designs in Chapter 5, would be presented before describing split plot experiments that are so commonplace in practice. I did this because I feel it is important to understand random factors before studying designs where there is more than one error term.

Acknowledgments This book is the culmination of many years of thought prompted by consulting and teaching. I would be remiss if I did not thank Melvin Carter, my advisor at Brigham Young University (BYU) who introduced me to the computer analysis of experimental data over forty years ago, and whose enthusiasm about the subject of designed experiments inspired my lifelong interest in this area. I would also like to thank John Erjavec, my boss and mentor at FMC Corp., for introducing me to the ideas of Box, Hunter and Hunter long before their original book *Statistics for Experimenters* was published. I also thank the many consulting clients over the years who have challenged me with interesting problems, and the many students who have asked me to explain things more clearly. Special thanks to my former students Willis Jensen at Gore and Michael Joner at Procter and Gamble for their careful review and comments on my manuscript, and Seyed Mottaghinejad for providing a solutions manual and finding many typos and unclear points in the text. Finally, I thank my wife Francesca for her never-ending support and encouragement during the writing of this book.

John Lawson
Department of Statistics
Brigham Young University

CHAPTER 1

Introduction

1.1 Statistics and Data Collection

Statistics is defined as the science of collecting, analyzing and drawing conclusions from data. Data is usually collected through sampling surveys, observational studies, or experiments.

Sampling surveys are normally used when the purpose of data collection is to estimate some property of a finite population without conducting a complete census of every item in the population. For example, if there were interest in finding the proportion of registered voters in a particular precinct that favor a proposal, this proportion could be estimated by polling a random sample of voters rather than questioning every registered voter in the precinct.

Observational studies and experiments, on the other hand, are normally used to determine the relationship between two or more measured quantities in a conceptual population. A conceptual population, unlike a finite population, may only exist in our minds. For example, if there were interest in the relationship between future greenhouse gas emissions and future average global temperature, the population, unlike registered voters in a precinct, cannot be sampled from because it does not yet exist.

To paraphrase the late W. Edwards Deming, the value of statistical methods is to make predictions which can form the basis for action. In order to make accurate future predictions of what will happen when the environment is controlled, cause and effect relationships must be assumed. For example, to predict future average global temperature given that greenhouse gas emissions will be controlled at a certain level, we must assume that the relationship between greenhouse gas emissions and global temperature is cause and effect. Herein lies the main difference in observational studies and experiments. In an observational study, data is observed in its natural environment, but in an experiment the environment is controlled. In observational studies it cannot be proven that the relationships detected are cause and effect. Correlations may be found between two observed variables because they are both affected by changes in a third variable that was not observed or recorded, and any future predictions made based on the relationships found in an observational study must assume the same interrelationships among variables that existed in the past will exist in the future. In an experiment, on the other hand, some variables are purposely changed while others are held constant. In that way the effect that is caused by the change in the purposely varied variable can be directly observed, and predictions can be made about the result of future changes to the purposely varied variable.

1.2 Beginnings of Statistically Planned Experiments

There are many purposes for experimentation. Some examples include: determining the cause for variation in measured responses observed in the past; finding conditions that give rise to the maximum or minimum response; comparing the response between different settings of controllable variables; and obtaining a mathematical model to predict future response values.

Presently planned experiments are used in many different fields of application such as engineering design, quality improvement, industrial research and manufacturing, basic research in physical and biological science, research in social sciences, psychology, business management and marketing research, and many more. However, the roots of modern experimental design methods stem from R. A. Fisher's work in agricultural experimentation at the Rothamsted Experimental Station near Harpenden, England.

Fisher was a gifted mathematician whose first paper as an undergraduate at Cambridge University introduced the theory of likelihood. He was later offered a position at University College, but turned it down to join the staff at Rothamsted in 1919. There, inspired by daily contact with agricultural research, he not only contributed to experimental studies in areas such as crop yields, field trials, and genetics, but also developed theoretical statistics at an astonishing rate. He also came up with the ideas for planning and analysis of experiments that have been used as the basis for valid inference and prediction in various fields of application to this day. Fisher (1926) first published his ideas on planning experiments in his paper "The arrangement of field experiments"; nine years later he published the first edition of his book *The Design of Experiments*, Fisher (1935).

The challenges that Fisher faced were the large amount of variation in agricultural and biological experiments that often confused the results, and the fact that experiments were time consuming and costly to carry out. This motivated him to find experimental techniques that could:

- eliminate as much of the natural variation as possible
- prevent unremoved variation from confusing or biasing the effects being tested
- detect cause and effect with the minimal amount of experimental effort necessary.

1.3 Definitions and Preliminaries

Before initiating an extended discussion of experimental designs and the planning of experiments, I will begin by defining the terms that will be used frequently.

DEFINITIONS AND PRELIMINARIES

- *Experiment* (also called a *Run*) is an action where the experimenter changes at least one of the variables being studied and then observes the effect of his or her actions(s). Note the passive collection of observational data is not experimentation.
- *Experimental Unit* is the item under study upon which something is changed. This could be raw materials, human subjects, or just a point in time.
- *Sub-sample, sub-unit or observational unit* When the experimental unit is split, after the action has been taken upon it, this is called a sub-sample or sub-unit. Sometimes it is only possible to measure a characteristic separately for each sub-unit; for that reason they are often called observational units. Measurements on sub-samples, or sub-units of the same experimental unit are usually correlated and should be averaged before analysis of data rather than being treated as independent outcomes. When sub-units can be considered independent and there is interest in determining the variance in sub-sample measurements, while not confusing the F-tests on the treatment factors, the mixed model described in Section 5.8 should be used instead of simply averaging the sub-samples.
- *Independent Variable* (*Factor* or *Treatment Factor*) is one of the variables under study that is being controlled at or near some target value, or *level*, during any given experiment. The level is being changed in some systematic way from run to run in order to determine what effect it has on the response(s).
- *Background Variable* (also called *Lurking variable*) is a variable that the experimenter is unaware of or cannot control, and which could have an effect on the outcome of the experiment. In a well-planned experimental design, the effect of these lurking variables should balance out so as to not alter the conclusion of a study.
- *Dependent Variable* (or the *Response* denoted by Y) is the characteristic of the experimental unit that is measured after each experiment or run. The magnitude of the response depends upon the settings of the independent variables or factors and lurking variables.
- *Effect* is the change in the response that is caused by a change in a factor or independent variable. After the runs in an experimental design are conducted, the effect can be estimated by calculating it from the observed response data. This estimate is called the *calculated effect*. Before the experiments are ever conducted, the researcher may know how large the effect should be to have practical importance. This is called a *practical effect* or the *size of a practical effect*.
- *Replicate* runs are two or more experiments conducted with the same settings of the factors or independent variables, but using different experimental units. The measured dependent variable may differ in replicate runs due to changes in lurking variables and inherent differences in experimental units.

- *Duplicates* refer to duplicate measurements of the same experimental unit from one run or experiment. The measured dependent variable may vary among duplicates due to measurement error, but in the analysis of data these duplicate measurements should be averaged and not treated as separate responses.

- *Experimental Design* is a collection of experiments or runs that is planned in advance of the actual execution. The particular runs selected in an experimental design will depend upon the purpose of the design.

- *Confounded Factors* arise when each change an experimenter makes for one factor, between runs, is coupled with an identical change to another factor. In this situation it is impossible to determine which factor causes any observed changes in the response or dependent variable.

- *Biased Factor* results when an experimenter makes changes to an independent variable at the precise time when changes in background or lurking variables occur. When a factor is biased it is impossible to determine if the resulting changes to the response were caused by changes in the factor or by changes in other background or lurking variables.

- *Experimental Error* is the difference between the observed response for a particular experiment and the long run average of all experiments conducted at the same settings of the independent variables or factors. The fact that it is called "error" should not lead one to assume that it is a mistake or blunder. Experimental errors are not all equal to zero because background or lurking variables cause them to change from run to run. Experimental errors can be broadly classified into two types: bias error and random error. Bias error tends to remain constant or change in a consistent pattern over the runs in an experimental design, while random error changes from one experiment to another in an unpredictable manner and average to be zero. The variance of random experimental errors can be obtained by including replicate runs in an experimental design.

With these definitions in mind, the difference between observational studies and experiments can be explained more clearly. In an observational study, variables (both independent and dependent) are observed without any attempt to change or control the value of the independent factors. Therefore any observed changes in the response, or dependent variable, cannot necessarily be attributed to observed changes in the independent variables because background or lurking variables might be the cause. In an experiment, however, the independent variables are purposely varied and the runs are conducted in a way to balance out the effect of any background variables that change. In this way the average change in the response can be attributed to the changes made in the independent variables.

1.4 Purposes of Experimental Design

The use of experimental designs is a prescription for successful application of the scientific method. The scientific method consists of iterative application of the following steps: (1) observing of the state of nature, (2) conjecturing or hypothesizing the mechanism for what has been observed, then (3) collecting data, and (4) analyzing the data to confirm or reject the conjecture. Statistical experimental designs provide a plan for collecting data in a way that they can be analyzed statistically to corroborate the conjecture in question. When an experimental design is used, the conjecture must be stated clearly and a list of experiments proposed in advance to provide the data to test the hypothesis. This is an organized approach which helps to avoid false starts and incomplete answers to research questions.

Another advantage to using the experimental design approach is ability to avoid confounding factor effects. When the research hypothesis is not clearly stated and a plan is not constructed to investigate it, researchers tend toward a trial and error approach wherein many variables are simultaneously changed in an attempt to achieve some goal. When this is the approach, the goal may sometimes be achieved, but it cannot be repeated because it is not known what changes actually caused the improvement.

One of Fisher's early contributions to the planning of experiments was popularizing a technique called randomization which helps to avoid confusion or biases due to changes in background or lurking variables. As an example of what we mean by bias is "The Biggest Health Experiment Ever," Meier (1972), wherein a trial of a polio vaccine was tested on over 1.8 million children. An initial plan was proposed to offer vaccination to all children in the second grade in participating schools, and to follow the polio experience of first through third graders. The first and third grade group would serve as a "control" group. This plan was rejected, however, because doctors would have been aware that the vaccine was only offered to second graders. There are vagaries in the diagnosis of the majority of polio cases, and the polio symptoms of fever and weakness are common to many other illnesses. A doctor's diagnosis could be unduly influenced by his knowledge of whether or not a patient had been vaccinated. In this plan the factor purposely varied, vaccinated or not, was biased by the lurking variable of doctors' knowledge of the treatment.

When conducting physical experiments, the response will normally vary over replicate runs due solely to the fact that the experimental units are different. This is what we defined to be experimental error in the last section. One of the main purposes for experimental designs is to minimize the effect of experimental error. Aspects of designs that do this, such as randomization, replication and blocking, are called methods of *error control*. Statistical methods are used to judge the average effect of varying experimental factors against the possibility that they may be due totally to experimental error. Another purpose for experimental designs is to accentuate the factor effects (or signal). Aspects of designs that do this, such as choice of the number and spacing of factor levels and factorial plans, are called methods of *treatment design*. How this is done will be explained in the following chapters.

1.5 Types of Experimental Designs

There are many types of experimental designs. The appropriate one to use depends upon the objectives of the experimentation. We can classify objectives into two main categories. The first category is to study the sources of variability, and the second is to establish cause and effect relationships. When variability is observed in a measured variable, one objective of experimentation might be to determine the cause of that variation. But before cause and effect relationships can be studied, a list of independent variables must be determined. By understanding the source of variability, researchers are often led to hypothesize what independent variables or factors to study. Thus experiments to study the source of variability are often a starting point for many research programs. The type of experimental design used to classify sources of variation will depend on the number of sources under study. These alternatives will be presented in Chapter 5.

The appropriate experimental design that should be used to study cause and effect relationships will depend on a number of things. Throughout the book the various designs are described in relation to the purpose for experimentation, the type and number of treatment factors, the degree of homogeneity of experimental units, the ease of randomization, and the ability to block experimental units into more homogeneous groups. After all the designs are presented, Chapter 13 describes how they can be used in sequential experimentation strategies where knowledge is increased through different stages of experimentation. Initial stages involve discovering what the important treatment factors are. Later, the effects of changing treatment factors are quantified, and in final stages, optimal operating conditions can be determined. Different types of experimental designs are appropriate for each of these phases.

Screening experiments are used when the researcher has little knowledge of the cause and effect relationships, and many potential independent variables are under study. This type of experimentation is usually conducted early in a research program to identify the important factors. This is a critical step, and if it is skipped, the later stages of many research programs run amuck because the important variables are not being controlled or recorded.

After identifying the most important factors in a screening stage, the researcher's next objective would be to choose between constrained optimization or unconstrained optimization (see Lawson, 2003). In constrained optimization there are usually six or fewer factors under study and the purpose is to quantify the effects of the factors, interaction or joint effects of factors, and to identify optimum conditions among the factor combinations actually tested.

When only a few quantitative factors are under study and curvilinear relationships with the response are possible, it may be possible to identify

PLANNING EXPERIMENTS

improved operating conditions by interpolating within the factor levels actually tested. If this is the goal, the objective of experimentation is called unconstrained optimization. With an unconstrained optimization objective, the researcher is normally trying to map the relationship between one or more responses and five or fewer quantitative factors.

Specific experimental design plans for each of the stages of experimentation will be presented as we progress through the book.

Figure 1.1 shows the relationship between the objectives of experimentation, the design of the experiment, and the conclusions that can be drawn. The objective of a research program dictates which type of experimental design should be utilized. The experimental design plan in turn specifies how the data should be collected and what mathematical model should be fit in order to analyze and interpret the data. Finally the type of data and the mathematical model will determine what possible conclusions can be drawn from the experiment. These steps are inseparable and dependent upon each other. Many mistakes are made in research by trying to dissever these steps. An appropriate analysis of data cannot be completed without knowledge of what experimental design was used and how the data was collected, and conclusions are not reliable if they are not justified by the proper modeling and analysis of the data.

Figure 1.1 *Objectives, Design and Conclusions from Experimentation*

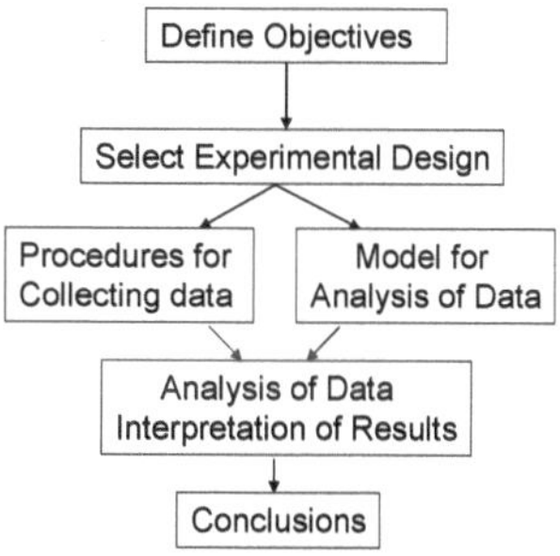

```
          Define Objectives
                 ↓
        Select Experimental Design
            ↙         ↘
  Procedures for    Model for
  Collecting data   Analysis of Data
            ↘         ↙
         Analysis of Data
      Interpretation of Results
                 ↓
            Conclusions
```

1.6 Planning Experiments

An effective experimental design plan should include the following items: (1) a clear description of the objectives, (2) an appropriate design plan that guarantees unconfounded factor effects and factor effects that are free of bias, (3) a provision for collecting data that will allow estimation of the variance of the experimental error, and (4) a stipulation to collect enough data to satisfy the objectives. Bisgaard (1999) recommends a formal proposal to ensure that a plan includes all of these elements. The proposal should include a checklist for planning the experiments. Below is a checklist that is similar to Bisgaard's.

Examples of some of the steps from this checklist will be illustrated in discussing a simple experiment in the next section.

1. *Define Objectives.* Define the objectives of the study. First, this statement should answer the question of why is the experiment to be performed. Second, determine if the experiment is conducted to classify sources of variability or if its purpose is to study cause and effect relationships. If it is the latter, determine if it is a screening or optimization experiment. For studies of cause and effect relationships, decide how large an effect should be in order to be meaningful to detect.

2. *Identify Experimental Units.* Declare the item upon which something will be changed. Is it an animal or human subject, raw material for some processing operation, or simply the conditions that exist at a point in time or *trial*? Identifying the experimental units will help in understanding the experimental error and variance of experimental error.

3. *Define a Meaningful and Measurable Response or Dependent Variable.* Define what characteristic of the experimental units can be measured and recorded after each run. This characteristic should best represent the expected differences to be caused by changes in the factors.

4. *List the Independent and Lurking Variables.* Declare which independent variables you wish to study. Ishikawa Cause and Effect Diagrams (see SAS Institute, 2004b) are often useful at this step to help organize variables thought to affect the experimental outcome. Be sure that the independent variables chosen to study can be controlled during a single run, and varied from run to run. If there is interest in a variable, but it cannot be controlled or varied, it cannot be included as a factor. Variables that are hypothesized to affect the response, but cannot be controlled, are lurking variables. The proper experimental design plan should prevent uncontrollable changes in these variables from biasing factor effects under study.

5. *Run Pilot Tests.* Make some pilot tests to be sure you can control and vary the factors that have been selected, that the response can be measured, and that the replicate measurements of the same or similar experimental units are consistent. Inability to measure the response accurately or to control the factor levels are the main reasons that experiments fail to produce desired results. If the pilot tests fail, go back to steps 2, 3 and 4. If these tests are successful, measurements of the response for a few replicate tests with the same levels of the factors under study will produce data that can be used to get a preliminary estimate of the variance of experimental error.

6. *Make a Flow Diagram of the Experimental Procedure for Each Run* This will make sure the procedure to be followed is understood and will be standardized for all runs in the design.

7. *Choose the Experimental Design.* Choose an experimental design that is suited for the objectives of your particular experiment. This will include a description of what factor levels will be studied and will determine how the

PERFORMING THE EXPERIMENTS 9

experimental units are to be assigned to the factor levels or combination of factor levels if there are more than one factor. One of the plans described in this book will almost always be appropriate. The choice of the experimental design will also determine what model should be used for analysis of the data.

8. *Determine the Number of Replicates Required* Based on the expected variance of the experimental error and the size of a practical difference, the number of replicate runs that will give the researcher a high probability of detecting an effect of practical importance.

9. *Randomize the Experimental Conditions to Experimental Units.* According to the particular experimental design being used, there is a proscribed method of randomly assigning experimental conditions to experimental units. The way this is done affects the way the data should be analyzed, and it is important to describe and record what is done. The best way to do this is to provide a data collection worksheet arranged in the random order in which the experiments are to be collected. For more complicated experimental designs Bisgaard (1999) recommends one sheet of paper describing the conditions of each run with blanks for entering the response data and recording observations about the run. All these sheets should then be stapled together in booklet form in the order they are to be performed.

10. *Describe a Method for Data Analysis.* This should be an outline of the steps of the analysis. An actual analysis of simulated data is often useful to verify that the proposed outline will work.

11. *Timetable and Budget for Resources Needed to Complete the Experiments.* Experimentation takes time and having a schedule to adhere to will improve the chances of completing the research on time. Bisgaard (1999) recommends a Gantt Chart (see SAS Institute, 2004a) which is a simple graphical display showing the steps of the process as well as calendar times. A budget should be outlined for expenses and resources that will be required.

1.7 Performing the Experiments

In experimentation, careful planning and execution of the plan are the most important steps. As we know from Murphy's Law, if anything can go wrong it will, and analysis of data can never compensate for botched experiments. To illustrate the potential problems that can occur, consider a simple experimenter conducted by an amateur gardener described by Box *et al.* (1978). The purpose was to determine whether a change in the fertilizer mixture would result in a change in the yield of his tomato plants. Eleven tomato plants were planted in a single row, and the fertilizer type (A or B) was varied. The experimental unit in this experiment is the tomato plant plus the soil it is planted in, and the treatment factor is the type of fertilizer applied. Easterling (2004) discusses some of the nuances that should be considered when planning and carrying out such a simple experiment. It is instructive to think about these in context with the checklist presented in the last section.

When defining the objectives for this experiment, the experimenter needs to think ahead to the possible implications of conclusions that he can draw. In this case, the possible conclusions are (1) deciding that the fertilizer has no effect on the yield of tomatoes, or (2) concluding that one fertilizer produces a greater yield. If the home gardener finds no difference in yield, he can choose to use the less expensive fertilizer. If he finds a difference, he will have to decide if the increase in yield offsets any increase in cost of the better fertilizer. This can help him determine how large a difference in yield he should look for and the number of tomato plants he should include in his study. The answer to this question, which is crucial in planning the experiment, would probably be much different for a commercial grower than for a backyard enthusiast.

The experimental units for this experiment were defined in the paragraph above, but in identifying them, the experimenter should consider the similarity or homogeneity of plants and how far apart he is going to place the tomato plants in the ground. Will it be far enough that the fertilizer applied to one plant does not bleed over and affect its neighbors?

Defining a meaningful response that can be measured may be tricky in this experiment. Not all the tomatoes on a single plant ripen at the same time. Thus, to measure the yield in terms of weight of tomatoes, the checklist and flow diagram describing how an experiment is conducted must be very precise. Is it the weight of all tomatoes on the plant at a certain date, or the cumulative weight of tomatoes picked over time as they ripen? Precision in the definition of the response and consistency in adherence to the definition when making the measurements are crucial.

There are many possible lurking variables to consider in this experiment. Any differences in watering, weeding, insect treatment, the method and timing of fertilizer application, and the amount of fertilizer applied may certainly affect the yield; hence the experimenter must pay careful attention to these variables to prevent bias. Easterling (2004) also pointed out that the row position seems to have affected the yield as well (as can be seen in Figure 1.2). The randomization of fertilizers to plants and row positions should equalize these differences for the two fertilizers. This was one of the things that Box et al. (1978) illustrated with this example. If a convenient method of applying the fertilizers (such as A at the beginning of the row followed by B) had been used in place of random assignment, the row position effect could have been mistaken for a treatment effect. Had this row position effect been known before the experiment was planned, the adjacent pairs of plots could have been grouped together in pairs, and one fertilizer assigned at random to one plot-plant in each pair to prevent bias from the row position effect. This technique is called blocking and will be discussed in detail in Chapter 4.

Easterling (2004) also raised the question: why were only eleven plants used in the study (five fertilized with fertilizer A and six with fertilizer B)? Normally flats of tomato plants purchased from a nursery come in flats of

Figure 1.2 *Plot of Yield by Row Position - Tomato Experiment*

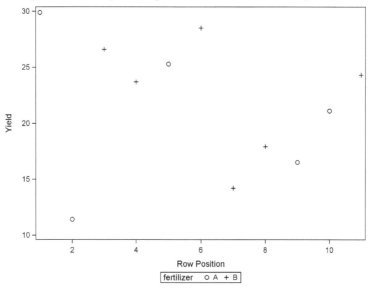

twelve. Was one plant removed from the study because it appeared unhealthy or got damaged in handling? The yield for the plant in the second row position (see Figure 1.2) of the eleven plants used was considerably lower than the others planted in neighboring row positions with the same fertilizer. Was this plant unhealthy or damaged as well?

Any problems that arise during the conduct of experiments should be carefully observed, noted, and recorded as comments on the data collection form described in step 9 of the checklist. Perhaps if this had been done for the tomato experiment, the low yield at row position two could be explained.

This discussion of a very simple experiment helps to emphasize the importance of carefully considering each step of the checklist presented in Section 11.6, and the importance of strict adherence to a flowchart for conducting the experiments, described in step 6 of that checklist. Failing to consider each point of the checklist, and inconsistency in conducting experiments and recording results may lead to the demise of an otherwise useful research project.

1.8 Use of SAS Software

Fisher's original book on Experimental Designs clearly laid the logical principles for experimentation, but users of experimental designs needed to have more detailed descriptions of the most useful designs along with accompanying plans. Consulting statisticians needed to have a systematic explanation of the relation between experimental designs and the statistical theory of least squares and linear hypotheses, and to have an enumeration of designs and descriptions of experimental conditions where each design was most appropriate.

These needs were satisfied by Cochran and Cox (1950)'s and Kempthorne (1952)'s books. However, Cochran and Cox and Kempthorne's books were published before the age of computers and they both emphasize extensive tables of designs, abundant formulas and numerical examples describing methods of manual analysis of experimental data and mathematical techniques for constructing certain types of designs. Since the publication of these books, use of experimental designs has gone far beyond agricultural research where it was initially employed, and a plethora of new books have been written on the subject. Even though computers and software (to both design and analyze data from experiments) are widely available, a high proportion of the more recent books on experimental design still follow the traditional pattern established by Cochran and Cox and Kempthorne by presenting extensive tables of designs and formulas for hand calculations and methods for constructing designs.

One of the objectives of this book is to break from the tradition and present computer code and output in place of voluminous formulas and tables. This will leave more room in the text to discuss the appropriateness of various design plans and ways to interpret and present results from experiments. The particular computer software illustrated in this book is SAS, which was originally developed in 1972. Its syntax is relatively stable and is widely used. SAS has probably the widest variety of general procedures for design of experiments and analysis of experimental data, including `proc plan`, `proc factex`, `proc optex`, and the menu driven SAS ADX for the design of experiments, and `proc glm`, `proc varcomp`, `proc mixed` and many other procedures for the analysis of experimental data. These procedures are available in the SAS/Stat and SAS/QC software.

1.9 Review of Important Concepts

This chapter describes the purpose for experimental designs. In order to determine if cause and effect relationships exist, an experimental design must be conducted. In an experimental design, the factors under study are purposely varied and the result is observed. This is different from observational studies or sampling surveys where data is collected with no attempt to control the environment. In order to predict what will happen in the future, when the environment is controlled, you must rely on cause and effect relationships. Relationships obtained from observational studies or sampling surveys are not reliable for predicting future results when the environment is to be controlled.

Experimental designs were first developed in agricultural research, but are now used in all situations where the scientific method is applied. The basic definitions and terminology used in experimental design are given in this chapter along with a checklist for planning experiments. In practice there are many different types of experimental designs that can be used. Which design

REVIEW OF IMPORTANT CONCEPTS 13

is used in a particular situation depends upon the research objectives and the experimental units. Figure 1.3 is a diagram that illustrates when the different experimental designs described in this book should be used. As different experimental designs are presented in chapters to follow, reference will be made back to this figure to describe when the designs should be used.

Figure 1.3 *Design Selection Roadmap*

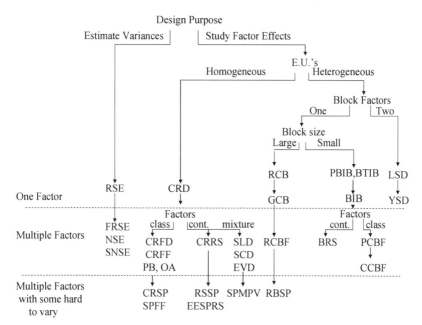

1.10 Exercises

1. A series of runs were performed to determine how the wash water temperature and the detergent concentration affect the bacterial count on the palms of subjects in a hand washing experiment.

    (a) Identify the experimental unit.
    (b) Identify the factors.
    (c) Identify the response.

2. Explain the difference between an experimental unit and a sub-sample or sub-unit in relation to the experiments described in 1.

3. Explain the difference between a sub-sample and a duplicate in relation to the experiment described in 1.

4. Describe a situation within your realm of experience (your work, your hobby, or school) where you might like to predict the result of some future action. Explain how an experimental design, rather than an observational study might enhance your ability to make this prediction.

5. Kerry and Bland (1998) describe the analysis of cluster randomized studies where a group of subjects are randomized to the same treatment. For example, when women in some randomly selected districts are offered breast cancer screening while women in other districts are not offered the screening, or when some general practitioners are randomly assigned to receive one or more courses of special training and the others are not offered the training. The response (some characteristic of the patients) in the cluster trials must be measured on each patient rather than the group as a whole. What is the experimental unit in this type of study? How would you describe the individual measurements on patients?

CHAPTER 2

Completely Randomized Designs with One Factor

2.1 Introduction

In a completely randomized design, abbreviated as CRD, with one treatment factor, n experimental units are divided randomly into t groups. Each group is then subject to one of the unique levels or values of the treatment factor. If $n = tr$ is a multiple of t, then each level of the factor will be applied to r unique experimental units, and there will be r replicates of each run with the same level of the treatment factor. If n is not a multiple of t, then there will be an unequal number of replicates of each factor level. All other known independent variables are held constant so that they will not bias the effects. This design should be used when there is only one factor under study and the experimental units are homogeneous.

For example, in an experiment to determine the effect of time to rise on the height of bread dough, one homogeneous batch of bread dough would be divided into n loaf pans with an equal amount of dough in each. The pans of dough would then be divided randomly into t groups. Each group would be allowed to rise for a unique time, and the height of the risen dough would be measured and recorded for each loaf. The treatment factor would be the rise time, the experimental unit would be an individual loaf of bread, and the response would be the measured height. Although other factors, such as temperature, are known to affect the height of the risen bread dough, they would be held constant and each loaf would be allowed to rise under the same conditions except for the differing rise times.

2.2 Replication and Randomization

Replication and randomization were popularized by Fisher. These are the first techniques that fall in the category of *error control* that was explained in Section 1.4.

The technique of replication dictates that r bread loaves are tested at each of the t rise times rather than a single loaf at each rise time. By having replicate experimental units in each level of the treatment factor, the variance of the experimental error can be calculated from the data, and this variance will be compared to the treatment effects. If the variability among the treatment means is not larger than the experimental error variance, the treatment differences are probably due to differences of the experimental units assigned to each treatment. Without replication it is impossible to tell if treatment differences are real or just a random manifestation of the particular experimental units used in the study. Sub-samples or duplicate measurements, described in Chapter 1, cannot substitute for replicates.

The random division of experimental units into groups is called randomization, and it is the procedure by which the validity of the experiment is guaranteed against biases caused by other lurking variables. In the bread rise experiment randomization would prevent lurking variables, such as variability in the yeast from loaf to loaf and trends in the measurement technique over time, from biasing the effect of the rise time.

When experimental units are randomized to treatment factor levels, an exact test of the hypothesis that the treatment effect is zero can be accomplished using a randomization test, and a test of parameters in the general linear model, normally used in the analysis of experimental data, is a good approximation to the randomization test.

A simple way of constructing a randomized data collection form, dividing n experimental units into t treatment groups, can be accomplished using the SAS data step commands. For example, in the bread rise experiment, if the experimenter wants to examine three different rise times (35 minutes, 40 minutes and 45 minutes) and test four replicate loaves of bread at each rise time, the following SAS commands will create the list.

```
/* The unrandomized design */
data unrand;
  do Loaf=1 to 12;
    if (Loaf <= 4) then time=35;
    if (5 <=Loaf <= 8) then time=40;
    if (Loaf >=9 ) then time=45;
    dough_height='_____';
    u=ranuni(0);
    output;
  end;
run;
/* Sort by random numbers to randomize */
proc sort data=unrand out=crd; by u;
/* Put Experimental units back in order */
data list; set crd;
  Loaf =_n_;
/* Print the randomized data collection form */
proc print double;  var time dough_height; id Loaf;
run;
```

These commands create the levels of the factor (rise time) in order, create an additional column of random numbers using the `ranuni()` function, create the variable dough_height that contains a space to record data in the output list, and then sorts the file by the random numbers to create a randomly ordered list. In the second data step the file crd is copied into a file called list, and the experimental units (loaves) are put back in sequential order. The proc print statement prints the SAS data file list using double spacing to allow room to write the response data after the experiments are run, and labels each

REPLICATION AND RANDOMIZATION

line with the variable `Loaf`. After running these commands the SAS output window contains a data collection form like the one shown below.

Loaf	time	dough_height
1	45	------------
2	45	------------
3	35	------------
4	45	------------
5	45	------------
6	35	------------
7	35	------------
8	40	------------
9	40	------------
10	35	------------
11	40	------------
12	40	------------

This list shows us that the first and second loaves, or experimental units, should be allowed to rise 45 minutes. The third loaf, or experimental unit, should be allowed to rise 35 minutes, etc. If you run the same commands, you may get a different random order due to the specific random numbers you obtain.

In addition to the data step, two SAS procedures (`proc plan`, and `proc factex`) can be conveniently used to create randomized data collection forms. These procedures will be used to create more complicated designs in forthcoming chapters, so just a simple introduction is presented here. For a completely randomized design SAS `proc plan` can be used to randomize a file containing the treatment indicators. The code listing below randomizes the list in the file `unrand` created in the first list above.

SAS `proc plan` automatically randomizes the order of the experimental

```
/* Randomize the design with proc plan*/
proc plan seed=27371;
   factors Loaf=12;
   output data=unrand out=crd;
run;
/* Put Experimental units back in order */
proc sort data=crd;
   by Loaf;
/* Print the randomized data collection form */
proc print double;  var time dough_height; id Loaf;
run;
```

18   COMPLETELY RANDOMIZED DESIGNS WITH ONE FACTOR

units in the file crd, so no sorting by the random numbers is required. The second data step puts the experimental units back in sequential order so that the treatment levels appear randomized in the printed data collection form.

The other alternative is to use proc factex. The lines below call proc factex to create the file crd.

```
/* Creates CRD in random order using proc factex  */
proc factex;
   factors time/nlev=3;
   output out=crd time nvals=(35 40 45) designrep=4 randomize;
/* Add experimental unit id and space for response*/
data list; set crd;
   Loaf =_n_; dough_height='_____';
/* Print the data Collection form                 */
proc print double;  var time dough_height; id Loaf;
run;
```

This procedure allows one to specify the level values for the factor and randomize the order of the list. The designrep=4 specifies that four replicate of each factor level be included in the output file. Alternatively, a randomized list could be constructed using the =rand() function and the sort menu in an Excel spreadsheet. This would result in an electronic worksheet that could additionally be used to record the response data during execution of the experiments and later read into SAS for analysis.

2.3 A Historical Example

To illustrate the checklist for planning an experiment, consider a historical example taken from the 1937 Rothamstead Experimental Station Report, unknown (1937). This illustrates some of the early work done by Fisher in developing the ideas of experimental design and analysis of variance for use on agricultural experiments at the research station.

Objectives The objective of the study was to compare the times of planting, and methods of applying mixed artificial fertilizers (NPK) prior to planting, on the yield of sugar beets. Normally fertilizer is applied and seeds planted as early as the soil can be worked.

Experimental Units The experimental units were the plots of ground in combination with specific seeds to be planted in each plot of ground.

Response or Dependent Variable The dependent variable would be the yield of sugar beets measured in cwt per acre.

Independent Variables and Lurking Variables The independent variables of interest were the time and method of applying mixed artificial fertilizers. Four levels of the treatment factor were chosen as listed below:

1. (A) no artificial fertilizers applied
2. (B) artificials applied in January (plowed)
3. (C) artificials applied in January (broadcast)
4. (D) artificials applied in April (broadcast)

LINEAR MODEL FOR CRD

Lurking variables that could cause differences in the sugar beet yields between plots were differences in the fertility of the plots themselves, differences in the beet seeds used in each plot, differences among plots in the level of weed infestation, differences in cultivation practices of thinning the beets, and hand harvesting the beets.

Pilot Tests Sugar beets had been grown routinely at Rothamstead, and artificial fertilizers had been used by both plowing and broadcast for many crop plants; therefore, it was known that the independent variable could be controlled and that the response was measurable.

Choose Experimental Design The completely randomized design (CRD) was chosen so that differences in lurking variables between plots would be unlikely to correspond to changes in the factor levels listed above.

Determine the Number of Replicates A difference in yield of 6 cwt per acre was considered to be of practical importance, and based on historical estimates of variability in sugar beet yields at Rothamstead, four or five replicates were determined to be sufficient.

Randomize Experimental Units to Treatment Levels Eighteen plots were chosen for the experiment, and a randomized list was constructed assigning four or five plots to each factor level.

2.4 Linear Model for CRD

The mathematical model for the data from a CRD, or completely randomized design, with an unequal number of replicates for each factor level can be written as:

$$Y_{ij} = \mu_i + \epsilon_{ij} \qquad (2.1)$$

where Y_{ij} is the response for the jth experimental unit subject to the ith level of the treatment factor, $i = 1, \ldots, t$, $j = 1, \ldots, r_i$ and r_i is the number of experimental units or replications in ith level of the treatment factor.

This is sometimes called the cell means model with a different mean, μ_i, for each level of the treatment factor. The distribution of the experimental errors, ϵ_{ij}, are mutually independent due to the randomization and assumed to be normally distributed. This model is graphically represented in Figure 2.1.

Figure 2.1 *Cell Means Model*

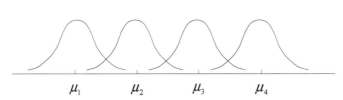

20   COMPLETELY RANDOMIZED DESIGNS WITH ONE FACTOR

An alternate way of writing a model for the data is

$$Y_{ij} = \mu + \tau_i + \epsilon_{ij}. \tag{2.2}$$

This is called the effects model and the τ_is are called the effects. τ_i represents the difference between the long-run average of all possible experiments at the ith level of the treatment factor and the overall average. With the normality assumption $Y_{ij} \sim N(\mu + \tau_i, \sigma^2)$ or $\epsilon_{ij} \sim N(0, \sigma^2)$. For equal number of replicates, the sample means of the data in the ith level of the treatment factor is represented by

$$\bar{y}_{i\cdot} = \frac{1}{r_i} \sum_{j=1}^{r_i} y_{ij} \tag{2.3}$$

and the grand mean is given by

$$\bar{y}_{\cdot\cdot} = \frac{1}{t} \sum_{i=1}^{t} \bar{y}_{i\cdot} = \frac{1}{n} \sum_{i=1}^{t} \sum_{j=1}^{r_i} y_{ij} \tag{2.4}$$

where $n = \sum r_i$. Using the method of least squares, the estimates of the cell means are found by choosing them to minimize the error sum of squares

$$ssE = \sum_{i=1}^{t} \sum_{j=1}^{r_i} (y_{ij} - \mu_i)^2. \tag{2.5}$$

This is done by taking partial derivatives of ssE with respect to each cell mean, setting the results equal to zero, and solving each equation

$$\frac{\partial ssE}{\partial \mu_i} = -2 \sum_{i=1}^{t} \sum_{j=1}^{r_i} (y_{ij} - \mu_i) = 0.$$

This results in the estimates:

$$\hat{\mu}_i = \bar{y}_{i\cdot}.$$

2.4.1 Matrix Representation

Consider a CRD with $t = 3$ factor levels and $r_i = 4$ replicates for $i = 1, \ldots, t$. We can write the effects model concisely using matrix notation as:

$$\boldsymbol{y} = \boldsymbol{X\beta} + \boldsymbol{\epsilon} \tag{2.6}$$

LINEAR MODEL FOR CRD

Where

$$y = \begin{pmatrix} y_{11} \\ y_{12} \\ y_{13} \\ y_{14} \\ y_{21} \\ y_{22} \\ y_{23} \\ y_{24} \\ y_{31} \\ y_{32} \\ y_{33} \\ y_{34} \end{pmatrix}, \quad X = \begin{pmatrix} 1 & 1 & 0 & 0 \\ 1 & 1 & 0 & 0 \\ 1 & 1 & 0 & 0 \\ 1 & 1 & 0 & 0 \\ 1 & 0 & 1 & 0 \\ 1 & 0 & 1 & 0 \\ 1 & 0 & 1 & 0 \\ 1 & 0 & 1 & 0 \\ 1 & 0 & 0 & 1 \\ 1 & 0 & 0 & 1 \\ 1 & 0 & 0 & 1 \\ 1 & 0 & 0 & 1 \end{pmatrix}, \quad \beta = \begin{pmatrix} \mu \\ \tau_1 \\ \tau_2 \\ \tau_3 \end{pmatrix}, \quad \epsilon = \begin{pmatrix} \epsilon_{11} \\ \epsilon_{12} \\ \epsilon_{13} \\ \epsilon_{14} \\ \epsilon_{21} \\ \epsilon_{22} \\ \epsilon_{23} \\ \epsilon_{24} \\ \epsilon_{31} \\ \epsilon_{32} \\ \epsilon_{33} \\ \epsilon_{34} \end{pmatrix},$$

and $\epsilon \sim MVN(\mathbf{0}, \sigma^2 \mathbf{I})$.

The least squares estimators for β are the solution to the normal equations $X'X\beta = X'y$. The problem with the normal equations is that $X'X$ is singular and cannot be inverted. The solution to this problem using the SAS `proc glm` is to partition the X matrix as:

$$X = \begin{pmatrix} 1 & 1 & 0 & | & 0 \\ 1 & 1 & 0 & | & 0 \\ 1 & 1 & 0 & | & 0 \\ 1 & 1 & 0 & | & 0 \\ 1 & 0 & 1 & | & 0 \\ 1 & 0 & 1 & | & 0 \\ 1 & 0 & 1 & | & 0 \\ 1 & 0 & 1 & | & 0 \\ 1 & 0 & 0 & | & 1 \\ 1 & 0 & 0 & | & 1 \\ 1 & 0 & 0 & | & 1 \\ 1 & 0 & 0 & | & 1 \end{pmatrix} = [X_1 | X_2]$$

Then $X'X$ can be written in partitioned form as:

$X'X = \begin{bmatrix} X_1'X_1 & X_1'X_2 \\ X_2'X_1 & X_2'X_2 \end{bmatrix}$, and $X_1'X_1$ is non-singular. Therefore $X'X^- = \begin{bmatrix} X_1'X_1^{-1} & 0 \\ 0 & 0 \end{bmatrix}$ is a generalized inverse for $X'X$, and $\hat{\beta} = (X'X)^- X'y$ is a solution that is equivalent to the estimates that would be obtained with the restriction that $\beta_t = 0$ (see Freund and Littell, 1981). For the example with $t = 3$ factor levels and $r_i = 4$ replicates for $i = 1, \ldots, t$,

$$\beta = \begin{pmatrix} \mu \\ \tau_1 \\ \tau_2 \\ \tau_3 \end{pmatrix}, \quad \hat{\beta} = \begin{pmatrix} \hat{\mu} + \hat{\tau}_3 \\ \hat{\tau}_1 - \hat{\tau}_3 \\ \hat{\tau}_2 - \hat{\tau}_3 \\ 0 \end{pmatrix}$$

2.4.2 L.S. Calculations with SAS proc glm

Table 2.1 shows the data from a CRD design for the bread rise experiment described earlier in this chapter.

Table 2.1 *Data from Bread Rise Experiment*

Rise Time	Loaf Heights
35 minutes	4.5, 5.0, 5.5, 6.75
40 minutes	6.5, 6.5, 10.5, 9.5
45 minutes	9.75, 8.75, 6.5, 8.25

Using these data we have

$$X'X = \begin{pmatrix} 12 & 4 & 4 & 4 \\ 4 & 4 & 0 & 0 \\ 4 & 0 & 4 & 0 \\ 4 & 0 & 0 & 4 \end{pmatrix}, \quad X'y = \begin{pmatrix} 88.0 \\ 21.75 \\ 33.0 \\ 33.25 \end{pmatrix},$$

and

$$X'X^- = \begin{pmatrix} 0.25 & -0.25 & -0.25 & 0 \\ -0.25 & 0.50 & 0.25 & 0 \\ -0.25 & 0.25 & 0.50 & 0 \\ 0 & 0 & 0 & 0 \end{pmatrix}, \quad \hat{\beta} = (X'X)^- X'y = \begin{pmatrix} 8.3125 \\ -2.8750 \\ -0.0625 \\ 0.0000 \end{pmatrix}.$$

The SAS commands to read in this data and compute these estimates are:

```
/* Reads the data from compact list    */
data bread;
  input time h1-h4;
  height=h1; output;
  height=h2; output;
  height=h3; output;
  height=h4; output;
  keep time height;
datalines;
35 4.5 5.0 5.5 6.75
40 6.5 6.5 10.5 9.5
45 9.75 8.75 6.5 8.25
run;
/* Fits model with proc glm           */
proc glm;
  class time;
  model height=time/solution;
run;
```

These commands read the data in a compact format, like Table 2.1, then use the output statement to create four lines in the file, bread, for each line in the input records. The result of the model /solution; option is:

LINEAR MODEL FOR CRD

Parameter		Estimate
Intercept		8.312500000 B
time	35	-2.875000000 B
time	40	-0.062500000 B
time	45	0.000000000 B

We can see the estimates are the same as those shown above. These parameter estimates are not unique because they depend on the ordering of the class variable time. proc glm recognizes this and designates the estimates as biased by following each value with a B.

2.4.3 Estimation of σ^2 and distribution of quadratic forms

The estimate of the variance of the experimental error, σ^2, is $ssE/(n-t)$. It is only possible to estimate this variance when there are replicate experiments at each level of the treatment factor. When measurements on sub-samples or duplicate measurements on the same experimental unit are treated as replicates, this estimate can be seriously biased.

In matrix form, ssE can be written as

$$ssE = y'y - \hat{\beta}'X'y = y'(I - X(X'X)^{-}X')y,$$

and from the theory of linear models it can be shown that the ratio of ssE to the variance of the experimental error, σ^2, follows a chi-square distribution with $n - t$ degrees of freedom, i.e., $ssE/\sigma^2 \sim \chi^2_{n-t}$.

2.4.4 Estimable Functions

A linear combination of the cell means is called an estimable function if it can be expressed as the expected value of a linear combination of the responses, i.e.,

$$\sum_{i=1}^{t} b_i(\mu + \tau_i) = E\left[\sum_{i=1}^{t}\sum_{j=1}^{r_i} a_{ij}Y_{ij}\right] \quad (2.7)$$

From this definition it can be seen that effects, τ_i, are not estimable, but a cell mean, $\mu + \tau_i$, or a contrast of effects, $\sum c_i \tau_i$, where $\sum c_i = 0$, is estimable.

In matrix notation $L\beta$ is a set of estimable functions if each row of L is a linear combination of the rows of X, and $L\hat{\beta}$ is its unbiased estimator. $L\hat{\beta}$ follows the multivariate normal distribution with covariance matrix $\sigma^2 L'(X'X)^{-}L$, and the estimator of the covariance matrix is $\hat{\sigma}^2 L'(X'X)^{-}L$. For example using the data from the bread rise experiment above,

$$L = \begin{pmatrix} 0 & 1 & -1 & 0 \\ 0 & 1 & 0 & -1 \end{pmatrix}. \quad (2.8)$$

$L\beta = \begin{pmatrix} \tau_1 - \tau_2 \\ \tau_1 - \tau_3 \end{pmatrix}$, and $L\hat{\beta} = \begin{pmatrix} \hat{\tau}_1 - \hat{\tau}_2 \\ \hat{\tau}_1 - \hat{\tau}_3 \end{pmatrix} = \begin{pmatrix} -2.8025 \\ -2.8750 \end{pmatrix}$ is a vector

24 COMPLETELY RANDOMIZED DESIGNS WITH ONE FACTOR

of contrasts of the effects. The number of degrees of freedom, or number of linearly independent contrasts of effects in a CRD, is always the number of levels of the treatment factor minus one, i.e., $t-1$. Whenever there is a set of $t-1$ linearly independent contrasts of the effects, they are called a *saturated set of estimable contrasts*.

It can be shown that $(L\hat{\beta})'(L(X'X)^-L')^{-1}(L\hat{\beta})$ follows the noncentral chi-square distribution, $\chi^2(p, \lambda)$ where the noncentrality parameter

$$\lambda = (\sigma^2)^{-1}(L\beta)'(L(X'X)^-L')^{-1}(L\beta),$$

and L is the coefficient matrix for an estimable contrast like (2.8), and the degrees of freedom p is equal to the rank of L.

Estimable contrasts can be obtained from SAS `proc glm` using the estimate command. For example to estimate the average difference in the cell means for the first and second levels of the treatment factor, $(\mu+\tau_1)-(\mu+\tau_2) = \tau_1-\tau_2$, use the command:

```
estimate '1 - 2' time 1 -1;
```

where the string in quotes is the label to be used in the output, `temp` is the name of the treatment factor, and the numbers 1 -1 are the contrast coefficients, c_i. The command:

```
estimate '1 -2 -3' time 1 -1 -1;
```

would produce no output and in the log file the message "1 -2 -3 is not estimable" would be printed, since $\sum c_i \neq 0$.

2.4.5 Hypothesis Test of No Treatment Effects

In the model for the CRD, the statistical hypothesis of interest is $H_0: \mu_1 = \mu_2 = \ldots \mu_t$ or $\tau_1 = \tau_2 = \ldots = \tau_t$ versus the alternative H_a: at least two of the τs differ. If the null hypothesis is true, the model $y_{ij} = \mu_i + \epsilon_{ij} = \mu + \tau_i + \epsilon_{ij}$ simplifies to $y_{ij} = \mu + \epsilon_{ij}$ which can be represented as a single normal distribution with mean μ and variance σ^2 rather than multiple normal distributions like those shown in Figure 2.1.

The sums of squares about the mean is $ssTotal = \sum_{i=1}^{t}\sum_{j=1}^{r_i}(y_{ij}-\bar{y}_{..})^2 = y'y - (1'y)^2/(1'1)$, where $\bar{y}_{..}$ is the grand mean and $\mathbf{1}$ is a column vector of 1s. This sum of squares can be partitioned as:

$$ssTotal = ssT + ssE \qquad (2.9)$$

where $ssT = \hat{\beta}'X'y - (1'y)^2/(1'1) = (L\hat{\beta})'(L(X'X)^-L')^{-1}(L\hat{\beta})$, and L is the coefficient matrix for a saturated set of estimable contrasts. This quantity is called the treatment sums of squares. Under the null hypothesis $H_0: \mu_1 = \mu_2 = \ldots \mu_t$, both ssT and ssE follow the chi-squared distribution. These sums of squares and their corresponding mean squares, which are formed by dividing each sum of squares by its degrees of freedom, are usually presented

LINEAR MODEL FOR CRD

in an Analysis of Variance or ANOVA table like that shown symbolically in Table 2.2.

Table 2.2 Analysis of Variance Table

Source	df	Sum of Squares	Mean Squares	F-ratio
Treatment	$t-1$	ssT	msT	$F = msT/msE$
Error	$n-t$	ssE	msE	
Total	$n-1$	$ssTotal$	$msTotal$	

Under the null hypothesis, the F-ratio msT/msE follows the F-distribution with $t-1$ and $n-t$ degrees of freedom, and under the alternative it follows the non-central F distribution with noncentrality parameter

$$\lambda = (\sigma^2)^{-1}(L\beta)'(L(X'X)^-L')^{-1}(L\beta) = \frac{r}{\sigma^2}\sum_{i=1}^{t}(\mu_i - \bar{\mu}.)^2.$$

It is the generalized likelihood ratio test statistic for H_0, and is the formal method of comparing the treatment effects to the experimental error variance described in Section 2.2.

The sums of squares, mean squares, degrees of freedom in the ANOVA table, and associated F-test statistic are automatically calculated by SAS `proc glm`. For example, the result of the `proc glm; class temp; model height=temp;` command for the bread experiment shown earlier is:

```
                 The GLM Procedure

Dependent Variable: height

                              Sum of
Source              DF       Squares    Mean Square   F Value   Pr > F

Model                2    21.57291667   10.78645833      4.60   0.0420
Error                9    21.09375000    2.34375000

Corrected Total     11    42.66666667
```

In this table the ssT and msT and the associated degrees of freedom are on the line labeled `Model`, the ssE is on the line labeled `Error` and the $ssTotal$ is on the line labeled `Corrected Total`. The F-value is the ratio msT/msE and the last column labeled `Pr > F` is the probability of exceeding the calculated F-value if the null hypothesis is true. This is called the P-value and is illustrated graphically in Figure 2.2. If the experimenter chooses the significance level, α, for his hypothesis test, he would reject the hypothesis if the `Pr > F` value on the `proc glm` output is less than the chosen value of α.

For the bread rise experiment there are significant differences among the mean risen dough heights for each rise time at the significance level $\alpha = 0.05$, since $0.042 < 0.05$.

26   COMPLETELY RANDOMIZED DESIGNS WITH ONE FACTOR

Figure 2.2 $Pr > F$

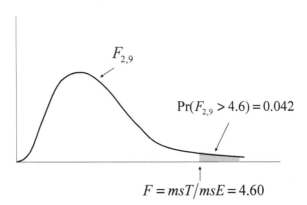

2.4.6 A Word of Caution

When a completely randomized design in one factor is conducted, the model for analysis is Equation (2.1) or (2.2) and the correct analysis is through the analysis of variance as shown symbolically in Table 2.2. The use of computer software like SAS makes it easy to analyze data and draw conclusions; however, if the experiment was not properly conducted even a sophisticated analysis of the data could be useless. The ϵ_{ij} term in the model (2.1) or (2.2), and its associated sums of squares, ssE, represents replicate experimental units. In many cases experimenters do not have replicate experimental units in each level of the treatment factor and substitute sub-samples or duplicates for them in the analysis. In other cases the experimental units are not properly randomized to treatment factor levels. When this is the situation, performing the analysis as if the design had been properly conducted may be completely wrong and misleading. Wrong conclusions can be drawn that do not hold up to later scrutiny, and a bad reputation is unfairly ascribed to statistically designed experiments and statistical analyses of data.

For example, consider an experiment where a professor would like to determine the effect of teaching methods on student test scores. If he uses one teaching method for the morning class, another for his evening class, and treats test scores for individual students as replicates, the results of his analysis may be totally wrong. This situation is similar to the cluster randomized studies described in exercise 5 of Chapter 1. The experimental unit is the class, since he applied the teaching method to a whole class simultaneously, and the individual students are sub-samples or observational units (since he must test individual students, not the class as a whole). The treatment effect should be judged against the variability in experimental units or classes. The variability among students in a class may be much different than variability from class

VERIFYING ASSUMPTIONS OF THE LINEAR MODEL

average to class average. Sub-sample observations should be averaged before analysis, as explained in Section 1.3. If this were done, he would only have one observation per class per teaching method and no replicates for use in calculating ssE in Table 2.2. There is no denominator for calculating the F-test statistic for teaching method. If he uses the variability in students within a class to calculate ssE, it may be too large or too small, causing him to reach the wrong conclusion about significance of the treatment effect. Further, if he did not randomize which teaching method was used in the morning and evening classes, and if he has no replicate classes that were taught with the same teaching method, his analysis is wide open to biases. Students in the morning classes may be fundamentally different than students in the evening classes, and any difference in average scores between the two teaching methods may be entirely due to differences among the two groups of students. In fact, if the professor knows there are differences in morning and evening students, he may purposely use the teaching method he wants to promote on the better class, thus ruining the objectivity of his research.

2.5 Verifying Assumptions of the Linear Model

The most critical assumption justifying the analysis based on the linear model presented in the last section is independence of the experimental error terms ϵ_{ij}. This assumption is justified if proper randomization of the experimental units to treatment factor levels has been performed. No further steps are needed to justify this assumption.

The other two assumptions for the linear model are constancy of the variance of the experimental error, σ^2, across all levels of the treatment factor, and normality of the experimental errors. Both of these assumptions can be visualized in Figure 2.1.

To verify the equal variance assumption, a simple scatter plot of the response data versus the factor levels can be constructed. Figure 2.3 shows an example of this type plot for the bread rise experiment. In this plot, similarity in the range of variability in dough heights between the different levels of rise time would justify the assumption of equal variance.

This figure was created in SAS using the commands:

```
proc sgplot data=bread;
    scatter y=Height x=Time;
run;
```

A similar plot that is used to check the equality of variance assumption is a plot of the residuals versus the cell means, or predicted values. The residuals are the differences of the response data and their respective cell means. The mean residual, for each level of the treatment factor, is zero. Therefore, in a plot of residuals it is easier to visualize whether the scatter or variability is equal. A common departure from equal variance occurs when the variance of the experimental errors increases as the mean response increases. By plotting the residuals versus the cell means, this departure can be detected if the vertical scatter in the residuals increases from left to right on the plot.

Figure 2.3 *Plot of Response versus Factor Level for Bread Rise Experiment*

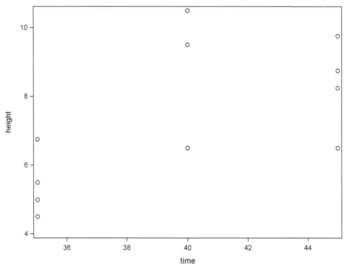

For the bread rise experiment, the SAS commands:

```
/* Fit model and Examine Residual Plots */
ods graphics on;
proc glm data=bread plots=diagnostics(unpack);
  class time;
  model height=time/solution;
run;
ods graphics off;
```

will fit the model for the CRD design, and when the ods graphics are turned on as shown in this example, the option plots=diagnostics(unpack) creates a panel of diagnostic plots to determine whether the assumptions of the model are satisfied. The (unpack) feature separates the plots so that they can be individually displayed. One of the plots created in this panel is labeled Residuals by Predicted and is shown in Figure 2.4. For the CRD design with one factor, the predicted values from the model, $\hat{y}_{ij} = \hat{\mu}_i$, are the cell means. As can be seen in Figure 2.4, there is a wider spread in the residuals for larger cell means or predicted values. This indicates that the equality of variance assumption may be violated. This fact is easier to see in Figure 2.4 than it was in Figure 2.3.

To verify the assumption of normality of the experimental error, a normal probability plot of the residuals can be examined. In the panel of diagnostic plots created by proc glm the normal probability plot is labeled Q-Q Plot of Residuals, and it is shown in Figure 2.5.

VERIFYING ASSUMPTIONS OF THE LINEAR MODEL

Figure 2.4 *Plot Residuals versus Cell Means for Bread Rise Experiment*

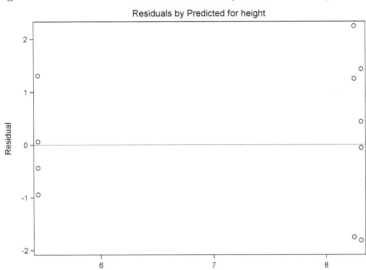

Figure 2.5 *Normal Plot of Residuals for Bread Rise Experiment*

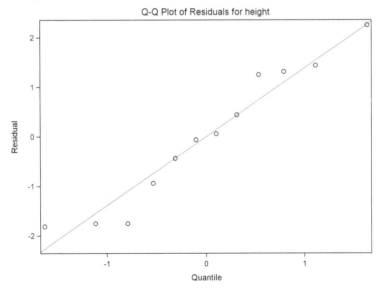

In a normal probability plot, if the points fall along a straight line, the normal distribution would be justified. The more data, and more points on the plot, the closer the points must lie to a straight line to justify the normality

30 COMPLETELY RANDOMIZED DESIGNS WITH ONE FACTOR

assumption. In Figure 2.5, the points lie reasonably close to a straight line, and the normality assumption would appear justified. The equal variance assumption is more critical than the normality assumption, but they sometimes go hand in hand. When the equal variance assumption is violated, the normality assumption is often violated as well, and the corrective measures used for modifying the analysis when there is heterogeneity of variance will often correct both problems.

2.6 Analysis Strategies When Assumptions Are Violated

One common cause of heterogeneity of variances between levels of the treatment factor is a non-linear relationship between the response and stimulus or treatment. For example, in the upper half of Figure 2.6, it can be seen that the response increases non-linearly in response to the treatment. The density functions, drawn on their sides at three treatment levels, represent how non-linearity often affects the distribution of the response. As the mean or center of the distribution increases, the variance or spread in the distribution also increases, and the distributions have long tails on the right. One way of correcting this situation is to transform the response data prior to analysis.

Figure 2.6 *Representation of Effect of Non-linearities on Distribution of Response*

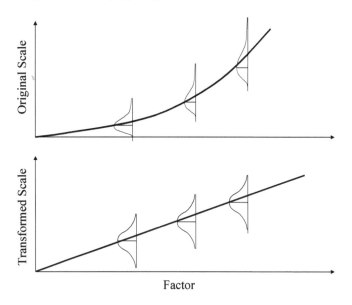

The bottom half of Figure 2.6 shows the potential result of a variance stabilizing transformation. On the transformed scale, the variance appears constant at different factor levels and the distribution appears more normal.

2.6.1 Box-Cox Power Transformations

One way to recognize the need for a variance stabilizing transformation is to examine the plot of residuals versus cell means described in the last section. If the spread in the residuals tends to increase proportionally as a function of the cell means, as illustrated in Figure 2.4, a transformation, $Y = f(y)$ can usually be found that will result in a more sensitive analysis. Box and Cox (1964) proposed a series of power transformations $Y = y^\lambda$ that normally work well. If the variance tends to increase as the mean increases, choose a value of λ less than one, and if the variance tends to decrease as the mean increases, choose a value of λ greater than one. Table 2.3 summarizes some common Box-Cox Power transformations. A common situation where the $\sigma \propto \mu$ is when the response is actually a measure of variability, like the sample variance s^2.

Table 2.3 Box-Cox Power Transformations

Relation Between σ and μ	λ	Transformation
$\sigma \propto \mu^2$	-1	Reciprocal
$\sigma \propto \mu^{3/2}$	$-\frac{1}{2}$	Square root of Reciprocal
$\sigma \propto \mu$	0	Log
$\sigma \propto \mu^{1/2}$	$\frac{1}{2}$	Square Root

In a CRD design with replicate experiments in each level of the treatment factor, one way to determine the most appropriate value of λ to use in the Box-Cox transformation is to fit a least squares regression line

$$\log(s_i) = a + b\log(\bar{y}_{i.})$$

to the summary statistics. If the straight line provides a reasonable fit to the data, use $\lambda = 1 - b$. For example, for the bread rise experiment the SAS commands below use **proc sort** and **proc means** to calculate and store the cell means and standard deviations in the SAS file, **means**.

```
proc sort data=bread; by time;
proc means data=bread; var height; by time;
  output out=means mean=meanht stddev=s;
data logs; set means;
  logmean=log(meanht);
  logs=log(s);
```

Logarithms of both the cell means and standard deviations are computed in the data step and stored in the file **logs**. In the commands shown at the top of the next page **proc reg** is used to calculate the slope and intercept of the regression line and **proc sgplot** is used to plot the log standard deviations by log means with the fitted regression line as shown in Figure 2.7.

```
proc reg data=logs;
model logs=logmean;
run;
proc sgplot data=logs;
  reg y=logs x=logmean/curvelabel="Slope=1.29487";
  xaxis label="log mean";
  yaxis label="log standard deviation";
run;
```

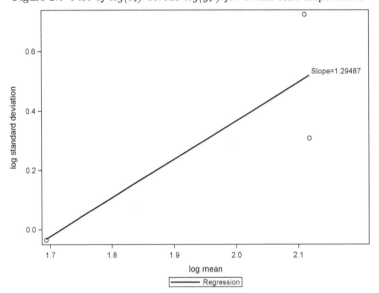

Figure 2.7 *Plot of $\log(s_i)$ versus $\log(\bar{y}_{i\cdot})$ for Bread Rise Experiment*

In this plot it can be seen that $\log(s_i)$ increases as $\log(\bar{y}_{i\cdot})$ increases, and that the linear fit provides a good representation of that relationship. Therefore, the appropriate λ to use for transforming the dough heights is $\lambda = 1 - 1.294869$. The SAS data step commands shown earlier to read in the data from the bread rise experiment can be modified to incorporate the transformation as follows:

```
data bread2;
  input time h1-h4;
  height=h1**(1-1.294869); output;
  height=h2**(1-1.294869); output;
  height=h3**(1-1.294869); output;
  height=h4**(1-1.294869); output;
  keep time height;
datalines;
```

ANALYSIS STRATEGIES WHEN ASSUMPTIONS ARE VIOLATED 33

After incorporating this transformation, the plot of residuals versus cell means is shown in Figure 2.8. It can be seen in this figure that the spread or variability of the residuals is nearly the same for each value of the predicted value or log cell mean.

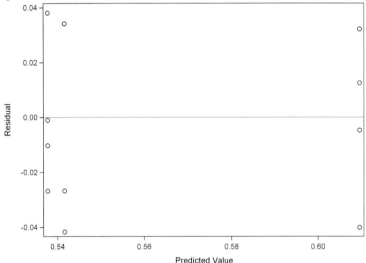

Figure 2.8 *Plot of Residuals versus Cell means after y^λ Transformation for Bread Rise Experiment*

This transformation also makes the F-test more sensitive. The ANOVA table for the transformed response is shown below, where it can be seen that the P-value has decreased from 0.042 (before transformation) to 0.0227 (after transformation).

```
                    The GLM Procedure

Dependent Variable: height

                              Sum of
Source              DF       Squares    Mean Square   F Value   Pr > F

Model                2     0.01305604    0.00652802      5.94   0.0227
Error                9     0.00989833    0.00109981

Corrected Total     11     0.02295437
```

Even when transformations are used to justify the assumptions of the analysis and make the F-test more sensitive, the results are normally reported on the original scale. For example, it would be confusing to talk about the (dough height)$^{-.294869}$ in a report.

2.6.2 Distribution-Based Transformations

The distribution assumption for the effects model for the CRD described in Section 2.3, was $Y_{ij} \sim N(\mu + \tau_i, \sigma^2)$. However, if it is known that the data follow some distribution other than the normal distribution, such as the Binomial, Poisson or Lognormal, then it would also be known that the standard deviation would not be constant. For example, if the response, Y, was a binomial count of the number of successes in n trials, then due to the central limit theorem, Y would be approximately normal, but $\mu_Y = np$ and $\sigma_Y = \sqrt{np(1-p)}$, where p is the probability of success. In situations like this where the distribution of the response is known to follow some specific form, then an appropriate transformation can be found to stabilize the variance. Table 2.4 shows the transformation for common situations often encountered.

Table 2.4 *Response Distribution-Based Transformations*

Response Distribution	Variance in Terms of Mean μ	Transformation $f(y)$
Binomial	$\frac{\mu(1-\mu)}{n}$	$\sin^{-1}\sqrt{y/n}$ (radians)
Poisson	μ	$\sqrt{y}$ or $\sqrt{y+\frac{1}{2}}$
Lognormal	$c\mu^2$	$\log(y)$

2.6.3 Alternatives to Least Squares Analysis

When the variance of the experimental error is not constant for all levels of the treatment factor, but it is not related to the cell means, a transformation will not be an appropriate way of equalizing or stabilizing the variances. A more general solution to the problem is to use weighted least squares. Using weighted least squares, $\hat{\boldsymbol{\beta}}$ is the solution to the normal equations $\boldsymbol{X'WX\beta} = \boldsymbol{X'Wy}$, where $\boldsymbol{W}$ is a diagonal matrix whose diagonal elements are the reciprocals of the standard deviation within each treatment level. As an illustration of this method, consider the SAS commands, at the top of the next page, for analyzing the data from the bread rise experiment.

As in the previous examples, the data is read in the same way. The sort and means procedures calculate the cell standard deviations and the next data step merges these with the original data and computes the weights, `w`, that are the reciprocals of the standard deviations. The `weight` statement, in the `proc glm` call, solves the weighted least squares normal equations. The results appear below the commands.

ANALYSIS STRATEGIES WHEN ASSUMPTIONS ARE VIOLATED

```
/* Calculates mean and standard deviations  */
proc sort data=bread; by time;
proc means data=bread; var height; by time;
  output out=means stddev=s;
/* Creates weights                          */
data bread3; merge bread means; by time;
  w=1/s;
proc glm data=bread3;
  class time;
  model height=time;
  weight w;
run;
```

```
                   The GLM Procedure

Dependent Variable: height

Weight: w

                                   Sum of
Source                   DF       Squares    Mean Square  F Value  Pr > F

Model                     2   18.20937757     9.10468879     6.23  0.0201
Error                     9   13.16060821     1.46228980

Corrected Total          11   31.36998578
```

With these results, it can be seen that F-test from the weighted least squares is more sensitive than the unweighted least squares, and the P-value is similar to what was obtained with the Box-Cox transformation shown in Section 2.5.1.

When the error distribution is not normal, an alternative to analyzing a transformation of the response is to use a generalized linear model (see McCullagh and Nelder, 1989). In fitting a generalized linear model, the user must specify the error distribution and a link function in addition to the model. The method of maximum likelihood is used to estimate the model parameters and the generalized likelihood ratio tests are used to test the hypotheses. When the link function is the identity and the distribution is normal, the generalized linear model analysis will result in the method of least squares and the ANOVA F-test. The SAS procedure **genmod** will fit the generalized linear model and compute appropriate likelihood ratio test statistics. This procedure allows the user to choose from the distributions Binomial, Poisson, Negative Binomial, Multinomial, Gamma, Inverse Gaussian, as well as Normal. For each distribution, **genmod** has a default link function.

To illustrate the use of **proc genmod** to analyze experimental data, consider the following example. A professor wanted to compare three different teaching methods to determine how the students would perceive the course. The treatment factor was the teaching method, the experimental unit was a class of students, and the response was the summary of student ratings for the course. The professor taught two sections of the course for three consecutive semesters resulting in a total of six experimental units or classes. He constructed a randomized list so that two classes were assigned to each teaching method. This would reduce the chance that other differences in the classes, or differences in his execution of the teaching methods, would bias the results. At the end of each semester, the students were asked to rate the course on a five-point scale, with 1 being the worst and 5 being the best. Therefore, the response from each class was not a single, normally distributed response, y, but a vector $(y_1, \ldots, y_5)$ response that followed the multinomial distribution. The summary data from the experiment is shown in Table 2.5.

Table 2.5 *Counts of Student Rating Scores*

Class	Method	1	2	3	4	5
1	1	2	14	12	8	6
2	3	1	11	15	15	10
3	2	3	8	18	14	10
4	3	1	9	17	15	12
5	2	4	12	19	9	7
6	1	3	16	13	10	4

The following SAS commands read in the data and call `proc genmod` to make the generalized linear model analysis.

```
data teaching;
   input count score $ class method;
datalines;
2 1 1 1
14 2 1 1
   ...
proc genmod;
   freq count;
   class method;
   model score=method/dist=multinomial aggregate=method type1;
run;
```

The `$` following `score` in the input statement causes it to be read as a character variable. The `dist=multinomial` option on the model statement specifies the distribution. `proc genmod` then assumes the default cumulative logit link function. The `aggregate=method` option requests that a likelihood ratio test of no method effect be printed. The results appear as shown on the next page.

The P-value for the likelihood ratio chi-square statistic is small indicating there is a significant difference between the teaching methods. Teaching method 1 had an average score of 2.98, teaching method 2 had an average score of 3.22, and teaching method 3 appeared to be the best with an average

DETERMINING THE NUMBER OF REPLICATES

```
          LR Statistics For Type 1 Analysis
                                      Chi-
Source           Deviance       DF    Square    Pr > ChiSq

Intercepts       13.7965
method            4.1486         2     9.65       0.0080
```

score of 3.47. This can also be visualized in the bar charts in Figure 2.9 which shows that the percentage of high scores given increases for teaching method 2 and 3.

Figure 2.9 *Percentage of Student Rating Scores by Teaching Method*

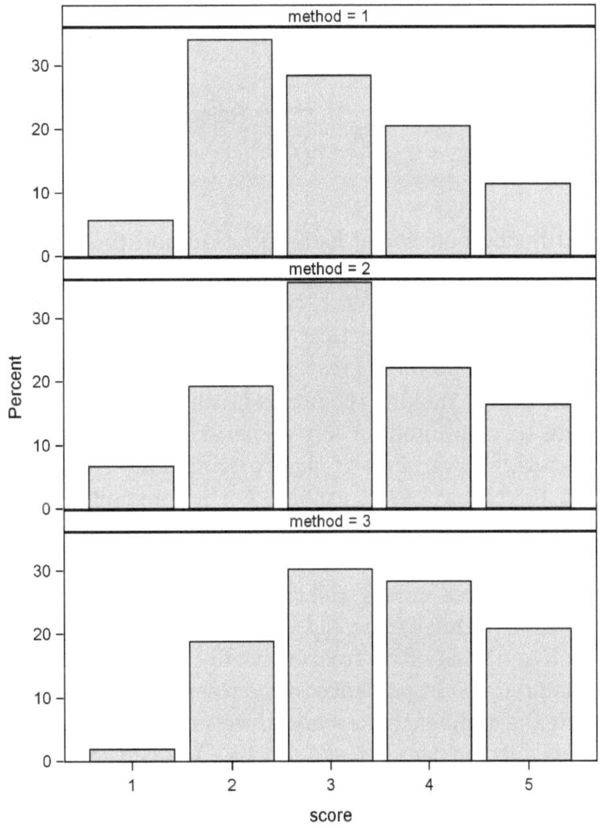

2.7 Determining the Number of Replicates

The significance level, α, of the ANOVA F-test of no treatment effect is the probability of rejecting the null hypothesis $H_0 : \mu_1 = \mu_2, \ldots, = \mu_t$, when it is

COMPLETELY RANDOMIZED DESIGNS WITH ONE FACTOR

true. The power of the test is the probability of rejecting the null hypothesis when it is false. The test statistic msT/msE follows the F-distribution when the null hypothesis is true, but when the null hypothesis is false it follows the noncentral F-distribution. The noncentral F-distribution has a wider spread than the central F-distribution, as shown in Figure 2.10.

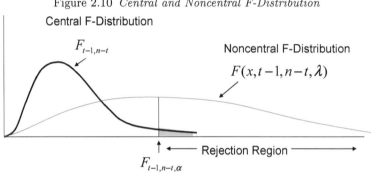

Figure 2.10 *Central and Noncentral F-Distribution*

The spread in the noncentral F-distribution and probability exceeding the critical limit from the central F-distribution is an increasing function of the non-centrality parameter, λ. When the distribution is the non-central F, the probability of exceeding the critical limit from the central F-distribution is called the power. The power is greater than the significance level, α, when the null hypothesis is false making the non-centrality parameter greater than zero. The power can be computed for any scenario of differing means, if the values of the cell means, the variance of the experimental error, and the number of replicates per factor level are specified. For a constant difference among cell means, represented by $\sum_{i=1}^{t}(\mu_i - \bar{\mu}.)^2$, the non-centrality parameter and the power increase as the number of replicates increase.

When the differences among cell means is large enough to have practical importance, the experimenter would like to have high power, or probability of rejecting the hypothesis of no treatment effects. When the difference among the means has practical importance to the researcher we call it *practical significance*. Practical significance does not always correspond to statistical significance as determined by the F-test from the ANOVA. Sometimes the number of replicates in the experiment is too few and the probability or power of detecting a difference of practical significance too low. Statistical significance can be made to coincide with practical significance by determining the appropriate number of replicates that result in the desired power. Doing this is the second technique that falls in the category of *error control* discussed in Chapter 1. The idea that increasing the number of replicates increases the sensitivity of the experiment is also due to Fisher (1935).

For example, if there is a difference among the cell means so that the cor-

DETERMINING THE NUMBER OF REPLICATES

rected sum of squares ($css = \sum_{i=1}^{t}(\mu_i - \bar{\mu}.)^2$) is greater than zero, then the power or probability of rejecting $H_0 : \mu_1 = \mu_2, \ldots, = \mu_t$ is given by

$$\pi(\lambda) = \int_{F_{t-1,t(r-1),\alpha}}^{\infty} F(x, t-1, t(r-1), \lambda) dx \qquad (2.10)$$

where $F_{t-1,t(r-1),\alpha}$ is the αth percentile of the central F distribution with $t-1$ and $t(r-1)$ degrees of freedom, $F(x, t-1, t(r-1), \lambda)$ is the noncentral F-distribution with non-centrality parameter $\lambda = \frac{r}{\sigma^2}\sum_{i=1}^{t}(\mu_i - \bar{\mu}.)^2$. For a fixed value of $\frac{1}{\sigma^2}\sum_{i=1}^{t}(\mu_i - \bar{\mu}.)^2$, the non-centrality parameter, and the power increase as a function of the number of replicates, r. This probability can be calculated for various values of r until a value is found with adequate power. In this way the appropriate number of replicates can be determined. The fprob function in SAS facilitates these computations.

In the bread rise experiment, suppose less than a three-inch difference in risen dough heights is of no consequence. However, if changing the rise time from 35 minutes to 45 minutes causes a difference of more than three inches in risen dough height, the experimenter would like to know about it, because he will need to monitor rise time closely in the future to produce loaves of consistent height. In this case we can regard $\Delta = 3.0$ as a practical difference in cell means. The smallest $css = \sum_{i=1}^{t}(\mu_i - \bar{\mu}.)^2$ could be, with at least two cell means differing by Δ, would be the case when one cell mean was $\Delta/2$ higher than the grand mean, a second was $\Delta/2$ less than the grand mean, and a third was equal to the grand mean. This would result in

$$css = \sum_{i=1}^{t}(\mu_i - \bar{\mu}.)^2 = \left(\frac{\Delta}{2}\right)^2 + 0^2 + \left(-\frac{\Delta}{2}\right)^2 = \left(\frac{\Delta^2}{2}\right) = \left(\frac{3^2}{2}\right) = 4.5$$

Assuming the variance of the experimental error $\hat{\sigma}^2 = 2.1$ was estimated from the sample variance in risen dough heights in a pilot experiment where several loaves were allowed to rise for the same length of time, then the noncentrality factor can be calculated as $\lambda = \frac{r}{2.1} \times (4.5)$. The power is calculated for $r = 2, \ldots, 6$ using the SAS data step commands shown at the top of the next page.

40   COMPLETELY RANDOMIZED DESIGNS WITH ONE FACTOR

```
*Example power computation in SAS data step using Bread Example;
data Power;
do r=2 to 6;
   nu1=3-1; * df for numerator;
   nu2=3*(r-1); * df for denomonator;
   alpha=.05;
   Fcrit=finv(1-alpha,nu1,nu2); *F critical value;
   sigma2=2.1;
   css=4.5;
   nc=r*(css)/sigma2;*noncentrality parameter for noncentral F;
   power=1-probf(Fcrit,nu1,nu2,nc);
   output;
end;
keep r nu1 nu2 nc power;
title Power Calculation in Data Step;
proc print; run;
```

Below are the results.

Power Calculation in Data Step

Obs	r	nu1	nu2	nc	power
1	2	2	3	4.2857	0.19480
2	3	2	6	6.4286	0.40419
3	4	2	9	8.5714	0.59034
4	5	2	12	10.7143	0.73289
5	6	2	15	12.8571	0.83299

From this we can see that with $r = 5$ replicates there would be a 73% chance of detecting a difference in cell means as large as 3.0, and with $r = 6$ there is a 83% chance. With fewer than five replicates there is at least a 40% chance this difference will be missed. As a rule of thumb, the number of replicates that result in power between 0.80 and 0.90 is usually sufficient for most experimental studies.

A similar power calculation can be accomplished with the SAS Analyst Sample Size tool for the One Way ANOVA, with proc power, with the proc glmpower, or with the SAS Power and Sample Size Application. The interactive input for the SAS Analyst Sample Size tool is similar to the data step commands above, and the $css = \sum_{i=1}^{t}(\mu_i - \bar{\mu}.)^2$, must be specified. SAS Analyst will no longer be available after SAS version 9.2.

The SAS proc power, proc glmpower procedure, and the SAS Power and Sample Size Application require the user to enter the values of the cell means, μ_i, that he believes exhibit a practical significance rather than the $css = \sum_{i=1}^{t}(\mu_i - \bar{\mu}.)^2$, and σ rather than σ^2. The following commands will produce an equivalent result to the data step commands above for the bread rise experiment using proc power.

COMPARISON OF TREATMENTS AFTER THE F-TEST

```
* Eample Power Calculation Using proc power;
proc power;
   OneWayANOVA
      Alpha = 0.05
      GroupMeans = (-1.5 0 1.5)
      StdDev = 1.449
      Power = .
      NPerGroup = 2 to 6 by 1;
run;
```

`proc glmpower` allows for more complicated models and will be illustrated in Chapter 3. The SAS Power and Sample Size Application is accessed by **Start ▶ Programs ▶ SAS ▶ SAS Power and Sample Size ▶ SAS Power and Sample Size 3.1**(or the latest release). It provides a GUI interface for `proc power` and `proc glmpower`.

2.8 Comparison of Treatments after the F-Test

When the F-test for the null hypothesis $H_0 : \mu_1 = \mu_2 = \ldots \mu_t$ is rejected, it tells us that there are significant differences between at least two of the cell means, but if there are several levels of the treatment factor, it does not necessarily mean that all cell means are significantly different from each other. When the null hypothesis is rejected, further investigation should be conducted to find out exactly which cell means differ. In some cases the investigator will have preplanned comparisons he would like to make; in other situations he may have no idea what differences to look for.

2.8.1 Preplanned Comparisons

Considering the treatment factor levels in the sugar beet yield experiment conducted at Rothamstead in 1937 and described in Section 2.3, some preplanned comparisons that might have been of interest are:

1. $H_0 : \mu_1 = \frac{1}{3}(\mu_2 + \mu_3 + \mu_4)$
2. $H_0 : \mu_2 = \mu_3$
3. $H_0 : \mu_3 = \mu_4$

The first comparison asks the question: Does a mix of artificial fertilizers change yield? The second comparison asks the question: Is there a difference in yields between plowed and broadcast application of artificial fertilizer? The third comparison asks the question: Does timing of the application change the yield?

These hypotheses can all be expressed in the general form $H_0 : \sum_{i=1}^{t} c_i \mu_i = 0$, where $\sum_{i=1}^{t} c_i = 0$. Since $\sum_{i=1}^{t} c_i \mu_i = 0$ are estimable functions, each of these hypotheses can be tested by computing the single estimable function $L\hat{\beta}$ and its standard error $s_{L\hat{\beta}} = \sqrt{\hat{\sigma}^2 L'(X'X)^- L}$. The ratio of the estimable function to its standard error follows the t-distribution. The `estimate` command in SAS `proc glm` performs this test. For the sugar beet experiment the commands on the next page produce the output that follows the commands.

42 COMPLETELY RANDOMIZED DESIGNS WITH ONE FACTOR

```
proc glm;
  class treat;
  model yield=treat;
  estimate 'fertilizer effect' treat 1 -.33333 -.33333 -.33333;
  estimate 'plowed vs broadcast' treat 0 1 -1 0;
  estimate 'January vs April' treat 0 0 1 -1;
  means treat/dunnettl('A');
run;
```

Parameter	Estimate	Standard Error	t Value	Pr > |t|
fertilizer effect	-8.80001200	0.82520231	-10.66	<.0001
plowed vs broadcast	-3.80000000	0.97518951	-3.90	0.0016
January vs April	0.10000000	0.91941749	0.11	0.9149

The P-values in the column labeled Pr >|t|, in the above output, can be interpreted the same way the P-values for the F-statistic were interpreted, and we can see that: (1) artificial fertilizers enhance yield, (2) broadcast application results in higher yields than plowed application, and (3) there is no significant difference in yield between April and January application time.

When factor levels are quantitative, such as the rise time in the bread dough rise experiment, preplanned comparisons often involve looking for the significance of linear or higher order polynomial trends in the response. Contrast coefficients, c_i for testing orthogonal polynomial trends, can be obtained from SAS proc iml orpol function. The required inputs for this function are a vector containing the levels of a quantitative factor and the degree of polynomial desired. The result is an orthogonal matrix with the contrast coefficients desired beginning in the second column. For example, for the bread dough rise experiment the commands

```
proc iml;
  t={35 40 45};
  C=orpol(t);
  print C;
quit;
```

produce the result

```
            C
0.5773503  -0.707107   0.4082483
0.5773503          0  -0.816497
0.5773503   0.7071068  0.4082483
```

The estimate commands in the following call to proc glm take the second and third columns as the coefficients for the linear and quadratic polynomials.

```
proc glm data=bread;
  class time;
  model height=time;
  estimate 'Linear Trend' time -.707107 0 .707107;
  estimate 'Quadratic Trend' time .4082483 -.816497 .4082483;
run;
```

COMPARISON OF TREATMENTS AFTER THE F-TEST

The result of the contrast statement is shown below, where we can see that there is a significant (at the $\alpha = 0.05$ level) linear trend, but no significant quadratic trend.

```
                              Standard
Parameter          Estimate      Error  t Value  Pr > |t|

Linear Trend     2.03293262  0.76546578    2.66    0.0262
Quadratic Trend -1.12268280  0.76546594   -1.47    0.1765
```

2.8.2 Unplanned Comparisons

When a set of preplanned comparisons can be expressed as a saturated set of orthogonal contrasts, like the examples shown in the last section, these comparisons are independent and equivalent to partitioning the overall F-test of $H_0 : \mu_1 = \ldots = \mu_t$. However, if the comparisons are not planned in advance of running the experiment, the analyst might be tempted to choose the comparisons he or she would like to make based on the means of the data. This implicitly means that all possible comparisons have been made. When testing all possible comparisons, each at the $\alpha=0.05$ significance level, the overall significance level can be much higher than 0.05, greater than 50% in some cases. This means that even when there is no difference in the cell means $\mu_1, \ldots, \mu_t$ there could be a high probability of finding one or more comparisons significant when each is tested individually. In order to reduce the overall (or experimentwise) chance of a type I error, an adjustment must be made.

For pairwise comparisons of the form $H_0 : \mu_i = \mu_j$ for $i \neq j$ Tukey's HSD (or honestly significant difference) method adjusts the critical region by using the studentized range statistic instead of the student's t-distribution. Using the HSD reject $H_0 : \mu_i = \mu_j$ in favor of the alternative $H_a : \mu_i \neq \mu_j$ if $|\hat{\mu}_i - \hat{\mu}_j| > q_{I,n-t,\alpha/2} s_{\hat{\mu}_i - \hat{\mu}_j}$ where $q_{I,n-t,\alpha/2}$ is the $\alpha/2$ percentile of the studentized range. If $X_1, \ldots, X_I$ are independent random variables following $N(\mu, \sigma^2)$ and $R = \max_i X_i - \min_i X_i$ then $R/\hat{\sigma}$ follows the studentized range distribution (see Tukey, 1949a).

There are two different commands in SAS **proc glm** that will do pairwise comparisons using Tukey's HSD method. Both produce differently formatted output. The first is the means statement with the **tukey** option as shown below.

```
proc glm data=Sugarbeet;
  class treat;
  model yield=treat;
  means treat/tukey;
run;
```

The output illustrates the results from the sugar beet experiment and is shown on the next page.

44 COMPLETELY RANDOMIZED DESIGNS WITH ONE FACTOR

Comparisons significant at the 0.05 level are indicated by ***.

treat Comparison	Difference Between Means	Simultaneous 95% Confidence Limits		
C - D	0.1000	-2.5723	2.7723	
C - B	3.8000	0.9655	6.6345	***
C - A	10.1000	7.2655	12.9345	***
D - C	-0.1000	-2.7723	2.5723	
D - B	3.7000	0.8655	6.5345	***
D - A	10.0000	7.1655	12.8345	***
B - C	-3.8000	-6.6345	-0.9655	***
B - D	-3.7000	-6.5345	-0.8655	***
B - A	6.3000	3.3122	9.2878	***
A - C	-10.1000	-12.9345	-7.2655	***
A - D	-10.0000	-12.8345	-7.1655	***
A - B	-6.3000	-9.2878	-3.3122	***

The first column of the output lists the comparison made, the next column lists the difference in cell means, and the next column is a 95% confidence interval on the difference of means of the form $|\mu_i - \mu_j| \pm q_{I,n-t,0.025} s_{\hat{\mu}_i.-\hat{\mu}_j.}$. The final column is an indicator whether the null hypothesis has been rejected. For example, the confidence interval for the first comparison, $\mu_C - \mu_D$ includes zero and therefore no indicator (***) for significance is present, and the sugar beet yield for treatment (C - artificial applied broadcast in January) is not significantly different than the yield for treatment (D - artificial applied broadcast in April). All other pairwise comparisons show a significant difference.

Another way of obtaining the same result is through using the lsmeans statement with the adjust=tukey option. More on the difference in the means statement and the lsmeans statement will be given in Chapter 3. The command to do this is:

```
proc glm data=Sugarbeet;
  class treat;
  model yield=treat;
  lsmeans treat/pdiff adjust=tukey;
run;
```

and the output is shown on the next page.

In this output the actual cell means are listed at the top, followed by a table of P-values for the pairwise comparisons using Tukey's HSD method. The P-value in the third column fourth row 0.9995>.0.05 indicating the mean for treatment C is not significantly different than the mean for treatment D. All other P-values are less than 0.05 providing the same information as the means statement.

A less conservative method of comparing all possible cell means was developed independently by Newman (1939) and Keuls (1952). This method is also

COMPARISON OF TREATMENTS AFTER THE F-TEST

```
              The GLM Procedure
              Least Squares Means
 Adjustment for Multiple Comparisons: Tukey-Kramer

                                    LSMEAN
           treat    yield LSMEAN    Number

            A        38.7000000       1
            B        45.0000000       2
            C        48.8000000       3
            D        48.7000000       4


       Least Squares Means for effect treat
       Pr > |t| for H0: LSMean(i)=LSMean(j)

             Dependent Variable: yield
```

i/j	1	2	3	4
1		0.0001	<.0001	<.0001
2	0.0001		0.0078	0.0094
3	<.0001	0.0078		0.9995
4	<.0001	0.0094	0.9995	

based on the studentized range statistic, but is based on the range of the particular pair of means being compared, within the entire set of ordered means, rather than the range of the largest - smallest as Tukey's HSD. The means comparison using the student Newman-Keuls method can be made using the **snk** option of the **means** statement as follows:

```
proc glm data=Sugarbeet;
  class treat;
  model yield=treat;
  means treat/snk;
run;
```

The output is shown below.

```
Means with the same letter are not significantly different.
```

SNK Grouping	Mean	N	treat
A	48.8000	5	C
A			
A	48.7000	5	D
B	45.0000	4	B
C	38.7000	4	A

In this form of presentation the cell means are listed, followed by the number of replications and treatment factor level, and preceded by a grouping indicator letter. Means preceded by the same grouping letter are not significantly different. Means that are preceded by different letters are significantly different. This is a concise way of indicating differences in treatment means that could be used to represent the results of any method of pairwise comparisons of means. Often in reports the means are written horizontally and underlines are used for the treatment grouping. For example, the results below show the means from an experiment to determine the effect of the download site upon the time to download a file.

$$\begin{array}{ccccc} B & D & A & C & E \\ 2.73 & 3.20 & 3.79 & 4.03 & 5.27 \end{array}$$

The results show that the download time is not significantly different between sites B and D, and not significantly different between sites D and A, but there is a significant difference in the download time between sites B and A. Likewise, there is no significant difference in download times for sites A and C, but the download time for site C is significantly longer than either site B or D. Finally, site E has a significantly longer download time than any of the other sites.

2.8.3 Comparison of All Means to a Control or the Best

In some experiments one of the treatment levels is the current or default level and the others are new or experimental levels. One of the main objectives in this type of experiment might be to compare the mean of each experimental level to the default, or sometimes called the *control* level. Dunnett (1955) developed a method to do this and control the experimentwise type I error rate. In the sugar beet yield experiment, treatment level (A - no artificial fertilizer) can be thought of as the control. To compare all other treatment levels to this one using SAS `proc glm`, the `dunnett` option for either the `means` or `lsmeans` can be used as shown below.

```
proc glm data=Sugarbeet;
  class treat;
  model yield=treat;
  means treat/dunnett('A');
  lsmeans treat/pdiff=control('A') adjust=dunnett;
run;
```

The results from `means` or `lsmeans` will be the same, but printed in a different format like the result of Tukey's HSD method shown in the last section.

When comparing all treatment levels to a control, the desired direction of the difference is often known. Therefore, a one-tailed test, rather than a two-tailed test, may be required. To get a one-tailed test to see if the experimental means are larger than the control means, change the **means** statement above to **means treat/dunnettu('A');** This will give a one-sided lower confidence bound on the difference in means (experimental-control). If the lower bound is greater than zero, the experimental mean is significantly higher than the control mean. The same result can be obtained from the **lsmeans** by changing the command to **lsmeans treat/pdiff=controlu('A') cl adjust=dunnett;**. To get a one-sided test of whether the experimental means are less than the control, change **dunnettu** to **dunnettl** in the **means** statement or **controlu('A')** to **controll('A')** in the **lsmeans** statements.

When there is no control level of the treatment factor, there may still be interest in comparing all treatment levels to the best level. For example in an experiment to see the effect of the download site on the time to download a file, it may be of interest to find those sites whose download times are not significantly higher than the site with the minimum observed mean download time. To compare all treatment levels to the observed best level and to control the experimentwise error rate, the MCB procedure (Hsu, 1984) turns out to be equivalent to Dunnett's procedure. Look at the observed cell means in the data, determine which is the best, label the best as the control level, and use the **dunnett** option. For example, in the means for the experiment to compare download times for different download sites shown at the end of Section 2.8.2, site B had the shortest observed download time. To compare all other sites to B use the command:

```
proc glm;
  class site;
  model dtime=site;
  means site/dunnett('B');
run;
```

Those sites that are not significantly different from B can be considered to be a subset of sites that have the fastest download times.

48 COMPLETELY RANDOMIZED DESIGNS WITH ONE FACTOR

2.9 Review of Important Concepts

In order to determine if cause and effect relationships exist and to make predictions about the results of future actions, experiments must be performed wherein certain factors are purposely varied while others are held constant. The one factor design is the simplest case where one factor is varied while all other known factors are held constant.

Figure 2.11 shows a roadmap for selecting an appropriate experimental design. When there is only one factor under study and experimental units are homogeneous, the CRD design should be used as indicated in black in the figure. This is the only situation presented in Chapter 2. As additional designs are presented in subsequent chapters the other branches in Figure 2.10 will be explained.

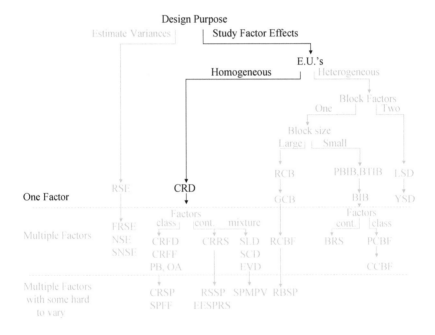

Figure 2.11 *Design Selection Roadmap*

Fisher's technique of randomizing experimental units to treatment levels guarantees the long run validity of the CRD and minimizes the chance that changes in unknown factors or lurking variables from biasing the results. The way a series of experiments is conducted dictates what model should be used for analysis.

The model for the analysis of the CRD or completely randomized for one factor design is $y_{ij} = \mu_i + \epsilon_{ij}$ or $y_{ij} = \mu + \tau_i + \epsilon_{ij}$, where y_{ij} is the observed response for the jth replicate of the ith treatment level, μ_i is the cell mean

REVIEW OF IMPORTANT CONCEPTS

for the ith level, and τ_i is the effect. ϵ_{ij} is the experimental error for the jth observation on treatment level i. ϵ_{ij}s are assumed to be independent and normally distributed with constant variance σ^2. The typical analysis is to fit the linear model by the method of least squares - maximum likelihood and perform a likelihood ratio F-test of the $H_0 : \mu_1 = \ldots = \mu_t$. If the data were not collected in a proper randomized design with replicates, analyzing data in this way may be totally misleading.

The credibility of the conclusions of analysis depends on the degree to which the assumptions are valid. The independence assumption is the most critical and it is guaranteed when replicate experimental units are randomized to treatment factor levels. The other assumptions should be checked. The constant variance and normality assumptions can be checked by plotting the residuals versus cell means and by making a normal probability plot of the residuals. If these assumptions are violated, the data should be analyzed on a transformed scale or by weighted least squares or the method of maximum likelihood for the generalized linear model.

If a significant difference in cell means is found with the overall F-test, further investigation of the differences can be made. If comparisons are planned in advance of running the experiment and can be described by as set of orthogonal comparisons, the overall F-test can be partitioned to test these comparisons. The experimentwise type I error rate for all possible comparisons of means can be controlled by using Tukey's HSD or the less conservative student Newman-Keuls method. For comparing all means to a control level, Dunnett's method should be used and for comparing all means to the best (largest or smallest), Hsu's MCB method should be used.

50 COMPLETELY RANDOMIZED DESIGNS WITH ONE FACTOR

2.10 Exercises

1. Consider the simple experiment described in Section 2.1 and used throughout this chapter to determine the effect of rise time on the height of bread dough. Describe the steps in the planning process for this experiment similar to what was done for this historical example in Section 2.3. Describe some of the potential problems that could occur in performing these experiments (similar to Section 1.7), and discuss how careful planning using the checklist from Chapter 1 could prevent these problems.

2. Paper helicopters can be cut from one half of an $8\frac{1}{2} \times 11$ sheet of paper as shown below.

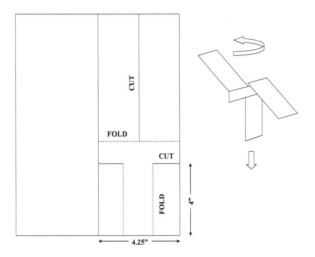

These helicopters can be made quickly and inexpensively, and can be used to demonstrate experimental design concepts. An experiment can be performed by constructing a helicopter, dropping it from a fixed height, and clocking the time it takes to rotate to the floor, as shown below, with a stopwatch. The wing length could be varied by trimming some paper off the top prior to folding the wings. Trimming some paper off would reduce the weight of the helicopter, but would also result in less surface area on the blades. You could experiment to determine if changing the wing length affects the flight time.

(a) Describe the experimental unit.
(b) Explain the difference in replicates and duplicates for this situation.
(c) Describe the treatment factor.
(d) Describe any lurking variables that might affect the results of experiments.
(e) Explain why randomization would be important.

EXERCISES

(f) Create a randomized list of experiments for examining four wing lengths of (4", 4.75", 5.5" and 6") with eight replicate experiments in each level.
(g) Carry out the experiments and collect the data.
(h) Perform an analysis of variance with your data.
(i) Check the equal variance and normality assumptions with residual plots.
(j) Test whether there is a significant linear or quadratic trend in flight times as the wing length increases.

3. In Section 2.8.2 an experiment for determining the effect of the download site selected upon the time to download a file was discussed. In this experiment:

(a) Describe the experimental unit.
(b) Describe the treatment factor.
(c) Describe the response.
(d) Discuss the causes for experimental error in this experiment and why the principles of replication and randomization would be important in reaching a valid conclusion.

4. In an experiment to study the effect of the amount of baking powder in a biscuit dough upon the rise heights of the biscuits, four levels of baking powder were tested and four replicate biscuits were made with each level in a random order. The results are shown in the table below.

.25 tsp	.5 tsp	.75 tsp	1 tsp
11.4	27.8	47.6	61.6
11.0	29.2	47.0	62.4
11.3	26.8	47.3	63.0
9.5	26.0	45.5	63.9

(a) What is the experimental unit?
(b) Perform the Analysis of Variance to test the hypothesis of no treatment effect.
(c) Formulate a contrast to test the hypothesis that increase in rise height is proportional to the increase in baking powder in the dough and test this hypothesis.
(d) Estimate the variance of the experimental error σ^2.
(e) Make a plot of residuals versus predicted values and normal plot of residuals and comment on whether the assumptions of the linear model are justified.
(f) If the dough were made in batches and the four replicate biscuit rise heights in each column (shown in the table above) were all from the same batch, would your answer to (a) be different? How could the data be analyzed if this were the case?

52   COMPLETELY RANDOMIZED DESIGNS WITH ONE FACTOR

5. The effect of plant growth regulators and spear bud scales on spear elongation in asparagus was investigated by Yang-Gyu and Woolley (2006). Elongation rate of spears is an important factor determining final yield of asparagus in many temperate climatic conditions. Spears were harvested from 6-year-old Jersey Giant asparagus plants grown in a commercial planting at Bulls (latitude 40.2S, longitude 175.4E), New Zealand. Spears were harvested randomly and transported from field to lab for investigation. After trimming to 80mm length, spears were immersed completely for 1 h in aqueous solutions of 10 mg l-1 concentration of indole-3-acetic acid (IAA), abscisic acid (ABA), GA3 or CPPU (Sitofex EC 2.0%; SKW, Trostberg, Germany) in test tubes. Control spears were submerged in distilled water for 1 h. The experiment was a completely randomized design with five replications (spears) per treatment. The resulting data (final spear length in mm) is shown below.

Control	IAA	ABA	GA3	CPPU
94.7	89.9	96.8	99.1	104.4
96.1	94.0	87.8	95.3	98.9
86.5	99.1	89.1	94.6	98.9
98.5	92.8	91.1	93.1	106.5
94.9	99.4	89.4	95.7	104.8

(a) Perform the Analysis of Variance to test the hypothesis of no treatment effect.

(b) Use the Tukey method to test all pairwise comparisons of treatment means.

(c) Use the Dunnett procedure to compare all treatment group means to the control mean.

6. Consider an experimental situation where the investigator was interested in detecting a maximum difference in treatment means that is twice the standard deviation of the response measured on replicate experimental units assigned to the same level of the treatment factor, i/e. $\Delta = 2.0\sigma$. If there are 4 levels of the treatment factor:

(a) Modify the SAS data step commands in Section 2.7 to calculate the power for various numbers of replicates r per treatment level.

(b) Use the SAS analyst tool and `proc glmpower` to reproduce your results.

(c) Calculate the number of replicates necessary to have 0.90 power of detecting a difference as large as $\Delta = 2.0\sigma$.

(d) How would the result you got in (c) change if the number of levels of the treatment factor increased to 8, or decreased to 2?

CHAPTER 3

Factorial Designs

3.1 Introduction

In Chapter 2 we examined one-factor designs. These are useful only when one factor is under study. When multiple factors are under study, one classical approach is to study each separately while holding all others constant. Fisher (1935) pointed out that this approach is useful for demonstrating known relationships to students in laboratory courses where the influence of other factors is known, but it is both inefficient and potentially misleading when it comes to discovering new knowledge through experimentation. A much better strategy for experimenting with multiple factors is to use a factorial design. In a factorial design the cells consist of all possible combinations of the levels of the factors under study. Factorial designs accentuate the factor effects, allow for estimation of interdependency of effects (or *interactions*) and are the first technique in the category of what is called *treatment design*.

By examining all possible combinations of factor levels, the number of replicates of a specific level of one factor is increased by the product of the number of levels of all other factors in the design, and thus the same power or precision can be obtained with fewer replicates. In addition, if the effect of one factor changes depending on the level of another factor, it will be seen in a factorial plan. This phenomenon will be missed in the classical approach where each factor is only varied at constant levels of the other factors. The example in the next section will illustrate these facts.

3.2 Classical One at a Time versus Factorial Plans

In the exercises for Chapter 2, a set of experiments with paper helicopters was described. In those experiments only one factor, the wing length, was under study. However, to maximize the flight time of paper helicopters, it would be advisable to consider more than one factor. For example, consider varying wing length over 4 levels as before, and the body width over four levels, such as 4.25", 4.0", 3.75" and 3.5". The left side of Figure 3.1 represents the classical plan in which one factor is varied at a time. The circles in the diagram represent experiments or runs. Using this approach, the experiments across the bottom of the figure would be completed by varying wing length while holding body width constant at 3.5". Next, the three additional experiments up the left side of the figure would be completed by varying body width while holding the wing length constant at its low level of 4.0". If eight replicate runs were to be made for each of these experiments (as suggested in exercise 1 of Chapter 2), a total of 56 experiments would be required.

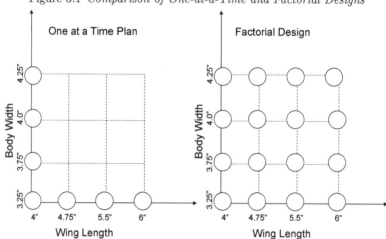

Figure 3.1 *Comparison of One-at-a-Time and Factorial Designs*

If the objective were to find the combination with the longest flight time, the classical approach would be to complete the experiments with one factor first. Next one would calculate the cell means and then select the level with the highest mean. Finally, the second factor would be varied while holding the first constant at its optimal level, not the lowest level as shown in Figure 3.1. However, Fisher's caution to randomize would tell you this is a bad strategy. If any unknown forces changed after the first set of experiments, the results could be biased. Additionally, the optimal level of one factor may depend upon the level of the other factor. Therefore, by varying one factor at a time, the overall optimum may be missed.

The diagram on the right side of Figure 3.1 represents a factorial plan for the helicopter experiments. Here it can be seen that experiments are run at all combinations of the levels of the two factors. In this plan, if two replicates of each cell are completed, there will be eight replicates of each level of wing length, and eight replicates of each level of body width which is equivalent to the one-at-a-time plan. Therefore the factorial plan would have the same precision or power for detecting factor effects as the one-at-a-time plan, but is more efficient since it requires only $2 \times 16 = 32$ total runs as opposed to the 56 required by the one-at-a-time plan. The number of replicates of each factor level in the factorial design is equal to the number of replicates per cell times the product of the levels of all other factors in the design. This multiplication is referred to as *hidden replication*. In the case shown in Figure 3.1, there are only two factors each at four levels; therefore, the number of replicates of each factor level is $2 \times 4 = 8$. In the factorial plan, the 32 treatment

INTERPRETING INTERACTIONS

combinations would be randomized to experimental units, thus preventing biases from unknown sources.

In more complicated research problems many treatment factors may be under study. The efficiency of factorial designs can be demonstrated even in the simplest case where each factor has only two levels. For example, consider a design with 4 factors. A factorial design would require all combinations of four factors at two levels, or $2^4 = 16$ cells. If two replicates were run for each cell, there would be a total of $2 \times 16 = 32$ experiments or runs. To examine the effect of any one of the four factors, half the runs (or $2 \times 2^3 = 16$ due to the hidden replication) would be at one level of the factor and half at the other level. Thus the treatment effect would consist of a difference of two averages of 16. Results from the same 32 experiments can be used to calculate the treatment effect for each of the four factors. To have equal precision for comparing treatment effects using a one-at-a-time plan, 32 runs would be required for comparing the levels of each factor while holding the others constant. This would result in $4 \times 32 = 128$ experiments, or four times the number required for a factorial design!

3.3 Interpreting Interactions

If there is an interaction or joint effect between two factors, the effect of one factor upon the response will differ depending on the level of the other factor. This can be illustrated graphically in Figure 3.2. On the left side of the figure is a contour plot representing the results of a series of experiments with paper helicopters. This plot can be interpreted like a topological map with the lines representing contours of equal flight time. You can simulate what would happen in a series of experiments where wing length was held constant

Figure 3.2 *Contour Plot of Flight Time for Helicopter Experiment*

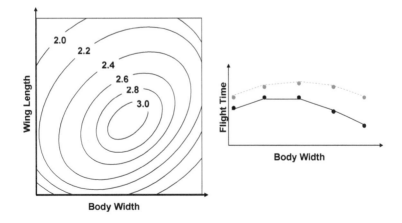

and body width varied by drawing a straight line parallel to the body width axis across the contour plot. The flight time for various runs can be read off as the label for the contour lines the straight line intersects. For example, if wing length were held constant a value below its mid-point on the left of the contour plot, the flight times resulting from five runs with varying body width are represented as the black line traced on the graph at the right in Figure 3.2. If the wing length were held constant at a higher value, the grey line indicates what the result of a series of experiments with body width might look like. The fact that the two lines or curves on the right side of the figure are not parallel indicates there is an interaction between wing length and body width. They show that the effect of body width depends upon the wing length.

Interactions are common in the real world, but using the classical one-at-a-time strategy of experimentation tacitly assumes that interactions do not exist. To see the fallacy that results from this assumption, examine Figure 3.3, which represents what would happen if one were to search for the optimum combination of wing length and body width.

Figure 3.3 *One-at-a-Time Optimization with Paper Helicopters*

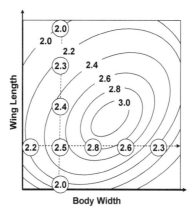

The vertical set of circles are drawn at the wing length, body width combinations for a series of experiments that vary wing length while holding body width constant. The numbers within the circles represent the resulting flight times. After examining the result of this series of experiments, the optimal wing length would be chosen and another series of experiments would be conducted by holding wing length constant at its optimal value and varying body width. The results of these experiments can be visualized as the horizontal

INTERPRETING INTERACTIONS

series of circles. The maximum result, 2.8, is not the overall optimum, because the optimal wing length depends on the body width and vice versa.

Table 3.1 *Mooney Viscosity of Silica B at* $100°C$

Naphthene Oil (phr)	Filler (phr)					
	0	12	24	36	48	60
0	25	30	35	40	50	60
10	18	21	24	28	33	41
20	13	15	17	20	24	29
30	11	14	15	17	18	25

When the effect of the factors is close to linear, the interaction is easier to explain in words. Table 3.1 shows the results of a factorial experiment conducted by Derringer (1974) to determine the effect of elastomer compounds on the viscosity silica at 100°C. The elastomer compounds were Naphthene Oil, studied at 4 levels, and Filler content, studied at 6 levels.

Figure 3.4 shows a graphical representation of the data in the table. This figure is called an interaction plot. As Filler is increased from 0 to 60, the viscosity increases along a fairly linear trend. However, the slope of the trend line depends upon the level of Naphthene Oil. When there is no Naphthene Oil added, increasing Filler from 0 to 60 causes viscosity to increase rapidly from 25 to 60; but when there is 30 phr of Naphthene Oil, increasing Filler from 0 to 60 causes a more gradual increase in viscosity from 11 to 25.

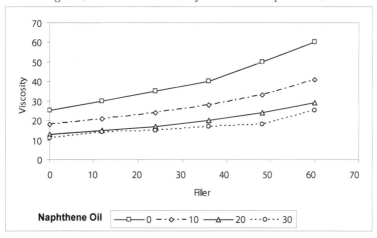

Figure 3.4 *Interaction Plot of Filler and Naphthene Oil*

Since interactions are common in factorial experiments, it is important to learn how to explain or interpret an interaction in order to clearly present the results of research studies. This is best done by describing the effect of one factor upon the response, and then contrasting or comparing how that effect changes depending on the level of the other factor. An interaction plot, like Figure 3.4, can guide the description or interpretation. Many more examples of interpreting interactions will be given throughout this chapter and the remainder of the book.

3.4 Creating a Two-Factor Factorial Plan in SAS

A factorial design can be easily created using nested do loops in the SAS data step. For example, to create a factorial design to study paper helicopters with three levels of body width (3.25", 3.75" and 4.25") and three levels of wing length (4.0", 5.0" and 6.0") the following SAS data step creates all possible combinations.

```
data copter;
  do BW = 3.25 to 4.25 by .5;
    do WL = 4 to 6 by 1;
      output;
    end;
  end;
run;
```

To add two replicates of each cell and a random number for later use in randomizing the list, add an additional nested do loop as shown below.

```
data copter;
  do BW = 3.25 to 4.25 by .5;
    do WL = 4 to 6 by 1;
      do rep = 1 to 2;
        u=ranuni(0);
        output;
      end;
    end;
  end;
run;
```

To randomize the order of the factorial design or CRFD, add the following data step.

```
proc sort data=copter; by u;
data list; set copter;
  experiment=_n_;
title Paper Helicopter Experiments;
proc print;
  var BW WL;
  id experiment;
run;
```

CREATING A TWO-FACTOR FACTORIAL PLAN IN SAS

This results in the following randomized list.

```
Paper Helicopter Experiments

experiment        BW        WL
    1            3.25        4
    2            3.75        6
    3            4.25        4
    4            4.25        6
    5            4.25        5
    6            3.25        5
    7            3.75        5
    8            3.75        6
    9            3.75        4
   10            3.75        4
   11            3.25        6
   12            4.25        4
   13            3.25        5
   14            3.75        5
   15            4.25        6
   16            3.25        4
   17            4.25        5
   18            3.25        6
```

This list shows that the first experiment consists of constructing a helicopter with a body width of 3.25" and a wing length of 4", dropping it from a fixed height and timing its flight. The second experiment consists of constructing a helicopter with a body width of 3.75" and a wing length of 6", dropping it from the same fixed height and timing its flight, etc. The randomization will help prevent biases from any lurking variables such as changes in air currents, changes in temperature or learning curves in dropping or timing helicopters. This randomized list could be converted to a double spaced data collection form as shown in the example in Section 2.2.

By utilizing **proc factex** a more compact set of commands can be obtained for creating a randomized list for a factorial design. The commands below will produce a randomized list in the file **copter** that is similar to that shown above.

```
proc factex;
  factors BW WL/nlev=3;
  model estimate=(BW|WL);
  output out=copter randomize designrep=2
  BW nvals=(3.25 3.75 4.25) WL nvals=(4 5 6);
run;
proc print data=copter;
run;
```

The **nlev=3** option on the factors statement specifies three levels for body width and wing length. The default number of levels in **proc factex** is 2. The **estimate=(BW|WL)** specifies that the design should allow estimation of the body width and wing length main effects and their interaction. Finally the **output** statement stores the list in the file. The **designrep** option specifies two replicates of each combination of factor levels, the **randomize** option randomizes the order of the list and the `nvals=` option specifies numerical values for the levels of the two factors. If the levels were qualitative like (low, mid, high), the `cvals=('low' 'mid' 'high')` option could be used instead of `nvals`. For a full factorial `proc factex` produces no output itself. Therefore `proc print` command is used to print the randomized list.

3.5 Analysis of a Two-Factor Factorial in SAS

The mathematical model for a completely randomized two-factor factorial design can be written as:

$$y_{ijk} = \mu_{ij} + \epsilon_{ijk} \tag{3.1}$$

where i represents the level of the first factor, j represents the level of the second factor, and k represents the replicate number. This model is called the cell means model and μ_{ij} represents the expected response in the ijth cell.

Another way of representing the model is the effects model

$$y_{ijk} = \mu + \alpha_i + \beta_j + \alpha\beta_{ij} + \epsilon_{ijk} \tag{3.2}$$

In this model α_i, β_j are the main effects and represent the difference between the marginal average of all experiments at the ith level of the first factor and the overall average, and the difference between the marginal average at the jth level of the second factor and the overall average, respectively. The interaction effects, $\alpha\beta_{ij}$, represent the difference between the cell mean, μ_{ij} and $\mu + \alpha_i + \beta_j$.

The usual assumptions are that the experimental errors are independent and $\epsilon_{ijk} \sim N(0, \sigma^2)$. The independence assumption is guaranteed if the treatment combinations are randomly assigned to the experimental units, and the equal variance and normality assumptions can be verified with a residual versus predicted plot and a normal probability plot of the residuals as described in Section 2.4.

Estimable functions, as described for the one-factor model in Section 2.3.4, are linear combinations of the cell means that can be expressed as a linear combination of the data. For the two-factor factorial the cell means, $\mu_{ij} = \mu + \alpha_i + \beta_j + \alpha\beta_{ij}$, are estimable functions but the individual effects, α_i, β_j, and $\alpha\beta_{ij}$ are not estimable functions. Contrasts among the effects such as $\sum_i c_i \alpha_i$, and $\sum_j c_j \beta_j$, where $\sum_i c_i = 0, \sum_j c_j = 0$ and $\sum_i \sum_j b_{ij} \alpha\beta_{ij}$, where $\sum_i b_{ij} = 0, \sum_j b_{ij} = 0$ are estimable. These estimable functions and their standard errors can be computed with the `estimate` command in SAS `proc glm`. The marginal means $\mu + \alpha_i + \overline{\alpha\beta}_{i.}$ and $\mu + \beta_j + \overline{\alpha\beta}_{.j}$ are estimable functions and they and the cell means can be computed using the `means` or `lsmeans` statement in SAS `proc glm`.

3.5.1 Matrix Representation of Model and Analysis

The effects model can be represented in matrix notation as

$$y = X\beta + \epsilon = \begin{pmatrix} 1 & | & X_A & | & X_B & | & X_{AB} \end{pmatrix} \begin{pmatrix} \mu \\ \beta_A \\ \beta_B \\ \beta_{AB} \end{pmatrix} + \epsilon \qquad (3.3)$$

For example, consider a case where the first factor has two levels, the second factor has three levels, and there are two replicates per cell. Then

$$\begin{pmatrix} y_{111} \\ y_{112} \\ y_{211} \\ y_{212} \\ y_{121} \\ y_{122} \\ y_{221} \\ y_{222} \\ y_{131} \\ y_{132} \\ y_{231} \\ y_{232} \end{pmatrix} = \begin{pmatrix} 1 & 1 & 0 & 1 & 0 & 0 & 1 & 0 & 0 & 0 & 0 & 0 \\ 1 & 1 & 0 & 1 & 0 & 0 & 1 & 0 & 0 & 0 & 0 & 0 \\ 1 & 0 & 1 & 1 & 0 & 0 & 0 & 1 & 0 & 0 & 0 & 0 \\ 1 & 0 & 1 & 1 & 0 & 0 & 0 & 1 & 0 & 0 & 0 & 0 \\ 1 & 1 & 0 & 0 & 1 & 0 & 0 & 0 & 1 & 0 & 0 & 0 \\ 1 & 1 & 0 & 0 & 1 & 0 & 0 & 0 & 1 & 0 & 0 & 0 \\ 1 & 0 & 1 & 0 & 1 & 0 & 0 & 0 & 0 & 1 & 0 & 0 \\ 1 & 0 & 1 & 0 & 1 & 0 & 0 & 0 & 0 & 1 & 0 & 0 \\ 1 & 1 & 0 & 0 & 0 & 1 & 0 & 0 & 0 & 0 & 1 & 0 \\ 1 & 1 & 0 & 0 & 0 & 1 & 0 & 0 & 0 & 0 & 1 & 0 \\ 1 & 0 & 1 & 0 & 0 & 1 & 0 & 0 & 0 & 0 & 0 & 1 \\ 1 & 0 & 1 & 0 & 0 & 1 & 0 & 0 & 0 & 0 & 0 & 1 \end{pmatrix} \begin{pmatrix} \mu \\ \alpha_1 \\ \alpha_2 \\ \beta_1 \\ \beta_2 \\ \beta_3 \\ \alpha\beta_{11} \\ \alpha\beta_{21} \\ \alpha\beta_{12} \\ \alpha\beta_{22} \\ \alpha\beta_{13} \\ \alpha\beta_{23} \end{pmatrix} + \begin{pmatrix} \epsilon_{111} \\ \epsilon_{112} \\ \epsilon_{211} \\ \epsilon_{212} \\ \epsilon_{121} \\ \epsilon_{122} \\ \epsilon_{221} \\ \epsilon_{222} \\ \epsilon_{131} \\ \epsilon_{132} \\ \epsilon_{231} \\ \epsilon_{232} \end{pmatrix}$$

The $X'X$ is singular and to solve the normal equations SAS `proc glm` uses a generalized inverse, $(X'X)^-$ which is equivalent to placing the restriction $\alpha_2 = \beta_3 = \alpha\beta_{21} = \alpha\beta_{22} = \alpha\beta_{13} = \alpha\beta_{23} = 0$. This is the same thing that was done to get a solution for the one-factor model in Sections 2.3.1 and 2.3.2.

The error sum of squares $ssE = y'y - \hat{\beta}'X'y = y'(I - X(X'X)^- X')y$, where $\hat{\beta} = (X'X)^- X'y$ are the estimates that `proc glm` designates as biased. To test the hypothesis $H_0 : \alpha_1 = \alpha_2 = 0$, $H_0 : \beta_1 = \beta_2 = \beta_3 = 0$ and $H_0 : \alpha\beta_{11} = \alpha\beta_{21} = \alpha\beta_{12} = \alpha\beta_{22} = \alpha\beta_{13} = \alpha\beta_{23} = 0$, the likelihood ratio F-tests are obtained by calculating ratios of the ANOVA means squares. What SAS `proc glm` designates as the type I sums of squares for factor A is $ssA = \hat{\beta}'X'y - (1'y)^2/(1'1)$, where the model is simplified to include only the effects for the first factor, i.e., $X = \begin{pmatrix} 1 & | & X_A \end{pmatrix}$. The error sums of squares for this simplified model is ssE_A. The type I sums of squares for factor A is denoted $R(\alpha|\mu)$. The type I sums of squares for factor B is denoted

$R(\beta|\alpha,\mu)=ssE_A - ssE_B$ where ssE_B is the error sums of squares from the reduced model where $X = (\ 1 \ | \ X_A \ | \ X_B \ )$. Finally, the type I sums of squares for the interaction AB is denoted $R(\alpha\beta|\beta,\alpha,\mu)=ssE_B - ssE$. In general when there are a levels of factor A, b levels of factor B, and r replicates per cell, the type I ANOVA Table for the two-factor factorial design can be represented symbolically as shown in Table 3.2.

Table 3.2 *Analysis of Variance Table*

Source	df	Sum of Squares	Mean Squares	F-ratio
A	$a-1$	$R(\alpha\|\mu)$	$\frac{ssA}{(a-1)}$	$F = \frac{msA}{msE}$
B	$b-1$	$R(\beta\|\alpha,\mu)$	$\frac{ssB}{(b-1)}$	$F = \frac{msB}{msE}$
AB	$(a-1)(b-1)$	$R(\alpha\beta\|\beta,\alpha,\mu)$	$\frac{ssAB}{(a-1)(b-1)}$	$F = \frac{msAB}{msE}$
Error	$ab(r-1)$	ssE	$\frac{ssE}{ab(r-1)}$	

The sums of squares ssA, ssB, and $ssAB$ can also be written in the form

$$ssA = (L_\alpha\hat{\beta})'(L_\alpha(X'X)^- L'_\alpha)^{-1}(L_\alpha\hat{\beta})$$
$$ssB = (L_\beta\hat{\beta})'(L_\beta(X'X)^- L'_\beta)^{-1}(L_\beta\hat{\beta})$$
$$ssAB = (L_{\alpha\beta}\hat{\beta})'(L_{\alpha\beta}(X'X)^- L_{\alpha\beta}')^{-1}(L_{\alpha\beta}\hat{\beta}),$$

where L_α, L_β and $L_{\alpha\beta}$ are contrast matrices computed internally by SAS proc glm, see Goodnight (1980). Under the null hypotheses the F-ratios msA/msE, msB/msE and $msAB/msE$ follow the F-distribution with the degrees of freedom shown in the table, and under the alternative they follow the noncentral F-distribution. The noncentrality parameter for $F = msA/msE$ is given by the expression $\lambda_\alpha = (\sigma^2)^{-1}(L_\alpha\hat{\beta})'(L_\alpha(X'X)^- L_\alpha')^{-1}(L_\alpha\hat{\beta})$. The non-centrality parameters for the F-ratios msB/msE and $msAB/msE$ are similarly given. When there is an equal number, r, of replicates in each cell, the noncentrality parameters can be shown to be equal to

$$\lambda_\alpha = br\sum_i \alpha_i^2/\sigma^2 \tag{3.4}$$

$$\lambda_\beta = ar\sum_j \beta_j^2/\sigma^2 \tag{3.5}$$

and

$$\lambda_{\alpha\beta} = r\sum_i\sum_j \alpha\beta_{ij}^2/\sigma^2 \tag{3.6}$$

To illustrate the analysis of a two-factor factorial experiment in using SAS proc glm consider the data in Table 3.3. These are the results of a two-factor experiment given by Hunter (1989). In this data, an experiment consisted

ANALYSIS OF A TWO-FACTOR FACTORIAL IN SAS 63

of burning an amount of fuel and determining the CO emissions released. The experimental unit is the portion of a standard fuel required for one run, and the response, y, is the carbon monoxide (CO) emissions concentration in grams/meter^3 determined from that run. Factor A is the amount of ethanol added to an experimental unit or portion of the standard fuel, and factor B is the fuel-to-air ratio used during the burn of that fuel.

Table 3.3 *Data from Ethanol Fuel Experiment*

A=ethanol additions	B=air/fuel ratio	y=CO emissions
0.1	14	66, 62
0.1	15	72, 67
0.1	16	68, 66
0.2	14	78, 81
0.2	15	80, 81
0.2	16	66, 69
0.3	14	90, 94
0.3	15	75, 78
0.3	16	60, 58

The SAS commands to read in this data and produce the analysis are shown below.

```
Data CO;
input Ethanol Ratio CO;
datalines;
.1 14 66
.1 14 62
... etc.
;
ods graphics on;
proc glm;
  class Ethanol Ratio;
  model CO=Ethanol Ratio Ethanol*Ratio/solution;
  lsmeans Ethanol Ratio Ethanol*Ratio;
  estimate '0.3 vs 0.1' Ethanol -1 0 1;
  estimate '16 vs 14' Ratio -1 0 1;
run;
ods graphics off;
```

The resulting ANOVA table is shown on the next page. There it can be seen that `proc glm` produces a table of the type I sums of squares, as described earlier, and a table of type III sums of squares. For the case where there is an equal number of replications in each cell, these two tables will always be identical. It can be seen from the tables that the two effects and their interaction are significant as indicated by the P-values to the right of the F-values.

The result of the `/solution` option on the model statement produces the table of parameters shown below the ANOVA table. SAS labels these estimates

FACTORIAL DESIGNS

Source	DF	Sum of Squares	Mean Square	F Value	Pr > F
Model	8	1654.000000	206.750000	40.02	<.0001
Error	9	46.500000	5.166667		
Corrected Total	17	1700.500000			

Source	DF	Type I SS	Mean Square	F Value	Pr > F
Ethanol	2	324.0000000	162.0000000	31.35	<.0001
Ratio	2	652.0000000	326.0000000	63.10	<.0001
Ethanol*Ratio	4	678.0000000	169.5000000	32.81	<.0001

Source	DF	Type III SS	Mean Square	F Value	Pr > F
Ethanol	2	324.0000000	162.0000000	31.35	<.0001
Ratio	2	652.0000000	326.0000000	63.10	<.0001
Ethanol*Ratio	4	678.0000000	169.5000000	32.81	<.0001

as not unique or biased as can be clearly seen. This is a result of the restrictions $\alpha_3 = \beta_3 = \alpha\beta_{13} = \alpha\beta_{23} = \alpha\beta_{31} = \alpha\beta_{32} = \alpha\beta_{33} = 0$ required to eliminate singularities in $X'X$.

Parameter			Estimate		Standard Error	t Value	Pr > |t|
Intercept			59.00000000	B	1.60727513	36.71	<.0001
Ethanol	0.1		8.00000000	B	2.27303028	3.52	0.0065
Ethanol	0.2		8.50000000	B	2.27303028	3.74	0.0046
Ethanol	0.3		0.00000000	B	.	.	.
Ratio	14		33.00000000	B	2.27303028	14.52	<.0001
Ratio	15		17.50000000	B	2.27303028	7.70	<.0001
Ratio	16		0.00000000	B	.	.	.
Ethanol*Ratio	0.1	14	-36.00000000	B	3.21455025	-11.20	<.0001
Ethanol*Ratio	0.1	15	-15.00000000	B	3.21455025	-4.67	0.0012
Ethanol*Ratio	0.1	16	0.00000000	B	.	.	.
Ethanol*Ratio	0.2	14	-21.00000000	B	3.21455025	-6.53	0.0001
Ethanol*Ratio	0.2	15	-4.50000000	B	3.21455025	-1.40	0.1951
Ethanol*Ratio	0.2	16	0.00000000	B	.	.	.
Ethanol*Ratio	0.3	14	0.00000000	B	.	.	.
Ethanol*Ratio	0.3	15	0.00000000	B	.	.	.
Ethanol*Ratio	0.3	16	0.00000000	B	.	.	.

NOTE: The X'X matrix has been found to be singular, and a generalized inverse was used to solve the normal equations. Terms whose estimates are followed by the letter 'B' are not uniquely estimable.

The result of the **means** statement is shown on the next section of output. The upper two sections of the table show the means for each factor along with the standard deviation of the values averaged in each mean. If the interaction was not significant, the marginal means would reveal the direction of the factor effects, but further preplanned comparisons or other multiple comparison procedures could be used to draw definite conclusions.

Finally, the result of the two estimate statements is shown next. These

ANALYSIS OF A TWO-FACTOR FACTORIAL IN SAS

Level of Ethanol	N	CO Mean	Std Dev
0.1	6	66.8333333	3.2506410
0.2	6	75.8333333	6.6156380
0.3	6	75.8333333	14.8649476

Level of Ratio	N	CO Mean	Std Dev
14	6	78.5000000	12.7082650
15	6	75.5000000	5.3197744
16	6	64.5000000	4.4609416

Level of Ethanol	Level of Ratio	N	CO Mean	Std Dev
0.1	14	2	64.0000000	2.82842712
0.1	15	2	69.5000000	3.53553391
0.1	16	2	67.0000000	1.41421356
0.2	14	2	79.5000000	2.12132034
0.2	15	2	80.5000000	0.70710678
0.2	16	2	67.5000000	2.12132034
0.3	14	2	92.0000000	2.82842712
0.3	15	2	76.5000000	2.12132034
0.3	16	2	59.0000000	1.41421356

are both estimable functions, and the estimates along with their respective standard errors and t-ratios for testing the hypotheses, $H_0 : \sum_i c_i \alpha_i = 0$ and $H_0 : \sum_j c_j \beta_j = 0$, are given.

Parameter	Estimate	Standard Error	t Value	Pr > |t|
0.3 vs 0.1	9.0000000	1.31233465	6.86	<.0001
16 vs 14	-14.0000000	1.31233465	-10.67	<.0001

These estimates would be meaningful if there were no significant interaction between ethanol addition level and air/fuel ratio, but in this case there is a significant interaction and the difference in CO emissions caused by changing the amount of ethanol addition will depend on the air/fuel ratio and the difference in CO emission caused by changing the air/fuel ratio will depend on the amount of ethanol added. An interaction graph is a better way of interpreting these results. By including the `ods graphics on` before using `proc glm` an interaction plot is automatically generated if there are only two factors in the model and an interaction is included. Figure 3.5 was produced by the SAS commands above.

In this plot we can see more clearly the dependence of effects. Increasing the amount of ethanol added to the fuel from 0.1 to 0.3 causes CO emissions

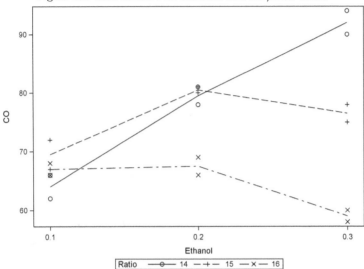

Figure 3.5 *Interaction Plot Ethanol and Air/Fuel Ratio*

to increase linearly from 64 grams/liter to 92 grams/liter when the air/fuel ratio is at its low level of 14. This is shown by the solid line with circles representing the data. However, when the air/fuel ratio is at its high level of 16 (illustrated by the dash-dot line with ×s representing the data points), increasing the ethanol added to the fuel from 0.1 to 0.3 actually causes a decrease in CO emissions from 67 grams/liter to 59 grams/liter along a nearly linear trend. Finally, when the air/fuel ratio is held constant at its mid-level of 15 (illustrated by the dashed line with +s representing the data points), increasing ethanol from 0.1 to 0.2 causes CO emissions to increase by 11 grams/liter; but a further increase in ethanol to 0.3 causes a decrease in CO emissions of 4 grams/liter to 76.5.

The interpretation above again illustrates the principle of comparing the effect of one factor across the levels of the other factor in order to describe an interaction. This was done by comparing the effect of changing the ethanol addition between the levels of air/fuel ratio. It could also be done in the opposite way. For example, the SAS code below reverses the interaction plot as shown in Figure 3.6.

```
ods graphics on;
proc glm data=CO;
   class    Ratio Ethanol ;
   model CO= Ratio Ethanol   Ratio*Ethanol;
run;
ods graphics off;
```

In this plot the dash-dot line, with ×s representing the data points, shows the effect of increasing air/fuel ratio when ethanol is added at the high rate

ANALYSIS OF A TWO-FACTOR FACTORIAL IN SAS

of 0.3. Carbon monoxide emissions decrease linearly from 92 grams/liter to 59 grams/liter. However, when ethanol is added at the low rate of 0.1, the CO emissions actually increase slightly from 64 grams/liter to 67 grams/liter as a result of increasing air/fuel ratio from 14 to 16. This can be seen on solid line with circles representing the data points. When ethanol is added at the mid-rate of 0.2, there is little change in CO emissions when air/fuel ratio is increased from 14 to 15, but there is a decrease in CO emissions of 13 grams/liter caused by increasing air/fuel ratio from 15 to 16. The latter result can be visualized on the dashed line with +s representing the data points.

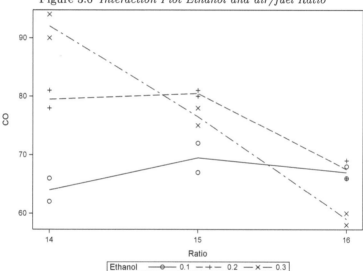

Figure 3.6 *Interaction Plot Ethanol and air/fuel Ratio*

Either way of presenting and interpreting the interaction is valid as long as one discusses how the effect of one factor changes depending upon the level of the other. The factor effects, that should be compared, depend on which one is of most interest in a particular research problem. Another thing to notice about the two interpretations is that cause and effect relationships are assumed. We say the change in the response is caused by the change in the factor or the change in the response is the result of changing the factor. This statement could not be made if discussing the results of an observational study rather than a controlled randomized experiment.

3.5.2 Determining the Number of Replicates

One of two possible methods can be followed to determine the number of replicates for a factorial experiment that will result in a power between 0.80 to 0.90 (for detecting differences that have practical significance). The first method is to consider detecting differences among the cell means. The second method is to consider detecting a practical size difference in the marginal means for the factors in the experiment.

When looking for differences among the cell means, the cells in the factorial are considered to be an unstructured group as in a one-factor design. Using the cell means model $y_{ijk} = \mu_{ij} + \epsilon_{ijk}$ the procedure is the same as it was described for the one-factor model $y_{ij} = \mu_i + \epsilon_{ij}$ in Section 2.6. The non-centrality parameter for the F-test is:

$$\lambda = (r/\sigma^2) \sum_{i=1}^{a} \sum_{j=1}^{b} (\bar{\mu}_{ij} - \bar{\mu}_{..})^2.$$

When looking for differences in the marginal means for the factors, the non-centrality factor for the first main effect is:

$$\lambda_a = br \sum_i \alpha_i^2/\sigma^2 = br \sum_i (\bar{\mu}_{i.} - \bar{\mu}_{..})^2/\sigma^2,$$

and for the second main effect the non-centrality factor is:

$$\lambda_b = ar \sum_j \beta_j^2/\sigma^2 = ar \sum_j (\bar{\mu}_{.j} - \bar{\mu}_{..})^2/\sigma^2.$$

If Δ is considered to be the size of a practical difference in cell means, then the smallest $\lambda = r/\sigma^2 \sum_{i=1}^{a} \sum_{j=1}^{b} (\bar{\mu}_{ij} - \bar{\mu}_{..})^2$ could be with two cells differing by at least Δ is $r\Delta^2/2\sigma^2$. Likewise if Δ is considered to be the size of a practical difference in marginal means for factor A, the smallest $\lambda_a = br \sum_i (\bar{\mu}_{i.} - \bar{\mu}_{..})^2/\sigma^2$ could be with two marginal means differing by at least Δ is $br\Delta^2/2\sigma^2$. Here again we can see the efficiency of factorial designs because the non-centrality factor for detecting differences in marginal factor A means is larger than the non-centrality factor for detecting differences of cell means by a factor of, b the number of levels of factor B.

Consider the following example. A paper helicopter experiment is planned to investigate the effects of four levels of factor A = wing length, and four levels of factor B = body width, upon the flight time. If pilot experiments with nine replicates of one design resulted in flight times of 2.8, 2.6, 3.5, 3.0, 3.1, 3.5, 3.2, 3.4, and 3.4 seconds. How many replicates would be required to detect a difference in flight times of 1 second with a power of .90?

From the pilot tests the variance of experimental error can be estimated as $s^2 = 0.1024$. If $\Delta = 1.0$ considered to be a practical size difference in cell means, the SAS commands in Section 2.6, can be modified to give the answer. In a 4-by-4 factorial there are 16 cells so the degrees of freedom for the numerator is $\nu_1 = 15$, and the degrees of freedom for the denominator is $\nu_2 = 16(r-1)$. The modified SAS code is shown on the next page.

ANALYSIS OF A TWO-FACTOR FACTORIAL IN SAS

```
data Power;
do r=2 to 8;
   nu1=16-1; * df for numerator;
   nu2=16*(r-1); * df for denominator;
   alpha=.05;
   Fcrit=finv(1-alpha,nu1,nu2); *F critical value;
   sigma2=.1024;
   css=0.5;
   nc=r*(css)/sigma2;*noncentrality parameter for
       noncentral F;
   power=1-probf(Fcrit,nu1,nu2,nc);
   output;
end;
keep r nu1 nu2 nc power;
title Power Calculation in Data Step;
proc print; run;
```

The results below show that 6 replicates per cell would be required to obtain a power of at least 0.90.

Power Calculation in Data Step

Obs	r	nu1	nu2	nc	power
1	2	15	16	9.7656	0.24173
2	3	15	32	14.6484	0.48174
3	4	15	48	19.5313	0.69246
4	5	15	64	24.4141	0.83829
5	6	15	80	29.2969	0.92326
6	7	15	96	34.1797	0.96664
7	8	15	112	39.0625	0.98655

If $\Delta = 1.0$ considered to be a practical size difference in marginal means for one of the factors, the results will be different. The degrees of freedom for the numerator would be $\nu_1 = 4 - 1$, the degrees of freedom for the denominator would be $\nu_2 = 16(r-1)$ and the non-centrality factor for a main effect would be $\lambda_a = br\Delta^2/2\sigma^2 = 4r1^2/2(0.1024)$. Modifying the SAS code above we get

FACTORIAL DESIGNS

```
data Power;
  do r=2 to 4;
    nu1=4-1; * df for numerator;
    nu2=16*(r-1); * df for denominator;
    alpha=.05;
    Fcrit=finv(1-alpha,nu1,nu2); *F critical value;
    sigma2=.1024;
    css=0.5;
    nc=4*r*(css)/sigma2;*noncentrality parameter for noncentral F;
    power=1-probf(Fcrit,nu1,nu2,nc);
    output;
  end;
  keep r nu1 nu2 nc power;
title Power Calculation in Data Step;
proc print; run;
```

The results appear below. Here it can be seen that with only two replicates per cell the power for detecting a $\Delta = 1.0$ difference in marginal means for factor A is greater than the power for detecting differences of $\Delta = 1.0$ in cell means with $r = 8$ replicates per cell. Again this demonstrates the efficiency of factorial experiments through hidden replication.

```
            Power Calculation in Data Step

    Obs   r    nu1    nu2       nc        power

     1    2     3      16     39.0625    0.99838
     2    3     3      32     58.5938    1.00000
     3    4     3      48     78.1250    1.00000
```

The power for detecting factorial main effects and interactions can also be estimated by SAS **proc glmpower** by specifying the cell means, the standard deviation of the experimental error, and the total number of observations $N = a \times b \times r$. Since **proc glmpower** can calculate the power for factorial models in addition to simple one-factor designs, the cell means must be read in the data step (in the same format that the data would be read with **proc glm**) rather than including them in the commands as with **proc power**. On the next page is an example of the input with three levels of body width and wing length for a proposed helicopter experiment. The cell means represent what might be thought to be differences of practical significance. The statement **ntotal=36 to 81 by 9** tells **proc glmpower** to calculate the power for three replicates through nine replicates per cell.

```
data powglm;
input BW WL cellmean;
datalines;
4.0   4.75  2.4
4.0   5.5   2.3
4.0   6.25  2.1
3.75  4.75  2.8
3.75  5.5   2.4
3.75  6.25  2.2
3.5   4.75  2.4
3.5   5.5   2.4
3.5   6.25  2.2
proc glmpower;
  class BW WL;
  model cellmean=BW WL BW*WL;
  power
    stddev=0.32
    ntotal=36 to 81 by 9
    power = .;
run;
```

The results are shown below. The power is given for each term in the model (i.e., main effects for body width, wing length and interaction) as a function of the total number of observations. With the cell means as specified in the input there would be adequate power for detecting the wing length main effect with six or more replicates per cell (i.e., ntotal ≥ 54).

```
              The GLMPOWER Procedure

            Fixed Scenario Elements
    Dependent Variable             cellmean
    Error Standard Deviation       0.32
    Alpha                          0.05
                  Computed Power
                    N      Test    Error
Index    Source   Total     DF      DF     Power

  1       BW       36        2      27     0.242
  2       BW       45        2      36     0.302
  3       BW       54        2      45     0.361
  4       BW       63        2      54     0.418
  5       BW       72        2      63     0.472
  6       BW       81        2      72     0.524
  7       WL       36        2      27     0.659
  8       WL       45        2      36     0.774
  9       WL       54        2      45     0.855
 10       WL       63        2      54     0.910
 11       WL       72        2      63     0.945
 12       WL       81        2      72     0.967
 13      BW*WL     36        4      27     0.165
 14      BW*WL     45        4      36     0.206
 15      BW*WL     54        4      45     0.248
 16      BW*WL     63        4      54     0.291
 17      BW*WL     72        4      63     0.335
 18      BW*WL     81        4      72     0.378
```

With the ability to calculate power quickly using the `probf` function in the data step or `proc glmpower`, it is possible to explore many potential designs before actually running the experiments. Questions such as, "Is it better to use three or four levels of the factors?" can be answered with respect to the effect on the power of detecting significant differences.

3.5.3 Analysis with an Unequal Number of Replicates per Cell

Although it would be unusual to plan a factorial experiment with an unequal number of replicates per cell, the data from a factorial experiment may end up with an unequal number of replicates due to experiments that could not be completed, or responses that could not be measured, or simply lost data. As long as the chance of losing an observation was not related to the treatment factor levels, the data from a factorial experiment with an unequal number of replicates per cell can still be analyzed and interpreted in a manner similar to the way it would be done for the equal replicate case. However, the computational formulas for analyzing the data differ for the case with an unequal number of replicates.

To illustrate why the analysis shown in Section 3.5.1 is inappropriate, consider again the data from the ethanol fuel experiment described in Section 3.5.1. This time assume one observation in the cell where air/fuel ratio = 16 and ethanol level = 0.3 was missing. Then Table 3.4 shows the data with each response value written above its symbolic expected value.

Table 3.4 *Fuel Experiment with Unequal Reps*

Ethanol	air/fuel 14	air/fuel 15	air/fuel 16
0.1	66 62 $\mu + \alpha_1 + \beta_1 + \alpha\beta_{11}$	72 67 $\mu + \alpha_1 + \beta_2 + \alpha\beta_{12}$	68 66 $\mu + \alpha_1 + \beta_3 + \alpha\beta_{13}$
0.2	78 81 $\mu + \alpha_2 + \beta_1 + \alpha\beta_{21}$	80 81 $\mu + \alpha_2 + \beta_2 + \alpha\beta_{22}$	66 69 $\mu + \alpha_2 + \beta_3 + \alpha\beta_{23}$
0.3	90 94 $\mu + \alpha_3 + \beta_1 + \alpha\beta_{31}$	75 78 $\mu + \alpha_3 + \beta_2 + \alpha\beta_{32}$	60 $\mu + \alpha_3 + \beta_3 + \alpha\beta_{33}$

The marginal column means for the levels of air/fuel ratio factor computed using the `means` statement as shown in Section 3.5.1 would be 78.5, 75.5 and 65.8, respectively. The expected value of the marginal means for the first two columns would be: $\mu + \beta_1$, $\mu + \beta_2$, since $(\alpha_1 + \alpha_2 + \alpha_3)/3 = 0$ and $(\alpha\beta_{1i} + \alpha\beta_{2i} + \alpha\beta_{3i})/3 = 0$ for $i = 1, 2$. However, the expected value of the last marginal column mean would be $\mu + \beta_3 + (2\alpha_1 + 2\alpha_2 + \alpha_3)/5 + (2\alpha\beta_{13} + 2\alpha\beta_{23} + \alpha\beta_{33})/5$ and is not an unbiased estimate of $\mu + \beta_3$. The comparison

ANALYSIS OF A TWO-FACTOR FACTORIAL IN SAS

between the first and third column means would not be an unbiased estimate of $\beta_1 - \beta_3$. Likewise the last marginal row mean would not be an unbiased estimate of $\mu + \alpha_3$.

In order to get means that have expectations $\mu + \beta_1$, $\mu + \beta_2$, $\mu + \beta_3$ when there are an unequal number of replicates per cell, use the lsmeans statement rather than the means statement in SAS proc glm. The means statement simply takes the averages of all observations in each level of a factor, whereas the lsmeans statement computes the marginal means of the predicted cell means $(\hat{\mu} + \hat{\alpha}_i + \hat{\beta}_j + \hat{\alpha\beta}_{ij})$ obtained from the least squares estimates of the model parameters. Remember that the cell means are estimable functions.

Another problem with the analysis shown in Section 3.5.1 is that the type I F-tests will not test the same hypotheses that they do in the case of equal number of replicates in the cells. When there is an unequal number of replicates in the cells, the non-centrality parameter for the type I F-test of $H_0: \alpha_1 = \cdots = \alpha_a$, that is based on $R(\alpha|\mu)$ will not be $\lambda_a = rb\sum_i \alpha_i^2$ but a quadratic form involving the elements of $\boldsymbol{\alpha}, \boldsymbol{\beta}$ as well as $\boldsymbol{\alpha\beta}$. The non-centrality for the type I F-test test of $H_0: \beta_1 = \cdots = \beta_b$ based on $R(\beta|\mu, \alpha)$ will be a quadratic form involving the elements of $\boldsymbol{\alpha}$ and $\boldsymbol{\alpha\beta}$.

To get the correct test of the null hypothesis for the main effects, use the F-tests from proc glm type III ANOVA table. The type III sums of squares for the factors A and B can be represented as $ssA_{III} = R(\alpha|\mu, \beta, \alpha\beta)$, and $ssB_{III} = R(\beta|\mu, \alpha, \alpha\beta)$. $R(\alpha|\mu, \beta, \alpha\beta)$ is the difference in the error sums of squares for the reduced model where $\boldsymbol{X} = (\boldsymbol{1}|\boldsymbol{X_B}|\boldsymbol{X_{AB}})$ and the full model where $\boldsymbol{X} = (\boldsymbol{1}|\boldsymbol{X_A}|\boldsymbol{X_B}|\boldsymbol{X_{AB}})$. Recall that when there is an equal number of replications per cell, the type I sums of squares are identical to the type III sums of squares. With the non-full rank $\boldsymbol{X}$ matrix, normally $R(\alpha|\mu, \beta, \alpha\beta)$ would be zero, since $\boldsymbol{X_A}$ is linearly dependent on $\boldsymbol{X_{AB}}$. However, using proc glm's generalized inverse, which is equivalent to the restriction that $\alpha_2 = \beta_3 = \alpha\beta_{13} = \alpha\beta_{23} = \alpha\beta_{31} = \alpha\beta_{32} = \alpha\beta_{33} = 0$, it can be computed and the corresponding non-centrality factor for the corresponding F-test will be a quadratic form that only involves $\boldsymbol{\alpha}' = (\alpha_1, \alpha_2, \alpha_3)$.

The proc glm analysis of the ethanol fuel experiment after removing the observation with the value of 58 from the cell air/fuel ratio $= 16$ and ethanol level $= 0.3$ is shown at the top of the next page.

There it can be seen that the type III sums of squares for Ethanol and Ratio are different than their type I sums of squares. The F-tests in the type III table test the correct hypothesis. The results of the means statement and the lsmeans statement are shown in the SAS output below the ANOVA tables on the next page, illustrating that the means differ in the third level of each factor. The unbiased means are given by the lsmeans statement.

In general the type III sums of squares and lsmeans should be used, because they will test the correct hypotheses and provide unbiased factor level means whether there is an equal or unequal number of replications per cell.

```
                         Sum of
Source            DF     Squares       Mean Square   F Value   Pr > F
Model             8      1423.029412   177.878676    31.98     <.0001
Error             8      44.500000     5.562500
Corrected
Total             16     1467.529412

Source            DF     Type I SS     Mean Square   F Value   Pr > F
Eth               2      472.6627451   236.3313725   42.49     <.0001
Ratio             2      395.3282051   197.6641026   35.54     0.0001
Eth*Ratio         4      555.0384615   138.7596154   24.95     0.0001

Source            DF     Type III SS   Mean Square   F Value   Pr > F
Eth               2      319.4545455   159.7272727   28.72     0.0002
Ratio             2      511.4545455   255.7272727   45.97     <.0001
Eth*Ratio         4      555.0384615   138.7596154   24.95     0.0001
```

```
              Results of means statement

                   The GLM Procedure

Level of                   ---------------CO---------------
Eth            N                Mean              Std Dev

0.1            6             66.8333333          3.2506410
0.2            6             75.8333333          6.6156380
0.3            5             79.4000000         13.4461891


Level of                   ---------------CO---------------
Ratio          N                Mean              Std Dev

14             6             78.5000000         12.7082650
15             6             75.5000000          5.3197744
16             5             65.8000000          3.4928498


              Results of lsmeans statement

                   The GLM Procedure
                  Least Squares Means

               Eth          CO LSMEAN

               0.1          66.8333333
               0.2          75.8333333
               0.3          76.1666667


               Ratio        CO LSMEAN

               14           78.5000000
               15           75.5000000
               16           64.8333333
```

3.5.4 Testing for Interaction with one Replicate per Cell

When there is adequate power for detecting main effects with $r = 1$ replicate per cell, then it would make sense to run a factorial design with only one observation per cell and $a \times b$ total observations. Adding one replicate to each cell, in this case, would double the effort required, and would usually not be required. However, with only one replicate per cell in a factorial design, there is no way to compute the ANOVA ssE and therefore no way to make F-tests on the main effects and interaction in the traditional way. If the interaction term is assumed to be zero, then F-tests on the main effects can be made by using the additive model $y_{ij} = \mu + \alpha_i + \beta_j + \epsilon_{ij}$. Even so, this could be dangerous if the interaction actually exists. There are ways to test whether the interaction is zero in this case.

If the levels of both factors are quantitative as in the proposed paper helicopter experiments or the ethanol fuel experiment, the sums of squares for the interaction term can be partitioned into orthogonal polynomial single degrees of freedom. For example, if there are three equally spaced quantitative levels of factor A, and three equally spaced quantitative levels for factor B, then the sums of squares for the interaction can be partitioned into four single degrees of freedom (namely: linear×linear, linear×quadratic, quadratic×linear and quadratic×quadratic). Using the Taylor Series philosophy that low order polynomials can approximate most functional relationships, the three higher order terms might be assumed to be negligible and pooled to estimate the ssE, which could then be used as an error term to test the linear×linear portion of the interaction. This will be illustrated with the data from the ethanol fuel experiment given in Table 3.4.

First, consider the averages of the two replicates in each cell of Table 3.3 to be the result of a single experiment. The SAS commands shown below average the data in each cell to produce the file `cells` with one observation per cell. Fitting the model 3.2 to this data with `proc glm` results in an ANOVA with zero degrees of freedom for ssE, and no F-tests.

```
proc sort data=CO; by Ethanol Ratio;
proc means noprint;
  by Ethanol Ratio; var CO;
  output out=cells mean=COmean;
proc glm data=cells;
  class Ethanol Ratio;
  model COmean=Ethanol Ratio Ethanol*Ratio:
run;
```

The type III portion of the output appears on the next page.

FACTORIAL DESIGNS

Source	DF	Type III SS	Mean Square	F Value
Ethanol	2	162.0000000	81.0000000	.
Ratio	2	326.0000000	163.0000000	.
Ethanol*Ratio	4	339.0000000	84.7500000	.

To get the sums of squares for the linear×linear portion of the interaction sums of squares, the `contrast` command is added to the `proc glm` commands above just before the `run;` statement:

```
contrast 'Linear x Linear' Eth*Ratio .500000309  0 -.500000309
                                     0           0  0
                                    -.500000309  0  .500000309;
```

The `contrast` command is similar to the `estimate` command that has been used earlier. Instead of producing an estimate, its standard error, and a t-statistic for testing the hypothesis $H_0 : L'\beta = 0$, the `contrast` produces the sums of squares for a contrast, $(L\hat{\beta})'(L(X'X)^-L')^{-1}(L\hat{\beta})$, and an F-test for testing the hypothesis $H_0 : L'\beta = 0$.

To determine the coefficients for obtaining the linear×linear sums of squares using the contrast statement, multiply the coefficients for the linear contrast on ethanol by the coefficients for the linear contrast on air/fuel ratio as shown in the table below.

Table 3.5 *Coefficients for Interaction Contrast*

		Ratio			Linear Coef.
		14	15	16	
	0.1	.500000309	0	-.500000309	-.707107
Ethanol	0.2	0	0	0	0
	0.3	-.500000309	0	.500000309	.707107
Linear Coef.		-.707107	0	.707107	

On the margins of the table are the coefficients for the linear contrasts for both ethanol and air/fuel ratio. These were obtained from the `proc iml` commands as shown in Section 2.8.1. The values in the body of the table were obtained as the product of the marginal values. For example, in the upper left cell $.500000309 = -.707107 \times -.707107$. Next the coefficients in the body of the table are entered row after row in the contrast statement since ethanol is the first factor in the model statement and moves slowest. In this case since both factors have three equally spaced levels, it doesn't matter if you enter the coefficients from the body of the table row-wise or column-wise. However, if the number of levels differ or if you are trying to get the sums of squares for the linear×quadratic contrast, the order will matter and the coefficients should be entered row-wise with the factor defining the rows being the first factor in the `proc glm` model statement. If you have any doubt whether you

ANALYSIS OF A TWO-FACTOR FACTORIAL IN SAS

have entered interaction contrast coefficients in the right order in a contrast statement command, add the option `/e;` to the command and it will print out a table showing which contrast coefficients `proc glm` has associated with each combination of factor levels. The output of the contrast statement is shown below, and using this one can partition the interaction sums of squares into the portion due to the linear×linear contrast and the remainder which will be treated as error.

Contrast	DF	Contrast SS	Mean Square	F Value
Linear x Linear	1	324.0000000	324.0000000	.

In other words, $ssE = ssInteraction - ssLinear \times Linear$ or $ssE = 339.0 - 324.0 = 15.0$ with $4 - 1 = 3$ degrees of freedom. Therefore F-test value for the linear×linear part of the interaction can be calculated as shown in the table below.

Table 3.6 *Test of Linear by Linear Portion of Interaction*

Source	df	SS	MS	F	Pr>F
Linear×Linear	1	324.0	324.0	64.8	<0.01
Remainder (error)	3	15.0	5.0		

The linear×linear part of the interaction is significant and can be visualized by drawing straight lines through the interaction plot rather than by connecting the cell means as was done in Section 3.5.1. To do this the `proc sgplot` commands, shown earlier, can be modified as shown below.

```
proc sgplot data=cells;
  scatter x=Ratio y=COmean/group=Ethanol;
  reg x=Ratio y=COmean/group=Ethanol;
  yaxis label="CO emmissions";
run;
```

and the result is shown in Figure 3.7, which should be compared to Figure 3.6 that was created earlier. In this figure it can be seen that the linear trend in the response, CO emissions, which are caused by a change in the air/fuel ratio depends upon the level of ethanol. When ethanol is at its high level (0.3) increasing air/fuel ratio from 14 to 16 causes a steep decrease in CO emissions. When ethanol is at its mid-level (0.2) increasing air/fuel ratio from 14 to 16 causes a slight decrease in CO emissions represented by the gentle negative sloping line. However, when ethanol is at its low level (0.1) increasing air/fuel ratio from 14 to 16 actually causes an increase in CO emissions illustrated by the positively sloped line.

When there is only one replicate per cell in a factorial experiment and the

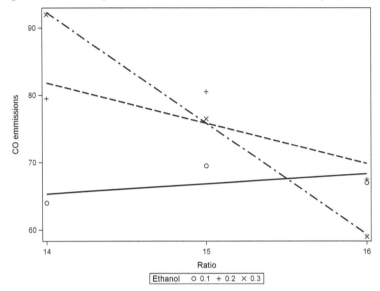

Figure 3.7 *Linear by Linear Interaction Plot Ethanol and Air/Fuel Ratio*

factors do not have quantitative levels, partitioning the interaction sums of squares into orthogonal polynomial contrasts and combining the higher order terms as an error sums of squares may not be appropriate. However, Tukey (1949b) has developed an alternate method for testing a single degree of freedom partitioned from interaction sums of squares. This method is equivalent to restricting the $\alpha\beta_{ij}$ in model 3.2 of Section 3.5 to be a second-degree polynomial function of the main effects α_i and β_j, (see Scheffé, 1959). By doing this, the sums of squares

$$ssAB = \frac{ab\left[\sum_i \sum_j y_{ij}\bar{y}_{i.}\bar{y}_{.j} - (ssA + ssB + ab\bar{y}_{..}^2)\bar{y}_{..}\right]^2}{(ssA)(ssB)} \qquad (3.7)$$

for testing the restricted hypothesis $H_0 : \alpha\beta_{ij} = 0$ for all i and j will have one degree of freedom and the difference between it and the error term for the additive model will form the error sums of squares similar to the example above with quantitative factor levels.

To illustrate the use of Tukey's single degree of freedom test for interaction, consider the data in Table 3.7, which is a portion of the data from a study to validate an assay of viral contamination reported by Lin and Stephenson (1998). Assays of viral contamination are used to determine the presence (and amount) of a specific virus in biological products such as blood clotting Factor Eight. An experiment, or run, consists of making a solution with a known viral contamination, allowing the virus in a contaminated solution to grow, then

ANALYSIS OF A TWO-FACTOR FACTORIAL IN SAS

measuring the result. The experimental unit is the specific viral sample in combination with the place and time where it is allowed to grow. Factor A represents the sample number, or solution with which the viral sample is mixed (or spiked). Factor B represents different dilutions of the spiked sample. The measured response is the $\log_{10}$ of the plaque forming units per mL of solution.

Table 3.7 $\log_{10}(PFU/mL)$ Assay of Viral Contamination

		Sample					
		1	2	3	4	5	6
	3	1.87506	1.74036	1.79934	2.02119	1.79934	1.59106
Dilution	4	1.38021	1.36173	1.25527	1.39794	1.20412	1.25527
	5	0.60206	0.90309	0.95424	1.00000	0.60206	0.60206

Since factor A (sample) is not a quantitative factor it would be inappropriate to use orthogonal polynomial contrasts to partition its sums of squares or the sums of squares of its interaction with factor B (Dilution). To determine if the additive model $y_{ij} = \mu + \alpha_i + \beta_j + \epsilon_{ij}$ is appropriate for this data, test to see whether there is a significant interaction using Tukey's method. A SAS macro (shown in the appendix to this chapter) calculates the non-additivity or interaction sums of squares, shown in Equation (3.7), and prints a report. The commands to read the data and call the macro are shown below. The code for the macro, shown in the appendix, should be included above the code shown below. When this code is executed, the macro is called with the %tukeysdf command. The macro looks in a file called file1 for the variables A, B, and y to perform the calculations.

```
data PFU;
input A B y;
datalines;
1 3 1.87506
1 4 1.38021
 ... etc.
data file1; set PFU;
%tukeysdf
```

The results are shown in the output shown below.

```
        Tukey's single df test for additivity

Source              df        SS          MS         F      PR>F
A                    5      0.19479     0.03896
B                    2      3.16639     1.5832
Error               10      0.12825     0.01283
NonAdditivity        1      0.00688     0.00688     0.51    0.493
Residual             9      0.12137     0.01349
```

Here it can be seen that the interaction (or non-additivity) is not significant.

Therefore, for this data, it would be appropriate to fit the additive model with the commands `proc glm; class A B; model y=A B;`.

3.6 Factorial Designs with Multiple Factors - CRFD

Two-factor factorial designs are more efficient than studying each factor separately in one-factor designs. Likewise, when many factors are under study, it is more efficient to study them together in a multi-factor factorial design than it is to study them separately in groups of two using two-factor factorial designs. When multiple factors are studied simultaneously, the power for detecting main effects is increased over what it would be in separate two-factor factorial designs. Also, the possibility of detecting interactions among any of the factors is possible. If the factors were studied separately in two-factor factorials, two-factor interactions could only be detected between factors studied together in the same design. In a multi-factor factorial not only is it possible to detect two-factor interactions between any pair of factors, but it is also possible to detect higher order interactions between groups of factors. A three-factor interaction between factors A, B and C, for example, means the effect of factor A differs depending on the combination of levels of factors B and C. Examples of higher order interactions will be presented in examples to follow.

The treatment combinations in a multi-factor factorial consist of all possible combinations of the levels of all factors. A design, along with randomization can be produced using nested do loops in the SAS data step or `proc factex` similar to the example shown in Section 3.4. The model for analysis is an extension of Equation (3.2), and the analysis can be made using `proc glm` or `proc genmod` similar to the examples shown earlier.

Consider an example of a multi-factor factorial design in marketing research. A company whose sales are made online through a Web page would like to increase the proportion of visitors to their Web site that sign up for their service by optimally configuring their Web page. In order to buy from the company, customers must sign up and fill out a form supplying their email address along with other required fields. Once a customer signs up, the company has contact information for their database and can email advertisements, special offers, etc. The company would like to experiment by testing different configurations of their Web page to see if they can increase the number of visitors to their site that actually sign up.

The experimental units in this study will be individuals who visit the company Web site. The response is binary; the customer either signs up or does not. The factors under study were characteristics that change the appearance of the Web page. For example, factor A was the background alternatives for the page with three options. Factor B was the font size in the main banner, with three levels; factor C was the text color with two alternatives; and factor D was a choice between a sign-up button or link. Based on these factors there were $3 \times 3 \times 2 \times 2 = 36$ possible configurations of the Web page when

FACTORIAL DESIGNS WITH MULTIPLE FACTORS - CRFD

considering all possible combinations of the levels of each factor. A four-factor factorial experiment would consist of randomly assigning visitors to the Web site to one of the possible configurations and recording their binary response. There are lurking variables that could affect the chance that a site visitor will sign up. For example, the position order that the link (for the company's Web site) comes up in a Web search for the products they sell, promotions offered by competitors, and attractiveness of competitors' Web sites. Random assignment of each sequential visitor to the site to one of the alternative configurations under study should minimize the chance of bias from changes in the lurking variables over time. The probability that a site visitor would sign up can be expressed by the model:

$$p_{ijkl} = \mu + \alpha_i + \beta_j + \alpha\beta_{ij} + \gamma_k + \alpha\gamma_{ik} + \beta\gamma_{jk} + \alpha\beta\gamma_{ijk} + \delta_l + \alpha\delta_{il} \\ + \beta\delta_{jl} + \alpha\beta\delta_{ijl} + \gamma\delta_{kl} + \alpha\gamma\delta_{ikl} + \beta\gamma\delta_{jkl} + \alpha\beta\gamma\delta_{ijkl}, \quad (3.8)$$

where α_i represents the effect of background choice, β_j represents the effect of font size in the main banner, γ_k represents the effect of text color, and δ_l represents the effect of sign-up link versus button.

The experiment was conducted by constructing thirty-six Web sites consisting of all possible combinations of the four factors described above. Each potential customer who visited the company's Web site during the trial period was randomly redirected to one of the thirty-six configurations. The number of visitors n_{ijkl} to the $ijkl$th configuration and the number that signed up x_{ijkl} was logged. x_{ijkl} is then binomially distributed

$$B(x_{ijkl}, n_{ijkl}, p_{ijkl}) = \binom{n_{ijkl}}{x_{ijkl}} p_{ijkl}^{x_{ijkl}} (1 - p_{ijkl})^{(n_{ijkl} - x_{ijkl})} \quad (3.9)$$

where n_{ijkl} is the number of visitors to the $ijkl$th configured Web page during the testing period.

The raw data from this experiment was read into SAS using the commands on the next page.

```
data raw;
input A B C D visitors signup;
datalines;
1    1    1    1    1016    22
1    1    1    2    1145    16
1    1    2    1    1145    17
1    1    2    2    1082    19
1    2    1    1    1121    28
1    2    1    2    1103    28
1    2    2    1    1072    15
1    2    2    2    1097    33
1    3    1    1    1145    18
1    3    1    2    1137    20
1    3    2    1    1091    22
1    3    2    2    1097    32
2    1    1    1    1062    21
2    1    1    2    1093    27
2    1    2    1    1020    28
2    1    2    2    1009    27
2    2    1    1    1091    19
2    2    1    2    1133    34
2    2    2    1    1037    17
2    2    2    2    1061    28
2    3    1    1    1082    32
2    3    1    2    1010    26
2    3    2    1    1028    38
2    3    2    2    1045    30
3    1    1    1    1014    17
3    1    1    2    1100    21
3    1    2    1    1031    19
3    1    2    2    1072    21
3    2    1    1    1035    15
3    2    1    2    1038    24
3    2    2    1    1097    26
3    2    2    2    1089    22
3    3    1    1    1108    19
3    3    1    2    1119    25
3    3    2    1    1141    28
3    3    2    2    1148    35
proc format;
   value Backg 1='type 1'
               2='type 2'
      3='type 3';
   value font  1='small'
               2='medium'
       3='large';
   value sgn   1='Sign-up Button'
               2='Sign-up Link';
   value txt   1='black'
               2='white';
run;
```

The proc format commands following the data assign labels to the levels of the factors A, B, C and D for use in clarifying the output.

The correct SAS procedure must be utilized to analyze the data, determine if any of the factor effects are significant, and to predict the optimal Web

FACTORIAL DESIGNS WITH MULTIPLE FACTORS - CRFD 83

page configuration. Since the responses for the individual visitors to each configuration of the Web site are Bernoulli, the aggregate response data, which is read into SAS, is Binomial with large and approximately equal sample sizes (i.e., number of visitors to each possible Web configuration). The arcsin square root transformation shown in Table 2.4 of Section 2.5.2 could be applied and **proc glm** could be used for analysis. However, the problem with using this procedure is that the individual responses were summed to get the aggregate responses, and when using these aggregate binomial responses there are no replicate observations in any of the cells, and thus no way to calculate ssE. This would be similar to summing or averaging the replicate responses in each cell if the data were normally distributed, leaving only one observation per cell and no way to compute ssE. The alternative is to use the method of maximum likelihood to fit model (3.8). This can be done using **proc genmod**, that was described briefly in Section 2.5.3. It will automatically set $\sigma^2 = 1.0$ and the type III sums of squares of the form $(L\hat{\beta})'(L(X'X)^{-}L')^{-1}(L\hat{\beta})$ will be asymptotically distributed as chi-squares under the null hypothesis. The commands to analyze the data using **proc genmod** are shown below.

```
proc genmod data=raw;
  class A B C D;
  model signup/visitors=A|B|C|D/ dist=bin type3 ;
run;
```

signup/visitors is the observed proportion signing up and the model specification A|B|C|D is a shorthand for A B A*B C A*C B*C A*B*C D A*D B*D A*B*D C*D A*C*D B*C*D A*B*C*D. The option **dist=bin** declares the response to be binomially distributed, and the option **type3** requests a table of the type III sums of squares and chi-square tests. The first output from **proc genmod** is a table of parameter estimates produced from the generalized inverse $\hat{\beta} = (X'X)^{-}X'y$ similar to that resulting from the **/solution** option in **proc glm**. It is not shown here. The individual coefficients are biased and should not be interpreted. The next section of the output is the table of type III analysis resulting from the **type3** option and is shown on the next page.

In this output we can see that (at the $\alpha = 0.05$ level of significance) factors A (background style) and factor D (sign-up button or link) were significant along with the three-way interaction ACD, where factor C represents the text color. Since there is a significant interaction, the main effects A and D cannot be interpreted separately. Whether it is better to use a sign-up button or link

84   FACTORIAL DESIGNS

```
       LR Statistics For Type 3 Analysis

                              Chi-
Source              DF       Square      Pr > ChiSq

A                    2        10.21        0.0061
B                    2         5.91        0.0520
A*B                  4         6.00        0.1994
C                    1         2.32        0.1275
A*C                  2         0.84        0.6576
B*C                  2         4.33        0.1146
A*B*C                4         2.43        0.6563
D                    1         5.05        0.0247
A*D                  2         0.04        0.9817
B*D                  2         4.11        0.1281
A*B*D                4         5.67        0.2249
C*D                  1         0.12        0.7259
A*C*D                2         6.06        0.0484
B*C*D                2         0.06        0.9725
A*B*C*D              4         2.01        0.7346
```

depends on whether the text color is black or white and what background style is chosen. To interpret the three-way interaction, it is necessary to make a table of the proportion signing up in each combination of factors A, C and D and a series of interaction graphs. The SAS commands to do this are shown below.

```
data plt; set raw;
  p=signup/visitors;
  rename A=Background D=Signup C=Text_Color;
run;
proc sort; by Background Text_Color Signup;
proc means noprint; by Background Text_Color Signup; var p;
  output out=pltint mean=pmean;
proc sort; by A;
proc print; run;
proc sgpanel data=pltint;
  panelby Background/columns=3;
  series x=Text_Color y=pmean/group=Signup markers;
  colaxis type=discrete;
  format Signup sgn. Text_Color txt.;
run;
```

proc sgpanel produces the three graphs in Figure 3.8 which show the effect of factor D (text color) for each combination of the levels of factors A (background type) and C (sign-up link or button).

The common way of interpreting the interaction is to compare the effect the variable represented on the horizontal axis between combinations of levels of the other factors. For this example an interpretation can be made as follows.

FACTORIAL DESIGNS WITH MULTIPLE FACTORS - CRFD

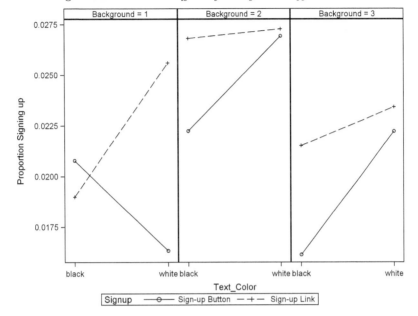

Figure 3.8 *Text Color Effect by Background Type and Link*

When using background type 2 or type 3, it can be seen that changing the text color from black to white causes an increase in the proportion of site visitors signing up. The increase (represented by the slope of the lines) is greater when a sign-up button is used rather than a sign-up link because the overall sign up rate is higher when a link is used regardless of the text color, and there is not that much room for improvement. However, when background type 1 is used, the effect of text color is altogether different. In this case changing the font from black to white actually causes a decrease in the proportion signing up when a sign-up button is used, and there is a big increase in the proportion signing up when changing from black to white font when using a sign-up link. This is the opposite of the effects seen for background types 2 and 3.

Any one of the three factors could be placed on the horizontal axis and an equivalent interpretation could be made. Sometimes the interpretation that results when placing the factor with the largest main effect on the horizontal axis is easiest to explain.

When interpreting a two-factor interaction, only one graph was necessary to illustrate the fact that the effect of one factor depended on the level of another factor. However, in this case more than one graph is required to illustrate how the effect of one factor depends on the combination of levels of the other two factors. The two lines on each graph show how the effect of text color changes when there is a sign-up link versus a button, and the different graphs show how the effect changes when the background is changed.

From inspection of the three graphs, or the table of averages that could be produced by removing the `noprint` option from `proc means`, it can be seen that the highest proportion signing up would be for the Web page with a sign-up button, white text, and background type 2. Here it is predicted that on the average, slightly more than 2.7% will sign up. The font size is insignificant, so it doesn't matter what font size is used.

3.7 Two-Level Factorials

As additional factors are added to a factorial design, the number of treatment combinations (runs) in the design increases exponentially. The example in the last section contained four factors and 36 treatment combinations. If there were five factors in a design each having four levels, the number of treatment combinations would be $4 \times 4 \times 4 \times 4 \times 4 = 4^5 = 1024$ runs in the design. It can be seen that it would not take too many factors to render the design impractical. In other words it would have too many treatment combinations to run in a reasonable period of time. However, it is better to reduce the number of levels of each factor and stay with the factorial design using all factors than it is to revert to one-at-a-time or two-at-a-time experiments and lose the efficiency of factorial experiments. With separate experiments the ability to detect higher order interactions, and the ability to detect interactions between any pair of factors is lost. If five factors in a factorial design were studied with only two levels each, the number of treatment combinations would be reduced to $2^5 = 32$. For this reason factorial designs with two levels for each factor, or a two-level factorials, are popular. A shorthand for a two-level factorial with k factors is a 2^k design.

In two-level factorials, if a factor has quantitative levels, the two levels are denoted symbolically by $(-)$ and $(+)$, where $(-)$ represents the lowest level the experimenter would consider, and $(+)$ represents the highest level the experimenter would consider. The high and low are usually spread out as far as feasibly possible in order to accentuate the signal or difference in response between the two levels. If a factor has qualitative levels, the $(-)$ and $(+)$ designations are arbitrary, but the two levels chosen normally would be two that the experimenter believes should result in the maximum difference in response.

3.7.1 Main Effects and Regression Slopes

The model for a factorial experiment with three factors can be written as:

$$y_{ijkl} = \mu + \alpha_i + \beta_j + \alpha\beta_{ij} + \gamma_k + \alpha\gamma_{ik} + \beta\gamma_{jk} + \alpha\beta\gamma_{ijk} + \epsilon_{ijkl}, \quad (3.10)$$

where α_i, β_j, etc. are the effects as defined earlier. However, in the case where each factor has only two levels represented by $(-)$ and $(+)$, i, j, k, and l can be replaced with either a $(-)$ or $(+)$, and $\alpha_- = -\alpha_+$, since $\alpha_- = \bar{y}_{-...} - \bar{y}_{....}$, $\alpha_+ = \bar{y}_{+...} - \bar{y}_{....}$, and $\bar{y}_{....} = (\bar{y}_{-...} + \bar{y}_{+...})/2$. A similar equality will be true

TWO-LEVEL FACTORIALS

for all the effects and interactions. Since the two effects for each factor are the same value with different signs, a more compact way of defining the main effects for a two-level factorial is $E_A = \bar{y}_{+\cdots} - \bar{y}_{-\cdots}$. This can be visualized on the left side of Figure 3.9 and represents the change in the average response caused by a change in the factor from its low $(-)$ level to its high $(+)$ level. This *effect* can then represented by the difference in two averages $\bar{y}_{+\cdots}$ and $\bar{y}_{-\cdots}$.

Figure 3.9 *Effect and Regression Coefficient for Two-Level Factorial*

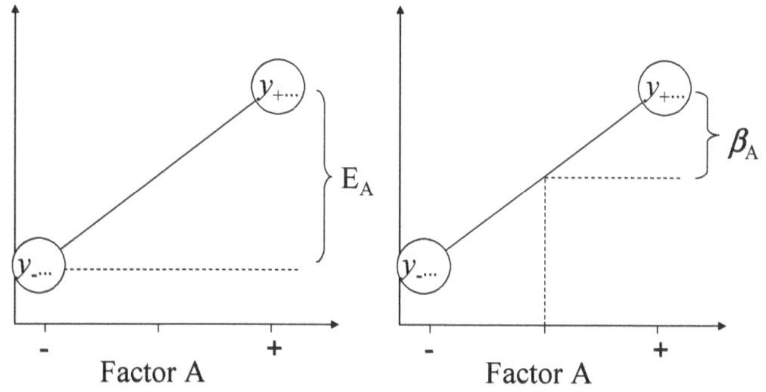

The regression slope β_A shown in the right side of Figure 3.9 is the vertical change in the average response for a one-unit change (i.e., from 0 to +1) in the factor level in symbolic units. Therefore the slope, β_A, is just one half the effect, E_A, or the difference in two averages divided by 2.

The treatment combinations in a two-level factorial can also be represented geometrically as the corners of a cube as shown in Figure 3.10. On the left side of this figure is a list of the treatment combinations or runs listed in standard or Yates' order with the first column changing fastest with alternating $-$ and $+$ signs, the second column changing in pairs of $-$ and $+$ signs, and the third column changing slowest in groups of four $-$ and $+$ signs. The treatment combinations in two-level factorial designs have traditionally been written in standard order to facilitate computation of main effects and interaction effects by hand using Yates' algorithm (see Daniel (1976)). The main effect for factor A can be visualized in the figure as the difference of the average of the responses on the right side of the cube in the grey-shaded circles and the average of the responses on the left side of the cube in the white circles. With modern computer programs such as SAS `proc glm` one half of the main effects, or regression coefficients (shown on the right side of Figure 3.9), can be computed by regression and we no longer need Yates' algorithm.

One of the desirable properties of a 2^k factorial plan is that factor effects are not obscured by planned changes in other factors. In the list of experiments for 2^k design, shown in Figure 3.10, this is evident by the fact that at the high level of each factor, there are an equal number of high and low levels of every other factor. Also at the low level of each factor, there are an equal number of high

Figure 3.10 *Geometric Representation of 2^3 Design and Main Effect Calculation*

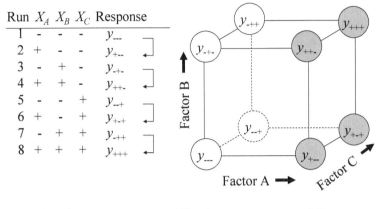

$$E_A = (y_{+--} + y_{++-} + y_{+-+} + y_{+++})/2 - (y_{---} + y_{-+-} + y_{--+} + y_{-++})/2$$

and low levels of every other factor. Thus the effect of a factor, or difference in average response between the high and low level of that factor, represents the effect of that factor alone, because the influence of all other factors has been averaged out. Mathematically this property is called orthogonality.

3.7.2 Interactions

When all the factors have only two levels, the interaction effect is defined as one-half the difference in the simple effect of factor A, $(\bar{y}_{++..} - \bar{y}_{-+..})$, when factor B is held constant at its high (+) level, and the simple effect of factor A, $(\bar{y}_{+-..} - \bar{y}_{---..})$, when factor B is held constant at its low (−) level, i.e., $((\bar{y}_{++..} - \bar{y}_{-+..}) - (\bar{y}_{+-..} - \bar{y}_{---..}))/2$. This is illustrated on the left side of Figure 3.11. The interaction effect could also be defined as one half the difference in the simple effect of factor B, $(\bar{y}_{++..} - \bar{y}_{+-..})$, when factor A is held constant at its high (+) level, and the simple effect of factor B, $(\bar{y}_{-+..} - \bar{y}_{---..})$, when factor A is held constant at its low (−) level. This is illustrated on the right side of Figure 3.11. Either way the interaction effect is $E_{AB} = (\bar{y}_{++..} + \bar{y}_{---..})/2 - (\bar{y}_{+-..} + \bar{y}_{-+..})/2$ is the difference of two averages.

It is easy to determine which responses should be averaged and which average should be subtracted from the other in order to calculate an interaction effect as illustrated in Figure 3.12. To calculate the AB interaction, we add a column of signs, $X_A \cdot X_B$ to the list of treatment combinations on the left side of the figure. The elements in this new column are just the elementwise products of signs in the column for X_A and X_B (i.e., $(-)(-) = +, (-)(+) = -$ etc.). Now the interaction effect can be visualized in the figure as the difference

TWO-LEVEL FACTORIALS

Figure 3.11 *Definition of an Interaction Effect for Two-level Factorial*

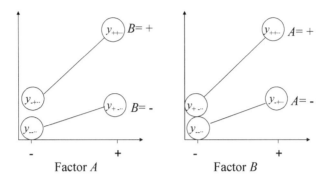

in the average response on one diagonal represented by grey circles and the average response on the other diagonal represented by white circles. From this

Figure 3.12 *Geometric Representation of 2^3 Design and Interaction Effect*

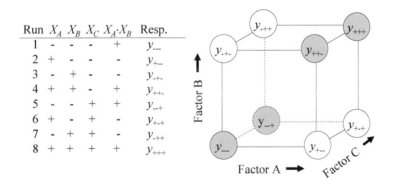

$$E_{AB}=(y_{---}+y_{++-}+y_{--+}+y_{+++})/2 - (y_{+--}+y_{-+-}+y_{++-+}+y_{-++})/2$$

representation, it can also be seen that interaction effects are not obscured by planned changes in other factors, or in other words they are orthogonal to main effects.

One-half of this interaction effect (or the regression coefficient) can be calculated using a regression program such as `proc glm` by adding a $X_A \times X_B$ term to the model. Higher order interaction effects can be similarly defined.

Therefore, a simpler way of writing the model for a two-level factorial is by using the familiar regression equation,

$$y = \beta_0 + \beta_A X_A + \beta_B X_B + \beta_{AB} X_A X_B + \beta_C X_C + \beta_{AC} X_A X_C + \beta_{BC} X_B X_C$$
$$+ \beta_{ABC} X_A X_B X_C + \epsilon$$
(3.11)

where, the βs are one-half of the effects and $X_A = -1$ if factor A is at its low level and $X_A = +1$ if factor A is at its high level. If we write this model in matrix terms, $\boldsymbol{y} = \boldsymbol{X\beta} + \boldsymbol{\epsilon}$, the orthogonality property of the design is expressed by the fact that the columns of the $\boldsymbol{X}$ matrix are orthogonal and the $\boldsymbol{X'X}$ matrix is diagonal with diagonal elements $\boldsymbol{r2^k}$, where r is the number of replicates of each cell.

3.7.3 Example of a 2^3 Factorial

To illustrate the design and analysis of a 2^3 factorial experiment, consider the following example (see Lawson and Erjavec, 2001). Students in a university electronics lab often complained that voltage measurements made on a circuit they constructed in class were inconsistent. The lab teaching assistant (TA) decided to conduct an experiment to try to identify the source of the variation. The three factors he varied were the A=the ambient temperature where the voltage measurement was made, B=the voltmeter warm-up time, and C=the time the power was connected to the circuit before the measurement was taken. The response was the measured voltage in millivolts. The two levels for factor A were $-$ = 22°C (room temperature) and $+$ = 32°C (close to the temperature in some industrial settings). An oven was used and the circuit was allowed to stabilize for at least five minutes prior to measurements. The settings for factors B and C were $-$ =30 seconds or less, and $+$ =5 minutes. The same circuit was measured for each combination of treatment factors so the experimental unit was nothing more than the trial or point in time at which the particular combination of treatment factor levels were applied to make the measurement. Two replicates of each of the eight experimental combinations were run in a random order to help prevent biases. The results of the experiment are shown in Table 3.8.

In this table, the actual factor settings are shown on the left, and the coded $-$ and $+$ levels are shown on the right. The actual settings on the left form a list of recipes or directions for performing each experiment. The order number on the far right next to the response was created with a random number generator and represents the order in which the experiments should be run. The coded factor levels are used as the independent variables in a regression program in order to calculate the regression coefficients or half effects.

The coded factor levels can be easily calculated from the actual factor settings using the coding and scaling formula. In this formula we subtract the midpoint of the two factor settings, then divide by half the range. For exam-

TWO-LEVEL FACTORIALS

Table 3.8 *Factor Settings and Response for Voltmeter Experiment*

Run	A	B	C	X_A	X_B	X_C	Rep	Order	y
1	22	0.5	0.5	−	−	−	1	5	705
2	32	0.5	0.5	+	−	−	1	14	620
3	22	5.0	0.5	−	+	−	1	15	700
4	32	5.0	0.5	+	+	−	1	1	629
5	22	0.5	5.0	−	−	+	1	8	672
6	32	0.5	5.0	+	−	+	1	12	668
7	22	5.0	5.0	−	+	+	1	10	715
8	32	5.0	5.0	+	+	+	1	9	647
1	22	0.5	0.5	−	−	−	1	4	680
2	32	0.5	0.5	+	−	−	1	7	651
3	22	5.0	0.5	−	+	−	1	2	685
4	32	5.0	0.5	+	+	−	1	3	635
5	22	0.5	5.0	−	−	+	1	11	654
6	32	0.5	5.0	+	−	+	1	16	691
7	22	5.0	5.0	−	+	+	1	6	672
8	32	5.0	5.0	+	+	+	1	13	673

ple, for factor A the midpoint between 22 and 32 is 27, and half the range is 5, thus

$$X_A = \left(\frac{Actual\,Factor\,Setting - 27}{5} \right).$$

The commands to read this data into SAS and analyze it with **proc glm** are shown below.

```
data volt;
input A B C y;
  XA=(A-27)/5;
  XB=(B-2.75)/2.25;
  XC=(C-2.75)/2.25;
datalines;
22 0.5 0.5 705
32 0.5 0.5 620
   ... etc.
proc glm;
   model y=XA XB XC XA*XB XA*XC XB*XC XA*XB*XC/
      solution;
run;
```

Here it can be seen that only the actual factor levels are read into SAS, and the coded factor levels are created in the data step. In order to do regression rather than ANOVA, no class statement is used in **proc glm** and the /solution option creates the table of the regression coefficients (or half effects) shown below.

Parameter	Estimate	Standard Error	t Value	Pr > |t|
Intercept	668.5625000	4.51776009	147.99	<.0001
XA	-16.8125000	4.51776009	-3.72	0.0059
XB	0.9375000	4.51776009	0.21	0.8408
XC	5.4375000	4.51776009	1.20	0.2632
XA*XB	-6.6875000	4.51776009	-1.48	0.1771
XA*XC	12.5625000	4.51776009	2.78	0.0239
XB*XC	1.8125000	4.51776009	0.40	0.6988
XA*XB*XC	-5.8125000	4.51776009	-1.29	0.2342

Here it can be seen that factor A (ambient temperature) and the A×C interaction, or interaction between the ambient temperature and the circuit warm-up time, are significant. The main effect has direct interpretation. The effect of factor A is twice the regression coefficient shown above or $E_A = 2 \times \hat{\beta}_A = 2(-16.8125) = -33.625$. This means that, on the average, when the ambient temperature is increased from 22° to 32°, the voltage measurement will decrease by 33.6 millivolts. However, since the interaction is significant in this example, it really is not meaningful to talk about the average main effect because the effect of ambient temperature depends on the circuit warm-up time.

Describing or interpreting the interaction is best done by looking at the interaction plot shown in Figure 3.13 on the next page. Here it can be seen that when the circuit warm-up time is short (0.5 minutes or 30 seconds) changing the ambient temperature from 22° to 32° causes a large (58.7 millivolt) decrease in the voltage reading. However, when the circuit warm-up time is long (5 minutes), changing the ambient temperature from 22° to 32° only causes a small (8.5 millivolt) decrease in the voltage reading. Therefore, to make voltage readings more consistent, the lab TA recommended that his students allow their circuits to warm up 5 minutes before making voltage measurements.

The regression was performed on the coded factor levels so that the regression coefficients produced by proc glm would be exactly half of the effects. However, the actual factor names and levels should be used for clarity when presenting the results graphically for inclusion in a report or presentation, as shown in Figure 3.13. Most readers or listeners will not remember what the − and + levels represent.

The SAS commands to produce the graph are shown below. Here an extra data step was employed to rename the factor C to Circuit_Warmup_Time in order to correctly label the legend on the graph. The axis1 statement was used to create a label for the horizontal axis and the plot is made of actual factor levels rather than coded factor levels so that the units on the horizontal axis will not be labeled as (-1) and (1).

By eliminating the insignificant terms, a simple prediction equation

TWO-LEVEL FACTORIALS

```
proc sort data=volt; by A C;
proc means noprint; by A C; var y;
  output out=s mean=meanvolt;
proc print; run;
data s; set s;
  rename C=Circuit_Warmup_Time;
proc sgplot data=s;
  series x=A y=meanvolt/ group=Circuit_Warmup_Time markers;
  xaxis label='Ambient Temperature' min=20 max=34;
  yaxis label='Voltage (millivolts)';
run;
```

Figure 3.13 *Interaction between Ambient Temperature and Circuit Warm-up Time*

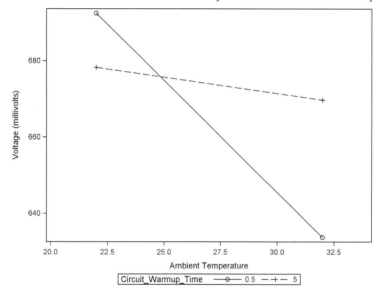

$$y = 668.563 - 16.813 \left(\frac{Temp - 27}{5}\right) - 6.688 \left(\frac{CWarm - 2.75}{2.25}\right) \left(\frac{Temp - 27}{5}\right)$$

can be written from the regression results and used to predict the voltage reading in millivolts for any ambient temperature between 22° and 32°, and any circuit warm-up time between 30 seconds and 5 minutes.

3.7.4 Shortcut Formula for Determining the Number of Replicates

Wheeler (1974) has developed a shortcut approximation formula for calculating the number of runs necessary to achieve power equal to 0.95 when the significance level for a two-level factorial is $\alpha = 0.05$. This formula is

$$N = ((8\sigma)/\Delta)^2 \qquad (3.12)$$

where σ is the standard deviation of the experimental error, Δ is the size of a practical effect. In this case the difference in average response between the low and high level of a factor and $N = r \times 2^k$ is the total number of experiments in the 2^k factorial. Since this formula is so compact, it is easy to use on the spot in meetings where experiments are being planned. As an illustration of its use, consider an example based on the voltage meter experiment presented in the last section.

The lab instructor felt that the standard deviation of the experimental error was about $\sigma = 15.0$ and the size of practical effect was about $\Delta = 30.0$. σ would be known by the lab instructors experience in making repeat voltage measurements of the same circuit under exactly the same conditions (i.e., factor levels), and Δ would be known from the amount of inconsistency in measurements claimed by the students who were getting inconsistent readings. Δ is the size of the effect the TA would like to detect in his experiments. Using the shortcut formula, this says that

$$N = ((8 \times 15.0)/30.0)^2 = 16$$

or that $r = 2$ replicates of each of the $2^3 = 8$ runs should result in a power of 0.95 for detecting effects of size 30.0 at significance level $\alpha = 0.05$. If you modify and use the SAS program for calculating the power in the data step for a factorial experiment presented in Section 3.5.2, the actual power for $r = 2$ replicates is closer to 0.94 than 0.95. However this approximate formula is accurate enough for most planning purposes.

The simple formula can also be used backwards by solving for Δ as a function of N, i.e., $\Delta = 8 \times \sigma/\sqrt{N}$. That way, if an experimenter knows his budget for experimentation, which dictates the largest N can be, he can calculate the size of the effect Δ that he is likely to be able to detect. If the experimenter does not have an accurate estimate of σ, the formula can still be used by talking about practical effect size in units of the unknown σ. For example, if an experimenters' budget allows him to make at most $N = 64$ experiments, he can hope to detect effects that are no more than one standard deviation of the experimental error, i.e., $\Delta = 8 \times \sigma/\sqrt{64} = \sigma$. This result will be true regardless of the number of factors in the two-level experiment. Consequently, with 64 runs he may have one factor with $r = 32$ replicates of each level, or six factors with $r = 1$ replicate of each of the $2^6 = 64$ treatment combinations.

TWO-LEVEL FACTORIALS

3.7.5 Analysis with One Replicate per Cell

Factorial designs with one replicate per cell are often referred to as *unreplicated* designs. When there is adequate power for detecting effects with $r = 1$ replication per cell, or treatment combination, there is no need to double the experimental work by replicating each experiment. However, in an un-replicated factorial, the same problem arises that was discussed in Section 3.5.4. There will be zero degrees of freedom for calculating ssE and thus no F-tests for the effects. However, when there are multiple factors in a two-level factorial, there are simple graphical tools that allow detection of the significant effects. Since not all main effects and interactions in a 2^k experiment are expected to be significant, the levels of insignificant factors and combinations of levels defined by the insignificant interactions are equivalent to having replicates in the design. Graphical tools allow the significant effects (or equivalently regression coefficients) to be recognized.

The most common graphical tool used to spot significant effects are normal or half normal plots that were first suggested by Daniel (1959). These are easy to produce using SAS, as were normal plots of residuals in Section 2.4. The additional graphical tools, Pareto Diagram, Lenth Plot and Bayes Plot are also useful for detecting significant effects and interactions and can be generated automatically using the interactive SAS ADX (Automated Design of Experiments) Tool. Examples of the input and output available for analysis of two-level factorials in ADX are shown in files on the web page (http://lawson.mooo.com) for this book.

To illustrate the analysis of an un-replicated two-level factorial, consider an example from the chemical industry. Experimental design principles were developed by Fisher in the early part of the twentieth century and were originally used in agricultural experiments. Within forty years there was extensive use of experimental design techniques in the chemical industry. Figure 3.14 is a diagram of a continuous chemical process. In this process continuous streams of two reactants, A and B, are combined at a juncture called the mixing-T where they begin to react. The mixture then flows into a reactor and is combined with solvent and a catalyst and the reaction is completed. The result of the reaction flows into a separator tank where the final product floats to the top in a solvent phase while the catalyst and water go to the bottom of the tank. The catalyst is concentrated and sent back into the reactor, while the product, byproducts and solvent are taken to a distillation column where the product is removed and the solvent is recycled to the reactor.

One of the problems experienced in this process was the production of byproduct (tars). Over time these tars would clog the reactor and force a shut-down of the process for cleaning. It also required an additional process step to purify the final product. Engineers decided to conduct experiments to see if they could increase the percent conversion which would lower the amount of byproducts. The factors they thought might affect the percent conversion were:

Figure 3.14 *Diagram of a Chemical Process*

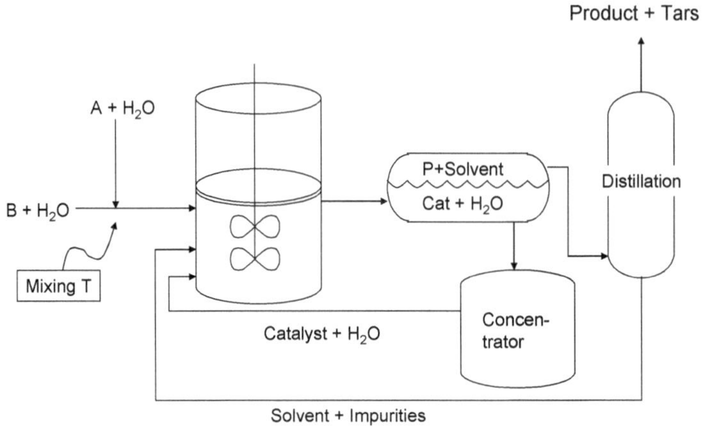

Symbol	Factor Name
A	Excess of Reactant A (over molar amount)
B	Catalyst Concentration
C	Pressure in the Reactor
D	Temperature of the Coated Mixing-T

Two levels of each factor were chosen that were spread apart as wide as the engineers thought feasible in order to maximize the chance of detecting factor effects with only two levels. During experimentation, the factor levels would be changed after a fixed interval of time. The experimental unit for this would be the particular reactants, catalyst and solvent entering the reaction zone during a given run, and the response, Y, would be the percent conversion calculated from the product produced during a run.

It was felt that if the percent conversion could be increased by $\Delta = 12\%$ (or more) it would substantially reduce the maintenance and extra processing currently required and would be worth detecting. From past experience with the process, the standard deviation in percent conversion on this process for product produced in the same length intervals as the runs in the proposed experimental design (with no changes in the factor levels) was $\sigma = 6\%$. Using the shortcut formula, the number of runs required to have a power of 0.95 for detecting factor effects of $\Delta = 12\%$ or more, was

$$N = (8\sigma/\Delta)^2 = ((8)(6)/12)^2 = 16$$

Sixteen runs with four factors means the design would be unreplicated. The

Table 3.9 *List of Experiments and Results for Chemical Process*

Random Run No.	A	B	C	D	Y
15	−	−	−	−	45
13	+	−	−	−	41
11	−	+	−	−	90
1	+	+	−	−	67
10	−	−	+	−	50
2	+	−	+	−	39
3	−	+	+	−	95
12	+	+	+	−	66
16	−	−	−	+	47
8	+	−	−	+	43
9	−	+	−	+	95
14	+	+	−	+	69
6	−	−	+	+	40
5	+	−	+	+	51
7	−	+	+	+	87
4	+	+	+	+	72

experimental conditions (in coded units), a list of random run orders, and results of the experiments are shown in Table 3.9.

The SAS commands to read the data and perform a regression analysis to estimate the half effects or regression coefficients are shown below.

```
data chem;
input A B C D y;
datalines;
-1 -1 -1 -1 45
 1 -1 -1 -1 41
 . . . etc.
proc glm data=chem;
   model y=A|B|C|D/solution;
   *this ouputs the parameter estimates to a file;
   ods output ParameterEstimates=sol;
run;
```

Unlike the example in Section 3.7.3, there will be no estimate of ssE and thus no t-tests on the regression coefficients in the `proc glm` output. The `ods` statement outputs the table of parameter estimates to a file called `sol` which can be read to produce graphs. The regression coefficients are shown in the `proc glm` output below. The regression coefficients for main effects A and B along with the $A \times B$ interaction are the largest effects, but a graph must be used to determine which are significant.

The effects in a two-level factorial are the difference of two averages. If changes in factor levels do not cause a change in the response, the effect will

98   FACTORIAL DESIGNS

Parameter	Estimate	Standard Error	t Value	Pr > |t|
Intercept	62.31250000	.	.	.
A	-6.31250000	.	.	.
B	17.81250000	.	.	.
A*B	-5.31250000	.	.	.
C	0.18750000	.	.	.
A*C	0.81250000	.	.	.
B*C	-0.31250000	.	.	.
A*B*C	-0.18750000	.	.	.
D	0.68750000	.	.	.
A*D	2.06250000	.	.	.
B*D	-0.06250000	.	.	.
A*B*D	-0.68750000	.	.	.
C*D	-0.68750000	.	.	.
A*C*D	2.43750000	.	.	.
B*C*D	-0.43750000	.	.	.
A*B*C*D	-0.31250000	.	.	.

be just the difference in averages of random data (due to random fluctuations in experimental error). If none of the factors or interactions cause changes in the response, the entire set of effects, or regression coefficients, should appear as a sample from the normal distribution with zero mean due to the Central Limit Theorem. Therefore if we make a normal probability plot of the effects, the insignificant effects should lie along a straight line and any significant effects or interactions should appear as outliers on the plot.

The SAS commands to create a normal plot of the regression coefficients in the file sol are shown on top of the next page. These commands drop the first row in the file nplot, containing the intercept term β_0 shown in Equation (3.11). Next, the variables StdErr tValue Probt are dropped from the file. proc rank is used to calculate the normal scores, and in the next data step a new file nplots is created where the variable parameter, which will be used to label the outlier effects, is set to a blank when the z-score is less than or equal to 1.2 in absolute value. Finally, proc sgplot is used to create the plot.

In this normal plot (Figure 3.15) most of the points lie along a straight line drawn through the origin at $(0,0)$. However, the points representing main effects A, B, and the $A \times B$ interaction tend to be above and to the right or below and to the left of the straight line. This indicates that these three effects are significant. The points along the straight line are insignificant and the slope of this line is an estimate of the standard error of an effect $\hat{\sigma}_\beta$.

Another way of looking at the effects graphically is to make a half normal plot. The half normal plot is created by essentially folding the normal plot at 45° along the line $y = -x$. This puts the origin of the graph in the lower left corner and makes the straight line of insignificant effects appear to come out of the origin. To make a half-normal plot, plot the absolute value of the effects or regression coefficients on the vertical axis and their corresponding half normal scores on the horizontal axis. Since proc rank does not compute

TWO-LEVEL FACTORIALS

```
data nplot; set sol;
  estimate=estimate;
  if _n_>1; drop StdErr tValue Probt;
* proc rank calculates normal scores for parameter estimates ;
proc rank data=nplot out=nplots normal=blom; var estimate;
  ranks zscore;
data nplots; set nplots;
if abs(zscore)<=1.2 then parameter='    ';
proc sgplot data=nplots;
scatter x=zscore y=estimate/datalabel=parameter;
xaxis label='Normal Scores';
run;
```

Figure 3.15 *Normal Probability Plot of Regression Coefficients*

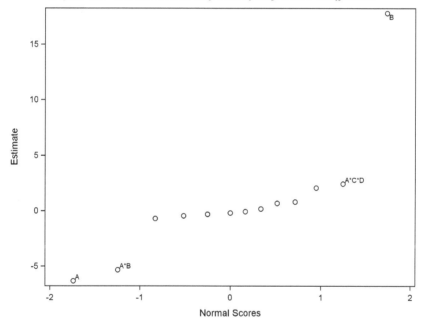

half normal scores directly, the half normal score for $|\hat{\beta}|_{(i)}$ can be approximated by the formula $\Phi^{-1}\left(((i-1/2)+1)/2\right)$ where $\Phi^{-1}()$ is the inverse CDF of the standard normal distribution, and i is the rank order of $|\hat{\beta}|_{(i)}$. In SAS the probit function computes $\Phi^{-1}()$ and the code for creating a half-normal plot is shown on the next page.

The half-normal plot shown in Figure 3.16 has a reference line coming out of the origin in the lower left, and an upper prediction limit line. Effects A, B, and the $A \times B$ interaction fall above the upper limit line and are clearly significant. The reference line and upper limit in Figure 3.16 were added automatically by the SAS macros described by Lawson (2008a) following the procedure outlined

```
data hnplot; set sol;
  estimate=abs(estimate);
  if _n_>1; drop StdErr tValue Probt;
* proc rank calculates ranks for parameter estimates ;
proc rank data=hnplot out=hnplots; var estimate; ranks rnk;
* data step calculates half normal scores for estimates;
data hnplots; set hnplots;
  zscore=probit((((rnk-.5)/15)+1)/2);
  if abs(zscore)<=1.0 then parameter='    ';
proc sgplot data=hnplots;
scatter x=zscore y=estimate/datalabel=parameter;
xaxis label='Half Normal Scores';
run;
```

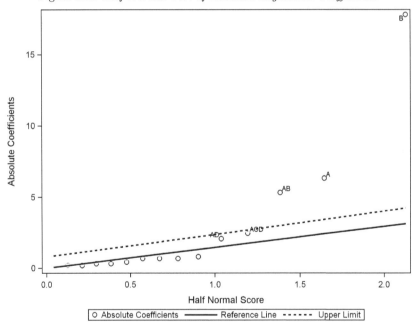

Figure 3.16 *Half-Normal Plot of Absolute Regression Coefficients*

by Lawson *et al.* (1998). When drawing the reference line by hand on a plot, it is easier to use the half-normal plot than the normal plot, because the first half to two thirds of the points trending from the lower left will almost always form a straight line. However, on the half-normal plot the signs of the coefficients are lost. For example, in the `proc glm` output and the normal plot it can be clearly seen that main effect A (the excess of reactant A) has a negative effect, and that increasing the excess of reactant A will on the average cause a decrease in the percent conversion. In the half-normal plot, it can be seen that

TWO-LEVEL FACTORIALS

main effect A is significant but one must refer back to the table of coefficients to see whether it has a positive or negative effect.

Of course, in this example there is a significant interaction, and therefore the main effects cannot be interpreted separately. To assist in interpreting the interaction, an interaction plot should be created. The SAS commands to create and label the interaction plot between factor A (excess of reactant A) and factor B (catalyst concentration) is shown below, and the resulting graph is shown in Figure 3.17.

```
proc format;
  value levels -1='Low Level' 0=' ' 1='High Level';
  run;
*makes table of means;
proc sort data=chem; by A B;
proc means data=chem noprint; var y; by A B;
  output out=intplt mean=yhat;
run;
proc print; run;
*makes plot of interaction;
proc sgplot data=intplt;
  series x=A y=yhat/ group=B markers;
  xaxis label='Excess of Reactant A' type=discrete;
  yaxis label='Percent Conversion';
  format B levels. A levels.;
run;
```

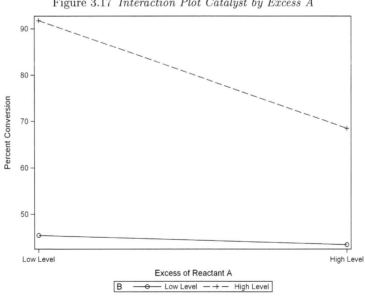

Figure 3.17 *Interaction Plot Catalyst by Excess A*

There it can be seen that increasing the level of catalyst increases conversion. Increasing the excess of reactant A has little effect on conversion when a low level of catalyst is used. However, if a high level of catalyst is used, increasing the excess of reactant A decreases conversion by more than 20 percent. Therefore, to achieve the highest level of conversion, a high level of catalyst and a low level of excess reactant A should be used.

3.8 Verifying Assumptions of the Model

When there are replicate experimental units in each cell of a factorial model, or when an interaction term can be assumed negligible and removed from the model (as result of a preliminary test like those described in Section 3.5.4 or 3.7.5), the normality and constant variance assumptions of the factorial model can be verified with residual plots as described in Section 2.4.

However, in the case of 2^k design with only one replicate per cell it is a little more difficult to check the assumption of normality. The normality assumption is most often violated by having one outlier or atypical value. The calculated main effects and interactions in a two-level factorial can always be represented as the difference of two averages, $\bar{y}_+ - \bar{y}_-$. When experimental errors follow a normal distribution, the calculated effects for factors and interactions that have a negligible influence on the response should be normally distributed with mean zero. The significance of potential influential factors are judged by their relation to a reference line of points on a normal or half-normal plot of effects formed by the negligible factors. However, one atypical value will bias each calculated effect positively or negatively away from zero. The variability of the calculated effects for the non-influential factors and interactions will be much larger and it will be more difficult to judge significance of effects, much less check the normality assumption of residuals.

Daniel (1960) proposed a manual method for detecting and correcting an outlier or atypical value in an unreplicated 2^k design. This method consists of three steps. First, the presence of an outlier is detected by a gap in the center of a normal plot of effects. Second, the outlier is identified by matching the signs of the insignificant effects with the signs of the coded factor levels and interactions of each observation. The third step is to estimate the magnitude of the discrepancy and correct the atypical value.

As an example, consider the normal plot of effects (Figure 3.18) from an unreplicated 2^4 experiment described by Box (1991). In this plot it appears that main effects B and C may be significant, but there is a vertical gap in the line of insignificant effects that indicates an outlier may be present.

Lawson and Gatlin (2006) automated Daniel's procedure identifying and correcting an atypical value, and a SAS macro for performing this procedure is available from Lawson (2008a). The SAS commands on the next page illustrate the use of this macro with the data described by Box (1991). The first data step creates a 2^4 design and stores it in a data set called **xmatrix16**. The next data step reads the response data into a data set called **box**. Next the macro variables **%nobs** and **%vname1** are defined. **%nobs** is the number of

VERIFYING ASSUMPTIONS OF THE MODEL

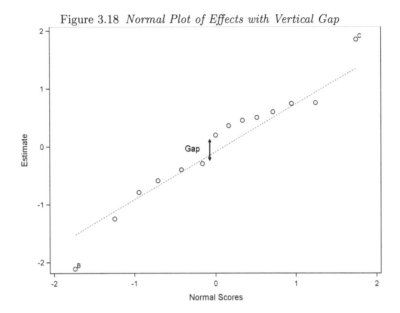

Figure 3.18 *Normal Plot of Effects with Vertical Gap*

runs in the design, and %vname1 is a list of the names of the columns in the X matrix. These macro variables are needed by the macro. Finally the data set representing the X matrix is copied into a data set called x1 and the data set representing the response variable is copied into the data set y. These are the names required by the macro. Finally the macro is called by the statement %gapmth. The output of the macro consists of a table of the regression coefficients and a test of their significance, a test for the significance of the gap in the normal plot, and when the gap is determined to be significant, a corrected response table and a table of regression coefficients calculated with the corrected data is produced. The corrected response table is shown on the next page below the SAS commands to produce it.

The method detected an outlier on the 13th run, and it corrected the response by changing 59.15 to 52.15. Reanalysis of the corrected data resulted in the table shown on the page that follows the table of corrected data.

The macro also produces half-normal plots of the regression coefficients calculated from the raw data and corrected data. Figure 3.19 shows the half-normal plot of the coefficients calculated with the corrected data. In this plot it can be seen that the AC interaction is clearly significant in addition to the main effects B and C, that were identified in Figure 3.18. They could not be detected in Figure 3.18 because the outlier inflated the estimates of the insignificant effects and the AC effect was buried in the noise.

Whenever an outlier is discovered, using this method or residual plots, and the conclusions of the analysis change when the outlier is removed or corrected, the experimenter should proceed cautiously. When there are more than two

```
*Create the design matrix of contrast coefficients for a 2^4
    design;
data xmatrix16;
  do x4=-1 to 1 by 2;
    do x3=-1 to 1 by 2;
      do x2=-1 to 1 by 2;
        do x1=-1 to 1 by 2;
        I=1;  A=x1;  B=x2;  C=x3;  D=x4;
        AB=A*B;  AC=A*C;  AD=A*D;  BC=B*C;  BD=B*D;  CD=C*D;
        ABC=A*B*C;  ABD=A*B*D;  ACD=A*C*D;  BCD=B*C*D;  ABCD=A*B*C*D;
        output;
        end;
      end;
    end;
  end;
drop x1-x4;
run;
*Read in the response data for 2^4 design;
data box;
input y @@;
datalines;
47.46 49.62 43.13 46.31 51.47 48.49 49.34 46.10 46.76 48.56 44.83
44.45 59.15 51.33 47.02 47.90
*Defines the nobs and vname1 macro variables needed by the macro;
%let nobs=16;
%let vname1={'I' 'A' 'B' 'C' 'D'
'AB' 'AC' 'AD' 'BC' 'BD' 'CD'
'ABC' 'ABD' 'ACD' 'BCD' 'ABCD'};
*Invokes the macro;
data y; set box;
data x1; set xmatrix16;
%gapmth
```

Corrected Data Report

Response	Corrected Response	Detect Outlier
47.46	47.46	no
49.62	49.62	no
43.13	43.13	no
46.31	46.31	no
51.47	51.47	no
48.49	48.49	no
49.34	49.34	no
46.1	46.1	no
46.76	46.76	no
48.56	48.56	no
44.83	44.83	no
44.45	44.45	no
59.15	52.75	yes
51.33	51.33	no
47.02	47.02	no
47.9	47.9	no

VERIFYING ASSUMPTIONS OF THE MODEL

```
                Effect Report

                Half
Label           Effect        Abs(Effect)    Sig(at 0.05)

A               -9E-16        89E-17         no
B               -1.71         1.71           yes
C                1.455        1.455          yes
D                0.105        0.105          no
AB               0.055        0.055          no
AC              -0.845        0.845          yes
AD               0.11         0.11           no
BC              -9E-16        89E-17         no
BD              -0.19         0.19           no
CD               0.345        0.345          no
ABC              0.2          0.2            no
ABD             -0.04         0.04           no
ACD              0.6          0.6            no
BCD             -0.39         0.39           no
ABCD             0.36         0.36           no

Lawson, Grimshaw & Burt Rn Statistic=1.626
95th percentile of Rn= 1.201
```

replicates at the same factor settings where the outlier was found, it may be clear that something is amiss. However, if there are two or less observations at factor settings where the outlier was found, it may be advisable to rerun the questionable experiment.

Figure 3.19 *half-normal Plot of Coefficients Calculated with Corrected Data*

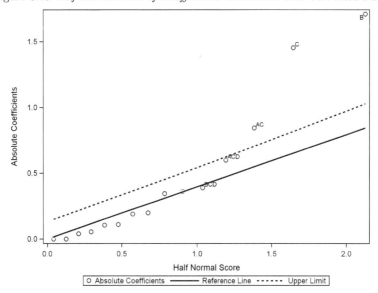

3.9 Review of Important Concepts

When experimenting with more than one treatment factor, factorial designs with all possible treatment combinations randomized to experimental units are much more efficient than separate one factor designs. When there are homogeneous experimental units and multiple factors under study the CRFD or completely randomized factorial design should be used.

Figure 3.20 shows when CRD and CRFD should be used. If there is one factor under study with homogeneous experimental units, use a CRD. However, if there are multiple experimental units under study, use the CRFD. In

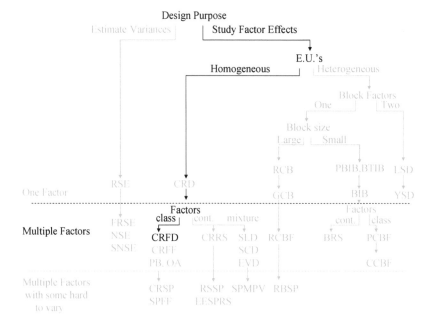

Figure 3.20 *Design Selection Roadmap*

a factorial design the total number of replicates of each level of one factor is the number of replicates per cell multiplied by the product of the number of levels of all other factors in the design. This *hidden replication* increases the power for detecting factorial effects or reduces the number of experiments needed to obtain the same power as a series of separate one-factor designs. The CRFD should be used when the experimental units are homogeneous and it is reasonable to run all combinations of levels of the factors.

By studying more than one treatment factor simultaneously in a factorial design, interaction or joint effects of the factors can be detected. Interactions occur when the effect of one factor is different depending on the level of another factor or on a combination of levels of other factors. Interactions are

REVIEW OF IMPORTANT CONCEPTS

common in the real world, and ignoring them by experimenting with one factor at a time can be very misleading. Examples in this chapter show that it is easy to describe or interpret the meaning of an interaction by comparing or contrasting the effect of one factor over the levels of another factor. Interaction graphs are helpful in describing interactions.

The model for analysis of a factorial with two factors can be written as

$$y_{ijk} = \mu + \alpha_i + \beta_j + \alpha\beta_{ij} + \epsilon_{ijk}$$

and can be easily extended to multiple factors. The assumptions for the analysis are homogeneity and normality of experimental errors, and can be checked using the residual plots described in Chapter 2. When there is an equal number of replications per cell or treatment combination, the proc glm type I or type III sums of squares can be used to test the main effects and interactions, and the means or lsmeans statements can be used to get unbiased estimates of cell means and marginal means. However, if there is an unequal number of replicates per cell, the type III sums of squares and lsmeans must be used.

When multiple factors are studied in a factorial experiment, often only two levels of each factor are used in order to reduce the total amount of experimentation required. In that case, the model simplifies to a regression model

$$y = \beta_0 + \beta_A X_A + \beta_B X_B + \beta_{AB} X_A X_B + \beta_C X_C + \beta_{AC} X_A X_C + \beta_{BC} X_B X_C + \beta_{ABC} X_A X_B X_C + \epsilon$$

and the regression coefficients are exactly one half the effect or difference in the average response between the low and high levels of the factors. Interaction effects are also easily defined as differences of averages in this case. There is a shortcut formula for approximating the power in two-level factorials, and when there is only one replicate per treatment combination, the significance of effects and interactions can be determined using normal or half-normal plots.

3.10 Exercises

1. A consultant was called to assist the police department of a large metropolitan city in evaluating its human relations course for new officers. He planned a 2-factor factorial experiment where the treatments were A the type beat to which officers were assigned, and B the length of the human relations course. A sample of 45 new officers was chosen, and 5 were randomly assigned to each of the 9 treatment combinations. A test was developed to measure officers' attitude toward minority groups and was administered to the participating officers after their training had ended and they had experienced two weeks on their beat. Better attitudes result in higher scores on this test. Analysis of the data revealed a significant A×B interaction effect between the type beat and length of human relations course. The table below shows the mean test scores for the 9 combinations of treatment levels.

| | Length of human relations course | | |
Type beat	5 hours	10 hours	15 hours
upper-class beat	34.4	35.5	39.2
middle-class beat	30.2	32.4	34.7
inner-city beat	20.1	39.4	54.3

(a) Construct an interaction graph.

(b) Write an interpretation of the interaction in a few complete sentences.

2. A wooden catapult can be used to flip a foam ball. The catapult has three factors that can be adjusted: the start angle, the stop angle, and the pivot height. The distance the ball travels can be measured with a tape measure.

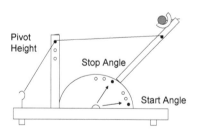

(a) If experiments were to be conducted with the catapult by flipping the ball and measuring the distance, what would the experimental unit be?

(b) Using the numbers 1, 2 and 3 to represent the levels of start angle and stop angle, and holding the pivot height constant at its high level, make a randomized list of experiments for a 3×3 factorial experiment with $r = 2$ replicates per cell.

EXERCISES

(c) If the variance of the experimental error in the measured distance was $\sigma^2 = 12$ inches, calculate the number of replicates you would need to have a power of 0.90 for detecting a difference in 10 inches in cell means.

(d) Calculate the number of replicates you would need to have a power of 0.90 for detecting a difference of 24 inches in marginal means for either factor.

(e) If a catapult is available, conduct the list of experiments you wrote in part (b).

(f) Calculate the ANOVA with your resulting data and test the main effects and interaction.

(g) Explain or interpret any significant effects (use graphs if necessary to aid in your explanation).

3. In an experiment to maximize the $Y=$ resolution of a peak on a gas Chromatograph, a significant interaction between $A =$ column temperature and $C =$ gas flow rate was found. The table below shoes the mean resolution in each combination of column temperature and gas flow rate.

| Column | Gas Flow Rate | |
Temperature	Low	High
120	10	13
180	12	18

(a) Construct an interaction graph.

(b) Write a sentence, or two, to interpret this interaction.

4. Consider performing experiments to determine the effect of popcorn brand, power level, and time on the percentage of edible popcorn (unpopped or burnt is not edible) kernels made in a microwave oven. The object is to maximize the proportion of edible kernels. Start with $\frac{1}{3}$ cup of kernels and do pilot experiments to determine the range of the factors you would like to study and an estimate of the standard deviation of replicates made under the same conditions.

(a) What is the experimental unit for this experiment?

(b) Determine how many replicates will be required to detect a maximum difference in marginal (main effect) means of 0.25.

(c) Determine the number of levels of the factors you would like to study, and create a randomized list of the experiments for a factorial design.

(d) Actually perform the experiments and collect the data.

(e) Analyze the data. Partition any significant effects of power level and time into orthogonal polynomial contrasts. Interpret the results, and determine the optimum combination of factor levels from among those you tested.

(f) How many experiments would it have taken to get the same power for main effects using a vary one-factor-at-a-time plan? Would you detect the same optimum using the vary one-factor-at-a-time plan?

5. Use `proc glmpower` to verify that the power would be 0.80 for detecting differences in marginal means of $\Delta = 1.0$ in. in the paper helicopter experiments with four levels of wing length, four levels of body width, and $r = 2$ replicates per cell (as shown in Section 3.5.2).

6. In a continuation of the experiments to investigate the effect of plant growth regulators and spear bud scales on spear elongation in asparagus, described in exercise 4 of Chapter 2, Yang-Gyu and Woolley (2006) conducted a 4×3 factorial experiment varying the concentration of the plant growth regulator CPPU in the solution, and the time the asparagus spears were dipped in the solution. The results are shown in the table below.

CPPU conc.	dipping time (sec)		
	30	60	90
0 (control)	92.5	92.9	91.3
0.5 ppm	97.8	94.9	101.3
1.0 ppm	97.0	98.5	101.6
10.0 ppm	103.4	102.9	98.6

(a) Partition the sums of squares for CPPU concentration and dipping time into orthogonal polynomial contrasts. Partition the interaction of CPPU conc. by dipping time into linear×linear, quadratic×linear, cubic×linear, linear×quadratic, quadratic×quadratic, and cubic×quadratic. Pool (sum) all sums of squares but the linear×linear portion to form an error sum of squares and use it to test the linear×linear part of the interaction and the polynomial contrasts you found for the main effects. Interpret your results.

(b) Use the %tukeysdf macro to test the significance of the interaction.

(c) Based on the results you obtained in (a) and (b) would you recommend the additive model for this data?

7. Kenett and Steinberg (1987) describe a two-level factorial experiment conducted by students to study the time required to boil 1 qt of water. Factors were A=flame level (low or high), B=pan size (small or large), C=pan cover (none or glass cover) and D=salt added to water (no or yes).

(a) If the standard deviation in boiling time (tested at the same conditions) was found to be $\hat{\sigma}=0.236$ minutes, use the shortcut formula to determine how many experiments you will need to perform in order to have power of 0.95 for detecting effects of size $\Delta=0.50$ minutes. Would this answer change if you decided to only perform an experiment with 3 of the 4 factors?

EXERCISES

(b) Create a list of experiments in random order for performing these experiments.

(c) Actually perform the experiments by boiling the water and collect the data.

(d) Analyze the data to determine which effects and interactions are significant.

(e) Interpret and explain any significant effects you find.

8. Consider the data in Table 3.9 with the third observation (90) replaced by an atypical value (78) for those factor settings.

   (a) Calculate the effects for this data.
   (b) Make a normal or half-normal plot of the effects.
   (c) What effects do you judge to be significant?
   (d) Run the SAS macro for Daniel's method of detecting and correcting an outlier with this data.

3.11 Appendix–SAS Macro for Tukey's Single df Test

```
%macro tukeysdf;
***********************************************************;
**   Input for this macro is a file called file1 that has a   ***;
**   variable A for Factor A levels, a variable B for Factor B **;
**   levels and a variable y for the response variable        **;
***********************************************************;
*computes ANOVA for additive model;
proc glm data=file1;
class A B;
model y=A B;
ods output ModelANOVA=SumS;
ods output OverallANOVA=ESums;
*gets Error SS and df from ANOVA;
data ssE; set Esums;
if Source='Error' ;
ssE=SS; dfe=df;
keep ssE dfe;
*gets sum of squares and df for Factor A from ANOVA;
data ssA; set SumS;
if HypothesisType=1; if Source='A';
ssA=SS; dfa=df;
keep ssA dfa;
*gets sum of squares and df for Factor B from ANOVA;
data ssB; set SumS;
if HypothesisType=1; if Source='B';
ssB=SS;
dfb=df;
keep ssB dfb;
*creates marginal means for factor A;
proc sort data=file1; by A;
proc means noprint; var y; by A;
output out=m1 mean=ybaridot;
*calculates number of levels of factor A;
proc means noprint data=m1; var ybaridot;
output out=mA n=a;
*creates marginal means for Factor B;
proc sort data=file1; by B;
proc means noprint; var y; by B;
output out=m2 mean=ybardotj;
*calculates number of levels of factor B;
proc means noprint data=m2; var ybardotj;
output out=mB n=b;
*note creates grand mean;
proc means noprint data=file1; var y;
output out=m3 mean=ybardd;
*note merges marginal means to original data;
proc sort data=file1; by A;
data B; merge file1 m1; by A;
proc sort data=B; by B;
data C; merge B m2; by B;
yijyidydj=y*ybaridot*ybardotj;
proc means data=C noprint; var yijyidydj;
output out=s1 sum=sp1;
*combines Summary info and calculates single df interaction SS;
data comb; merge s1 ssA ssB ssE mA mB m3;
msA=ssA/dfa; msB=ssB/dfB; msE=ssE/dfe;
SSAB=(a*b*(sp1-(ssA+ssB+a*b*ybardd**2)*ybardd)**2)/(ssA*ssB);
ssR=ssE-SSAB; dfn=1; dfr=dfe-1; msR=ssR/dfr;
FA=msA/msR; FB=msB/msR; FAB=SSAB/msR;
probA=1-probf(FA,dfa,dfr); probB=1-probF(FB,dfb,dfr);
```

APPENDIX–SAS MACRO FOR TUKEY'S SINGLE DF TEST

```
data _null_; set comb end=eof;
file print notitles;
put @20 ' Tukey''s single df test for additivity'
    /
    /
    / @15 'Source              df          SS          MS          F
    PR>F'
    /;
put @ 15 'A'              @32 dfa @38 ssA best7. @50 msA best7.;
put @ 15 'B'              @32 dfb @38 ssB best7. @50 msB best7.;
put @ 15 'Error'          @32 dfe @38 ssE best7. @50 msE best7.;
put @ 15 'NonAdditivity'  @32 dfn @38 SSAB best7. @50 SSAB best7.
                          @62 FAB best5.@71 probAB best5.;
put @ 15 'Residual'       @32 dfr @38 ssR best7. @50 msR best7.;
run;
%mend tukeysdf;
```

CHAPTER 4

Randomized Block Designs

4.1 Introduction

In order to eliminate as much of the natural variation as possible and increase the sensitivity of experiments, it would be advisable to choose the experimental units for a study to be as homogeneous as possible. In mathematical terms this would reduce the variance, σ^2, of the experimental error and increase the power for detecting treatment factor effects. On the other hand, most experimenters would like the conclusions of their work to have wide applicability. Consider the following example. An experimenter would like to compare several methods of aerobic exercise to see how they affect the stress and anxiety level of experimental subjects. Since there is wide variability in stress and anxiety levels in the general population, as measured by standardized test scores, it would be difficult to see any difference among various methods of exercise unless the subjects recruited to the study were a homogeneous group each similar in their level of stress. However, the experimenter would like to make general conclusions from his study to people of all stress levels in the general population.

Blocking can be used in this situation to achieve both objectives. Blocking is the second technique that falls in the category of error control defined in Section 1.4. In a randomized block design, a group of heterogeneous experimental units is used so that the conclusions can be more general; however, these heterogeneous experimental units are grouped into homogeneous subgroups before they are randomly assigned to treatment factor levels. The act of grouping the experimental units together in homogeneous groups is called blocking. Randomly assigning treatment factor levels to experimental units within the smaller homogeneous subgroups of experimental units, or blocks, has the same effect as using only homogeneous units, yet it allows the conclusions to be generalized to the entire class of heterogeneous experimental units used in the study.

If experimental units represent physical entities, blocking or grouping by similar physical characteristics often results in more homogeneous groups. For example, plots of land in agricultural experiments are usually blocked by proximity because plots in close proximity normally have similar soil characteristics. When experimental units are animals, the grouping of genetically similar animals, such as litter-mates, often reduces variability within groups. When experimental units are simply trials, or points in time where treatments will be applied, they are often blocked by time since many lurking variables may change over time and trials in close temporal proximity are more alike.

In a *randomized complete block design*, or RCB, with one treatment factor, when the factor has t levels there will be b blocks (or subgroups of homogeneous experimental units) that each contain exactly t experimental units for a total of $t \times b$ experimental units. The t experimental units within each block are as similar as possible, and the groups of experimental units vary enough from block to block to allow general conclusions to be drawn. The randomization of experimental units to treatment factor levels, described in Chapter 2, is performed within each block. Fisher first proposed block designs for agricultural experiments where the experimental units were plots of ground. Blocks represented compact plots in close proximity which were similar. Variability among blocks represented the range of conditions over which conclusions would be generalized.

If there are more than t experimental units within each block, so that treatment factor levels are replicated r times within each block, there will be a total of $r \times t \times b$ experimental units. In this situation we call the design a *general complete block design*. Normally the RCB would be preferred over the general complete block design because smaller blocks of experimental units allow for greater homogeneity within the blocks and thus smaller experimental error and more precise tests and estimates.

4.2 Creating an RCB in SAS

The randomization of experimental units to treatment factor levels in a randomized block design can be accomplished using the SAS data step similar to what was shown in Chapter 2, or using SAS `proc plan`. To illustrate how this is done, consider the following experimental situation. A student wanted to investigate wives' tales of methods for extending the life of cut flowers. The treatment factor was the liquid to fill the vase. The levels were:

1. Tap water
2. Tap water with one spoonful of sugar added
3. Tap water with one cup of carbonated water
4. Tap water with one cup of 7-up

The experimental units were single flowers and the response was the time in days until the flower wilted. The student wanted the conclusions of his study to apply to many types of flowers, so she used an RCB design. The blocks were:

1. Rose
2. Carnation
3. Daisy
4. Tulip

CREATING AN RCB IN SAS

```
data sequence;
input block$ 1-9 v1-v4;
  flower=v1; treat=v1; output;
  flower=v2; treat=v2; output;
  flower=v3; treat=v3; output;
  flower=v4; treat=v4; output;
datalines;
rose       1 2 3 4
carnation  1 2 3 4
daisy      1 2 3 4
tulip      1 2 3 4
;
data flowers; set sequence;
  keep block flower;
data rand; set sequence;
  u=ranuni(0);
  keep block treat u;
proc sort; by block u;
data rcb;
  merge flowers rand;
  keep block flower treat;
proc print; id block; var flower treat; run;
```

The SAS commands above create the list of treatment indicators randomized within each block. The first data file **sequence** has sixteen observations and variables for the flower number within a block and treatment level. This file is not randomized. Next, the files **flowers** and **rand** are created by copying the file **sequence**. The variable treat was dropped from the file **flowers**, the variable flower was dropped from the file **rand**, and a random uniform variable u was added to **rand**. Next, the file **rand** was sorted by block and then by the random variables within the block to create a random sequence of treatment levels within each block. Finally, the two files **flowers** and **rand** were merged by block to create the randomized list which appears on the next page.

This randomized list can be easily expanded to a double spaced data collection plan, like the example in Section 2.2 by including the following data step commands:

```
data list; set rcb;
  time_days='_____';
proc print double; id block; var flower treat time_days; run;
```

RANDOMIZED BLOCK DESIGNS

```
block           flower      treat
carnation       1           1
carnation       2           4
carnation       3           2
carnation       4           3
daisy           1           2
daisy           2           1
daisy           3           3
daisy           4           4
rose            1           3
rose            2           2
rose            3           1
rose            4           4
tulip           1           3
tulip           2           2
tulip           3           4
tulip           4           1
```

To use this list, the student would number the flowers in each block from 1 to 4. Then using the above list the carnation number 1 would be placed in a vase with treatment factor level 1 or plain tap water. Carnation number 2 would be placed in a vase with treatment level four or tap water with one cup of 7-up, etc.

To create the randomization for a general complete block design, repeat the sequence 1 2 3 4 in `datalines` for the file `sequence` as many times as you have replicates, and expand the list `v1-v4` and the following output statements accordingly.

SAS `proc plan` can also be used as a quick way of randomizing treatment levels within a block. For example, a plan similar to the one given above can be created using the SAS commands:

```
proc plan seed=27321 ordered;
  factors block=4 flower=4;
  treatments treatment=4 random;
  output out=rcb block
    cvals=('carnation' 'daisy' 'rose' 'tulip');
run;
proc print data=rcb; id block; run;
```

The output from `proc plan` appears on the next page. Here it can be seen that the first flower in block 1 receives treatment level 2, the second flower in block 1 receives treatment level 4, etc. When printed, the output data set (`rcb`) created by proc plan looks similar to the list shown above.

```
                The PLAN Procedure

                  Plot Factors

Factor       Select        Levels        Order

block          4             4          Ordered
flower         4             4          Ordered


               Treatment Factors

Factor       Select        Levels        Order

treatment      4             4          Random


   block        --flower--        --treatment--

     1         1  2  3  4         2  4  1  3
     2         1  2  3  4         1  3  4  2
     3         1  2  3  4         2  3  4  1
     4         1  2  3  4         3  1  4  2
```

4.3 Model for RCB

The model for the analysis of an RCB design is

$$y_{ij} = \mu + b_i + \tau_j + \epsilon_{ij}, \qquad (4.1)$$

where, b_i represent the block effects, τ_j represent the treatment effects. The usual assumptions of normality of experimental error and homogeneity of variance of experimental error across levels of the treatment factors and blocks are required for this model. These assumptions can be checked with residual plots as shown in Section 2.4.

Notice that this is an additive model which does not include the interaction between block and treatment. Since there are only $t \times b$ experimental units, there would be zero degrees of freedom for the error term ssE if a block by treatment interaction term were included in the model. However, the block by treatment interaction is in fact the correct error term for testing the treatment effects. The experimenter wants to generalize his conclusions about treatment effects over all the experimental units, so the average treatment effects should be larger than any differences in treatment effects among blocks of experimental units. The difference in treatment effects among blocks is exactly what the interaction measures and is therefore the correct error term. By leaving the interaction out of the model, the ssE becomes identical to the interaction sums of squares.

The ANOVA table for an RCB design is shown symbolically in Table 4.1. Representations for the type I sums of squares for blocks and treatments are shown in the table, similar to what was shown in Section 3.5.1, but they will be identical to the type III sums of squares for this design. The error sums of squares $ssE = y'y - \hat{\beta}'X'y = y'(I - X(X'X)^-X')y$, where $\hat{\beta} = (X'X)^-X'y$.

Table 4.1 *Analysis of Variance Table*

Source	df	Sum of Squares	Mean Squares	F-ratio
Blocks	$b-1$	$ssBlk$ $R(b\|\mu)$	$ssBlk/(b-1)$	
Treatments	$t-1$	ssT $R(\tau\|b,\mu)$	$ssT/(t-1)$	msT/msE
Error	$(b-1)(t-1)$	ssE	$ssE/(b-1)(t-1)$	

The degrees of freedom for the error $(b-1)(t-1)$ is smaller than it would be in a completely randomized design with b replicates of each treatment level; however, if the groups of experimental units within the blocks are more homogeneous, the msE should be smaller and the power for detecting differences among the treatment levels higher.

The estimate of the variance of the homogeneous experimental units within each block is given by

$$\hat{\sigma}^2_{rcb} = \frac{ssE}{(b-1)(t-1)} \qquad (4.2)$$

An estimate of the variance of the entire group of heterogenous experimental units can be made from the mean squares in the RCB ANOVA. It is given by the formula

$$\hat{\sigma}^2_{crd} = \frac{(b-1)msBlk + b(t-1)msE}{tb-1} \qquad (4.3)$$

which is a weighted average of the mean square for blocks and the mean square for error. However, the weights are not simply the degrees of freedom for each mean square. If the $msBlk$ is zero, it can be seen that $\hat{\sigma}^2_{crd} < \hat{\sigma}^2_{rcb}$. The ratio of $\hat{\sigma}^2_{crd}$ and $\hat{\sigma}^2_{rcb}$ is a measure of the efficacy of blocking.

The error degrees of freedom for the RCB is $\nu_{rcb} = (b-1)(t-1)$, and the error degrees of freedom for a completely randomized design (CRD) with the same number of experimental units would be $\nu_{crd} = t(b-1)$. The relative efficiency of the RCB is then given by the formula:

$$RE = \frac{(\nu_{rcb}+1)(\nu_{crd}+3)}{(\nu_{rcb}+3)(\nu_{crd}+1)} \frac{\hat{\sigma}^2_{crd}}{\hat{\sigma}^2_{rcb}} \qquad (4.4)$$

RE can be used to determine the number of observations that would be required in a CRD, with heterogeneous experimental units, in order to have the

AN EXAMPLE OF AN RCB

variances for the treatment means equivalent to that achieved with the RCB. If $b \times t$ experimental units were used in the RCB design, then $RE \times (b \times t)$ experimental units would be required in a CRD design, without blocking, to attain equivalent variances of treatment means.

4.4 An Example of an RCB

Consider the data in Table 4.2 from Lim and Wolfe (1997), partially modified from Heffner et al. (1974). The effect of the drug d-amphetamine sulfate on the behavior of rats was the object of the experiment. The behavior under study

Table 4.2 *Rat Behavior Experiment*

Rat	0.0	0.5	1.0	1.5	2.0
1	0.60	0.80	0.82	0.81	0.50
2	0.51	0.61	0.79	0.78	0.77
3	0.62	0.82	0.83	0.80	0.52
4	0.60	0.95	0.91	0.95	0.70
5	0.92	0.82	1.04	1.13	1.03
6	0.63	0.93	1.02	0.96	0.63
7	0.84	0.74	0.98	0.98	1.00
8	0.96	1.24	1.27	1.20	1.06
9	1.01	1.23	1.30	1.25	1.24
10	0.95	1.20	1.18	1.23	1.05

was the rate at which water-deprived rats pressed a lever to obtain water. The response was the lever press rate defined as the number of lever presses divided by the elapsed time of the session. The treatment factor levels were five different dosages of the drug in milligrams per kilogram of body weight, including a control dosage consisting of saline solution. An experiment, or run, consisted of injecting a rat with a drug dosage, and after one hour an experimental session began where a rat would receive water each time after a second lever was pressed. The experimental unit in these experiments was not a rat, but the state of a single rat during one experiment or run, since an individual rat could be used in many experiments by repeatedly injecting it with different doses of the drug (after an appropriate washout period) and by observing the lever pressing behavior. Because there was wide variability in lever pressing rate between rats, a RCB design was used, and a rat represented the blocking factor. Each rat received all five doses in a random order with an appropriate wash out period in between.

To utilize SAS `proc glm` to produce the ANOVA, the commands shown on the next page are used. The output statement in the `proc glm` call produces a file s that contains the predicted and residual values so that the assumptions can be checked with residual plots.

In this case, the rat is represented by the term b_i in the model

```
data drug;
input rat r1-r5;
  dose=0.0; rate=r1; output; dose=0.5; rate=r2; output;
  dose=1.0; rate=r3; output; dose=1.5; rate=r4; output;
  dose=2.0; rate=r5; output; keep rat dose rate;
datalines;
1   .60   .80   .82   .81   .50
2   .51   .61   .79   .78   .77
    . . . etc.
proc glm;
  class rat dose;
  model rate=rat dose;
  output out=s p=prate r=rrate;
run;
```

$y_{ij} = \mu + b_i + \tau_j + \epsilon_{ij}$. The experimental error, represented by ϵ_{ij}, is the effect of the state of rat i during the run when it received dose j. If data were presented without describing the experimental unit and the randomization process, the model could be easily misspecified as $y_{ij} = \mu + \tau_i + \epsilon_{ij}$ resulting in the wrong analysis and conclusions.

The ANOVA table resulting from the correct model specification in SAS appears below. The table of type III sums of squares is not shown because it is identical to the type I sums of squares. The F-tests show that there is a significant difference in treatment factor levels.

Source	DF	Sum of Squares	Mean Square	F Value	Pr > F
Model	13	2.12866600	0.16374354	19.61	<.0001
Error	36	0.30055200	0.00834867		
Corrected Total	49	2.42921800			

Source	DF	Type I SS	Mean Square	F Value	Pr > F
rat	9	1.66845800	0.18538422	22.21	<.0001
dose	4	0.46020800	0.11505200	13.78	<.0001

To interpret the differences in treatment factor levels comparisons of means should be made. Since the factor levels are quantitative, orthogonal polynomial contrasts as described in Section 2.8 are useful. The results of the estimate statement calculating the linear, quadratic, and cubic contrasts are shown at the top of the next page. They show that there is a quadratic trend in lever press rate over the dose of the drug. This trend can be visualized by plotting the means as a function of dose.

The SAS commands below the output table on the next page produce the graph in Figure 4.1. Here the quadratic trend line, produced by the **degree=2** option on the **reg** statement, is added to the plot and shows that the lever press rate increases as a function of dose until it reaches a maximum somewhere between 1.0 and 1.5 milligrams per kilogram of body weight.

AN EXAMPLE OF AN RCB

Parameter	Estimate	Standard Error	t Value	Pr > |t|
Linear	0.07810832	0.02889408	2.70	0.0104
Quadratic	-0.19857492	0.02889403	-6.87	<.0001
Cubic	-0.02023863	0.02889408	-0.70	0.4882

```
ods graphics on;
proc sort data=drug; by dose;
proc means noprint; by dose; var rate;
   output out=means mean=meanrate;
proc sgplot data=means noautolegend;
   scatter x=dose y=meanrate;
   reg x=dose y=meanrate/degree=2;
   yaxis label='average lever press rate';
run;
ods graphics off;
```

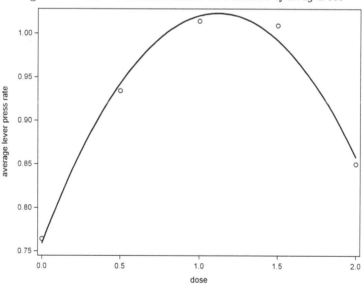

Figure 4.1 *Lever Press Rate Means as Function of Drug Dose*

The estimated variance of the experimental units (trials) within a block (or rat) is the mean square error $\hat{\sigma}^2_{rcb} = 0.00834867$. The variance of the heterogeneous experimental units is given by

$$\hat{\sigma}^2_{crd} = \frac{(b-1)msBlk + b(t-1)msE}{tb-1}$$

$$= \frac{9(0.18538422) + 10(5-1)(0.00834867)}{(5)(10) - 1} = 0.040865. \quad (4.5)$$

This is approximately five times larger than the variance within a rat and demonstrates the effectiveness of blocking by rat in the experiment. The relative efficiency is given by

$$RE = \frac{(\nu_{rcb}+1)(\nu_{crd}+3)}{(\nu_{rcb}+3)(\nu_{crd}+1)} \frac{\hat{\sigma}^2_{crd}}{\hat{\sigma}^2_{rcb}} = \frac{(37)(48)}{(39)(46)} \cdot \frac{0.040865}{0.0083487} = 4.8457. \quad (4.6)$$

This means that blocking has reduced the variance of experimental units approximately 80%= $1 - \frac{0.0083487}{0.040865}$, and that it would take approximately 5 times as many trials to have the equivalent variances for treatment means if each rat had been used for only one trial in a CRD design, and the rat-to-rat variability had not been removed from the error term.

4.5 Determining the Number of Blocks

The F-test for treatment or dose effect in the last example was highly significant ($P < .0001$). If the experiment were to be repeated in a situation where the variability of the response (lever press rate within a rat) and differences in treatment means were to remain approximately the same, fewer blocks or rats would be required to detect significant differences in treatments.

The non-centrality parameter for the F-test of treatment effects in the randomized complete block design is $\lambda = b\sum_j \tau_j^2/\sigma^2$, and the degrees of freedom are $\nu_1 = t-1$, and $\nu_2 = (b-1)(t-1)$. Therefore in order to calculate the number of blocks that will result in a power between 0.8 and 0.9 for detecting a difference in treatment means, the SAS code in Section 3.5.2 can be modified by changing the formula for the denominator degrees of freedom and the non-centrality factor.

Using the results from the last experiment, the estimate of $\sigma^2_{rcb} = 0.00834867$ and $css = \sum_j \tau_j^2$ can be estimated to be

$$(.764 - .9142)^2 + \ldots + (0.850 - .9142)^2 = 0.0460208$$

The SAS code on the next page calculates the power for $b = 2$ to 5. Running this code results in the output shown below the commands where it can be seen that power between 0.8 and 0.9 can be achieved with $b = 4$ or 5 blocks or rats in the experiment.

If an estimate σ^2_{crd} were available from previous experiments or pilot studies, Hinkelmann and Kempthorne (1994) have shown the relative efficiency (RE) σ^2_{crd} can also be used to get a rough estimate of the number of blocks required for a RCB design. For example, suppose σ^2_{crd} were estimated to be 0.040865 from previous experiments and the number of replicates of each treatment

FACTORIAL DESIGNS IN BLOCKS

```
data powerRCB;
  do b=2 to 5;
    t=5;
    nu1=(t-1);
    nu2=(b-1)*(t-1);
    alpha=.05;
    sigma2=0.0083487;
    css=0.0460208;
    Fcrit=finv(1-alpha,nu1,nu2);
    nc=b*(css)/sigma2;
    power=1-probf(Fcrit,nu1,nu2,nc);
    output;
    keep b nu2 nc power;
  end;
proc print; run;
```

Obs	b	nu2	nc	power
1	2	4	11.0247	0.32078
2	3	8	16.5370	0.68276
3	4	12	22.0493	0.88778
4	5	16	27.5617	0.96711

required for a CRD design to achieve adequate power for detecting a practical difference in means was $r = 20$, determined by the methods of Section 3.5.2. If blocking was expected to reduce the variance by 80% (i.e., $\sigma^2_{rcb} = 0.20 \times \sigma^2_{crd}$, or $RE = 5.0$). Then the number of blocks required to achieve the same power with a RCB design is $b = \frac{r}{RE} = \frac{20}{5} = 4$.

4.6 Factorial Designs in Blocks

Blocking is even more effective when combined with a factorial design in treatment factors. In the case where more than one treatment factor is studied, the number of experimental units in each block must be equal to the product of levels of all the factors. This is called a randomized complete block factorial or RCBF. As an example of a blocked factorial experiment consider the data in Table 4.3. This experiment was conducted by Festing (2003) to determine whether BHA (a common antioxidant used in processed foods) induced activity of the liver enzyme EROD in mice and whether this activity was independent of strain of mice. It was part of a larger study to determine if antioxidants help protect against cancer.

The factors in this experiment were A, whether a mouse was treated with BHA or not, and B, the strain of the mouse. Since you can't assign a particular mouse to be of a certain strain, the experimental unit is not the mouse but the trial or conditions existing in the lab when a particular experiment was run. One run consisted of selecting a mouse from a specified strain; then either incorporating BHA in the diet (for a three-week period) or not (depending on

Table 4.3 *Activity of EROD Liver Enzyme in Control and BHA-Treated Mice*

Strain	Block 1		Block 2	
	Treated	Control	Treated	Control
A/J	18.7	7.7	16.7	6.4
129/Ola	17.9	8.4	14.4	6.7
NIH	19.2	9.8	12.0	8.1
BALB/c	26.3	9.7	19.8	6.0

what was specified); and finally humanely sacrificing the mouse and performing an autopsy to determine the activity level of the enzyme EROD in the liver. Since the results of in vivo experiments like this can vary substantially from time to time in the same laboratory due to differences in reagents used for the enzyme analysis, calibration of instruments, and environmental factors in the animal husbandry, the experiments were blocked in time. Two mice were selected from each of four strains, one was randomly chosen to receive BHA in the diet, and eight trials or runs were conducted simultaneously. This represented one block. Three months later, the whole process was repeated for a second block of runs. Since there are two levels of the treatment factor (BHA treated or control) and four levels of the factor strain, there were $2 \times 4 = 8$ experimental units or runs per block.

The model for analysis of a two-factor factorial in a randomized block design, like that shown in Table 4.3, is

$$y_{ijk} = \mu + b_i + \alpha_j + \beta_k + \alpha\beta_{jk} + \epsilon_{ijk}, \qquad (4.7)$$

where b_i represents the block effect, α_j represents the treatment factor effect and β_k represents the strain factor effect. This model is easily generalized to multi-factor factorial designs in randomized blocks. Notice that in the model (4.7) there is an interaction $\alpha\beta_{jk}$ between the two factorial factors, but there is no interaction between the block factor, b_i, and the factorial factors, α_j and β_k. The interactions with blocks are the error term for the analysis and if they are included in the model there will be zero degrees of freedom for ssE.

The SAS commands to read this data and perform the analysis are shown on the next page with the resulting ANOVA table. There it can be seen that BHA treatment, strain, and the interaction are significant. The block sums of squares is also large resulting in a relative efficiency $RE = 2.11$. This means that it would take approximately three times as many mice to have the same power or sensitivity if the experiments were not blocked by time.

Figure 4.2 helps in interpreting the factorial effects and interaction. It can be seen that on the average, BHA added to the diet increases activity of the enzyme EROD in the mouse liver. However, this increase is nearly doubled for mice of strain BALB/c.

FACTORIAL DESIGNS IN BLOCKS

```
data bha;
input strain $ bha $ block erod;
datalines;
A/J      Yes   1 18.7
A/J      No    1  7.7
   ... etc.
proc glm; class strain bha block;
  model erod=block bha strain bha*strain;
run;
```

Source	DF	Sum of Squares	Mean Square	F Value	Pr > F
Model	8	543.2175000	67.9021875	26.20	0.0002
Error	7	18.1400000	2.5914286		
Corrected Total	15	561.3575000			

Source	DF	Type I SS	Mean Square	F Value	Pr > F
block	1	47.6100000	47.6100000	18.37	0.0036
bha	1	422.3025000	422.3025000	162.96	<.0001
strain	3	32.9625000	10.9875000	4.24	0.0527
strain*bha	3	40.3425000	13.4475000	5.19	0.0337

Figure 4.2 *BHA Effect for Each Strain*

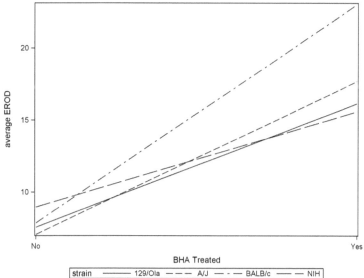

4.7 Generalized Complete Block Design

When experimental units represent physical entities, smaller groups or blocks of experimental units usually result in greater homogeneity. The larger the group, the more likely it is to have many experimental units that are widely different than the norm or average. For that reason it is unadvisable to have blocked designs with more than the minimum, t, experimental units per block, where t is the number of levels or combination of levels of treatment factors. However, in some cases where experimental units represent trials rather than physical entities and the experimental runs can be made quickly, larger block sizes may not increase the variability of experimental units within a block. In that case, a design with replicates of each treatment level within a block (called a generalized complete block design or GCB) can be used.

Consider the following example from *Golf Magazine* (Bastable, 2006, June). An experiment was conducted to determine the ideal tee height for driving a golf ball as far as possible. The purpose was to recommend to all readers of the article what tee height they should use. The treatment factor was the tee height as shown in Table 4.4.

Table 4.4 *Treatment Factor Levels for Golf Experiment*

Level	Tee Height
1	Entire ball below crown
2	Half the ball above the crown
3	Bottom of ball at top of club-face

An experiment consisted of a golfer hitting a golf ball from a specified height, and the response was the distance the ball traveled. To make a general conclusion, a representative group of golfers had to be used rather than one golfer. Since the ability to drive the ball by the golfers used in the study differed, it made sense to group or block the trials by golfer and randomize the order that each golfer hit a ball from each of the tee heights. However, since hitting more than three golf balls would not be likely to fatigue a golfer and cause more variability in his driving distance, there was no need to restrict a golfer to hitting just three balls. Instead, each golfer hit $r = 5$ golf balls from each of $t = 3$ tee heights. The results from this experiment are shown in the appendix. Nine golfers were used in this part of the study; each golfer hit five balls from each tee height (15 balls total) in a random order.

Since there are replicate experimental units for each treatment in each block, it is possible to fit the model

$$y_{ijk} = \mu + b_i + \tau_j + b\tau_{ij} + \epsilon_{ijk} \qquad (4.8)$$

GENERALIZED COMPLETE BLOCK DESIGN

However, this leaves a dilemma. The msE in the traditional ANOVA and the denominator for the F-tests for treatment effect and the block by treatment interaction is based on the variability of experimental units within the same treatment and block (in this case golfer). If the interaction between block (golfer) and treatment factor (tee height) were significant, its interpretation would imply that the optimal tee height could be different for different golfers. The golfers in the study were just a sample of golfers, and if the optimal tee height were different for them, there would be no way to recommend an optimal tee height for all readers of the *Golf Magazine* article. To make a general recommendation, the treatment factor should be tested using the block by treatment interaction mean square as the denominator of the F-test (another justification for using the block by treatment interaction as the denominator of the F-test for treatment will be given in Chapter 5, Section 8). An F-test for treatments constructed in this way is not made automatically in proc glm, but can be specified using the test statement as shown in the SAS commands below where id represents block or golfer id.

```
data rcb;
input id teehgt cdistance;
datalines;
1 1 142.0
 ... etc.
proc glm data=rcb;
  class id teehgt;
  model cdistance=id teehgt id*teehgt;
  test h=teehgt e=id*teehgt;
  lsmeans teehgt/pdiff adjust=tukey e=id*teehgt;
run;
```

The h=teehgt and e=id*teehgt in the test statement specify the hypothesis to be $H_0 : \tau_1 = \tau_2 = \tau_3$ and the error term to be $b\tau_{ij}$. The e=id*teehgt in the lsmeans statement also specifies the error term to be $b\tau_{ij}$. The results of this analysis are shown on the next page. The table of type I sums of squares was left out since it was identical to the type III sums of squares.

There it can be seen that the tee height is significant at the $\alpha = 0.0124$ level. This is much different than the incorrect F-test in the table of type III sums of squares. There the tee height effect is significant at $\alpha < 0.0001$ level. Sometimes the F-test for treatments may be significant in the type III table but insignificant in the result of the test statement. In that case, the result of the test statement should be used if you want to generalize conclusions.

Sometimes there is a more powerful test of treatment effects than can be obtained using the test statement to create F-test. If the block by treatment interaction term is insignificant, the additive model $y_{ijk} = \mu + b_i + \tau_j + \epsilon_{ijk}$ can be fit to the data and the default F-test for treatments will use an msE term that is a pooled or weighted average of the interaction and error term from model (4.8). If the interaction is negligible, the interaction mean square

```
                         Sum of
Source              DF   Squares        Mean Square   F Value   Pr > F
Model               26   128821.4813    4954.6724     72.90     <.0001
Error               108  7340.7520      67.9699
Corrected
Total               134  136162.2333

Source              DF   Type III SS    Mean Square   F Value   Pr > F
ID                  8    124741.4493    15592.6812    229.41    <.0001
TEEHGT              2    1723.9320      861.9660      12.68     <.0001
ID*TEEHGT           16   2356.1000      147.2563      2.17      0.0102

Tests of Hypotheses Using the Type III MS for ID*TEEHGT as an
Error Term

Source              DF   Type III SS    Mean Square   F Value   Pr > F
TEEHGT              2    1723.932000    861.966000    5.85      0.0124
```

is estimating the same experimental error as the error mean square. In that case, pooling these two mean squares increases the degrees of freedom for the error term and increases the power or sensitivity of the F-test for treatment effects. Normally the preliminary F-test of the interaction, which is used to decide whether to fit the additive model or not, is conducted at a higher significance level like $\alpha = 0.25$. If the interaction is significant at the $\alpha = 0.25$ level, use the **test** statement to create the appropriate F-test for treatment effects. If the interaction is not significant at $\alpha = 0.25$, fit the additive model and use the F-test from the type I or type III table of sums of squares. This procedure is called the "Pool or not to Pool" procedure.

For the golf experiment, the interaction sum of squares is significant at the $\alpha = 0.0102 < 0.25$ level, and thus the additive model should not be fit, and there is no more powerful F-test for treatments than the one that can be made using the **test** statement.

To interpret the meaning of the significant treatment effect in the golf experiment, the **lsmeans teehgt/pdiff adjust=tukey e=id*teehgt;** statement was used. This results in the table of means and the Tukey pairwise comparisons of means shown on the next page. The **e=id*teehgt** option specifies the error term to be the means square for block by treatment interaction.

The least squares means show the average driving distance for the tee height $= 1$ (entire ball below crown) is 171.45 yards. It is significantly less than the average driving distance (177.8 yds.) for tee height $= 3$ (bottom of ball at top of club-face) at the $\alpha = 0.0125$ level and marginally significantly less than the distance (179.8 yds.) for tee height $= 2$ (half the ball above the crown). The average distance for tee heights 2 and 3 were not significantly different $\alpha = 0.7192$. Therefore the recommendation to readers of the article was to tee their golf ball up so that at least half the ball is above the crown of the driver club-face in order to maximize their driving distance.

TWO BLOCK FACTORS LSD

```
                  Least Squares Means
         Adjustment for Multiple Comparisons: Tukey
     Standard Errors and Probabilities Calculated Using
       the Type III MS for ID*TEEHGT as an Error Term
             CDISTANCE          LSMEAN
         TEEHGT                 LSMEAN          Number
         1                   171.457778           1
         2                   177.837778           2
         3                   179.837778           3

         Least Squares Means for effect TEEHGT
         Pr > |t| for H0: LSMean(i)=LSMean(j)
            Dependent Variable: CDISTANCE
     i/j              1                 2                3
     1                               0.0590           0.0125
     2             0.0590                             0.7192
     3             0.0125            0.7192
```

4.8 Two Block Factors LSD

It was first shown in agricultural experiments (Fisher, 1935) that the process of grouping experimental plots into homogeneous blocks might profitably be duplicated. For example, on the left side of Figure 4.3 we see a representation of a randomized block design RCB laid out in a field. In this design, one field, roughly square in shape, is divided into four rectangular blocks. Each block is further divided into four plots and the treatment levels (A, B, C, D) are randomly assigned to one plot within each block. In the figure the random assignment within the block is represented by each letter only appearing once in each row that represents a block. If there is a fertility gradient running from top to bottom in this field, the plots within a block will be more alike in fertility than plots in different blocks or rows, and the randomized block design would work well to reduce the variance of experimental errors within a block.

If there were no clear gradient in the field, but adjacent plots tended to be

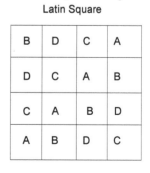

Figure 4.3 *Comparison of RCB and Latin Square Designs*

more alike than plots at different ends of the same block, a design like the one shown on the right side of Figure 4.3 assigns each treatment level only once to each row and once to each column. In that way, the variability from column to column can also be removed from the error sums of squares, further increasing the sensitivity for detecting treatment effects. The design shown on the right side of Figure 4.3 is called a Latin square design or LSD, and it is blocked both horizontally and vertically. The restriction with a LSD is that the number of row blocks equals the number of column blocks, which equals the number of levels of the treatment factor. This restriction will be relaxed in Chapter 7, where more general row-column blocking schemes will be discussed.

The model for a LSD is written

$$y_{ijk} = \mu + r_i + c_j + \tau_k + \epsilon_{ijk} \qquad (4.9)$$

where r_i represents the row blocking factor, c_j represents the column blocking factor, and τ_k represents the treatment factor. Like model (4.1) for the RCB design, no interactions are included in the model so that any differences in treatment factor levels can be generalized over rows and columns.

Latin square designs can be used whenever there are two independent blocking factors that can be used to group experimental units. For example, if an experiment were being conducted to determine the effect of tread design on the wear life of automobile tires, the experimental unit would be a wheel on a car and the treatment factor would be the tread design of the tire mounted on that wheel. It would make sense to block the experimental units by type of automobile, since tires may wear faster on heavier cars than they do on lighter cars. It would also make sense to block by position of the tire on a car since front right tires wear at a different rate than left rear tires, etc. These are independent blocking factors because all four wheel positions exist on any type car. To use a LSD, the number of tread types compared and the number of car types used in the study must be four in order to equal the number of wheel positions on a car. The row blocking factor in the Latin square would represent the type of car, with four alternatives ranging over the class of cars to which the experimenter would like to make inference. The column blocking factor would represent the position of a tire on the car (FL, FR, RL, RR), and the treatment factor would represent the four different tread designs being tested.

4.8.1 Creating and Randomizing Latin Square Designs

Latin square designs are easy to create by cyclically rotating the letters or symbols used to represent the treatment factor levels. For example, for a 5×5 Latin square, let the letters A, B, C, D, E, represent the levels of the treatment factor. Then the design is created as:

TWO BLOCK FACTORS LSD

```
A  B  C  D  E
B  C  D  E  A
C  D  E  A  B
D  E  A  B  C
E  A  B  C  D
```

To prevent biases from unknown lurking variables, randomization should be used in LSDs. However, care must be taken so that after randomization each treatment level still occurs once in each row and once in each column. This can be accomplished by first randomizing the order of the rows (keeping the columns positions fixed in each row), then randomizing the columns (keeping row positions fixed within each column), and finally randomizing the order of the treatment labels. This can be easily accomplished with SAS `proc plan`.

The commands below show the creation and randomization of a plan to study the effect of the number of shelf facings on the sales of toothpaste in drug stores. The treatment factor is the number of shelf facings (1-4), the column blocking factor is the store (to account for store-to-store differences), and the row blocking factor is the calendar week (to account for seasonal factors). The response would be the weekly sales in dollars. In this particular case the rows or weeks were not randomized because they will occur in sequential order.

```
proc plan seed=37430;
  factors rows=4 ordered cols=4 ordered / noprint;
  treatments tmts=4 cyclic;
  output out=lsd
  rows cvals=('Week 1' 'Week 2' 'Week 3' 'Week 4') ordered
  cols cvals=('Store 1' 'Store 2' 'Store 3' 'Store 4') random
  tmts nvals=( 1 2 3 4  ) random;
proc print;
run;
```

The resulting randomized data file `lsd` looks like the result shown on the next page and could be used to make a double-spaced data collection form as illustrated in Section 2.2.

```
Obs      rows       cols      tmts

 1     Week 1    Store 3      4
 2     Week 1    Store 4      2
 3     Week 1    Store 2      1
 4     Week 1    Store 1      3
 5     Week 2    Store 3      2
 6     Week 2    Store 4      1
 7     Week 2    Store 2      3
 8     Week 2    Store 1      4
 9     Week 3    Store 3      1
10     Week 3    Store 4      3
11     Week 3    Store 2      4
12     Week 3    Store 1      2
13     Week 4    Store 3      3
14     Week 4    Store 4      4
15     Week 4    Store 2      2
16     Week 4    Store 1      1
```

To verify that each treatment occurs once in each row and once in each column, we can print this design in the familiar Latin square layout using proc tabulate as shown below.

```
proc tabulate data=lsd;
  class rows cols;
  var tmts;
  table rows, cols*(tmts*f=8.) / rts=10;
run;
```

	cols			
	Store 1	Store 2	Store 3	Store 4
	tmts	tmts	tmts	tmts
	Sum	Sum	Sum	Sum
rows				
Week 1	3	1	4	2
Week 2	4	3	2	1
Week 3	2	4	1	3
Week 4	1	2	3	4

4.8.2 Analysis of a Latin Square Design

Latin square designs are frequently used in steer or dairy cow feeding experiments and in bioequivalence studies to compare different formulations of a drug in phase II clinical trials. In these studies the column blocking factor is time and the row blocking factor is animal or human subject. In some cases the treatment administered in one time period may have a carry-over effect on the response in the next period. However, if there is a sufficient washout period between column blocks, there will be no carry-over effects and the data can be analyzed as a traditional Latin square.

To illustrate the analysis of data from a Latin Square, consider the following bioequivalence study. The data is shown in Table 4.5 (taken from Selwyn and Hall, 1984).

Table 4.5 *Treatment and Resulting AUC for Bioequivalence Study*

Subject	Period		
	1	2	3
1	A 1186	B 642	C 1183
2	B 984	C 1135	A 1305
3	C 1426	A 1540	B 873

A=solution, B=tablet, C=capsule

The purpose was to test the bioequivalence of three formulations (A=solution, B=tablet, C=capsule) of a drug as measured by the AUC or area under the curve, which relates the concentration of the drug in the blood as a function of the time since dosing. Three volunteer subjects took each formulation in succession with a sufficient washout period between. After dosing, blood samples were obtained every half-hour for four hours and analyzed for drug concentration. AUC was calculated with the resulting data. Since there may be large variation in metabolism of the drug from subject to subject, subject was used as a row blocking factor. Since the absorbtion and metabolism of a drug will vary from time to time for a particular subject, time was used as a column blocking factor.

The SAS commands to read the data and fit model (4.9) and the resulting output are shown on the next page. The results indicate that there is no difference in the three formulations.

```
data bioeqv;
input subject  period   treat $  auc;
datalines;
1 1 A 1186
1 2 B  642
1 3 C 1183
2 1 B 1135
2 2 C 1305
2 3 A  984
3 1 C  873
3 2 A 1426
3 3 B 1540
proc glm;
  class subject period treat;
  model auc = subject period treat;
  lsmeans treat/ pdiff adjust=tukey;
run;
```

```
                         Sum of
Source          DF      Squares     Mean Square   F Value   Pr > F
Model            6   174460.6667    29076.7778      0.13    0.9774
Error            2   442157.5556   221078.7778
Corrected
Total            8   616618.2222


Source          DF    Type I SS     Mean Square   F Value   Pr > F
subject          2   114264.2222    57132.1111      0.26    0.7946
period           2    45196.2222    22598.1111      0.10    0.9073
treat            2    15000.2222     7500.1111      0.03    0.9672
```

```
              The GLM Procedure
             Least Squares Means
    Adjustment for Multiple Comparisons: Tukey
                                  LSMEAN
         treat      auc LSMEAN    Number
          A         1198.66667       1
          B         1105.66667       2
          C         1120.33333       3

      Least Squares Means for effect treat
        Pr > |t| for H0: LSMean(i)=LSMean(j)
              Dependent Variable: auc
   i/j            1             2             3
    1                        0.9687        0.9776
    2          0.9687                      0.9992
    3          0.9776        0.9992
```

4.8.3 Determining the Number of Replicates

The number of replicates of each treatment factor level, in a LSD with t rows and t columns must be equal to t. The only way the power for detecting differences in treatment means can be increased would be to increase the number of rows or columns. In the last example the row blocking factor represented subjects, and the column blocking factor represented periods. One way of increasing the power for detecting differences in treatment means would be to increase the number of subjects. If the number of subjects, r, were doubled, i.e., $r = 2t$, it would be essentially the same as replicating the Latin square with t additional subjects.

In general if we consider a replicated Latin square to have $r = nt$ rows (where n is an integer), t columns and t levels of the treatment factor, the model for the data will still be Equation (4.9), but the degrees of freedom for the error term will be $\nu_2 = (r-2)(t-1)$. The non-centrality factor for the F-test of no treatment effects is $\lambda = nt \sum_k \tau_k^2/\sigma^2$ where n is the number of times the square has been repeated. Therefore in order to calculate the number of replicates of the square, n, that will result in a power between 0.8 and 0.9 for detecting a difference in treatment means, the SAS code in Section 3.5.2 can be modified by changing the formula for the denominator degrees of freedom and the non-centrality factor.

4.9 Review of Important Concepts

When experimental units are not homogeneous and the experimenter does not want to restrict the conclusions of his study to a homogeneous subset of experimental units, the randomized block design can be used. In the randomized block design, heterogenous experimental units are grouped into homogeneous subgroups or blocks prior to assigning them to treatment levels. This allows the variation between blocks to be removed from the error sums of squares and increases the power or sensitivity for detecting treatment effects. The relative efficiency (RE) is a measure of the efficacy of blocking, and the higher the value for RE the more effective the blocking.

Figure 4.4 illustrates when the various types of blocked designs should be used. When there is one treatment factor and the heterogeneous experimental units can be grouped into categories based on one blocking factor the randomized complete block or RCB design is used. If the block size, or number of experimental units in a block, can be larger without increasing the variance of experimental units within a block the generalized complete block or GCB design can be used. If there are multiple factors under study, then the randomized complete block factorial or RCBF should be used. If the heterogeneous experimental units can be grouped into classifications based on two independent blocking factors and there is only one treatment factor, the Latin square design or LSD should be used.

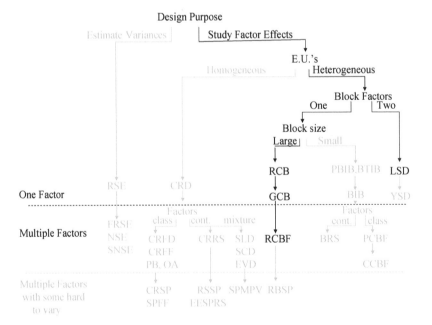

Figure 4.4 *Design Selection Roadmap*

REVIEW OF IMPORTANT CONCEPTS

A randomized complete block design or RCB has only one experimental unit per treatment level per block. The model for the RCB design is $y_{ij}=\mu+b_i+\tau_j+\epsilon_{ij}$, which does not include the interaction between block and treatment because this is the correct error for testing treatment effects if the experimenter wants general conclusions. In a generalized complete block design, replicates of each treatment level within each block are included. Because of the replicates in the GCB, the interaction between block and treatment can be included in the model, $y_{ijk}=\mu+b_i+\tau_j+b\tau_{ij}+\epsilon_{ijk}$, but it should still be used as the error term for testing the treatment effects and this can be accomplished by using the **test** statement in SAS **proc glm**.

In RCBF every combination of levels of the factors is randomized to experimental units within a block. The model for analysis is the usual factorial model with the addition of a block term, $y_{ijk}=\mu+b_i+\alpha_j+\beta_k+\alpha\beta_{jk}+\epsilon_{ijk}$. No interaction between blocks and factorial effects or interactions should be included in the model if the experimenter wants general conclusions. Writing a model that does not represent the way the experiment was conducted can result in using the wrong mean squares in the denominators of the F-tests and may cause the wrong conclusions to be drawn.

With two independent blocking factors, LSDs can be used to further decrease the variance of experimental error. The model for the Latin square is $y_{ijk}=\mu+r_i+c_j+\tau_k+\epsilon_{ijk}$, and again no interaction between row blocks, column blocks or treatment factors is included in the model. When randomizing a list of experiments for a LSD, care must be taken so that after randomization each treatment factor level still occurs exactly once in each row block and each column block.

4.10 Exercises

1. Consider planning an experiment to determine how the flavor duration of gum is affected by the characteristics of the gum such as Factor A (type of gum stick or tablet), Factor B (flavor of gum), and Factor C (regular or sugar-free), see Rogness and Richardson (2003).

   (a) If several subjects were recruited to participate in the study and each was asked to chew and record the flavor duration (in minutes) of one or more types of gum, what would the experimental unit be?
   (b) How could experimental units be blocked to reduce the variance of experimental error and increase the power for detecting differences?
   (c) Choose at least two of the factors listed above and provide a randomized list for an RCB design.
   (d) Recruit subjects and actually perform the experiment within a week.
   (e) Analyze your data. Did treatment factors and interactions affect the flavor duration? Was the blocking scheme you employed effective at reducing experimental error?

2. Consider room temperature as an additional blocking factor for the cut-flower experiment described in Section 4.2.

   (a) What design could you use to incorporate the two blocking factors of flower type and room temperature?
   (b) Set up a randomization plan to conduct this experiment.
   (c) What is the model you would use to analyze the data resulting from this experiment?

3. The data on the next page show the yields of five varieties of barley in a randomized complete block experiment carried out in Minnesota reported in *The Journal of the American Society of Agronomy* by Immer et al. (1934).

   (a) What would be the purpose of running these experiments in different locations and years?
   (b) How was variance of experimental error reduced in these experiments?
   (c) Compute the ANOVA for the RCB design model (4.1) and determine if there is a significant difference in varieties.
   (d) Calculate the Relative Efficiency, RE, and determine how many experiments would have been needed to have the same power using a completely randomized (CRD) design.
   (e) Use the Tukey method to make pairwise comparisons of the variety mean yields.

4. Horiuchi et al. (2005) conducted an experiment during Japan's 2004 Upper House election. The purpose of the experiment was to test the hypothesis that voter turnout is influenced by the amount of information available

EXERCISES

Table 4.6 *Total Yields of Barley Varieties in Twelve Independent Trials*

			Variety			
Place	Year	Manchuria	Svansota	Velvet	Trebi	Peatland
1	1931	81.0	105.4	119.7	109.7	98.3
1	1932	80.7	82.3	80.4	87.2	84.2
2	1931	146.6	142.0	150.7	191.5	145.7
2	1932	100.4	115.5	112.2	147.7	108.1
3	1931	82.3	77.3	78.4	131.3	89.6
3	1932	103.1	105.1	116.5	139.9	129.6
4	1931	119.8	121.4	124.0	140.8	124.8
4	1932	98.9	61.9	96.2	125.5	75.7
5	1931	98.9	89.0	69.1	89.3	104.1
5	1932	66.4	49.9	96.7	61.9	80.3
6	1931	86.9	77.1	78.9	101.8	96.0
6	1932	67.7	66.7	67.4	91.8	94.1

to voters. Potential voters were contacted by email and asked to fill out a screening survey and participate in the study. In the screening survey potential voters were asked what their gender was and whether they were planning to vote in the upcoming election (yes, no, or undecided). This information is known to be correlated with voter participation. Next, the respondents to the screening survey were randomly assigned to one of four groups. The first group was asked via email to view the official web site of the LDP party that shows their policy proposal. The second group was asked to view the web site of the DPJ party that shows their policy proposal. A third group was asked to view both web sites, and a fourth group was not asked to visit any web site prior to the election. Each participant was again contacted after the election to find out whether he or she had voted in the election or not.

(a) What was the treatment factor in this study?
(b) What was the experimental unit?
(c) What is the response?
(d) If potential voters were classified into the following six groups based on the screening survey

- Male - plans to vote
- Female - plans to vote
- Male - does not plan to vote
- Female - does not plan to vote
- Male - undecided
- Female - undecided

should this information be used in assigning respondents to the four groups? Why or why not?

(e) Explain how you would analyze the resulting data. What model would you use, what is the distribution of the dependent variable, etc.?

5. Lew (2007) presents the data from an experiment to determine whether cultured cells respond to two drugs. The experiment was conducted using a stable cell line plated onto Petri dishes, with each experimental run involving assays of responses in three Petri dishes: one treated with drug 1, one treated with drug 2, and one untreated serving as a control. The data are shown in the table below:

	Control	Drug 1	Drug 2
Experiment 1	1147	1169	1009
Experiment 2	1273	1323	1260
Experiment 3	1216	1276	1143
Experiment 4	1046	1240	1099
Experiment 5	1108	1432	1385
Experiment 6	1265	1562	1164

(a) Analyze the data as if it came from a completely randomized design using the model $y_{ij} = \mu + \tau_i + \epsilon_{ij}$. Is there a significant difference between the treatment groups?

(b) Analyze the data as a RCB design, where experiment number represents a blocking factor.

(c) Is there any difference in the results you obtain in (a) and (b)? If so explain what may be the cause of the difference in the results and which method would you recommend?

6. le Riche and Csima (1964) evaluated four hypnotic drugs and a placebo to determine their effect on quality of sleep in elderly patients. The treatment levels were labeled (A=Placebo, B=Ethchlorvynol, C=Glutethimide, D=Chloral hydrate and E=Secobarbitol sodium). Elderly patients were given one of the capsules for five nights in succession and their quality of sleep was rated by a trained nurse on a four-point scale (0=poor to 3=excellent) each night. An average score was calculated for each patient over the five nights in a week. Each patient received all five treatments in successive weeks. A Latin-square design was used to account for patient-to-patient differences and week-to-week effects. The design and the response (mean quality of sleep rating) are shown in the table on the next page.

(a) What is the appropriate model for this data?
(b) Complete the ANOVA and determine if there are any significant differences among the treatments.
(c) Use an appropriate method to determine if there is a significant difference between the placebo and the average of the other drugs, and if there are significant differences among the four drugs.

EXERCISES

			Week		
Patient	1	2	3	4	5
1	B (2.92)	E (2.43)	A (2.19)	C (2.71)	D (2.71)
2	D (2.86)	A (1.64)	E (3.02)	B (3.03)	C (3.03)
3	E (1.97)	B (2.50)	C (2.47)	D (2.65)	A (1.89)
4	A (1.99)	C (2.39)	D (2.37)	E (2.33)	B (2.71)
5	C (2.64)	D (2.31)	B (2.44)	A (1.89)	E (2.78)

(d) Use residual plots to check the assumptions for the model you fit.

7. Woodward (1970) conducted an experiment to determine the fastest path to second base in baseball. The three paths investigated were the round-out, narrow angle and wide angle shown in the figure below. The best path is defined to be the one that minimizes the time to reach second base.

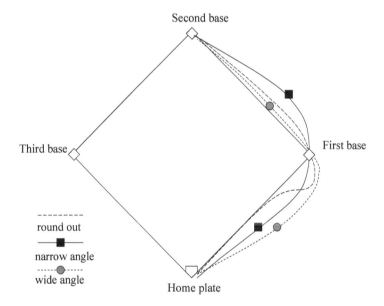

He used a stopwatch to time a runner going from home to second. He started the watch when the runner crossed a point 35 feet from home plate and stopped the watch at a point 15 feet short of second base. This eliminated the variability in times caused by stopping and starting. Finally he timed a random sample of 22 different runners, so that his conclusions could be generalized to all runners. In addition, after an appropriate rest period, he had each runner take each path (in a random order). In that way he could use the runners as blocks and eliminate the runner-to-runner variability from the error. The data is shown in the table below.

(a) What is the appropriate model for this data?

Player	Round-out	Narrow angle	Wide angle
1	5.40	5.50	5.55
2	5.85	5.70	5.75
3	5.20	5.60	5.50
4	5.55	5.50	5.40
5	5.90	5.85	5.70
6	5.45	5.55	5.60
7	5.40	5.40	5.35
8	5.45	5.50	5.35
9	5.25	5.15	5.00
10	5.85	5.80	5.70
11	5.25	5.20	5.10
12	5.65	5.55	5.45
13	5.60	5.35	5.45
14	5.05	5.00	4.95
15	5.50	5.50	5.40
16	5.45	5.55	5.50
17	5.55	5.55	5.35
18	5.45	5.50	5.55
19	5.50	5.45	5.25
20	5.65	5.60	5.40
21	5.70	5.65	5.55
22	6.30	6.30	6.25

(b) Complete the ANOVA and determine if there are any significant differences among the three paths.

(c) Use an appropriate multiple comparison procedure to determine what paths are different from each other.

(d) Was blocking effective in this experiment? Why or why not?

(e) Use residual plots to check the assumptions for the model you fit. If the assumptions do not hold, what would you recommend doing?

8. Consider the situation described in Exercise 6 of Chapter 2 with $t = 4$ levels of the treatment factor, and $\Delta = 2.0\sigma$.

   (a) If by blocking the experimental units into blocks it was believed that the variance of the experimental error, σ^2 could be reduced by 50%, calculate the number of blocks that would be required to have power of 0.90 for detecting a maximum difference in treatment means as large as Δ.

   (b) If by using a Latin square design the variance of the experimental error, σ^2, could be reduced another 10%, determine the power for detecting treatment differences when using a 4×4 Latin square design.

4.11 Appendix–Data from Golf Experiment

Table 4.7 *Data from Golf Experiment*

Golfer ID

Tee Hgt.	1	2	3	4	5	6	7	8	9
1	142.0	169.5	142.7	185.4	222.2	133.6	165.2	174.3	229.7
1	141.8	177.0	136.2	164.8	201.9	132.6	173.2	160.1	220.7
1	153.7	169.1	140.2	173.9	192.5	135.0	174.2	162.8	240.4
1	130.6	176.5	143.3	191.9	182.0	147.6	176.9	174.6	219.5
1	147.8	173.8	145.8	164.5	224.8	136.7	166.4	172.6	225.6
2	142.7	185.6	137.8	184.7	197.7	145.5	178.8	184.4	241.6
2	136.2	164.8	159.0	172.8	229.8	154.5	163.4	181.8	242.1
2	140.2	173.9	151.1	175.8	203.3	150.5	160.2	185.0	243.4
2	143.3	191.9	154.1	184.7	214.3	137.9	160.6	192.4	240.8
2	145.8	164.5	135.0	172.2	220.9	154.4	169.3	193.3	240.7
3	137.8	184.7	142.0	176.0	221.8	145.9	172.8	180.6	243.3
3	159.0	183.0	141.8	177.0	240.0	146.0	183.2	172.5	242.1
3	151.1	195.9	153.7	175.3	221.4	149.2	170.2	181.2	236.1
3	154.1	194.4	130.6	176.5	234.9	145.2	169.6	178.4	248.3
3	135.0	182.2	147.8	173.8	213.2	147.2	169.9	167.6	240.4

CHAPTER 5

Designs to Study Variances

5.1 Introduction

In the experiments described in Chapters 2 through 4, the purpose was to compare the response between different levels of controllable variables or factors in order to predict what the response might be in the future at specific levels of the factors, or to recommend the best factor level for future use. Another purpose of experimentation is to study sources of variability in the response. For example, cholesterol, blood glucose, and other diagnostic tests made by doctors are known to vary with the procedures and the equipment used to make the measurements. Experiments might be conducted to find out how much of the variability is due to equipment and how much is due to procedure. Symbolically $\sigma_T^2 = \sigma_p^2 + \sigma_e^2$ where σ_T^2 is the total variance, σ_p^2 and σ_e^2 are the portions of the total due to procedure and equipment, respectively. σ_p^2 and σ_e^2 are called the components of variance or variance components. An experiment can be conducted to collect data so that the variance components can be estimated. In this type of experiment, there might not be any interest in the difference in average diagnostic readings between specific pieces of equipment because there are (and will continue to be) many in use.

There are at least three reasons for conducting experiments to study the sources of variability. In some cases the purpose may be descriptive, and the variance components have value in themselves. A second reason for quantifying the sources of variability is to gain insight into how to reduce the variance of the response. A third reason for studying the sources of variability is to stimulate ideas about the causes of variability that could be tested in further experiments.

Two examples of where variance components are useful as descriptive measures are in genetics and in educational and psychological testing (see Searle et al., 1992). In dairy cow breeding, the variability in milk production can be partitioned into the amount due to the sire and the daughter, i.e., $\sigma_T^2 = \sigma_s^2 + \sigma_d^2$. The ratio of $h = 4\sigma_s^2/(\sigma_s^2 + \sigma_d^2)$ is called the heritability and is highly important to dairy farmers. In psychological and educational testing the variability in test scores can be partitioned into the person-to-person variability and the repeat test scores for the same person, i.e., $\sigma_T^2 = \sigma_p^2 + \sigma_r^2$. In this case, $\sigma_p^2/(\sigma_p^2 + \sigma_r^2)$ is called the intra-class correlation, and high values of it imply reliability of the testing procedure.

In industrial quality control there is a need to reduce variability in in-process measurements of key product and process characteristics. If one cannot measure accurately, there is no hope of controlling or improving quality. Total measurement variability within a plant can be attributed to the measurement equipment (or gage) and the operator (or inspector) making the measurement, i.e., $\sigma_T^2 = \sigma_g^2 + \sigma_o^2$. In order to reduce measurement variability, management needs to know where to concentrate its efforts. If the major proportion of variability is σ_g^2, perhaps effort needs to be placed on re-calibrating measurement equipment or buying more new and more precise and consistent equipment. On the other hand, if the major source of measurement variability is σ_o^2, perhaps better training of operator-inspectors may solve the problem.

In some cases a researcher would like to conduct an experiment like those described in Chapters 2-4 to compare the average response caused by different levels of controllable factors; however, he or she has such limited knowledge about the mechanism under study that it is difficult to hypothesize what factors or levels of factors to study. In this situation, determining the sources of variability in the response may prompt ideas about what factors would be most profitable to study. For example, knowing whether the majority of variability in an industrial process is batch-to-batch or within a batch would give insight as to whether factors that could be varied within a batch, or factors that could be varied from one batch to another, should be studied in optimization experiments. In cases where one factor experiment or a factorial experiment has been conducted and nothing was found to be significant, Leitnaker and Cooper (2005) suggest that a follow-up sampling study (to classify the sources of variability in the response) may explain why no significant factors were found.

5.2 Random Factors and Random Sampling Experiments

When the purpose of experimentation is to study differences in the average response caused by differences in factor levels (like the experiments described in Chapters 2-4), the factors in the experiment are called *fixed factors*. The levels of these factors are specifically selected by the experimenter. On the other hand, when the purpose of experimentation is to study the variance caused by changing levels of a factor, the factor is called a *random factor*. For example, the model $y_{ij} = \mu + \tau_i + \epsilon_{ij}$ in Chapter 2, the factor τ_i would be considered a fixed factor. Although we haven't called it a factor before, the term ϵ_{ij} in this model would be considered a random factor. ϵ_{ij} represents the effect of the jth experimental unit within the ith level of the treatment factor. In the bread rise experiment described in Chapter 2, the experimental unit was the loaf of bread. Replicate loaves (i.e., $j = 1, \ldots, 4$) were used in this experiment so that σ^2, the variance of the experimental units, could be estimated and used to judge the significance of the fixed factor (rise time). Since the purpose of including multiple levels (or loaves) within each rise time was to estimate σ^2, ϵ_{ij} is considered a random factor.

Whereas the levels of fixed factors are specifically chosen by the experimenter, the levels of random factors are just samples of possible levels that

RANDOM FACTORS AND RANDOM SAMPLING EXPERIMENTS   149

could have been used. For example, in the bread rise experiment, the experimenter chose 35, 40 and 45 minutes to study for the levels of rise time. However, the four replicate loaves used for each rise time represent only a sample of the loaves that could have been used in the experiment. For this reason experiments that are used to study variances can be thought of as random sampling experiments or RSE since the factor levels are just a sample of possible levels. For example, consider the data in Table 5.1 patterned after the international survey of apolipoproteins conducted in 1984-85 (see Henderson et al., 1987). Apo A-I is known to help clear cholesterol from arteries. However, the apo genotype yields poor predictive values when screening for clinically defined atherosclerosis. This may be in part due to difficulty in measurement.

Table 5.1 *Measured Apo A-I Concentrations by Laboratory Code*

Lab	A	B	C	D
	1.195	1.155	1.021	1.163
	1.144	1.173	1.037	1.171
	1.167	1.171	1.022	1.182
	1.249	1.175	1.064	1.184
	1.177	1.153	1.094	1.175
	1.217	1.139	0.992	1.134
	1.187	1.185	1.072	1.169
		1.144		1.136

The purpose of the study was to examine and quantify the variation among laboratories with respect to their measurement of an international reference material for apo A-I and B. Several measurements of the relative concentrations of preparations of the reference material were made by twenty-eight selected laboratories. Table 5.1 shows data from four representative labs. The model for the data can be written as:

$$y_{ij} = \mu + t_i + \epsilon_{ij}, \tag{5.1}$$

where y_{ij} is the jth measurement of apo A-I concentration at the ith laboratory, μ is the overall average measured concentration, t_i is the laboratory effect, and ϵ_{ij} is the effect of the jth measurement in the ith lab. There was no interest in comparing measured apo A-I concentrations among the specific laboratories in the study since they can be considered a random sample, or representative sample, of several labs around the world. Since the purpose of including several labs was to estimate the component of variance, σ_t^2, in measured concentration due to lab, t_i can be considered a random factor and the experiment can be thought of as a sampling experiment. Note that the Roman letter t_i was used to represent the random region effect, whereas in model (2.2) in Chapter 2 the Greek letter τ_i was used to represent the fixed treatment factor. The replicate measurements of apo A-I made in each lab are only samples of the possible measurements that could be made in that lab. Since multiple measurements were made in each lab in order to estimate σ^2,

150   DESIGNS TO STUDY VARIANCES

ϵ_{ij} can also be considered a random factor. Except for the ϵ used to represent the random experimental unit, the convention in this book will be to use Roman letters to represent random factors and Greek letters to represent fixed factors.

The usual assumptions regarding model (5.1) are that the random effects, t_i and ϵ_{ij}, are independent and normally distributed with zero means and variances equal to the variance components σ^2 and σ_t^2, respectively. Since the variance of a sum of independent random variables is the sum of the variances, $\sigma_y^2 = \sigma_t^2 + \sigma^2$. Data from sampling experiments can be used to partition the variance in the response, σ_y^2, into the two variance components of which it is composed.

5.3 One-Factor Sampling Designs

One-factor sampling experiments can be used to partition variability into two sources. As an example of partitioning variance into two sources, consider the paper helicopter experiments described in Exercise 2.1. In that exercise there might have been some differences in the description of the experimental unit among students. Some might argue that the experimental unit was a sheet of paper from which a helicopter design was cut. Others might argue that it was the trial, or air conditions, at the time a helicopter was dropped and timed. If the first definition was used, then replicate experiments would consist of making, dropping, and timing several helicopters of the same design once, where each one was made from a different sheet of paper. If the second definition was used, replicate experiments would consist of repeatedly dropping and timing one helicopter. One practical way to decide how to define the experimental unit would be to partition the variability in drop times into helicopter-to-helicopter variability and variability among repeat drops of the same helicopter. If a substantial part of the variability was among helicopters of the same design cut from different pieces of paper, there would be reason to make multiple helicopters for replicates. If, on the other hand, all of the variability was among drop times of the same helicopter, there would be no reason to make multiple helicopters of the same design for replicates. In that case, repeat drops of the same helicopter could be considered replicates.

The variability in helicopter drop times can be partitioned into the helicopter-to-helicopter variability and within helicopter variability using a one-factor sampling experiment. To do this, first randomly select six sheets of paper. From each of these sheets, cut and fold one standard helicopter with body width = 4.25", tail length = 4.0", and wing length = 6.5". Drop and time each of the six helicopters three times each according to a randomized order like that created in Section 2.2.

As a second example of partitioning variability into two sources, consider the following example presented by Davies (1949). A dye manufacturer wanted to know if there was an appreciable contribution to variability in dyestuff color yields owing to the quality of the intermediate acid batch used. It was a two-

ESTIMATING VARIANCE COMPONENTS

step process. Intermediate acid batches were produced in one step, and in a later step the acid was used to produce the dyestuff. The goal was to keep dyestuff color yields consistently high. If the majority of variation was caused by differences among the intermediate acid batches, then improvement efforts should be concentrated on the process that makes the acid batches. If the majority of variation was within preparations of the dyestuff made from the same acid batch, improvement efforts should be focused on the process step of making the dyestuff. A sampling experiment was run wherein six representative samples of H acid intermediate were taken from the step manufacturing process that produces it. From each acid sample, five preparations of the dyestuff Naphthalene 12B were made in a laboratory, and these were representative of the preparations that could be made with each sample. The data from the sampling experiment is shown in Table 5.2. The yields are given in grams of standard color.

Table 5.2 Yields of Naphthalene Black 12B

Sample of H acid	1	2	3	4	5	6
Individual yields in	1440	1490	1510	1440	1515	1445
grams of standard	1440	1495	1550	1445	1595	1450
color	1520	1540	1560	1465	1625	1455
	1545	1555	1595	1545	1630	1480
	1580	1560	1605	1595	1635	1520

5.4 Estimating Variance Components

The model (5.1) can be expressed in matrix terms as:

$$y = X\beta + \epsilon, \tag{5.2}$$

where $\beta' = (\mu, t')$ and t' is the vector of random effects, and the independence and normality assumptions can be expressed as:

$$\begin{pmatrix} t \\ \epsilon \end{pmatrix} \sim MVN\left(\begin{pmatrix} 0 \\ 0 \end{pmatrix}, \begin{pmatrix} \sigma_t^2 I_t & 0 \\ 0 & \sigma^2 I_n \end{pmatrix}\right), \tag{5.3}$$

where MVN represents the multivariate normal distribution, I_t is a $t \times t$ identity matrix, t is the number of levels of the random factor t_i, and $n = tr$ is the total number of experiments or runs. With the independence and normality assumptions, there are several methods of estimating variance components from the data obtained in sampling experiments.

5.4.1 Method of Moments Estimators

Fisher was the first to show how to estimate the variance components from an analysis of variance. To do this, the method of moments is used. The model for the one-factor sampling experiment is given by Equation (5.1), and the analysis of variance table is identical to Table 2.2 in Chapter 2. The F-test in the analysis of variance table can be used to test the null hypothesis $H_0 : \sigma_t^2 = 0$ against the alternative $H_a : \sigma_t^2 > 0$.

To estimate the variance components from the ANOVA, the mean squares are equated to their expected values and the simultaneous equations are solved. When the treatment factor is a random factor, as in model (5.1), and there are an equal number of replicates in each level, the msT term in the ANOVA table follows a distribution that is a multiple of the central chi-square distribution. This is different than the model with a fixed effect treatment factor. In that case, described in Section 2.3.5, the distribution of msT was a non-central chi-square. In either case, the msT from the ANOVA can be represented as the quadratic form $\boldsymbol{y}'\boldsymbol{A}\boldsymbol{y}$; and in the random effects model (5.1), Hartley (1967) has shown that its expected value can be written as $\sigma^2 + c\sigma_t^2$, where $c = \sum_i \boldsymbol{x}_i' \boldsymbol{A} \boldsymbol{x}_i$, and $\boldsymbol{x}_i$ is an indicator of the ith level of t_i (that is the $(i+1)$th column in the $\boldsymbol{X}$ matrix as shown in Equation (2.6). When there is an equal number of replicates, r, in each level of the random factor, the coefficient in the expected mean square simplifies to $c = r$. The method of moments estimators for the variance components can be used when there are equal number replicates in each level of the random factor as shown in Table 5.2, or an unequal number as shown in Table 5.1. For the equal replicates case, the estimates turn out to be uniformly best unbiased estimators, but for the unequal case estimators that are uniformly best do not exist (see Searle et al., 1992).

To illustrate the estimation of the variance components using SAS, the commands to read in the data from the apo measurement sampling experiment, shown in Table 5.1, along with the `proc glm` commands to produce the ANOVA are shown below.

```
data labs;
input lab $ conc;
datalines;
A 1.195
  . . .
proc glm data=labs; class lab;
  model conc=lab;
  random lab;
run;
```

The output is shown at the top of the next page.

The statement `random lab` tells `proc glm` that the lab factor is a random factor. By adding this statement the `proc glm` output includes the additional

table shown below that contains the expected value of msT calculated by Hartley's method.

As shown in Section 2.3.3 the expected value of the msE in the ANOVA Table is σ^2, therefore two simultaneous equations, in the two unknown vari-

ESTIMATING VARIANCE COMPONENTS

Source	DF	Sum of Squares	Mean Square
Model	3	0.09223328	0.03074443
Error	26	0.01898409	0.00073016
Corrected Total	29	0.11121737	

Source	Type III Expected Mean Square
lab	Var(Error) + 7.4889 Var(lab)

ance components σ^2 and σ_t^2, can be created by equating the mean squares for model and error to their expected values as shown below.

$$0.03074443 = \sigma^2 + 7.4889\sigma_t^2$$
$$0.00073016 = \sigma^2$$
$$\Rightarrow \hat{\sigma}^2 = 0.00073016$$
$$\text{and} \Rightarrow \hat{\sigma}_t^2 = (0.03074443 - 0.00073016)/7.4889 = 0.0040078$$

Since $\hat{\sigma}^2 < \hat{\sigma}_t^2$, these estimates show there is much or more variability among labs than within a lab.

The SAS procedure **proc varcomp** is more appropriate for estimating variance components from models with random effects. It produces the ANOVA table and the expected mean squares, like **proc glm**, and it also solves the equations to estimate the variance components. The commands to call **proc varcomp** with the data from the apo measurement sampling experiment are:

```
proc varcomp data=labs method=type1; class lab;
   model conc=lab;
run;
```

There are several different methods that can be used to estimate variance components, and they are available as options in **proc varcomp**. The option **method=type1** shown above tells the procedure to use the method of moments-analysis of variance method of estimating the variance components by equating the type I ANOVA mean squares to their expected values. If this option is left out, **proc varcomp** by default uses the MIVQUE0 method. The output produced by the above commands is shown on the next page.

5.4.2 Interval Estimates

When the normality assumptions apply and there is an equal number of replicates for each level of the random factor in model (5.1), exact interval estimates exist for σ^2, $\sigma_t^2/(\sigma^2 + \sigma_t^2)$, $\sigma^2/(\sigma^2 + \sigma_t^2)$, and σ_t^2/σ^2 based on the distributions of the means squares. Neither SAS **proc glm** nor **proc varcomp** produce

```
Source             DF  Sum of    Mean Square  Expected Mean Square
                       Squares

lab                3   0.092233  0.030744     Var(Error) + 7.4889 Var(lab)
Error              26  0.018984  0.000730     Var(Error)
Corrected Total    29  0.111217                 .        .

                       Type 1 Estimates
                   Variance Component    Estimate

                   Var(lab)              0.0040078
                   Var(Error)            0.0007302
```

these confidence intervals, but they can be easily calculated by hand from the statistics produced in the ANOVA table. If there are $i = 1, \ldots, T$ levels of the random factor and $j = 1, \ldots, r$ replicates in each level of the random factor, then Table 5.3 (that was taken from Searle et al. (1992)) contains exact formulas for confidence intervals on lines 1, 3-5, and an approximate formula on line 2.

Table 5.3 *Confidence Intervals for Functions of Variance Components in One-Factor Random Model with Equal Replication*

		Confidence Interval		
Line	Parameter	Lower Limit	Upper Limit	Confidence Coefficient
1	σ^2	$\dfrac{ssE}{\chi^2_{T(r-1),U}}$	$\dfrac{ssE}{\chi^2_{T(r-1),L}}$	$1-\alpha$
2	σ_t^2	$\dfrac{ssT(1-F_U/F)}{r\chi^2_{T-1,U}}$	$\dfrac{ssT(1-F_L/F)}{r\chi^2_{T-1,L}}$	$1-2\alpha$
3	$\dfrac{\sigma_t^2}{\sigma_t^2+\sigma^2}$	$\dfrac{F/F_U-1}{r+F/F_U-1}$	$\dfrac{F/F_L-1}{r+F/F_L-1}$	$1-\alpha$
4	$\dfrac{\sigma^2}{\sigma_t^2+\sigma^2}$	$\dfrac{r}{r+F/F_L-1}$	$\dfrac{r}{r+F/F_U-1}$	$1-\alpha$
5	$\dfrac{\sigma_t^2}{\sigma^2}$	$\dfrac{F/F_U-1}{r}$	$\dfrac{F/F_L-1}{r}$	$1-\alpha$

Notation: $F = msT/msE$,

$$\Pr\{\chi^2_{\nu,L} \leq \chi^2_\nu \leq \chi^2_{\nu,U}\} = 1-\alpha$$

$$\Pr\{F_L \leq F_{\nu_1,\nu_2} \leq F_U\} = 1-\alpha$$

As an example of calculating the confidence intervals, consider the data in the sampling experiment shown in Table 5.2 to study the variability in dyestuff

ESTIMATING VARIANCE COMPONENTS

color yields. From the ANOVA $ssT = 56{,}358$, $ssE = 58{,}830$, $T = 6$, $r = 5$, and $F = msT/msE = 4.59847$. The upper 0.975 percentile of the chi-square distribution with 24 degrees of freedom is 39.36. This can be obtained in a SAS data step using the cinv function as chiU=cinv(.975,24);. The upper 0.975 percentile of the F-distribution with 5 and 24 degrees of freedom is 3.15482. This can be obtained from the SAS data step using the finv function as FU=finv(.975,5,24);. The other percentiles of the F and chi-square can be obtained similarly. The exact 95% confidence interval on σ^2 is given by:

$$\left(\frac{ssE}{\chi^2_{24,0.975}}, \frac{ssE}{\chi^2_{24,0.025}}\right) = \left(\frac{58830}{39.6}, \frac{58830}{12.4}\right)$$
$$= (1485.61, 4744.35).$$

The approximate 90% confidence interval on σ_t^2 is given by:

$$\left(\frac{ssT(1 - F_{5,24,0.975}/F)}{r\chi^2_{5,0.975}}, \frac{ssT(1 - F_{5,24,0.025}/F)}{r\chi^2_{5,0.025}}\right)$$
$$= \left(\frac{56358(1 - 3.15482/4.60)}{5(12.8325)}, \frac{56385(1 - 0.15929/4.60)}{5(0.83121)}\right)$$
$$= (275.9551, 10{,}914.1765),$$

and the exact 95% confidence interval on $\sigma_t^2/(\sigma_t^2 + \sigma^2)$ is given by:

$$\left(\frac{F/F_{5,24,0.975} - 1}{r + F/F_{5,24,0.975} - 1}, \frac{F/F_{5,24,0.025} - 1}{r + F/F_{5,24,0.025} - 1}\right)$$
$$= \left(\frac{4.59847/3.15482 - 1}{6 + 4.59847/3.15482 - 1}, \frac{4.59847/0.15929 - 1}{6 + 4.59847/0.15929 - 1}\right)$$
$$= (0.07086, 0.822844).$$

Notice the interval estimate of σ_t^2 is much wider than the interval estimate of σ^2. This is because ssT only has 5 degrees of freedom in the ANOVA, while ssE has 24 degrees of freedom.

5.4.3 Maximum Likelihood and REML Estimators

Although the method of moments estimators are uniformly best unbiased estimators, they have one unfortunate property. When msT is less than msE in the analysis of variance, the estimator of σ_t^2 will be negative. This can happen quite frequently if $\sigma_t^2/\sigma^2 \leq 0.10$, and there are less than $T = 10$ levels of the random factor t_i. Maximum likelihood (ML) and (reduced or restricted) maximum likelihood (REML) are preferred methods of estimation that avoid this problem. REML is an adaptation of the maximum likelihood technique that maximizes part of the likelihood. The fact that maximum likelihood estimators cannot lie outside their parameter space prevents both the ML and REML methods from obtaining negative estimates of σ_t^2. To understand how maximum likelihood and REML work, we will consider the equal replication case. Given the model and assumptions in Equations (5.2) and (5.3), the distribution of y can be written as:

$$y \sim MVN(\mu \mathbf{1}, V), \tag{5.4}$$

where V is a block diagonal matrix with T blocks of $(\sigma_t^2 J_r + \sigma^2 I_r)$ along the diagonal. The likelihood function is

$$L(\mu, V|y) = \frac{\exp\left[-\frac{1}{2}(y - \mu \mathbf{1}_n)' V^{-1} (y - \mu \mathbf{1}_n)\right]}{(2\pi)^{\frac{1}{2}n} |V|^{\frac{1}{2}}}. \tag{5.5}$$

For the equal replication case, this can be simplified to:

$$L(\mu, \sigma^2, \lambda | y) = \frac{\exp\left\{-\frac{1}{2}\left[\frac{ssE}{\sigma^2} + \frac{ssT}{\lambda} + \frac{(\bar{y}_{..} - \mu)^2}{\lambda/n}\right]\right\}}{(2\pi)^{\frac{1}{2}n} \sigma^{2[\frac{1}{2}n]} \lambda^{\frac{1}{2}T}}, \tag{5.6}$$

where $\lambda = \sigma^2 + r\sigma_t^2$. The maximum likelihood estimates are obtained by maximizing the likelihood with respect to μ, σ^2, and λ. The REML estimates of σ^2, and σ_t^2 are obtained by maximizing $L(\sigma^2, \sigma_t^2 | ssT, ssE)$ with respect to σ^2, and σ_t^2, where

$$L(\sigma^2, \sigma_t^2 | ssT, ssE) = \frac{L(\mu, \sigma^2, \lambda | y)}{L(\mu | \bar{y}_{..})} \tag{5.7}$$

can be obtained by factoring $L(\mu, \sigma^2, \lambda | y)$ using the fact that ssE and ssT are independent of $\bar{y}_{..}$. The maximization can be done analytically for the equal replication case as shown by Searle et al. (1992), and can be done numerically for the unbalanced case.

A desirable property of the REML estimates is that they are the same as the method of moments (analysis of variance estimates) when there is equal replication in each level of the random factor and $msT > msE$. SAS **proc varcomp** has options to compute variance component estimates using the method of moments, maximum likelihood and REML. The maximum likelihood estimators and REML estimators are calculated by **proc varcomp** using a numerical solution for both the unbalanced and balanced cases.

To illustrate the REML estimators, consider the following case. A manufacturer of packaged dry soup mixes was experiencing excessive variability in the package weights of a dry soup mix component called the "intermix." The intermix is a mixture of flavorful ingredients such as vegetable oil, salt, etc. Too much intermix in a soup packet gives it too strong a flavor, and not enough gives too weak a flavor. It was a two-step process to make the packaged soup mix. The first step was to make a large batch of soup and dry it on a rotary dryer. Next, the dried soup batch was placed into a mixer, where the intermix was added through ports as it was mixed. Then it was packaged in sealed bags of uniform weight. There were several factors that could be changed in the first step (production of the dried soup batch), and several factors that could be changed in the second step (adding the intermix and mixing) that could possibly affect the variability of the weight of the intermix

ESTIMATING VARIANCE COMPONENTS

in each sealed bag. A factorial experiment was to be planned to find out which factors affected variability in intermix weights. In order to determine which factors to include in a factorial experiment, a reasonable first step would be to partition the variability in intermix weight into the variability from soup batch to soup batch and the variability within a batch caused by the process to mix and add intermix to the dried soup. If there was little variability from batch to batch, the experiment would only need to consider factors involved in the mixing step.

In order to partition the variability in package weights into batch-to-batch and within batch, a sampling experiment was performed. A random sample of four batches was selected over a month of production runs. From each batch, three 1-lb. samples of the finished soup mix were taken from the mixing tank as it was being packaged. The weight of the intermix was determined for each sample. The results of the sampling experiment are shown in Table 5.4.

Table 5.4 *Variability in Dry Soup Intermix Weights*

batch	weight
1	0.52, 2.94, 2.03
2	4.59, 1.26, 2.78
3	2.87, 1.77, 2.68
4	1.38, 1.57, 4.10

The SAS commands to read the data and produce the REML estimates of σ_t^2 and σ^2 are shown on the next page, and the results are shown below the commands.

In this output, the familiar ANOVA table is missing because it is not needed to obtain the REML estimates. The output instead shows the iteration history of the numerical process used to maximize the likelihood and the resulting estimators. It can be seen that $\hat{\sigma}_t^2$, the batch-to-batch variability, is estimated to be zero, while $\hat{\sigma}^2$, the variability caused by the mixing process within a batch, is estimated to be 1.41. If the method of moments estimators is used with this data (left as an exercise), the estimator of σ_t^2 turns out to be negative.

```
data soupmx;
input batch sample intweight;
datalines;
 1 1 0.52
 1 2 2.94
 1 3 2.03
  ... etc.
proc varcomp method=reml;   class batch;
   model intweight=batch;
run;
```

The conclusions from this sampling experiment indicate that further experiments to identify the causes of variability in the intermix should concentrate

```
                    REML Iterations

Iteration           Objective      Var(batch)         Var(Error)
        0        3.7798354425               0       1.4100446970
        1        3.7798354425               0       1.4100446970

Convergence criteria met.
                    REML Estimates

                    Variance
                    Component          Estimate
                    Var(batch)                0
                    Var(Error)          1.41004
```

on factors that can be varied in the second step of the process where the intermix is added and mixed with the batch. These further experiments were performed and will be shown in the next chapter.

In processes that contain more than two steps, Leitnaker and Cooper (2005) show that multistage sampling experiments, to be described in Section 5.6, are very useful for identifying process steps and factors within those steps that would be good candidates for further study with factorial type experiments.

When using the option `method=reml` or `method=ml`, SAS `proc varcomp` gives the asymptotic variances of the estimates in addition to the maximum likelihood or REML estimates of the variance components. For example, using the data from dyestuff color yields in Table 5.2, the output table shown at the top of the next page was produced by `proc varcomp`.

The asymptotic variances on the diagonal of the covariance matrix shown in the `proc varcomp` output can be used to get an asymptotic confidence interval on the variance components. For example, an asymptotic 95% confidence interval for σ^2 can be constructed using the estimates on the next page as shown below.

$$\hat{\sigma}^2 \pm 1.96\sqrt{Var(\hat{\sigma}^2)} = 2451.3 \pm 1.96\sqrt{500,718.9} = (1064.37, 3838.225)$$

A similar interval estimate can be obtained for σ_t^2. The asymptotic confidence intervals will be reasonably close to the exact confidence intervals produced using the formulas in Table 5.3, when the degrees of freedom for the term corresponding to the variance component being estimated is greater than 45.

ESTIMATING VARIANCE COMPONENTS

```
            REML Estimates
       Variance
       Component           Estimate
       Var(sample)          1764.1
       Var(Error)           2451.3

 Asymptotic Covariance Matrix of Estimates
                   Var(sample)          Var(Error)
Var(sample)         2052776.2           -100143.8
Var(Error)          -100143.8            500718.9
```

5.4.4 Determining the Sample Size For One-Factor Sampling Studies

Since there are two variance components being estimated in a one-factor sampling design, there are two things to consider when determining the sample size. In order to accurately estimate the replicate variance, σ^2, the important thing to consider is the number of degrees of freedom for error $\nu_2 = t(r-1)$. The accuracy of the estimate of the random factor, σ_t^2, will always be relative to the accuracy in estimating σ^2.

The accuracy for estimating σ^2 can be expressed in terms of the width of the confidence interval given on line 1 of Table 5.3. Since $E(ssE) = t(r-1)\sigma^2$, the expected width of the 95% confidence interval can be written as:

$$\sigma^2 \left[\frac{t(r-1) \times \left(\chi^2_{t(r-1),0.975} - \chi^2_{t(r-1),0.025} \right)}{\chi^2_{t(r-1),0.975} \times \chi^2_{t(r-1),0.025}} \right] \tag{5.8}$$

Therefore, if you would like the half-width of the confidence interval to be 50% of σ^2, search for the number of levels of the random factor, t, and the number of replicates, r, such that the multiplier of σ^2 in Equation (5.8) is 1.0. This can be done easily by enumerating various cases using the function cinv in the data step of SAS. The example on the next page shows the calculation of the multiplier of σ^2 that determines the expected width of the confidence interval. The resulting output, that is below the code on the next page, shows that any combination of t and r that result in $\nu_2 = t(r-1)$ in the range of 36 to 38 will give the desired accuracy in estimating σ^2.

A simple rule of thumb can be used to get an idea as to how many levels of the random factor, t, to include in the sampling experiment. When σ_t^2 is expected to be larger than σ^2, t should be as large as possible, so $t = \nu_2$, and $r = 2$ would be reasonable.

```
data ss;
  do nu=36 to 44;
    chiu=cinv(.975,nu);
    chil=cinv(.025,nu);
    widthf=nu*(chiu-chil)/(chil*chiu);
    halfwf=widthf/2;
    output;
  end;
proc print;
run;
```

Obs	nu	chiu	chil	widthf	halfwf
1	36	54.4373	21.3359	1.02599	0.51299
2	37	55.6680	22.1056	1.00913	0.50456
3	38	56.8955	22.8785	0.99306	0.49653
4	39	58.1201	23.6543	0.97772	0.48886
5	40	59.3417	24.4330	0.96307	0.48153
6	41	60.5606	25.2145	0.94904	0.47452
7	42	61.7768	25.9987	0.93560	0.46780
8	43	62.9904	26.7854	0.92271	0.46135
9	44	64.2015	27.5746	0.91033	0.45517

Another way of determining both t and r would be to consider the power of the F-test for testing the hypothesis $H_0 : \sigma_t^2 = 0$. Under the alternative hypothesis $H_a : \sigma_t^2 > 0$, the statistic $F = msT/msE$ follows a multiple of the central F-distribution and the power or probability of exceeding the critical limit can be expressed by

$$1 - \beta = Pr\left(F_{t-1,t(r-1)} > \frac{1}{1 + r \times R} F_{t-1,t(r-1),\alpha}\right), \qquad (5.9)$$

where $R = \sigma_t^2/\sigma^2$.

Again the sample sizes t and r that give adequate power for specified alternatives can be determined by enumerating several cases in a SAS data step. In this case the use of the SAS functions finv and probf make it easy. For example, if you wanted to have a power greater than $1 - \beta = 0.90$ for rejecting $H_0 : \sigma_t^2 = 0$ the SAS data step, shown on the next page, will find some alternatives. The resulting output list is below the code, and it shows several combinations of T and r that result in power greater than 0.90.

Of course this method does not consider the accuracy of the estimate of σ^2. One might consider using the first method shown above to determine $\nu_2 = t(r-1)$ to have the desired accuracy in estimating σ^2; and then with $\nu_2 = t(r-1)$ fixed at the number determined, use the second method shown above to determine how large t should be for adequate power in rejecting $H_0 : \sigma_t^2 = 0$ when $\sigma_t^2/\sigma^2 = R$ is at a specified level.

TWO-FACTOR SAMPLING DESIGNS

```
data ss2;
  Ratio=3.0;
    do t=5 to 7;
      do r=2 to 4;
        nu1=T-1;
        nu2=T*(r-1);
        fcrit=finv(.95,nu1,nu2);
        factor=1/(1+r*Ratio);
        plimit=factor*fcrit;
        power=1-probf(plimit,nu1,nu2);
        output;
      end;
    end;
proc print; var t r nu1 nu2 power;
run;
```

Obs	t	r	nu1	nu2	power
1	5	2	4	5	0.60253
2	5	3	4	10	0.83975
3	5	4	4	15	0.91424
4	6	2	5	6	0.68763
5	6	3	5	12	0.89721
6	6	4	5	18	0.95237
7	7	2	6	7	0.75659
8	7	3	6	14	0.93460
9	7	4	6	21	0.97375

5.5 Two-Factor Sampling Designs

When the purpose of experimentation is to study the variance in the response caused by varying the levels of two independent factors, the design is similar to the two-factor factorial design presented in Chapter 3. However, in the two-factor factorial designs presented in Chapter 3, the levels of the factors would be specifically selected by the experimenter because he would be interested in comparing the average response between these levels. In the two-factor sampling experiment, on the other hand, the levels of the factors are just a random or representative sample of possible levels, and the purpose is to determine how much of the variance in the response can be attributed to varying levels of the factors. In general these designs are called factorial random sampling experiments or FRSE.

Consider the example data presented in Table 5.5 taken from Sower *et al.* (1999). These are data from a Gage R&R study commonly performed in industrial quality assurance departments.

Table 5.5 *Data from Gage R & R Study*

Part	Operator 1	Operator 2	Operator 3
1	0.71	0.56	0.52
	0.69	0.57	0.54
2	0.98	1.03	1.04
	1.00	0.96	1.01
3	0.77	0.76	0.81
	0.77	0.76	0.81
4	0.86	0.82	0.82
	0.94	0.78	0.82
5	0.51	0.42	0.46
	0.51	0.42	0.49
6	0.71	1.00	1.04
	0.59	1.04	1.00
7	0.96	0.94	0.97
	0.96	0.91	0.95
8	0.86	0.72	0.78
	0.86	0.74	0.78
9	0.96	0.97	0.84
	0.96	0.94	0.81
10	0.64	0.56	1.01
	0.72	0.52	1.01

In these studies the purpose is to classify the variability in measured features of manufactured products or product components. Assuming the gage or measuring instrument is properly calibrated, a measured value determined during a quality control inspection can be considered to be a function of the true feature dimension, the gage repeatability and the gage reproducibility. Gage repeatability is the ability of a single operator to obtain the same measurement value multiple times using the same measuring instrument (or gage) on the same feature of a single manufactured component (or part). Gage reproducibility is the ability of different operators to obtain the same measured value multiple times using the same gage on the same part. If the variability in measurements caused by the gage repeatability plus the gage reproducibility is more than 10% of the tolerance range, the measurements may not be accurate enough to be used in monitoring product quality.

The Gage R&R sampling experiment consists of selecting a set of manufactured parts or components that are representative of the part-to-part variability in normal manufacturing. In Table 5.5, a sample of ten parts was selected and these parts represent the levels of the first factor in the sampling experiment. Next, a random or representative sample of inspectors is selected. The inspectors or operators represent the levels of the second factor in the sam-

TWO-FACTOR SAMPLING DESIGNS   163

pling experiment. Finally each inspector measures each part twice in a random order and the results are assembled in a table like Table 5.5. The replicate measurements represent the replicates in each cell.

Since each operator or inspector measured each part, the model for the data in Table 5.5 can be written in the form of a factorial model

$$y_{ijk} = \mu + a_i + b_j + ab_{ij} + \epsilon_{ijk}, \qquad (5.10)$$

where y_{ijk} is the kth measurement ($k = 1, \ldots, r$) made by the jth operator ($j = 1, \ldots, B$) on the ith part ($i = 1, \ldots, A$), a_i is the part effect, b_j is the operator or inspector effect, and ab_{ij} is the interaction effect.

The difference between this model and model (3.2) in Chapter 3 is the fact that the effects a_i, b_j and ab_{ij} are now assumed to be independent, normally distributed random variables with zero means and variances σ_a^2, σ_b^2, and σ_{ab}^2. Since the variance of a sum of independent random variables is the sum of the variances, the total variance in the response $\mathrm{Var}(y) = \sigma_y^2 = \sigma_a^2 + \sigma_b^2 + \sigma_{ab}^2 + \sigma^2$. σ_a^2 represents the portion of the total variance due to actual differences in part features, σ_b^2 is the portion of the variance caused by differences among operators, σ_{ab}^2 is the portion of the variance caused by the interaction of operator and part, and σ^2 is the portion of the variance caused by replicate measurements or gage repeatability. The sum of $\sigma_b^2 + \sigma_{ab}^2$ is the gage reproducibility. The repeatability plus reproducibility $\sigma_b^2 + \sigma_{ab}^2 + \sigma^2$ is a measure of the variance attributable to measurement error.

5.5.1 Estimating Variance Components

For the case with an equal number of replicates per subclass, like the example in Table 5.5, it is convenient to use either the method of moments or REML estimators of the variance components. For example the SAS commands to read the gage R&R data and compute the estimates are shown below:

```
options ls=64;
data gagerr;
input part oper y;
datalines;
1 1 .71
1 1 .69
... etc.
proc varcomp data=gagerr method=type1;
  class part oper;
  model y=part oper part*oper;
run;
proc varcomp data=gagerr method=reml;
  class part oper;
  model y=part oper part*oper;
run;
```

For this example the method of moments estimator of σ_b^2 is negative, as shown in the type 1 analysis on the next page. One or two atypical observations can result in an estimated variance component that is too large, too small, or even negative. Graphical techniques, like those described in Section 5.9, can often reveal atypical values and increase the chance of a useful interpretation of the data.

```
                   Type 1 Analysis of Variance

                                   Sum of
    Source              DF         Squares        Mean Square

    part                 9         1.448915        0.160991
    oper                 2         0.029703        0.014852
    part*oper           18         0.483930        0.026885
    Error               30         0.022550        0.000752
    Corrected Total     59         1.985098

                   Type 1 Analysis of Variance

    Source          Expected Mean Square

    part            Var(Error) + 2 Var(part*oper) + 6 Var(part)
    oper            Var(Error) + 2 Var(part*oper) + 20 Var(oper)
    part*oper       Var(Error) + 2 Var(part*oper)
    Error           Var(Error)
    Corrected Total

                        Type 1 Estimates

              Variance Component              Estimate

              Var(part)                        0.02235
              Var(oper)                       -0.0006017
              Var(part*oper)                   0.01307
              Var(Error)                       0.0007517
```

In cases where atypical values are not the cause of negative variance component estimates, the REML estimators avoid the negative values shown in the output containing the REML estimates on the next page. From these results it can be seen that most $94.3\% = 100 \times [0.01247/(0.01247+0.0007517)]$ of the measurement error is due to reproducibility. Therefore to reduce measurement error, efforts should be concentrated on better training of operator inspectors rather than investing in more precise gages.

5.5.2 Confidence Intervals on Variance Components in Two-Factor Designs

When there is an equal number of replicates per subclass, and the normality assumptions hold, the ANOVA mean squares are independently distributed as multiples of chi-square random variables. Therefore confidence intervals on any expected mean square can be obtained similar to line 1 in Table 5.3. However, except for $E(msE) = \sigma^2$, the expected value of all other mean

TWO-FACTOR SAMPLING DESIGNS

```
                   REML Estimates

        Variance
        Component              Estimate

        Var(part)              0.02255
        Var(oper)              0
        Var(part*oper)         0.01247
        Var(Error)             0.0007517
```

squares are linear combinations of two or more variance components as shown in the SAS output above. Although exact confidence intervals can be obtained on individual variance components in the balanced two-factor design following the formula on line 2 of Table 5.3, they are not applicable to all designs or unbalanced data. Burdick and Graybill (1992) show a method of computing approximate confidence intervals that is more generally applicable.

Whenever a variance component can be expressed in the form $\delta = c_1 E(ms1) - c_2 E(ms2)$ where $ms1$ and $ms2$ are mean squares in the ANOVA table and c_1 and c_2 are positive, the approximate confidence interval shown by Burdick and Graybill is applicable. For example in the ANOVA table shown above for the gage R&R study,

$$E(msPartOperator) = \sigma^2 + 2\sigma_{ab}^2$$

and

$$E(msOperator) = \sigma^2 + 2\sigma_{ab}^2 + 20\sigma_b^2,$$

therefore

$$\delta = 0.05 \times E(msOperator) - 0.05 \times E(msPartOperator) = \sigma_b^2.$$

An approximate $1 - \alpha$ confidence interval on δ is given by:

$$\left(\hat{\delta} - \sqrt{V_L}, \hat{\delta} + \sqrt{V_U}\right), \tag{5.11}$$

where $\hat{\delta} = c_1 ms1 - c_2 ms2$, ν_1 is the degrees of freedom for $ms1$, ν_2 is the degrees of freedom for $ms2$,

$$V_L = G_1^2 c_1^2 ms1^2 + H_2^2 c_2^2 ms2^2 + G_{12} c_1 c_2 (ms1)(ms2)$$
$$V_U = H_1^2 c_1^2 ms1^2 + G_2^2 c_2^2 ms2^2 + H_{12} c_1 c_2 (ms1)(ms2),$$

$$G_1 = 1 - \frac{1}{F_{\alpha,\nu_1,\infty}}$$

$$H_2 = \frac{1}{F_{1-\alpha,\nu_2,\infty}} - 1$$

$$G_{12} = \frac{(F_{\alpha,\nu_1,\nu_2} - 1)^2 - G_1^2 F_{\alpha,\nu_1,\nu_2}^2 - H_2^2}{F_{\alpha,\nu_1,\nu_2}}$$

$$H_1 = \frac{1}{F_{1-\alpha,\nu_1,\infty}} - 1$$

$$G_2 = 1 - \frac{1}{F_{\alpha,\nu_2,\infty}}$$

and

$$H_{12} = \frac{(1 - F_{1-\alpha,\nu_1,\nu_2})^2 - H_1^2 F_{1-\alpha,\nu_1,\nu_2}^2 - G_2^2}{F_{1-\alpha,\nu_1,\nu_2}}.$$

Although these formulas look formidable, they can easily be evaluated in a the SAS data step. A SAS, macro %vci, listed in the appendix of this chapter, will compute confidence intervals using these formulas. The input to the macro is a SAS data set called **meansq** that contains the variables confl=1 − α, c1=c_1, ms1=$ms1$, nu1=ν_1, c2=c_2, ms2=$ms2$, and nu2=ν_2. To see how to use this macro consider making confidence intervals on the variance component for operator, and the variance component for the part by operator interaction in the gage R&R study data.

To get a confidence interval on the variance component for operator, σ_b^2, $ms1$=0.014852 with 2 degrees of freedom, $ms2$=0.026885 with 18 degrees of freedom. These are read into a SAS data set as follows:

```
data meansq;
input confl c1 ms1 nu1 c2 ms2 nu2;
datalines;
.90 .05 .014852 2 .05 .026885 18
%vci;
```

The %vci statement calls the macro and produces the following results:

Obs	delta	L	U
1	-.00060165	-.001579783	.005717457

This shows that even though the method of moments estimator for σ_b^2 is negative, the upper 90% confidence bound is positive.

The 90% confidence interval for σ_{ab}^2 is found to be (0.008936 − 0.021895) using the same approximation formula.

TWO-FACTOR SAMPLING DESIGNS

5.5.3 Determining Sample Sizes for Two-Factor Sampling Experiments

In a two-factor sampling experiment, there are three sample sizes to be considered. First the number of levels of factor A, a, second the number of levels of factor B, b, and finally the number of replicates within each cell, r. The degrees of freedom for the replicate mean square is $ab(r-1)$. by replacing $t(r-1)$ by $ab(r-1)$ in formula (5.8), it can be used to determine the value of $ab(r-1)$ that will result in the desired width of the confidence interval for σ^2. Next using a rule of thumb like that expressed in Section 5.4.4 the levels of factor A and Factor B can be determined.

5.5.4 Two-Factor Studies with Unequal Replication

When there are unequal numbers of replicates in each subclass, the method of moments or analysis of variance estimators of the variance components for the two-factor sampling experiment are not unique. Their values will depend on whether the type I or type III sums of squares are used to get the estimates. On the other hand, the maximum likelihood estimators and REML estimators are unique and may be preferred in this situation.

Table 5.6 *Calcium in Blood Serum Solutions with Unknown Concentrations*

Laboratory	Standard Solution			
	1	2	3	4
A	87	92	179	177
		84	83	173
		80	76	166
B	80	69	138	151
		70	46	138
				132
C	70	67	173	176
		60	63	166
		44	48	

To illustrate the calculations, consider the data in Table 5.6 taken from a sampling study to estimate the sources of variability in an inter-laboratory assay of calcium in blood serum that was shown by Rao and Rao (1997). The commands to read the data and compute the variance components using `proc varcomp` are similar to those shown in the last section for balanced data. Using the method of moments-analysis of variance estimators `proc varcomp` only gives the option to use the type I sums of squares. The results of this analysis are shown on the next page.

```
              Type 1 Analysis of Variance

                             Sum of
  Source             DF     Squares      Mean Square

  lab                 2    2919.974074   1459.987037
  sol                 3      34869         11623
  lab*sol             6    1384.326453    230.721075
  Error              15      20787       1385.811111
  Corrected Total    26      59960

              Type 1 Analysis of Variance

Source          Expected Mean Square

lab             Var(Error) +2.525 Var(lab*sol) +0.0806 Var(sol)
                +8.963 Var(lab)
sol             Var(Error) +2.1979 Var(lab*sol) +6.4648 Var(sol)
lab*sol         Var(Error) +2.1335 Var(lab*sol)
Error           Var(Error)
Corrected Total

                  Type 1 Estimates

          Variance Component        Estimate

          Var(lab)                  144.91348
          Var(sol)                   1767.6
          Var(lab*sol)              -541.41218
          Var(Error)                 1385.8
```

There it can be seen that the unequal replication in the cells causes the coefficients in the expected mean squares to be non-integer values. In addition, the mean square for the lab by solution interaction is smaller than the error mean square, so the estimated variance component is negative.

The REML estimates avoid the problem of negative estimates and the results of **proc varcomp** with the option **method=reml** is shown on the page. There it can be seen that the majority of the variability in the assay for calcium in blood serum is due to differences among the standard solutions and among the repeat analyses using the same solution in the same lab.

Notice that while the variance component, σ_a^2, for lab is positive, its asymptotic variance, 23212.8, is quite large. However, since there are only three labs, an asymptotic confidence interval for σ_a^2 would not be accurate. With unequal replication the ANOVA mean squares are no longer distributed as multiples of chi-square random variables, so the confidence intervals using line 2 in Table 5.3 are also not appropriate.

Approximate confidence intervals can still be obtained if unweighted mean squares are substituted for the mean squares in Formula (5.11). The unweighted mean squares for factor A, B and the interaction can be obtained

```
                         REML Estimates
              Variance
              Component            Estimate

              Var(lab)             36.09073
              Var(sol)              1476.7
              Var(lab*sol)               0
              Var(Error)            1049.7

       Asymptotic Covariance Matrix of Estimates
                         Var(lab)             Var(sol)

Var(lab)                 23212.8               1435.9
Var(sol)                  1435.9            1772651.8
Var(lab*sol)                   0                    0
Var(Error)              -10609.4             -12397.2

       Asymptotic Covariance Matrix of Estimates
                         Var(lab*sol)         Var(Error)

Var(lab)                       0              -10609.4
Var(sol)                       0              -12397.2
Var(lab*sol)                   0                     0
Var(Error)                     0              103735.5
```

by performing an ANOVA on the cell means. Table 5.7 is a symbolic representation of the ANOVA with unweighted mean squares (and their expected values) from a two-factor sampling design.

Table 5.7 *Symbolic ANOVA with Unweighted Mean Squares*

Source	df	MS	EMS
Factor A	$a-1$	msA_U	$\sigma^2 + \bar{c}\sigma_{ab}^2 + b\bar{c}\sigma_a^2$
Factor B	$b-1$	msB_U	$\sigma^2 + \bar{c}\sigma_{ab}^2 + a\bar{c}\sigma_b^2$
Interaction AB	$(a-1)(b-1)$	$msAB_U$	$\sigma^2 + \bar{c}\sigma_{ab}^2$
Error	$\sum_i \sum_j r_{ij} - ab$	msE	σ^2

The symbol $\bar{c} = ab/\sum_i \sum_j (1/r_{ij})$ and r_{ij} is the number of replicates in the ijth cell. From this table it can be seen that $\sigma_a^2 = \left(\frac{1}{b\bar{c}}\right) msA_U - \left(\frac{1}{b\bar{c}}\right) msAB_U$ which is of the form $\delta = c_1 E(ms1) - c_2 E(ms2)$ shown in Section 5.5.2.

For the data in Table 5.6 the unweighted mean squares are obtained by the commands at the top of the next page.

```
proc sort data=bloods; by lab sol;
proc means noprint; by lab sol;
   var conc;
   output out=uss mean=yijbar;
proc glm data=uss;
   class lab sol;
   model yijbar=lab sol lab*sol;
run;
```

which result in $msA_U = 872.68056$, $msAB_U = 638.93056$. The factor

$$\bar{c} = ab / \sum_i \sum_j (1/r_{ij})$$

$$= \frac{(3)(4)}{\left(\frac{1}{1} + \frac{1}{3} + \frac{1}{3} + \frac{1}{3} + \frac{1}{1} + \frac{1}{2} + \frac{1}{2} + \frac{1}{3} + \frac{1}{1} + \frac{1}{3} + \frac{1}{3} + \frac{1}{2}\right)}$$

$$= 1.846,$$

and $c_1 = c_2 = \frac{1}{b\bar{c}} = \frac{1}{4(1.846)} = 0.13541$. Thus the data needed in the SAS data file `meansq` to calculate a 90% confidence interval for σ_a^2 is:

```
data meansq;
input confl c1 ms1 nu1 c2 ms2 nu2;
datalines;
.90 .1354166 872.68056 2 .1354166 638.93056 6
%vci;
```

and the resulting confidence interval is (-124.862, 1035.82). Therefore, although the point estimates of σ_a^2 are positive, the confidence interval shows it is not significantly different than zero.

5.6 Nested Sampling Experiments (NSE)

Many sampling experiments with more than one factor use nested factors or a hierarchical design. The levels of a *nested factor* are physically different depending on the level of factor it is nested within. That wasn't the case for the factors described in the last section. For example, in the gage R&R study, each operator measured each part; therefore, the operator number was uniquely defined and referred to the same operator regardless of which part he measured. We would call the operator and part in the gage R&R study *crossed factors*. To change the design so that operator was a nested factor, consider an experiment where n parts were selected and each part was measured by two operators, although it doesn't have to be the same two operators measuring each part. This might be a more convenient way to conduct the sampling experiment if parts were selected over a long period of time and the same

NESTED SAMPLING EXPERIMENTS (NSE)

operators were not always present to make the measurements. The fact that the operators differ depending upon the part number being measured makes the operator a nested factor (nested within part). The first operator measuring the first part is not physically the same person as the first operator measuring a subsequent part. Another common example where a nested design occurs is when the measurements are destructive. In that case, each operator must measure a different set of parts, and the part becomes the nested factor (nested within operator) because the first part measured by the first operator is not physically the same as the first part measured by subsequent operators. This type of sampling design is called a nested sampling experiment or NSE.

One example where we have already seen nested factors is in the term ϵ_{ij} in the models we have used thus far. It represents the effect of the jth replicate experimental unit, and since different experimental units are used for each factor level or combination of factor levels, the experimental unit is always nested within another factor level or cell in the design.

When two factors are crossed factors we can include their interaction in the model to represent the extra variability caused by changes in the level of one factor between levels of the other factor. However, if a factor is nested within another factor we cannot include an interaction between them because the nested factor includes the degrees of freedom that could be taken by the interaction. The model for a two-stage nested design with factor B nested within factor A is written as:

$$y_{ijk} = \mu + a_i + b_{(i)j} + \epsilon_{ijk} \tag{5.12}$$

and if there is an equal number of replicates, r, per cell, the ANOVA table for the nested model can be represented as:

Table 5.8 Symbolic ANOVA for Two-Factor Nested Design

Source	df	MS	EMS
Factor A	$a-1$	msA	$\sigma^2 + r\sigma_b^2 + br\sigma_a^2$
Factor B	$a(b-1)$	msB	$\sigma^2 + r\sigma_b^2$
Error	$ab(r-1)$	msE	σ^2

Here it can be seen that the degrees of freedom for Factor B,

$$a(b-1) = (b-1) + (a-1)(b-1)$$

is equal to the degrees of freedom for a crossed factor plus the degrees for the interaction AB.

Nested or hierarchical designs can easily be extended to include several stages or factors. For example Table 5.9 shows the results of a four-stage nested sampling study on the variability of properties of crude rubber, taken from Bennett and Franklin (1954). In this study a sample of four batches of rubber was taken from each of four suppliers. Since the first batch obtained from the first supplier is not physically the same as the first batch taken from the second supplier, batch is nested within supplier. The next two sample mixes were made from each batch, and since the two sample mixes for one batch are physically different than the sample mixes for any other batch, the sample mix is nested within batch. Finally three replicate tests were performed on each sample mix to determine the elasticity.

The model for the data can be written as:

$$y_{ijkl} = \mu + a_i + b_{(i)j} + c_{(ij)k} + \epsilon_{ijkl}$$

where y_{ijkl} is the lth elasticity determination made from the kth sample mix, taken from the jth batch from the ith supplier, a_i is the random supplier effect, $b_{(i)j}$ is the random batch effect, $c_{(ij)k}$ is the random sample mix effect and ϵ_{ijkl} is the random replicate determination effect, $i = 1, \ldots, 4$, $j = 1, \ldots, 4$, $k = 1, \ldots, 2$ and $l = 1, \ldots, 3$.

This model can be written in the notation of SAS `proc glm` or `proc varcomp` as

```
class supplier batch sample;
model elasticity = supplier batch(supplier) sample(batch);
```

and the variance components σ_a^2, σ_b^2, σ_c^2, and σ^2 can be estimated using the method of moments or REML.

In order to increase confidence in estimates of variance components, the number of levels of a random factor should be increased. However, in hierarchical designs with several stages, increasing the number of levels of the topmost factor greatly increases the overall sample size, even if all the other nested factors have only two levels. For example in the design shown in Table 5.9, if the number of suppliers was increased from 4 to 20 in order to get a more precise estimate of σ_a^2, the number of determinations that would have to be made would increase from the 96 shown in Table 5.9 to 480. Even if the number of batches per supplier and the number of sample mixes per batch and determinations per mix were reduced to 2 each, there would still be $20 \times 2 \times 2 \times 2 = 160$ determinations. If the sampling study had been done in this way there would be $a - 1 = 19$ degrees of freedom for the supplier effect, $a(b-1) = 20(2-1) = 20$ degrees of freedom for the batch effect, $ab(c-1) = 20 \times 2(2-1) = 40$ degrees of freedom for the sample effect and $abc(r-1) = 20(2)(2)(2-1) = 80$ degrees of freedom for the random replicate effect. Therefore the majority of the 160 observations are used to increase the precision of the bottom two variance components σ^2 and σ_c^2. For this reason balanced hierarchical designs are usually not recommended if there are more than three stages or sources of variability being studied. Staggered nested designs presented in the next section allow the convenience of nested factors in sampling studies, but allow the various variance components to be estimated with more uniform precision.

Table 5.9 *Modulus of Elasticity at 700% Elongation of 96 Prepared Specimens of Smoked Sheet Rubber*

Supplier	A		B		C		D	
	Sample Mix		Sample Mix		Sample Mix		Sample Mix	
	1	2	1	2	1	2	1	2
Batch I	211	171	196	196	200	240	323	262
	215	198	186	210	221	229	279	234
	197	268	190	156	198	217	251	249
Batch II	229	234	209	200	191	196	255	249
	196	210	193	186	189	198	235	247
	200	226	204	196	186	175	223	239
Batch III	204	225	204	174	211	196	228	262
	221	215	165	172	197	184	250	227
	238	196	194	171	210	190	260	272
Batch IV	229	248	198	202	196	180	273	273
	250	249	209	211	197	166	241	256
	238	249	221	204	186	172	221	230

5.7 Staggered Nested Designs

Staggered nested sampling experiments, or SNSE, were developed independently by Bainbridge (1965) and Prairie and Anderson (1962). In a completely nested design as discussed in the last section, each level of the topmost factor leads down into two (or more) levels of each succeeding factor or stage. In a staggered nested design, on the other hand, only one of the two levels of the succeeding factor leads to the next two-level stage. Figure 5.1 illustrates the difference between a nested and staggered nested design for three stages. If there are a levels of the topmost factor, the nested design requires $4a$ total observations while the staggered nested design only requires $3a$ observations.

The savings in observations are multiplied as the number of stages in a nested design increases. Figure 5.2 shows the schematic for staggered nested designs from three through six stages.

While the information or degrees of freedom available for estimating variance components in a completely nested design is concentrated in the lower tier factors, the information is balanced in a staggered nested design. Table 5.10 compares the degrees of freedom distribution between a staggered nested

Figure 5.1 *Comparison of 3 Stage Nested and 3 Stage Staggered Nested Design*

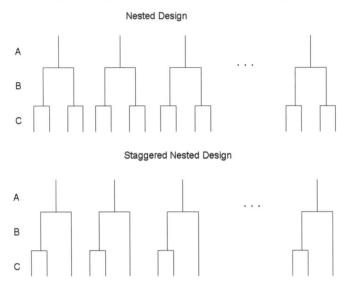

Figure 5.2 *Staggered Nested Designs for 3 to 6 Stages*

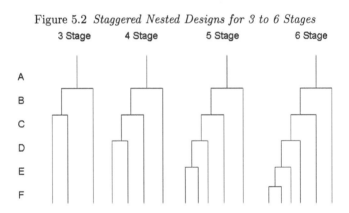

design and a completely nested design where each factor except the topmost has only two levels.

STAGGERED NESTED DESIGNS

Table 5.10 *Comparison of Degrees of Freedom between Staggered Nested and Nested Designs*

Source	Staggered Nested df	Nested df
A	$a-1$	$a-1$
B in A	a	a
C in B	a	$2a$
D in C	a	$4a$
E in D	a	$8a$
F in E	a	$16a$

Mason *et al.* (1989) described a study where a staggered nested design was used to estimate the sources of variability in a continuous polymerization process. In this process polyethylene pellets are produced in lots of one hundred thousand pounds. A four-stage design was used to partition the source of variability in tensile strength between lots, within lots and due to the measurement process. Thirty lots were sampled at random. Lot represented the topmost factor or source of variability A. From each lot two boxes of pellets were randomly selected. This represented the second stage or source of variability B. From the first box selected from each lot, two preparations were made for strength testing, but from the second box selected from each lot only one preparation was made. This represented the third stage or source of variability C. Finally two repeat strength tests were made from the first preparation from box one, while only one strength test was made from the other three preparations. This sampling scheme is diagramed in Figure 5.3, and the data is shown in Table 5.11 on the next page.

Figure 5.3 *Diagram of Sampling Scheme for Polymerization Study*

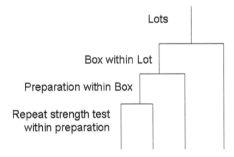

Table 5.11 *Data from Polymerization Strength Variability Study*

	Box 1			Box 2
	Preparation			Preparation
	1		2	1
Lot	test 1	test 2	test 1	test 1
1	9.76	9.24	11.91	9.02
2	10.65	7.77	10.00	13.69
3	6.50	6.26	8.02	7.95
4	8.08	5.28	9.15	7.46
5	7.84	5.91	7.43	6.11
6	9.00	8.38	7.01	8.58
7	12.81	13.58	11.13	10.00
8	10.62	11.71	14.07	14.56
9	4.88	4.96	4.08	4.76
10	9.38	8.02	6.73	6.99
11	5.91	5.79	6.59	6.55
12	7.19	7.22	5.77	8.33
13	7.93	6.48	8.12	7.43
14	3.70	2.86	3.95	5.92
15	4.64	5.70	5.96	5.88
16	5.94	6.28	4.18	5.24
17	9.50	8.00	11.25	11.14
18	10.93	12.16	9.51	12.71
19	11.95	10.58	16.79	13.08
20	4.34	5.45	7.51	5.21
21	7.60	6.72	6.51	6.35
22	5.12	5.85	6.31	8.74
23	5.28	5.73	4.53	5.07
24	5.44	5.38	4.35	7.04
25	3.50	3.88	2.57	3.76
26	4.80	4.46	3.48	3.18
27	5.35	6.39	4.38	5.50
28	3.09	3.19	3.79	2.59
29	5.30	4.72	4.39	6.13
30	7.09	7.82	5.96	7.14

The SAS code to read the data and compute the variance components is shown on the next page, and the output from SAS `proc varcomp` is below it.

STAGGERED NESTED DESIGNS

```
data polymer;
input lot box prep strength;
datalines;
 1 1 1 9.76
 1 1 1 9.24
 ... etc.
proc varcomp method=type1;
  class lot box prep;
  model strength=lot box(lot) prep(box lot);
run;
```

Type 1 Analysis of Variance

Source	DF	Sum of Squares	Mean Square
lot	29	855.957534	29.515777
box(lot)	30	50.094525	1.669818
prep(lot*box)	30	68.437450	2.281248
Error	30	19.438750	0.647958
Corrected Total	119	993.928259	

Type 1 Analysis of Variance

Source	Expected Mean Square
lot	Var(Error) + 1.5 Var(prep(lot*box)) + 2.5 Var(box(lot)) + 4 Var(lot)
box(lot)	Var(Error) + 1.1667 Var(prep(lot*box)) + 1.5 Var(box(lot))
prep(lot*box)	Var(Error) + 1.3333 Var(prep(lot*box))
Error	Var(Error)
Corrected Total	

Type 1 Estimates

Variance Component	Estimate
Var(lot)	6.92729
Var(box(lot))	-0.27151
Var(prep(lot*box))	1.22497
Var(Error)	0.64796

Once again the method of moments procedure produces a negative estimate for one of the sources. Changing the option method=type1 to method=reml results in the more reasonable REML estimates shown on the next page.

```
                    REML Estimates
           Variance Component          Estimate
                Var(lot)                7.24270
                Var(box(lot))                 0
                Var(prep(lot*box))      1.02957
                Var(Error)              0.65680

           Asymptotic Covariance Matrix of Estimates
                                    Var(lot)          Var(box(lot))
Var(lot)                             4.16206                      0
Var(box(lot))                              0                      0
Var(prep(lot*box))                  -0.01093                      0
Var(Error)                        -0.0080837                      0

           Asymptotic Covariance Matrix of Estimates
                              Var(prep(lot*box))         Var(Error)
Var(lot)                               -0.01093         -0.0080837
Var(box(lot))                                 0                  0
Var(prep(lot*box))                      0.10053           -0.02395
Var(Error)                             -0.02395            0.02933
```

From these results we see that 81% $=(100\times(7.2427)/(7.2427+1.0296+0.6568))$ of the total variation is due to variability among lots, while the within lot or box-to-box variability is negligible. Therefore, if the manufacturers would like to decrease the variability in tensile strength, they should focus on reducing the variation on influential factors that change between lots.

While no exact closed form confidence intervals have been developed for variance components estimated from data in staggered nested designs, when the number of levels of the topmost factor is greater than 30, approximate asymptotic estimates can be created using elements of the asymptotic covariance matrix of estimates produced by `proc varcomp` when the `method=ml` or `method=reml` estimates are used. For example, a 90% confidence interval on σ_a^2 the variance component due to lots is $7.2427 \pm 1.645\sqrt{4.16206} =$ (3.887; 10.598). Bayesian interval estimates of variance components are also useful for these designs (see Lawson, 2008b).

To determine the sample size for a staggered nested design, follow the procedure outlined in Section 5.4.4 for determining the sample size for estimating the replicate variance in a one-factor sampling design. Since the degrees of freedom for all factors or stages in the staggered nested design are near equal, following this procedure will give you approximately the same precision on all variance components.

5.8 Designs with Fixed and Random Factors

In some cases where fixed treatment factors are being studied in an experimental design like those discussed in Chapters 2 through 4, random factors are also introduced in the model by the way the experiment is conducted. For example consider the data in Table 5.12 that resulted from an experiment comparing different formulations and methods of applying a pesticide to the leaves of cotton plants. The goal was to increase the amount of active pesticide remaining on cotton plant leaves one week after application. The pesticide being studied degrades in sunlight and a certain additive to the formulation retards this process. Different application techniques may differ in the amount of pesticide delivered to the plant leaves. The treatment factors in this experiment were two different formulations of the pesticide and two different application methods, resulting in a 2^2 factorial experiment.

The experimental unit was a 20' row of cotton plants called a plot, because this was a convenient area within which the application of pesticide could be controlled. Eight plots were selected and two were randomly assigned to each of the four treatment combinations, resulting in two replicates per treatment combination. One week after application, the experimenters were ready to determine the pesticide residue remaining on the plant leaves. However, there was too much plant material in an entire plot to send to the lab for analysis. Therefore, two samples of leaves in an amount convenient for laboratory analysis of pesticide residues were selected from each plot. Each sample was sent to the lab resulting in the data shown in Table 5.12.

Formulation, application technique and their interaction are fixed factors because the experimenters were interested in comparing the average response between levels of these factors. The plot, on the other hand, is a random factor that represents differences in experimental units. It is nested within the combinations of formulation by application technique. There is no interest in comparing experimental units within each combination of formulation and application. Instead multiple plots per treatment combination were included in the design so that the variance caused by differing plots could be estimated and used to judge significance of formulation and application effects. The replicate samples taken from each plot were for convenience in conducting the experiment. They would be classified as sub-samples or observational units defined in Chapter 1.

The simplest way of analyzing the data would be to average the two sub-samples and proceed as illustrated in Section 3.5. However, if the sub-samples can be assumed independent and it is desirable to include all the data (shown in Table 5.12) in the analysis, then an additional term for sample must be included in the model. Sample is another random effect since there is no specific interest in comparing the response between the two samples from each plot.

The model for the data can be written in the form

$$y_{ijkl} = \mu + \alpha_i + \beta_j + \alpha\beta_{ij} + p_{(ij)k} + \epsilon_{ijkl}, \qquad (5.13)$$

Table 5.12 *Pesticide Residue on Cotton Plants*

Formulation	Application Technique	Plot	Sample 1	Sample 2
A	1	1	0.237	0.252
A	1	2	0.281	0.274
B	1	1	0.247	0.294
B	1	2	0.321	0.267
A	2	1	0.392	0.378
A	2	2	0.381	0.346
B	2	1	0.351	0.362
B	2	2	0.334	0.348

where y_{ijkl} is the pesticide residue found on the lth sample taken from the kth plot, treated with formulation level i and application technique j. In general $i = 1, \ldots, a$, $j = 1, \ldots, b$, $k = 1, \ldots, r$, and $l = 1, \ldots, s$. In this specific example, $a = 2$, $b = 2$, $r = 2$, and $s = 2$, α_i is the formulation effect, β_j is the application effect, $\alpha\beta_{ij}$ is the interaction effect, $p_{(ij)k}$ is the random plot effect, and ϵ_{ijkl} is the random sample effect.

The model can be written in matrix notation as:

$$y = X\beta + Z\gamma + \epsilon, \tag{5.14}$$

where

$$y = \begin{pmatrix} y_{1111} \\ y_{1112} \\ y_{1121} \\ y_{1122} \\ y_{2111} \\ y_{2112} \\ y_{2121} \\ y_{2122} \\ y_{1211} \\ y_{1212} \\ y_{1221} \\ y_{1222} \\ y_{2211} \\ y_{2212} \\ y_{2221} \\ y_{2222} \end{pmatrix}, \quad X = \begin{pmatrix} 1 & 1 & 0 & 1 & 0 & 1 & 0 & 0 & 0 \\ 1 & 1 & 0 & 1 & 0 & 1 & 0 & 0 & 0 \\ 1 & 1 & 0 & 1 & 0 & 1 & 0 & 0 & 0 \\ 1 & 1 & 0 & 1 & 0 & 1 & 0 & 0 & 0 \\ 1 & 0 & 1 & 1 & 0 & 0 & 1 & 0 & 0 \\ 1 & 0 & 1 & 1 & 0 & 0 & 1 & 0 & 0 \\ 1 & 0 & 1 & 1 & 0 & 0 & 1 & 0 & 0 \\ 1 & 0 & 1 & 1 & 0 & 0 & 1 & 0 & 0 \\ 1 & 1 & 0 & 0 & 1 & 0 & 0 & 1 & 0 \\ 1 & 1 & 0 & 0 & 1 & 0 & 0 & 1 & 0 \\ 1 & 1 & 0 & 0 & 1 & 0 & 0 & 1 & 0 \\ 1 & 1 & 0 & 0 & 1 & 0 & 0 & 1 & 0 \\ 1 & 0 & 1 & 0 & 1 & 0 & 0 & 0 & 1 \\ 1 & 0 & 1 & 0 & 1 & 0 & 0 & 0 & 1 \\ 1 & 0 & 1 & 0 & 1 & 0 & 0 & 0 & 1 \\ 1 & 0 & 1 & 0 & 1 & 0 & 0 & 0 & 1 \end{pmatrix}, \quad \beta = \begin{pmatrix} \mu \\ \alpha_1 \\ \alpha_2 \\ \beta_1 \\ \beta_2 \\ \alpha\beta_{11} \\ \alpha\beta_{21} \\ \alpha\beta_{12} \\ \alpha\beta_{22} \end{pmatrix},$$

DESIGNS WITH FIXED AND RANDOM FACTORS

$$Z = \begin{pmatrix} 1 & 0 & 0 & 0 & 0 & 0 & 0 & 0 \\ 1 & 0 & 0 & 0 & 0 & 0 & 0 & 0 \\ 0 & 1 & 0 & 0 & 0 & 0 & 0 & 0 \\ 0 & 1 & 0 & 0 & 0 & 0 & 0 & 0 \\ 0 & 0 & 1 & 0 & 0 & 0 & 0 & 0 \\ 0 & 0 & 1 & 0 & 0 & 0 & 0 & 0 \\ 0 & 0 & 0 & 1 & 0 & 0 & 0 & 0 \\ 0 & 0 & 0 & 1 & 0 & 0 & 0 & 0 \\ 0 & 0 & 0 & 0 & 1 & 0 & 0 & 0 \\ 0 & 0 & 0 & 0 & 1 & 0 & 0 & 0 \\ 0 & 0 & 0 & 0 & 0 & 1 & 0 & 0 \\ 0 & 0 & 0 & 0 & 0 & 1 & 0 & 0 \\ 0 & 0 & 0 & 0 & 0 & 0 & 1 & 0 \\ 0 & 0 & 0 & 0 & 0 & 0 & 1 & 0 \\ 0 & 0 & 0 & 0 & 0 & 0 & 0 & 1 \\ 0 & 0 & 0 & 0 & 0 & 0 & 0 & 1 \end{pmatrix}, \quad \gamma = \begin{pmatrix} p_{(11)1} \\ p_{(11)2} \\ p_{(21)1} \\ p_{(21)2} \\ p_{(12)1} \\ p_{(12)2} \\ p_{(22)1} \\ p_{(22)2} \end{pmatrix}, \quad \epsilon = \begin{pmatrix} \epsilon_{1111} \\ \epsilon_{1112} \\ \epsilon_{1121} \\ \epsilon_{1122} \\ \epsilon_{2111} \\ \epsilon_{2112} \\ \epsilon_{2121} \\ \epsilon_{2122} \\ \epsilon_{1211} \\ \epsilon_{1212} \\ \epsilon_{1221} \\ \epsilon_{1222} \\ \epsilon_{2211} \\ \epsilon_{2212} \\ \epsilon_{2221} \\ \epsilon_{2222} \end{pmatrix}.$$

β represents the vector of fixed effects, while γ and ϵ represent the vectors of random effects. The assumption of independent and normally distributed random effects can be expressed by $\gamma \sim MVN(\mathbf{0}, \sigma_p^2 \mathbf{I}_{abr})$, and $\epsilon \sim MVN(\mathbf{0}, \sigma^2 \mathbf{I}_{abrs})$; therefore, $y \sim MVN(X\beta, V)$, where $V = Z(\sigma_p^2 \mathbf{I}_{abr})Z' + \sigma^2 \mathbf{I}_{abrs}$. The least squares estimator of β would be $\hat{\beta} = (X'V^{-1}X)^{-}X'V^{-1}y$; however, σ_p^2, σ^2 are unknown, and therefore V is unknown.

SAS `proc glm` solves the problem by treating both β and γ as fixed. Then the least squares estimates are given by

$$\begin{pmatrix} \hat{\beta} \\ \hat{\gamma} \end{pmatrix} = \begin{pmatrix} X'X & X'Z \\ Z'X & Z'Z \end{pmatrix}^{-} \begin{pmatrix} X'y \\ Z'y \end{pmatrix}. \qquad (5.15)$$

The type I ANOVA sums of squares for each term in the model are similar to those shown in Table 3.2. The commands to read the data from Table 5.12, and analyze it using SAS `proc glm` are shown below. For this example the type I and type III sums of squares will be the same since there is equal replication and sub-samples.

```
data pesticide;
input form $ app plot sample residue;
datalines;
A 1 1 1 .237
A 1 1 2 .252
  ... etc.
proc glm;
  class form app plot;
  model residue=form app form*app plot(form*app);
  random plot(form*app)/test;
run;
```

The `random plot(form*app)/test;` statement tells `glm` that plot is a random effect, and the result is a table of the expected mean squares shown below. Here the coefficients for `Var(Error)`=σ^2 and `Var(plot(form*app)`=σ_p^2 were determined using Hartley's method described in Section 5.4.1, and the terms of the form `Q(form,form*app)` are the way `proc glm` indicates a quadratic form involving the fixed effects that are functions of the non-centrality parameters shown in Section 3.5.2.

```
Source            Type III Expected Mean Square
form              Var(Error) +2Var(plot(form*app)) +Q(form,form*app)
app               Var(Error) +2Var(plot(form*app)) +Q(app,form*app)
form*app          Var(Error) +2Var(plot(form*app)) +Q(form*app)
plot(form*app)    Var(Error) +2Var(plot(form*app))
```

Since the expected mean squares for formulation, application and their interaction all contain $\sigma^2 + 2\sigma_p^2$ in addition to a quadratic form involving the fixed effects, the correct mean square to use for the denominator of the F-ratio for testing these fixed effects is `plot(form*app)` whose expectation is $\sigma^2+2\sigma_p^2$. The expected value of the error mean square is σ^2, and it is too small for use as a denominator in the F-ratio for testing the fixed effects. However, this is exactly what is used by the `proc glm` type I and type III ANOVA tables. They are incorrect and they are not shown here.

Using the plot mean square to test the fixed effects is justified and is equivalent to averaging the subsamples and analyzing as illustrated in Section 3.5. The `/test` option on the random statement shown above causes `proc glm` to compute the correct F-ratios as shown in the output below. There we see that application method did have a significant effect on the amount of active pesticide residue on the cotton plant leaves one week after application. However, neither the formulation nor the interaction of formulation and application had a significant effect.

```
  Source          DF    Type III SS    Mean Square    F Value    Pr > F
* form             1    0.000018063    0.000018063       0.03    0.8692
* app              1       0.032310       0.032310      55.14    0.0018
  form*app         1       0.002186       0.002186       3.73    0.1256

  Error            4       0.002344       0.000586
Error: MS(plot(form*app))
* This test assumes one or more other fixed effects are zero.


  Source                 DF    Type III SS    Mean Square    F Value    Pr > F
  plot(form*app)          4       0.002344       0.000586       1.31    0.3432
  Error: MS(Error)        8       0.003569       0.000446
```

In many cases where observational units are different than the experimental units, the observational units will not be independent. In this example, common application of the pesticide to each plot might induce a correlation,

$E(\epsilon_{ijkl} \times \epsilon_{ijkl'}) = \rho$, between sub-samples from the same plot. Even though the independence assumption is violated, Casella (2008) shows the F-test on fixed effects using the plot mean square as the denominator is still valid.

In this example where the application method was found significant, we should look at the least squares means or in this case the marginal means for application (since there is equal replication) in order to determine which application method was best. The marginal means can be expressed as $\bar{y}_{.1..}$ and $\bar{y}_{.2..}$. The expected value $E(\bar{y}_{.j..}) = \mu + \beta_j + \bar{\alpha\beta}_{.j}$, the variance $Var(\bar{y}_{.j..}) = \sigma_p^2/ars + \sigma^2/ars$, and the variance of the difference in two marginal means would be $2(\sigma_p^2/ars + \sigma^2/ars)$. However, the standard error of the difference in means reported by the estimate statements in proc glm is $\sqrt{2\sigma^2/ars}$. proc glm does not estimate the variance component for σ_p^2, since plot was treated as a fixed effect in the calculations. A similar problem occurs with all the standard errors of estimable functions of fixed effects in the model when using the estimate statements in proc glm.

When there are both fixed and random effects in the model, due to the way the experiment was conducted, we call it a mixed model and SAS proc mixed is a better option for analysis. By default proc mixed uses the REML method to estimate the variance components for the random effects in the model, and then estimates the fixed effects using the formula $\hat{\boldsymbol{\beta}} = (\boldsymbol{X}'\hat{\boldsymbol{V}}^{-1}\boldsymbol{X})^{-}\boldsymbol{X}'\hat{\boldsymbol{V}}^{-1}\boldsymbol{y}$. The correct standard errors for estimable functions are produced by proc mixed, and the correct F-ratios for testing the fixed effects; however, as when using proc varcomp with the REML option, no ANOVA table is produced.

The commands to run proc mixed with the data in Table 5.12 are shown below.

```
proc mixed data=pesticide;
  class form app plot;
  model residue=form app form*app;
  random plot(form*app)/solution;
  lsmeans app;
  estimate 'application effect' app -1 1;
run;
```

These commands are similar to the commands for proc glm shown earlier, except the random effect plot(form*app) is removed from the model statement. It is still shown in the random statement. The /test option is also removed from the random statement since proc mixed automatically calculates the correct F-ratios for the fixed effects.

The result of the REML estimation is shown on the next page, and we see that $\hat{\sigma}_p^2 = 0.000070$ and $\hat{\sigma}^2 = 0.000446$.

```
          The Mixed Procedure

         Covariance Parameter
              Estimates

       Cov Parm          Estimate

       plot(form*app)    0.000070
       Residual          0.000446
```

The F-ratios for testing the fixed effects are shown below, and as with the `proc glm` output we see the application effect is significant.

```
              Type 3 Tests of Fixed Effects

                    Num    Den
       Effect       DF     DF     F Value    Pr > F

       form         1      4      0.03       0.8692
       app          1      4      55.14      0.0018
       form*app     1      4      3.73       0.1256
```

The result of the `estimate` and `lsmeans` statement is shown below. Here we see the standard error of the difference in the two application means is estimated correctly as $\sqrt{2(\hat{\sigma}_p^2/ars + \hat{\sigma}^2/ars)} = \sqrt{2 \times (0.000070/8 + 0.000446/8)} = 0.08987$, and that application method 2 caused a higher residue of active pesticide on the cotton plant leaves.

```
                            Estimates

                                Standard
Label                 Estimate  Error     DF   t Value   Pr > |t|

application effect    0.08987   0.01210   4    7.43      0.0018


                        Least Squares Means

                                Standard
Effect    app    Estimate       Error     DF   t Value   Pr > |t|

app       1      0.2716         0.008558  4    31.74     <.0001
app       2      0.3615         0.008558  4    42.24     <.0001
```

DESIGNS WITH FIXED AND RANDOM FACTORS

Another design whose model contains both fixed and random effects is the randomized block designs discussed in the last chapter. For example consider the example from *Golf Magazine* discussed in Section 4.6. In that experiment the treatment factor (tee height) can be considered to be a fixed effect, since the object of the experiment was to determine if there is any difference in driving distance caused by different tee heights. However, there was no interest in comparing the specific golfers used in the experiment since they just denoted a sample of representative golfers. Therefore the block factor, or golfer, can be considered to be a random effect. The interaction between a random and fixed effect is also defined to be a random effect. Therefore the interaction of golfer and tee height should also be considered a random effect, and the model for golf experiment can be written as:

$$y_{ijk} = \mu + b_i + \tau_j + b\tau_{ij} + \epsilon_{ijk}, \qquad (5.16)$$

where b_i represents the random golfer effect, τ_j represents the fixed tee height effect, $b\tau_{ij}$ represents the random interaction effect, and ϵ_{ijk} represents the random repeat hit effect, $i = 1, \ldots, 3$, $j = 1, \ldots, 9$, and $k = 1, \ldots, 5$. By assuming the random golfer, interaction, and repeat hit effects to be independent and normally distributed with variances σ_b^2, $\sigma_{b\tau}^2$ and σ^2, respectively, the expected value and variance of y_{ijk} are given by:

$$E(y_{ijk}) = \mu + \tau_i$$
$$Var(y_{ijk}) = \sigma_b^2 + \sigma_{b\tau}^2 + \sigma^2$$

The expected values for the ANOVA mean squares are given in Table 5.13, where $Q(\tau) = 8\sum_{i=1}^{3}(\tau_i - \bar{\tau}_.)^2/(3-1)$.

Table 5.13 *Expected Mean Squares for Randomized Block Design with Replicates within a Block*

Source	df	Expected Mean Square
Golfer (b_i)	(9-1)=8	$\sigma^2 + 5\sigma_{b\tau}^2 + 3\sigma_b^2$
Tee Height (τ_j)	(3-1)=2	$\sigma^2 + 5\sigma_{b\tau}^2 + Q(\tau)$
Golfer×Tee Height ($b\tau_{ij}$)	8×2=16	$\sigma^2 + 5\sigma_{b\tau}^2$
Repeat (ϵ_{ijk}=Error)	(5-1)×16=64	σ^2

These expected mean squares show that to test the hypothesis of no difference in tee height effect, i.e., $H_0 : \tau_1 = \tau_2 = \tau_3$, the correct denominator for the F-ratio would be the interaction Golfer×Tee Height, as used in Section 4.6, and not the error mean square that is used by default in the `proc glm` type I and type III ANOVA tables. Also the variance of a difference in marginal or least squares means for different tee heights $Var(\bar{y}_{i..} - \bar{y}_{i'..}) = 2\left(\frac{\sigma_{b\tau}^2}{45} + \frac{\sigma^2}{45}\right)$. By default `proc glm` does not estimate $\sigma_{b\tau}^2$. Therefore in Chapter 4, the `test h=teehgt e=id*teehgt` statement was used in `proc glm` was used in order to get the correct F-test for tee height. Another way of getting the correct F-test would be to use the `random id id*teehgt/test;` statement in `proc glm` or use `proc mixed`. In Chapter 4, the `lsmeans teehgt/pdiff adjust=tukey e=id*teehgt` statement was used in order to get correct standard error for Tukey comparison of means after the ANOVA. Using `proc mixed` the `lsmeans teehgt/pdiff adjust=tukey;` automatically uses the correct standard error for the difference in tee height means.

5.9 Graphical Methods to Check Model Assumptions

Graphical analysis of data from sampling experiments to study variances are useful for checking assumptions of the statistical models and identifying atypical observations which may have a heavy influence on the resulting variance component estimates. Snee (1983) explains that another important advantage of graphical analysis is that it forces the analyst to become more familiar with the data, and think critically about the mechanisms to explain the observed variation. He suggests simple graphical tools such as half normal plots, gamma plots in the analysis of nested sampling designs. Normal plots can also be used to check the normality assumption for random effects.

5.9.1 Simple Plots

Consider again the data in Table 5.2. Model (5.1) is appropriate for the data from this experiment, where the random sample effects, t_i, and the random experimental errors, ϵ_{ij}, are assumed to be normally distributed. The assumption of constant variance of ϵ_{ij}s across the various samples is also implied by $\epsilon_{ij} \sim N(0, \sigma^2)$. A simple way to verify these assumptions is to make a simple plot of the data like that shown in Figure 2.3. There, nonconstant variances or outlier observations can be quickly detected. The normality of the ϵ_{ij} can also be checked with a normal probability plot like that shown in Figure 2.5.

For the gage R&R two-factor sampling experiment presented in Section 5.5 using the method of moments resulted in a negative variance component estimate for σ_b^2 the variance due to operators. This estimate was negative because the mean square for interaction of part by operator was larger than the mean square for operator. A simple interaction plot like Figures 3.4 or 3.6 can facilitate an explanation of this result. Figure 5.4 shows the interaction plot for part by operator. An interaction is characterized by the fact that the trace of the average response across the levels of one factor plotted separately for each level of the other factor will not be parallel. In Figure 5.4, it can be seen that the line segments are close to parallel, except for the segments joining parts 5 through 7 and 9 through 10. This is due to the fact that operator 1 had a much lower average measurement on part number 6 than the other two operators, and operator 3 had a much higher measurement on part number 10 than the other two operators.

When the interaction mean square is largely due to one or two observations,

GRAPHICAL METHODS TO CHECK MODEL ASSUMPTIONS 187

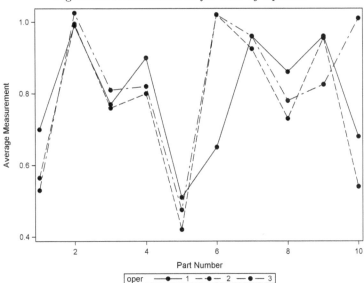

Figure 5.4 *Interaction Plot for Part by Operator*

Table 5.14 *Comparison of Method of Moments and REML Estimates on Gage R&R Study after Eliminating Parts 6 and 10*

Component	Method of Moments Estimator	REML Estimator
part (σ_a^2)	0.03191	0.02808
oper (σ_b^2)	0.0008601	0.0008089
part*oper (σ_{ab}^2)	0.0020045	0.0020082
Error (σ^2)	0.0004062	0.0004063

as it is in this case, it might be wise to question the results. Table 5.14 shows the variance component estimates from the data from the gage R&R study after eliminating parts 6 and 10. With these parts eliminated, the method of moments does not produce a negative variance component estimate, and the results obtained from the method of moments and REML are quite similar. In a situation like this, if the parts are still available, it might be wise to have the operators remeasure parts 6 and 10 to see if there were mistakes.

5.9.2 Gamma and Half-Normal Plots to Check Constant Variance Assumption

Wilk et al. (1962) proposed gamma probability plots as effective tools for investigating homogeneity of groups of variances. Sample of variances, with ν degrees of freedom, follow the gamma distribution with shape parameter $\nu/2$,

188   DESIGNS TO STUDY VARIANCES

and thus the points on a gamma probability plot sample variances that are estimating the same constant σ^2 should appear as a straight line. For many nested sampling designs and staggered nested sampling designs involve only samples of size two. In this case half-normal plots of standard deviations can be used in the same way. To illustrate this, consider again the data in Table 5.2. The SAS commands below compute the variances within each sample, output them to a SAS file s, use `proc rank` and the data step function `gaminv` to compute the gamma quantiles gam. Next `proc sgplot` is used to make the gamma probability plot. It is shown in Figure 5.5.

```
ods graphics on;
proc sort data=naph; by sample;
proc means noprint; by sample; var cYield;
   output out=s var=cvar;
proc rank data=s out=ranks; var cvar; ranks r;
data gam; set ranks;
   gscore=gaminv((r-.5)/6,2);
proc sgplot data=gam;
  scatter x=gscore y=cvar;
  xaxis label='Gamma Score';
  yaxis label='Sample Variance';
run;
ods graphics off;
```

Figure 5.5 *Gamma Plot of within Sample Variances from Data in Table 5.2*

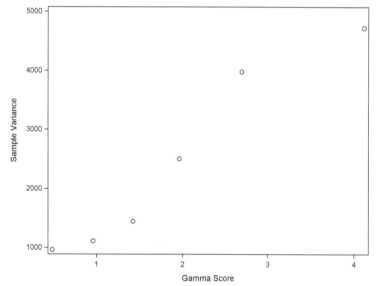

If the variances are homogeneous, the points on the gamma probability plot will lie roughly along a straight line. Individual points above the line to the

GRAPHICAL METHODS TO CHECK MODEL ASSUMPTIONS   189

far right would indicate variances that are larger than the norm. In Figure 5.5 with only 6 points it appears that the points fall fairly close to a straight line indicating constant variances.

For an example where there are more variances to plot, consider the polymerization study shown in Table 5.11. The methods of moments estimator for the box within lot for this staggered nested sampling study was negative. As with the gage R&R study, one or two atypical values may be the cause of this result, and a graphical analysis of the data may reveal that. The mean squares in the type I ANOVA presented in Section 5.7 are pooled estimates of the variances for each source in the design. If the four observations for lot number i are denoted by Y_{1i}, Y_{2i}, Y_{3i}, and Y_{4i} as shown below,

	Box 1			Box 2
	Preparation			Preparation
	1		2	1
Lot	test 1	test 2	test 1	test 1
i	Y_{1i}	Y_{2i}	Y_{3i}	Y_{4i}

Snee (1983) shows that the variances to be pooled from each source to create the ANOVA mean squares are given by:

Source	Variance s_i^2
Error or test(prep)	$(Y_{2i} - Y_{1i})^2/2$
prep(box)	$\frac{2}{3}\left(Y_{3i} - \frac{(Y_{1i}+Y_{2i})}{2}\right)^2$
box	$\frac{3}{4}\left(Y_{4i} - \frac{(Y_{1i}+Y_{2i}+Y_{3i})}{3}\right)^2$

The SAS commands to string the data out as shown above and compute the standard deviation within each source are shown on the next page along with the commands to make a half-normal plot of the standard deviations for prep within box.

The method of moments estimator of the box within lot was negative because the mean square for prep within box and lot was larger than the mean square for box. This would suggest an investigation of the variances that are pooled to form the mean square for prep within box. The standard deviations due to prep within box for each lot are labeled sd2 in the SAS commands on the next page.

In the plot, shown in Figure 5.6, we see a relatively straight line of points extending from the lower left of the plot, except there is one standard deviation that sticks out above the line at the right side. Examination of the `proc print` results of the file of standard deviations, on the page following the graph, shows that this large standard deviation is in lot 19 and is primarily due to the high result for y3=16.79, the first test of the second preparation for box 1.

If lot number 19 is removed from the data, the method of moments estimate

190   DESIGNS TO STUDY VARIANCES

```
proc sort data=polymer; by lot;
data string (keep = lot y1-y4);
   array y(4) y1-y4;
   do i=1 to 4;
     set polymer; by lot;
     y(i)=strength;
     if last.lot then return;
   end;
data vars; set string;
  sd1=sqrt((y2-y1)**2/2);
  sd2=sqrt((2/3)*(y3-(y1+y2)/2)**2);
  sd3=sqrt((3/4)*(y4-(y1+y2+y3)/3)**2);
proc print; run;
* proc rank calculates normal scores for
    parameter estimates ;
ods graphics on;
proc rank data=vars out=hnplots; var sd2;
    ranks rnk;
data hnplots; set hnplots;
zscore=probit((((rnk-.5)/30)+1)/2);
proc sgplot data=hnplots;
   scatter x=zscore y=sd2;
   xaxis label='Half-Normal Score';
   yaxis label='Std. due to prep within box';
run;
ods listing off;
```

Figure 5.6 *Half-Normal Plot of Standard Deviations of Prep(Box)*

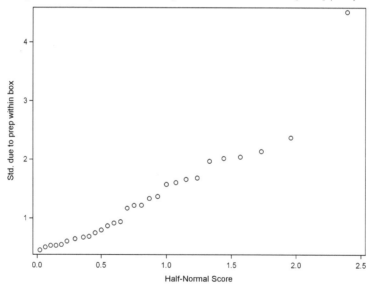

of the variance component for box within lot is no longer negative and method of moments and REML estimators are quite consistent, as can be seen in the

GRAPHICAL METHODS TO CHECK MODEL ASSUMPTIONS 191

Obs	y1	y2	y3	y4	lot	sd1	sd2	sd3
1	9.76	9.24	11.91	9.02	1	0.36770	1.96776	1.11140
2	10.65	7.77	10.00	13.69	2	2.03647	0.64503	3.65174
3	6.50	6.26	8.02	7.95	3	0.16971	1.33905	0.88623
4	8.08	5.28	9.15	7.46	4	1.97990	2.01675	0.03753
5	7.84	5.91	7.43	6.11	5	1.36472	0.45316	0.82272
6	9.00	8.38	7.01	8.58	6	0.43841	1.37171	0.38971
7	12.81	13.58	11.13	10.00	7	0.54447	1.68607	2.17084
8	10.62	11.71	14.07	14.56	8	0.77075	2.37192	2.10155
9	4.88	4.96	4.08	4.76	9	0.05657	0.68586	0.10392
10	9.38	8.02	6.73	6.99	10	0.96167	1.60850	0.91221
11	5.91	5.79	6.59	6.55	11	0.08485	0.60421	0.39260
12	7.19	7.22	5.77	8.33	12	0.02121	1.17167	1.38853
13	7.93	6.48	8.12	7.43	13	1.02530	0.74709	0.06928
14	3.70	2.86	3.95	5.92	14	0.59397	0.54705	2.09289
15	4.64	5.70	5.96	5.88	15	0.74953	0.64503	0.38682
16	5.94	6.28	4.18	5.24	16	0.24042	1.57584	0.19630
17	9.50	8.00	11.25	11.14	17	1.06066	2.04124	1.34811
18	10.93	12.16	9.51	12.71	18	0.86974	1.66157	1.59637
19	11.95	10.58	16.79	13.08	19	0.96874	4.51114	0.02309
20	4.34	5.45	7.51	5.21	20	0.78489	2.13514	0.48209
21	7.60	6.72	6.51	6.35	21	0.62225	0.53072	0.51384
22	5.12	5.85	6.31	8.74	22	0.51619	0.67361	2.58076
23	5.28	5.73	4.53	5.07	23	0.31820	0.79608	0.09526
24	5.44	5.38	4.35	7.04	24	0.04243	0.86549	1.71762
25	3.50	3.88	2.57	3.76	25	0.26870	0.91448	0.38394
26	4.80	4.46	3.48	3.18	26	0.24042	0.93897	0.92376
27	5.35	6.39	4.38	5.50	27	0.73539	1.21658	0.10970
28	3.09	3.19	3.79	2.59	28	0.07071	0.53072	0.66395
29	5.30	4.72	4.39	6.13	29	0.41012	0.50623	1.14893
30	7.09	7.82	5.96	7.14	30	0.51619	1.22066	0.15877

results shown in Table 5.15. This might lead one to question the result of test 1 of preparation 2 of box from lot 19. If the material was still available, a repeat test would be in order.

Table 5.15 *Comparison of Method of Moments and REML Estimates for Polymerization Study after Removing Lot 19*

Component	Method of Moments Estimator	REML Estimator
Lot (σ_a^2)	5.81864	6.09918
Box(Lot) (σ_b^2)	0.13116	0.04279
Prep(Box) (σ_c^2)	0.76517	0.79604
Error (σ^2)	0.63794	0.64364

5.9.3 Probability Plots of Empirical Best Linear Unbiased Estimates of Random Effects

In Section 2.4 the assumption that the experimental errors follow a normal distribution, a normal probability plot of the residuals was made. In matrix notation, the vector of residuals can be written as:

$$\hat{\epsilon} = y - X\hat{\beta}. \tag{5.17}$$

When there are random terms in the model in addition to the experimental error term, ϵ, we can check the normality of these random effects by making normal probability plots of the estimated random effects. A straight line of points in this plot would justify the assumption of normality for the random effect in question. This graphical technique will be useful in detecting atypical observations or departures from normality when there are at least 12 to 15 points on the plot. In the model

$$y = X\beta + Z\gamma + \epsilon \tag{5.18}$$

the empirical best linear unbiased predictors (EBLUPs) of the random effects γ are given by the equation

$$\hat{\gamma} = \hat{G}Z'\hat{V}^{-1}(y - X\hat{\beta}) \tag{5.19}$$

where $\hat{G}$ is the estimated variance covariance matrix of γ. SAS `proc varcomp` does not calculate the EBLUPs but `proc mixed` does and it can be used for models with all random effects as well as for models with both fixed and random effects. On the next page are the commands for fitting the nested random effects model for the polymerization data using proc mixed. Notice no terms are specified in the model statement since the model statement in `proc mixed` is used for specifying the fixed effects. The random statement includes all the terms in the model. The `/solution` option on the random statement causes `proc mixed` to calculate and print a table of EBLUPs. The `ods output` statement creates a SAS data file containing the table of printed EBLUPs. In the next data step, the SAS file `prep` is created that contains a subset of the EBLUPs for prep within box. Next the calls to `proc rank` and `proc sgplot` are similar to the ones used to produce a normal plot of the residuals in Section 2.4.

The points on the normal plot of the EBLUPs for prep within box shown in Figure 5.7 nearly follow a straight line. However, there are two apparent outliers on the upper right tail of the plot. When comparing the values of these two EBLUPs to the table printed out by `proc mixed` it can be seen that the largest EBLUP (2.2488) corresponds to the first test of the second preparation for box 1 in lot 19.

GRAPHICAL METHODS TO CHECK MODEL ASSUMPTIONS

```
proc mixed data=polymer;
  class lot box prep;
  model strength=;
  random lot box(lot) prep(box lot)/solution;
  ods output SolutionR=Eblup;
run;
proc print data=Eblup; run;
data prep; set Eblup;
  if Effect='prep(lot*box)';
  keep Estimate;
proc rank data=prep normal=vw; var Estimate;
    ranks zscore;
ods graphics on;
proc sgplot;
  scatter x=zscore y=Estimate;
  xaxis label="Normal Score";
  yaxis label="EBLUP";
run;
ods graphics off;
```

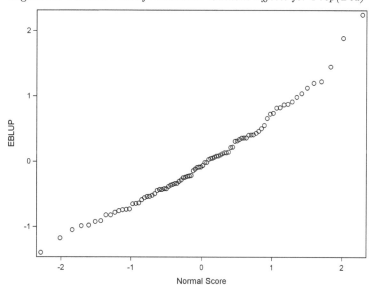

Figure 5.7 *Normal Plot of Estimated Random Effects for Prep(Box)*

This is the same point identified by the half-normal plot in the last section and warrants further investigation. The second largest EBLUP (1.8905) is associated with the test on preparation 1 in box 2 of lot 2. This value (13.69) is higher than any other value for lot 2 and again might warrant further investigation.

5.10 Review of Important Concepts

Sampling experiments are conducted to estimate variances and to partition variances into various sources called variance components. Sometimes estimating variances or partitioning variances into various sources may be the end goal. In other cases partitioning the variance may be an intermediate step in a study trying to reduce the variance or an intermediate step in a study seeking to determine the causes of variability. The factors in a sampling experiment are called random factors as opposed to fixed factors described in chapters 1-4. The purpose for including random factors in a design is to estimate the variability caused by varying levels of the random factor. Levels of the random factor are a representative sample of possible levels.

When the goal is to partition variance into two sources, one factor random sampling experiments, or RSE, are useful. Two sample sampling experiments or factorial random sampling experiments (FRSE), like those used in classical gage R&R studies, are used to partition variance into three independent sources. Nested sampling experiments, or NSE, are used when it is more convenient to use a different set of levels of one factor for each level of another factor. Several stages of nesting can be utilized in one design to partition variance into several sources. Staggered nested sampling experiments, or SNSE, are a variation of nested designs that balance the information collected over the various sources. Figure 5.8 illustrates where these designs should be used in relation to the designs described in earlier chapters.

Variance components can be estimated using the analysis of variance method of moments or the maximum likelihood or REML methods. This can be accomplished using SAS `proc varcomp`. Formulas were presented in the chapter for exact confidence intervals on variance components estimated from one factor random sampling experiments RSE. Formulas were presented for approximate confidence limits on variance components from two-factor FRSE, or nested designs NSE, and a SAS macro was introduced for evaluating these formulas. Asymptotic confidence intervals can be calculated for variance components estimated by maximum likelihood or REML. These intervals will be reasonably accurate when there are 30-40 degrees of freedom associated with the term in question.

Sample sizes for sampling experiments can be determined to make the width of the confidence interval for σ^2 the desired multiple of $\hat{\sigma}^2$. The number of levels for other random factors in the design should be chosen to partition the degrees of freedom among the various sources in the design according to their perceived importance relative to the replicate variance σ^2.

In designs to study the effects of fixed factors, random factors can be introduced into the model by the way the experiment is conducted. For example, when response measurements are made on sub-samples of experimental units an additional nested error term must be included in the model for analysis. Block terms and block by treatment interactions in randomized block experiments would normally be considered to be random factors. When both fixed

REVIEW OF IMPORTANT CONCEPTS

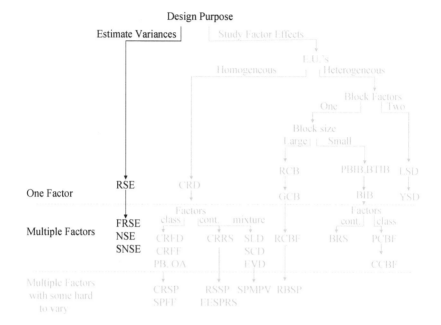

Figure 5.8 *Design Selection Roadmap*

and random effects are included in the model, the expected values of the ANOVA mean squares (EMS) show which mean square should be used in the denominator of the F-test for each fixed effect or interaction in the model. SAS `proc mixed` is most useful for analyzing data from designs that include fixed and random factors. The REML method is used for estimating the variance components of the random effects by `proc mixed` and it automatically computes the correct standard error for estimable functions of fixed effects.

Atypical response values can have a heavy influence on variance component estimates from sampling experiments. Simple plots, such as plots of the response versus the level in one-factor designs or interaction plots for two-factor designs, can help to identify atypical response values. For nested designs (or staggered nested designs), gamma plots of sample variances (or half-normal plots of standard deviations) within each source are useful for checking homogeneity of variances and identifying cells where atypical values may occur. Normal probability plots of empirical best linear unbiased estimators of random effects (EBLUPs) are useful for checking normality assumptions and again for identifying cells where atypical response values may lie.

5.11 Exercises

1. Carry out the sampling experiment described in Section 5.3 to partition the variability in paper helicopter flight times.

   (a) Read your data into SAS and estimate the variance components using the method of moments.

   (b) Using `proc varcomp`, estimate the variance components using the REML method.

   (c) Using the formulas in Table 5.3, calculate 90% confidence intervals for the two variance components.

   (d) Write your conclusions. What should you use as the experimental unit in future studies to optimize flight time?

2. Read the data from Table 5.2 into SAS and

   (a) Calculate the ANOVA sums of squares and mean squares used in calculating the confidence limits in Section 5.4.2.

   (b) Use `proc varcomp` to calculate and compare the method of moments and REML estimates of the variance components.

3. Using the mean squares and expected mean squares from the gage R&R study presented in Section 5.5, use the macro in the appendix of this chapter to compute a 90% confidence interval on the variance component for the interaction of operator and part.

4. Plan a sampling experiment to partition the source of variability in auto gas milage in your community between drivers, cars, and replicates.

   (a) Describe how you would plan the study if drivers and cars were crossed factors.

   (b) Describe how you would plan the study if drivers were nested within cars.

   (c) Determine the number of degrees of freedom you would need for the error or replicate term in your model if you desire the width of the confidence interval on the variance component for replicate to be 75% of σ^2.

5. Consider the data in Table 5.9.

   (a) Enter the data into SAS and use `proc varcomp` to estimate the variance components σ_a^2, σ_b^2, σ_c^2, and σ^2.

   (b) What are the major source or sources of variability?

   (c) Is there any difference between the method of moments or REML estimators of the variance components?

6. Consider the data (Anderson and McLean, 1974) in the following table to come from a sampling study to determine the source of variability in the average amount spent on healthcare per year in thousands of dollars. Towns are nested within state and households are nested within town.

EXERCISES

	State					
	1		2		3	
	Town		Town		Town	
	1	2	1	2	1	2
Household						
1	10	7	6	6	15	12
2	13	12	5	12	18	15
3	16	11	9	7	20	18
4	12	9	3	10	19	16

(a) Use the method of moments and REML methods to determine the variance components for state, town and household.

(b) What percent of the total variance is due to households? Towns and households?

(c) Using formula (5.11) obtain a 90% confidence interval on the variance component due to towns.

7. Consider the data from Smith and Beverly (1981) taken from a staggered nested design to investigate the sources of variability in impurities in raw materials received at a plant in trailer loads. Two samples of material were taken from each of 9 trailer loads of pellets. Two measurement of impurities was made on the first sample from each trailer but only on measurement for the second sample from each trailer.

Table 5.16 *Variability in Impurities in Received Pellets*

	Sample		
	1		2
	Measurement		Measurement
Trailer	1	2	1
1	47.06	44.37	49.3
2	47.43	50.35	50.42
3	48.90	48.05	50.64
4	52.32	52.26	53.47
5	46.53	45.60	53.98
6	46.99	50.87	51.87
7	47.49	51.55	58.57
8	47.41	47.63	48.63
9	48.37	51.03	50.15
10	54.80	51.57	54.52

(a) Write the model for the data.

(b) Analyze the data using `proc varcomp` using the option `method=type1` and estimate the three variance components.

(c) Analyze the data using `proc varcomp` using the option `method=reml` and check to see if your estimates remain the same.

(d) Make half-normal plots of the square root of the variances pooled to get the mean squares for sample(trailer) and measurement(sample). Does the assumption of homogeneous variances appear reasonable?

(e) Re-analyze the data using `proc mixed`; calculate the EBLUPs for the random trailer effect and make a normal plot to check the normality assumption. What is your conclusion?

8. Reanalyze the data from the golf experiment, presented in the appendix of Chapter 4, using proc mixed. Check to see if you get the same P-values and conclusions shown in Section 4.7.

5.12 Appendix

```
%macro vci;
*********************************************************************;
**        Macro for computing confidence interval for              **;
**           delta = c1*E(ms1) - c2*E(ms2)                         **;
**                 where c1, c2 >=0                                **;
** Input for this macro is a file called meansq that includes     **;
*one line containing variables confl, c1, ms1, nu1, c2, ms2, nu2;
data conf; set meansq;
delta=c1*ms1-c2*ms2;
alpha=1-confl;
Falpha1=finv(confl,nu1,10000000);
Falpha12=finv(confl,nu1,nu2);
Fconf2=finv(alpha,nu2,10000000);
Fconf12=finv(alpha,nu1,nu2);
Falpha2=finv(confl,nu2,10000000);
Fconf1=finv(alpha,nu1,10000000);
Fconf12=finv(alpha,nu1,nu2);
G1=1-(1/Falpha1);
H2=(1/Fconf2)-1;
G12=((Falpha12-1)**2-G1**2*Falpha12**2-H2**2)/Falpha12;
VL=G1**2*c1**2*ms1**2+H2**2*c2**2*ms2**2+G12*c1*c2*ms1*ms2;
H1=(1/Fconf1)-1;
G2=1-(1/Falpha2);
H12=((1-Fconf12)**2-H1**2*Fconf12**2-G2**2)/Fconf12;
VU=H1**2*c1**2*ms1**2+G2**2*c2**2*ms2**2;
L=delta-sqrt(VL);
U=delta+sqrt(VU);
proc print; var delta L U; run;
%mend vci;
```

CHAPTER 6

Fractional Factorial Designs

6.1 Introduction

There are two benefits to studying several treatment factors simultaneously in a factorial design. First, the interaction or joint effects of the factors can be detected. Second, the experiments are more efficient. In other words, the same precision of effects can be achieved with fewer experiments than would be required if each of the factors was studied one-at-a-time in separate experiments. The more factors included in a factorial design, the greater the efficiency and the greater the number of interactions that may be detected. However, the more factors included in a factorial experiment, the greater the number of runs that must be performed. When many factors are included in a factorial experiment, one way to reduce the number of runs is to use only two levels of each factor and run only one experiment per cell or treatment combination. These ideas were discussed in Sections 3.7 and 3.7.5.

In the preliminary stage of experimentation, where the objective may be to determine which factors are important from a long list of candidates, a factorial design may require too many experiments to perform even when there are only two levels of each factor and only one replicate per cell. Table 6.1 shows the number of experiments required for a 2^k factorial design as a function of the number of factors, k. With $k = 7$ or more factors, the large number of runs required for a 2^k is usually impractical.

Table 6.1 *Number of Experiments Required for 2^k Design*

Number of Factors (k)	Number of Experiments (2^k)
4	16
5	32
6	64
7	128
8	256
9	512

When the number of factors under study is large, researchers will often abandon the efficiency of factorial experiments altogether and revert to a "seat of the pants" approach or vary one-factor-at-a-time plan. Others will run a

200   FRACTIONAL FACTORIAL DESIGNS

factorial experiment with a subset of the factors, chosen from the longer list by guessing which ones may be more important. However, these approaches are less than optimal and do not retain the benefits of factorial experiments with a large number of factors. A better solution to this problem is to use a fraction of the experiments, or runs, required for a full factorial experiment. To be effective, the fraction of runs used must be carefully selected in order to preserve some of the benefits of a full factorial experiment.

One of the desirable properties of a 2^k factorial plan is that factor effects are not obscured by (or correlated with) planned changes in other factors. This property was called orthogonality in Section 3.7.

6.2 Half-Fractions of 2^k Designs

Consider first choosing a half-fraction of a 2^k factorial experiment. Careless choice of half the $n = 2^k$ runs may not retain the desirable orthogonality property of a 2^k design. One way to preserve this property, when selecting a one-half fraction of a 2^k factorial experiment, is to choose the runs where the coded factor levels for an interaction term (preferably the highest order interaction) are constant.

Table 6.2 illustrates this procedure. On the left side of the table is a representation of the coded factor levels for a 2^4 design. On the right side of the table are the runs that have a constant (+) value for $X_A \times X_B \times X_C \times X_D$. These runs represent the half-fraction. The order of the runs on the right side of the table have been reordered so that it can be easily seen that the standard factorial pattern is present in the first three columns. Therefore it can be seen that the orthogonality property is preserved for the first three columns. By further inspection, it can be seen that fourth factor is also orthogonal to the other three.

With 16 runs in a full 2^4 factorial, 15 effects can be estimated in addition to the grand mean. The 15 effects consist of the four main effects, six two-factor interaction effects, four three-factor interactions and one four-factor interaction. In a half-fraction of a 2^4 experiment, however, there are only 8 runs. Thus only 7 effects can be estimated in addition to the grand mean. By choosing the runs from a full factorial that have a constant value for an interaction, we automatically lose the ability to estimate that interaction effect. By studying the right half of Table 6.2, it can also be seen that the coded factor levels or column of signs for X_D is exactly the product of signs in the first three columns, i.e., $X_D = X_A \times X_B \times X_C$. This means that the effect we can estimate for X_D will be completely obscured or confused by the three-factor interaction $X_A X_B X_C$. This is not all. As will be shown later, each main effect and interaction in the design is obscured or *confounded* with one other interaction. This is the price we pay for running one-half the total number of experiments. However, in preliminary experiments where a large number of factors are included in order to find out which ones are truly important, this may not be a serious price to pay.

HALF-FRACTIONS OF 2^k DESIGNS

Table 6.2 *Creating a Half-Fraction by Choosing the Runs in a Full Fraction with Constant Values for an Interaction*

	Full Factorial						Half-Fraction			
run	X_A	X_B	X_C	X_D	$X_A X_B X_C X_D$	run	X_A	X_B	X_C	X_D
1	-	-	-	-	+	1	-	-	-	-
2	+	-	-	-	-	10	+	-	-	+
3	-	+	-	-	-	11	-	+	-	+
4	+	+	-	-	+	4	+	+	-	-
5	-	-	+	-	-	13	-	-	+	+
6	+	-	+	-	+	6	+	-	+	-
7	-	+	+	-	+	7	-	+	+	-
8	+	+	+	-	-	16	+	+	+	+
9	-	-	-	+	-					
10	+	-	-	+	+					
11	-	+	-	+	+					
12	+	+	-	+	-					
13	-	-	+	+	+					
14	+	-	+	+	-					
15	-	+	+	+	-					
16	+	+	+	+	+					

In preliminary experiments involving a large number of factors, usually only a small proportion of the factors will have significant effects. This fact has been called the *effect sparsity principle* by Box and Meyer (1986a). Just as two planets will line up with the moon in the night sky more often than three planets will, main effects are more likely to be important than two-factor interaction, and two-factor interactions are more likely to be important than three-factor interactions, etc. This general rule has been called the *hierarchical ordering principle* by Wu and Hamada (2000). Therefore, if fractions of factorial experiments can be planned in a way that main effects are confounded with three-factor and higher-order interactions, the amount of information lost by fractionating the number of runs will be small in comparison to the benefit of a reduced number of runs.

The way a fractional factorial of a 2^k is created in practice is actually the opposite order of what was shown above. Instead of starting with a full factorial and eliminating runs to get the desired fraction, start with a full factorial containing the desired number of runs and add additional factors to the design.

For example, to construct a one-half fraction of a 2^k design, denoted by $\frac{1}{2}2^k$ or 2^{k-1}, the procedure is as follows:

1. Write down the *base design*, a full factorial plan in $k-1$ factors using the coded factor levels $(-)$ and $(+)$.

2. Add the kth factor to the design by making its coded factor levels equal to the product of the other factor levels (i.e., the highest order interaction).

3. Use these k columns to define the design.

A complete list of interactions confounded with each main effect and interaction in a half-fractional factorial is called the *confounding pattern* or *alias structure* of the design. This list is easy to construct based on the assignment of the kth factor in item 2 of the list above. For example, in the 2^{4-1} design, if the levels of the fourth factor are equal to the product of the levels of the first three factors in the design, we write symbolically $D = ABC$. This is called the *generator* of the design. Multiplying on both sides of the generator we get:

$$D^2 = ABCD$$

or

$$I = ABCD$$

where I represents a column of pluses and is the multiplicative identity for elementwise products of columns of coded factor levels. The equation, $I = ABCD$, is called the *defining relation* for the fractional factorial design, and by multiplying on both sides of this equation, the interaction confounded with any main effect or interaction can be determined. For example, multiplying by the first factor on both sides of the defining relation we see

$$A(I) = A(ABCD)$$

or

$$A = BCD$$

This means that the effect of the first factor A is confounded with the three-factor interaction BCD. When data is collected, the effect of factor A is estimated as the difference in the average response at the high and low levels of factor A. However, that effect really estimates the sum of the effects of factor A and the three-factor interaction. Therefore we write it as $A + BCD$. The complete alias pattern for this 2^{4-1} design can be determined by multiplying the defining relation by each main effect and interaction resulting in:

HALF-FRACTIONS OF 2^k DESIGNS

$$I + ABCD$$
$$A + BCD$$
$$B + ACD$$
$$C + ABD$$
$$D + ABC$$
$$AB + CD$$
$$AC + BD$$
$$AD + BC$$

There are only eight unique results, and they represent the eight effects (I being the overall mean) that can be estimated from the 8-run fractional factorial design.

Replicates are not included in a one-half fractional factorial design because replicates would take as many additional experiments as it would to complete the full factorial. Therefore, there is no estimate of σ^2, the variance of experimental error, when running a fractional factorial. In order to judge the significance of the effects in a fractional factorial design, graphical methods like those described in Section 3.7.5 should be used. The effects found to be significant should be interpreted with the hierarchical ordering principle in mind. For example, if the effect for $B + ACD$ were found to be significant, it will be assumed to represent the effect of factor B rather than the three-way interaction.

Using the procedure outlined on the last page, it is easy to create a 2^{k-1} fractional factorial design in the data step of SAS. For example, the SAS code below creates a list of treatment combinations for a 2^{5-1} experiment.

```
*Create 2^(5-1) design in standard order;
data design;
do x4=-1 to 1 by 2;
 do x3=-1 to 1 by 2;
  do x2=-1 to 1 by 2;
   do x1=-1 to 1 by 2;
   A=x1; B=x2; C=x3; D=x4; E=A*B*C*D;
   output;
   end;
  end;
 end;
end;
keep A B C D E;
proc print; run;
```

In practice the treatment combinations should be randomized to the experimental units, and a randomized list created like the one shown in Section 3.4

should be used. This design is then called a completely randomized fractional factorial design of CRFF. The design resulting from the code above is shown below in standard order (not randomized).

Obs	A	B	C	D	E
1	-1	-1	-1	-1	1
2	1	-1	-1	-1	-1
3	-1	1	-1	-1	-1
4	1	1	-1	-1	1
5	-1	-1	1	-1	-1
6	1	-1	1	-1	1
7	-1	1	1	-1	1
8	1	1	1	-1	-1
9	-1	-1	-1	1	-1
10	1	-1	-1	1	1
11	-1	1	-1	1	1
12	1	1	-1	1	-1
13	-1	-1	1	1	1
14	1	-1	1	1	-1
15	-1	1	1	1	-1
16	1	1	1	1	1

To illustrate the analysis of a 2^{k-1} design, consider a continuation of the dry soup mix example presented in Section 5.4.3. In that example, the majority of variability in the soup "intermix" was found to be within a batch rather than between batches. The researchers responsible for the project made a list of factors that they thought might influence the variability within a batch (see Hare, 1988). These were options that could be changed on the mixer where a batch was mixed and the intermix was added through ports. The list consisted of: (1) the number of ports where intermix was added, (2) the temperature of the mixer (that could be controlled by adding cooling water to the jacket surrounding it), (3) the mixing time, (4) weight of the batch, and (5) the delay time between mixing and packaging. The response or variability in fill weights of the intermix was obtained by taking five consecutive samples of the soup mix every 15 minutes as the batch was being packaged. The factor labels and levels are shown in Table 6.3.

Table 6.3 *Factors and Levels for Soup Mix 2^{5-1} Experiment*

Factor Label	Name	Low Level	High Level
A	Number of Ports	1	3
B	Temperature	Cooling Water	Ambient
C	Mixing Time	60 sec.	80 sec.
D	Batch Weight	1500 lbs	2000 lbs
E	Delay Days	7	1

HALF-FRACTIONS OF 2^k DESIGNS

The normal batch size was 2000 lbs and the normal mixing time was 60 seconds. Since this experiment was to be run in a production facility, the research and production staff both had to agree to the plan. The factor levels could be changed between batches with relatively little effort and they were not a problem. The requested reduction in batch weight and the increase in mixing time for some batches in the planned list of experiments would not seriously interfere with the production schedule if the list was short, but it would slow production if a 32-run design were used. For this reason a 16-run fractional factorial plan was agreed upon. The alias structure for the 2^{5-1} plan shown above is:

$$A + BCDE$$
$$B + ACDE$$
$$C + ABDE$$
$$D + ABCE$$
$$E + ABCD$$
$$AB + CDE$$
$$AC + BDE$$
$$AD + BCE$$
$$AE + BCD$$
$$BC + ADE$$
$$BD + ACE$$
$$BE + ACD$$
$$CD + ABE$$
$$CE + ABD$$
$$DE + ABC$$

Therefore, if three-factor and four-factor interactions could be assumed negligible, estimates of all main effects and two-factor interactions could be made. The experimental unit for this experiment was the batch of dried soup mix put in the mixer, and Table 6.4 shows a randomized list of the 2^{5-1} experiments in actual factor levels. The response $\hat{\sigma}_p$ was an estimate of the standard deviation of fill weight within a batch, computed from the sampling study made during packaging of each batch.

The SAS commands, to read in the data (in standard order) and to merge it with the design in coded factor levels created earlier, are shown on the next page.

```
data process;
input s @@;
datalines;
1.13 1.25 .97 1.70 1.47 1.28 1.18 .98
.78 1.36 1.85 .62 1.09 1.10 .76 2.1
data soup; merge design process;
```

Table 6.4 2^{5-1} *Experiment To Determine Which Factors Are Associated with Fill Variation in Random Order*

Batch	(A) Number of Ports	(B) Temperature	(C) Mixing Time (sec)	(D) Batch Weight (lbs)	(E) Delay (days)	Response $\hat{\sigma}_p$
1	1	Cool Water	60	2000	7	0.78
2	3	Cool Water	80	2000	7	1.10
3	3	Ambient	60	1500	1	1.70
4	3	Cool Water	80	1500	1	1.28
5	1	Ambient	60	1500	7	0.97
6	1	Cool Water	80	1500	7	1.47
7	1	Ambient	60	2000	1	1.85
8	3	Ambient	80	2000	1	2.10
9	1	Ambient	80	2000	7	0.76
10	3	Ambient	60	2000	7	0.62
11	1	Cool Water	80	2000	1	1.09
12	1	Cool Water	60	1500	1	1.13
13	3	Cool Water	60	1500	7	1.25
14	3	Ambient	80	1500	7	0.98
15	3	Cool Water	60	2000	1	1.36
16	1	Ambient	80	1500	1	1.18

The SAS commands, to fit the model in the main effects and two-factor interactions and to make a normal probability plot of the regression coefficients, are shown on the next page. This is the same thing that was done in Section 3.7.5, except only 15 terms are included in the model rather than the full model which would contain main effects and interactions up to order 5. Each effect in this model is confounded with one other interaction as shown in the alias pattern on the previous page. Therefore only 15 effects can be estimated.

As can be seen in the normal plot of effects, the main effect E (Delay Time between mixing and packaging), BE (the interaction between Temperature and Delay Time), and DE (the interaction between Batch Weight and Delay Time) appear to be significant. If the hierarchical effect ordering can be as-

HALF-FRACTIONS OF 2^k DESIGNS

sumed, this is the correct interpretation, and the three-factor and four-factor interactions can be assumed negligible.

```
proc glm data=soup;
  model s=A B C D E A*B A*C A*D A*E B*C B*D B*E C*D
  C*E D*E/solution;
  ods output ParameterEstimates=sol;
run;
data nplot; set sol;
  estimate=estimate;
  if _n_>1; drop StdErr tValue Probt;
*proc rank calculates normal scores for parameter estimates;
proc rank data=nplot out=nplots normal=blom; var estimate;
  ranks zscore;
data nplots; set nplots;
  if abs(zscore)<=.9 then parameter='    ';
run;
ods graphics on;
proc sgplot data=nplots;
  scatter x=zscore y=estimate/datalabel=parameter;
  xaxis label='Normal Scores';
  yaxis label='Regression Coefficient';
run;
ods graphics off;
```

Figure 6.1 *Normal Plot of Effects from Fill Variability Experiment*

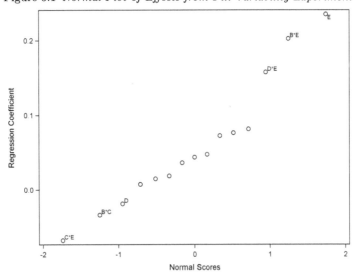

E=Delay Time has a positive effect (as seen on the normal plot); this would normally mean that increasing the Delay Time between mixing and packaging would increase the response (or within batch variation). However, due to the unconventional assignment of 7 to the low (coded) level of factor E and 1 to the high (coded) level in Table 6.3, it actually tells us that on the average increasing the delay time between mixing and packaging decreases the within batch variability. Since the interactions BE and DE also appear to be significant, the average main effect of factor E does not tell the whole story. Figure 6.2 shows the interaction plot for Temperature and Delay Time. Here it can be seen that Delay Time has little effect on the within batch variability if the mixer is cooled with cooling water during mixing. However, the within batch variability is decreased substantially by increasing delay time between mixing and packaging if the mixing was done at ambient temperature.

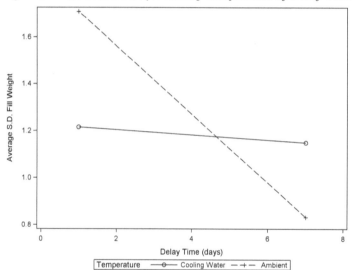

Figure 6.2 *Interaction Plot for Mixing Temperature by Delay Time*

Figure 6.3 shows the interaction plot for Batch Weight and Delay Time. Here it can be seen that increasing the delay time between mixing and packaging has little effect on the variability within a batch for small (1500 lbs) batches, while increasing the delay time between mixing and packaging decreases within batch variability for large (2000 lbs) batches.

Based on the results of this study, the minimum variation of intermix within a batch could be obtained by using the larger batch size (2000 lbs), the ambient temperature at mixing, and a 7-day delay between mixing and packaging. It did not answer a new question prompted by the results: does it take a full 7-day delay between mixing and packaging or could a 3-day or 4-day delay reduce variability just as much? However, the production and research staff agreed that the current results made sense, since the uniformity of intermix related to the hardness of the vegetable oil which is affected by temperature and

QUARTER AND HIGHER FRACTIONS OF 2^k DESIGNS

agitation. Production staff implemented the optimal conditions while research continued investigating the new question.

It is interesting to think that neither of the interactions would have been detected if one-factor-a-time experiments had been conducted, rather than the fractional factorial plan. The conclusions of one-factor-at-a-time type experiments may not be reproducible because it would not be realized that the effects (or absence of effects) seen could depend on other factor settings that are deemed to be insignificant.

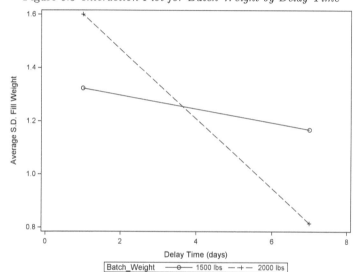

Figure 6.3 *Interaction Plot for Batch Weight by Delay Time*

6.3 Quarter and Higher Fractions of 2^k Designs

In 2^{k-1} designs only half the experimental runs are made, and each effect that can be estimated is confounded with one interaction. Likewise, in a quarter fraction 2^{k-2} design, only one-fourth the experimental runs from the full 2^k design are run, and each effect that can be estimated is confounded with three other interactions. In a one-eighth fraction, only one-eighth of the runs in the full factorial are made and each estimated effect will be confounded with seven other interactions, and so on. These designs may sound confusing at first because of the large number of effects confounded with each estimable effect. However, they are used quite frequently in practice, and by following the effect sparsity principle and the hierarchical ordering principle useful conclusions can usually be reached after just a fraction of the total runs required for a full factorial.

To construct a quarter fraction of a 2^k design, start with a base design in 2^{k-2} factors, then add two additional factors by making their coded factor levels equal to two interactions among the first $k-2$ columns, as illustrated below.

X_A	X_B	X_C	$\overbrace{X_AX_B}^{X_D}$	$\overbrace{X_AX_C}^{X_E}$	X_BX_C	$X_AX_BX_C$
−	−	−	+	+	+	−
+	−	−	−	−	+	+
−	+	−	−	+	−	+
+	+	−	+	−	−	−
−	−	+	+	−	−	+
+	−	+	−	+	−	−
−	+	+	−	−	+	−
+	+	+	+	+	+	+

$\Downarrow$

X_A	X_B	X_C	X_D	X_E
−	−	−	+	+
+	−	−	−	−
−	+	−	−	+
+	+	−	+	−
−	−	+	+	−
+	−	+	−	+
−	+	+	−	−
+	+	+	+	+

There are two generators for the design created as above, $D = AB$ and $E = AC$. From these it can be seen that $I = ABD$ and $I = ACE$. Also, since $I^2 = I$ a third equality, called the generalized interaction, is $I = ABD(ACE)$ or $I = BCDE$. Combining the three equations obtained from the two generators and the generalized interaction results in the defining relation for the design $I = ABD = ACE = BCDE$. The confounding pattern, or alias structure, for the design is found by multiplying through the defining relation by each effect that can be estimated:

$$A + BD + CE + ABCDE$$
$$B + AD + ABCE + CDE$$
$$C + ABCD + AE + BDE$$
$$D + AB + ACDE + BCE$$
$$E + ABDE + AC + BCD$$
$$BC + ACD + ABE + DE$$
$$BE + ADE + ABC + CD$$

Construction of one-eighth and higher fractions is similar. To construct a one-eighth fraction of a 2^6 design or 2^{6-3} design, start with a base design in $6 - 3 = 3$ factors, then add three additional factors by confounding them with

CRITERIA FOR CHOOSING GENERATORS FOR 2^{k-p} DESIGNS

interactions. For example, if we choose the generators $D = AB$, $E = AC$, and $F = BC$, the SAS code below could be used to generate the design.

```
data frac;
do x1=-1 to 1 by 2;
 do x2=-1 to 1 by 2;
  do x3=-1 to 1 by 2;
   A=x3; B=x2; C=x1;
   D=A*B; E=A*C; F=B*C; output;
  end;
 end;
end;
keep A B C D E F;
proc print; run;
```

There are eight runs in this design, and seven effects in addition to the overall mean can be estimated. Each effect will be aliased with seven interactions. To find what interactions are aliased with each effect that can be estimated, first find the defining relation. From the generators $I = ABD = ACE = BCF$. The two-factor generalized interactions are $ABD(ACE) = BCDE$, $ABD(BCF) = ACDF$, and $ACE(BCF) = ABEF$. The three-factor generalized interaction is $ABD(ACE)(BCF) = DEF$. Combining the equations found from the generators and the generalized interactions the defining relation is:

$$I = ABD = ACE = BCF = BCDE = ACDF = ABEF = DEF$$

Multiplying through the defining relation by the main effects $(A - F)$, and the three-factor interaction (ABC), the aliases for the seven effects that may be estimated can be determined. One quick way to do this in SAS is to use the automated ADX system. This is a point-and-click interface which automates creation and analysis 2^{k-p} designs and other types of designs. Examples of the completed input forms for ADX creation and analysis of 2^{k-p} designs are shown in the lecture slides accompanying this book.

6.4 Criteria for Choosing Generators for 2^{k-p} Designs

There is more than one alternative when selecting the generators for a 2^{k-p} design. For example, to create a quarter fraction of a 2^6 design the generators $E = ABC$ and $F = ABD$ could be used, or the generators $E = AB$ and $F = ACD$ could be used. The first selection results in the defining relation and alias structure (for the main effects only) shown below:

$$I = ABCE = ABDF = CDEF$$
$$A + BCE + BDF + ACDEF$$
$$B + ACE + ADF + BCDEF$$
$$C + ABE + ABCDF + DEF$$
$$D + ABCDE + ABF + CEF$$
$$E + ABC + ABDEF + CDF$$
$$F + ABCEF + ABD + CDE$$

The second set of generators results in the defining relation and alias structure (for the main effects only) below:

$$I = ABE = ACDF = BCDEF$$
$$A + BE + CDF + ABCDEF$$
$$B + AE + ABCDF + CDEF$$
$$C + ABCE + ADF + BDEF$$
$$D + ABDE + ABF + CEF$$
$$E + ABC + ACF + ABEF$$
$$F + ABEF + ACD + BCDE$$

Both generators result in sixteen run fractional factorials, but the first set of generators might be preferable, since each main effect is confounded with three-factor interactions and higher order interactions, while using the second set of generators results in a design where main effects A and B are confounded with one two-factor interaction each. The first design has a smaller chance for confusion since the hierarchical ordering principle tells us that three-factor interactions are less likely to be important than two-factor interactions.

Three general criteria have been proposed for guiding the choice among the various possible sets of generators for any possible 2^{k-p} design. These criteria are the resolution criteria, the aberration criteria and the clear effects criteria.

Box and Hunter (1961) first proposed the resolution criteria. The resolution of the design is defined to be the length of the shortest word in the defining relation. For example, in the first defining relation for a 2^{6-2} design shown above, the shortest word has length 4. Thus it is a resolution IV design. The shortest word for the second defining relation for a 2^{6-2} design shown above has length 3, and it is therefore a resolution III design. In general, if the number of runs in two designs is the same, the design with the higher resolution is preferred.

In a resolution R design no effect involving i factors is aliased with effects of order less than $R - i$. For example, in designs with resolution V, main effects are aliased with four-factor interactions and higher order interactions, and two-factor interactions are aliased with three-factor interactions and higher order interactions. Therefore if all three-factor and higher order interactions can be assumed negligible, all main effects and two-factor interactions can be estimated from a resolution V design. In resolution IV designs, main effects are aliased with three-factor and higher order interactions. While in a resolution III design, main effects are confounded with two-factor interactions. Resolution III designs are normally used only in screening experiments where the purpose is to discover which factors are important enough to be studied further in follow-up experiments.

The projective property of a fractional factorial is another attribute that can be determined from its resolution. In a resolution R design, any subset of $k = R - 1$ factors will form a full 2^k design (with possible replication of some runs). Therefore, if an experiment is started with a resolution R fractional factorial design and only $R - 1$ of the factors appear to be significant, then the data can be reanalyzed including only the $R - 1$ significant factors. Since the design in these factors is a full factorial, interactions of all orders possible among the $R - 1$ factors can be examined.

When two or more designs, created with different sets of generators, have the same number of runs and the same resolution, Fries and Hunter (1980) proposed another criteria for deciding which design is preferable. They called this criteria the *minimum aberration criteria*. If the number of words of length r in the defining relation of a design is defined to be A_r, then a design d_1 is said to have less aberration than a design d_2 if r is the smallest integer such that $A_r(d_1) \neq A_r(d_2)$ and $A_r(d_1) < A_r(d_2)$. For example, if d_1 is a resolution IV 2^{7-2} design created with the generators $F = ABCD$ and $G = ABCE$, it has less aberration than design d_2, created with the generators $F = ABC$ and $G = ADE$, since the defining relation for d_1 ($I = ABCDF = ABCEG = DEFG$) has only 1 word of length 4, while the defining relation for d_2 ($I = ABCF = ADEG = BCDEFG$) has two words of length 4. For any k and p there is always a minimum aberration 2^{k-p} design that has less aberration than any other 2^{k-p} design. For two designs of the same resolution, the minimum aberration design will have less confounding of main effects with low order interactions and is generally preferred.

A final criterion that is useful selecting the generators for a 2^{k-p} design is the number of clear effects. Chen et al. (1993) define an effect to be clear if none of its aliases are main effects or two-factor interactions. In some cases a design that is not the minimum aberration design may have more clear effects than the minimum aberration design. For example, Wu and Hamada (2000) explain that the 2^{6-2} design with defining relation $I = ABCE = ABDF = CDEF$ has all six main effects clear, while the 2^{6-2} design with defining relation $I = ABE = ACDF = BCDEF$ has three main effects (C, D, and F) clear along with six two-factor interactions BC, BD, BF, CE, DE, and EF. In cases where some two-factor interactions are believed to be important *a priori*, the second design may be preferred over the first. Wu and Hamada (2000)'s Table 4A lists the generators for the minimum aberration design and the design with the maximum number of clear effects (if different) for 8-run to 64-run designs.

When the SAS ADX automated system is used to create a 2^{k-p} fractional factorial design, the user first specifies the number of factors k. Next, the program presents the user with a limited set of choices of p. Once the user chooses the value of p, the program recalls the generators from an internally stored table. These stored generators are the ones for the highest resolution minimum aberration design. If the user would like to choose a different set of generators, he can do this by choosing the Create User Design option. The lecture slides for this book show examples of how to do that.

SAS proc factex can also create minimum aberration designs. proc factex does not allow the user specify the generators for the design, as can be done in ADX or in the data step like the examples in Sections 6.2 and 6.3. However, when the resolution is specified in the model statement along with the minimum aberration option, as shown in the commands below, proc factex uses an algorithm, similar to that proposed by Franklin (1985), to search for a set of generators for the design. The output statement causes the design to be stored in the SAS file ff. The examine aliasing(3) confounding;

```
proc factex;
  factors A B C D E F G H;
  size design=16;
  model resolution=4/ minabs;
  examine aliasing(3) confounding;
  output out=ff;
run;
```

will print the generators the procedure found and the aliases up to order 3, as shown in the output on the next page. If no design can be found that satisfies the criteria specified, the message WARNING: No such design found. is printed in the SAS log. This will happen if the factors statement above is changed to factors A B C D E F G H I;, since there is no resolution IV 2^{9-5} design.

Consider the following example of the design and analysis of a 2^{8-4} fractional factorial. AlmeidaeSilva et al. (1998) conducted an experiment to find the optimal conditions for culturing *Paecilomyces variotii* (a fungus commonly found in the air and soils of tropical countries) on eucalyptus hemicellulosic hydrolyzate with a view to producing microbial protein. Only 51.6% of total dry mass of eucalyptus wood is utilized by Brazilian industry while the rest (branches leaves, small trees, etc.) is left in the fields. The hemicellulose fraction of this waste can be easily removed by acid treatment, and the resulting hydrolyzate is rich in fermentable sugars. The fungus *P. variotii* was selected from among 21 species of yeasts and filamentous fungus for its performance on eucalyptus hemicellulose hydrolyzate. Protein biomass produced by this fungus during 72 hours of fermentation has an amino acid profile that is equal to or exceeds conventional products used for animal feed. The purpose of the

CRITERIA FOR CHOOSING GENERATORS FOR 2^{k-p} DESIGNS

```
              Factor Confounding Rules

                    E = B*C*D
                    F = A*C*D
                    G = A*B*D
                    H = A*B*C
```

Aliasing Structure

```
A = B*C*H = B*D*G = B*E*F = C*D*F = C*E*G = D*E*H = F*G*H
B = A*C*H = A*D*G = A*E*F = C*D*E = C*F*G = D*F*H = E*G*H
C = A*B*H = A*D*F = A*E*G = B*D*E = B*F*G = D*G*H = E*F*H
D = A*B*G = A*C*F = A*E*H = B*C*E = B*F*H = C*G*H = E*F*G
E = A*B*F = A*C*G = A*D*H = B*C*D = B*G*H = C*F*H = D*F*G
F = A*B*E = A*C*D = A*G*H = B*C*G = B*D*H = C*E*H = D*E*G
G = A*B*D = A*C*E = A*F*H = B*C*F = B*E*H = C*D*H = D*E*F
H = A*B*C = A*D*E = A*F*G = B*D*F = B*E*G = C*D*G = C*E*F
A*B = C*H = D*G = E*F
A*C = B*H = D*F = E*G
A*D = B*G = C*F = E*H
A*E = B*F = C*G = D*H
A*F = B*E = C*D = G*H
A*G = B*D = C*E = F*H
A*H = B*C = D*E = F*G
```

experiments was to study the influence of inhibitors, nutrients and fermentation time on the biomass growth produced by *P. variotii*. Table 6.5 shows the factors and levels that were to be studied.

Table 6.5 *Factors and Levels for Biomass Experiment*

Label	Factors	Levels −	+
A	Inhibitors (Furfural and Acetic Acid)	1.25g/L	7.8g/L
B	Rice Bran	10.0g/L	30.0g/L
C	Urea	0.0g/L	2.0g/L
D	Magnesium Sulfate	0.0g/L	1.5g/L
E	Ammonium Sulfate	0.0g/L	2.0g/L
F	Potassium Nitrate	0.0g/L	2.0g/L
G	Sodium Phosphate	0.0g/L	2.0g/L
H	Fermentation Time	72 hrs.	96 hrs.

A 2^{8-4} fractional factorial design was used with generators $E = BCD$, $F = ACD$, $G = ABC$, and $H = ABD$. This is the minimum aberration resolution IV design, and the clear effects in this design are the eight main effects plus the seven aliased strings of two-factor interactions shown below.

$$CG + DH + AB + EF$$
$$AC + BG + DF + EH$$
$$CF + AD + EG + BH$$
$$CH + DG + AE + BF$$
$$CD + GH + AF + BE$$
$$BC + AG + DE + FH$$
$$CE + FG + AH + BD$$

The design as created in the SAS ADX window appears in Figure 6.4. The list can be ordered randomly by clicking on the Edit Responses button and then choosing the Design→Randomize Design...

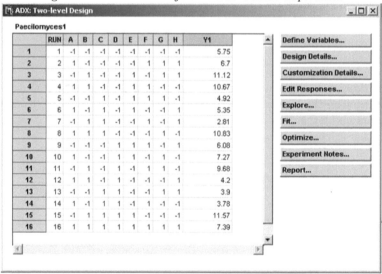

Figure 6.4 2^{8-4} Paecilomyces variotii Culture Experiment

Figure 6.5 shows the effects (average response at high factor level minus average response at low factor levels) computed by the SAS ADX tool.

Figure 6.6 shows a normal plot of effects, also produced by the SAS ADX. A reference line is automatically added to the normal plot by ADX. In this plot, there are no effects that obviously stick out from the reference line. However, main effects for B-Rice Bran, E-Ammonium Sulfate, H-Fermentation time, and the confounded string of interactions $CF + AD + EG + BH$ are largest in absolute value and are in the tails of the normal distribution of estimated

CRITERIA FOR CHOOSING GENERATORS FOR 2^{k-p} DESIGNS

Figure 6.5 *Effects from Paecilomyces variotii Culture Experiment*

Effect	Estimate	Std Error	t Ratio	P Value
A	0.045	.	.	.
B	3.065	.	.	.
C	-1.365	.	.	.
D	-0.535	.	.	.
E	2.09	.	.	.
F	-0.995	.	.	.
G	1.45	.	.	.
H	-2.115	.	.	.
A*B + ...	-0.5675	.	.	.
B*G + ...	0.9925	.	.	.
B*H + ...	-2.1925	.	.	.
B*F + ...	-0.7975	.	.	.
B*E + ...	1.2175	.	.	.
B*C + ...	0.5975	.	.	.
B*D + ...	-0.1125	.	.	.

Figure 6.6 *Normal Plot of Effects from 2^{8-4} Paecilomyces variotii Culture Experiment*

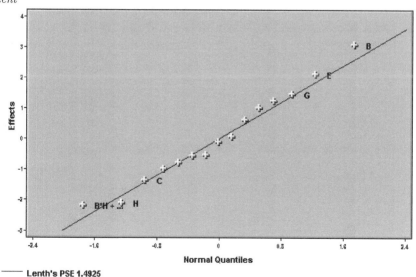

effects. If the three-factor interactions are assumed to be negligible, the effects for the three main effects could be interpreted as the cause of the large effects, but it is unclear what the string of confounded two-factor interactions represents.

The authors of the article felt that the experiment had given them evidence that factor A (the inhibitor) had little effect, and they confirmed this by citing other published reports. They also felt the experiment showed main effects D (magnesium sulfate) and F (potassium nitrite) were insignificant. However, due to the confounding of two-factor interactions with the second largest effect (in absolute value) and the fact that nothing clearly stuck out on the normal plot of effects, no definite conclusions could be drawn. They decided to run another resolution V 2^{5-1} follow-up experiment using factors B, C, E, G and H with defining relation $I = BCEGH$ and factors A, D and F held constant at the mid-point of the levels used in the first experiment. This was a 16-run design, but if it could be safely assumed that main effects A, D and F were negligible in the first set of sixteen experiments, then eight of the sixteen runs for the proposed follow-up design were already completed.

Figure 6.7 shows the 2^{5-1} design as created in SAS ADX. Run numbers 3, 4, 7, 8, 11, 12, 15 and 16 were already completed in the 2^{8-4} design as runs 14, 7, 6, 15, 13, 8, 12 and 16, respectively. The remaining eight experiments were completed in random order to get the results shown. Since this design is resolution V, all main effects and two-factor interactions are clear and can be estimated if three-factor and higher order interactions can be assumed negligible.

Figure 6.7 2^{5-1} *Paecilomyces variotii* Culture Experiment

RUN	B	C	E	G	H	Y1
1	-1	-1	-1	-1	1	3.37
2	1	-1	-1	-1	-1	3.55
3	-1	1	-1	-1	-1	3.78
4	1	1	-1	-1	1	2.81
5	-1	-1	1	-1	-1	5.53
6	1	-1	1	-1	1	10.43
7	-1	1	1	-1	1	5.35
8	1	1	1	-1	-1	11.57
9	-1	-1	-1	1	-1	2.93
10	1	-1	-1	1	1	7.23
11	-1	1	-1	1	1	3.9
12	1	1	-1	1	-1	10.83
13	-1	-1	1	1	1	11.69
14	1	-1	1	1	-1	10.59
15	-1	1	1	1	-1	4.92
16	1	1	1	1	1	7.39

Figure 6.8 shows the normal plot of effect from the follow-up design produced by SAS ADX. Here it can be seen that the main effects for B-Rice Bran, E-Ammonium Sulfate, G-Sodium Phosphate along with the interactions BH

CRITERIA FOR CHOOSING GENERATORS FOR 2^{k-p} DESIGNS   219

(Rice Bran×Fermentation Time), EG (Ammonium Sulfate×Sodium Phosphate), and CH (Urea×Fermentation Time) stick out from the reference line and appear to be significant.

Figure 6.8 *Normal Plot of Effects from 2^{5-1} Paecilomyces variotii Culture Experiment*

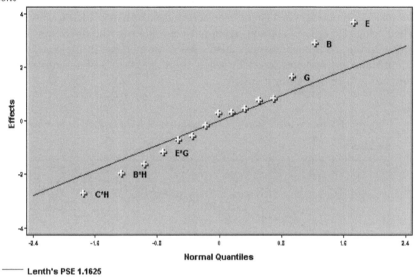

Since there appear to be significant interactions the main effects should not be interpreted in isolation. Figure 6.9 is an interaction plot created by SAS ADX. It shows how the effect of fermentation time depends upon the level of rice bran. When there is only 10 g/L of rice bran in the growth

Figure 6.9 *Interaction Plot for Fermentation Time and Level of Rice Bran*

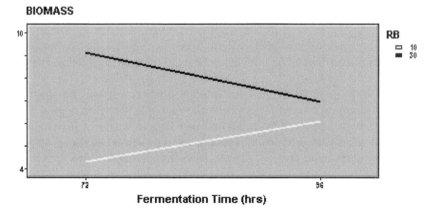

medium, it can be seen that increasing the fermentation time from 72 to 96 hrs increases the biomass produced. However, if there are 30 g/L of rice bran in the growth medium, increasing the fermentation time actually decreases the biomass produced.

Figure 6.10 shows how the effect of ammonium sulfate upon biomass production depends upon the level of sodium phosphate. Here it can be seen that increasing the level of ammonium sulfate from 0.0 to 2.0 g/L causes a greater increase in biomass when there is no sodium phosphate added to the nutrients than when 2.0 g/L of sodium phosphate is added to the nutrients. However, since adding sodium phosphate also causes an increase in biomass, the maximum biomass appears to result when there is 2.0 g/L of sodium phosphate and 2.0 g/L of ammonium sulfate.

Figure 6.10 *Interaction Plot for Ammonium Sulfate and Sodium Phosphate*

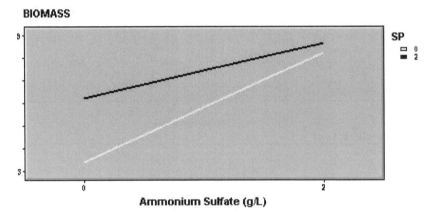

Figure 6.11 shows the interaction plot of fermentation time and urea, CH. On the average, it can be seen that increasing fermentation time seems to have little effect on biomass production. Also, on the average, adding urea to the growth medium seems to have little effect on the biomass produced. However, as can be seen in the graph, the effect of fermentation time upon biomass depends upon whether urea is present, and its effect appears to be exactly opposite depending on whether 2.0 g/L of urea is added to the growth medium.

While it is possible to have an interaction between two factors that do not have significant main effects (like the example shown in Figure 6.11), it is rare. Usually interactions occur between factors where at least one of the two main effects are significant. This has been described as the *effect heredity principle* by Hamada and Wu (1992). In this experiment since the two-factor interaction between fermentation time and urea is confounded with the three-factor interaction between rice bran, ammonium sulfate and sodium phosphate (i.e., $CH = BEG$), and all three of the latter factors have significant main

CRITERIA FOR CHOOSING GENERATORS FOR 2^{k-p} DESIGNS

Figure 6.11 *Interaction Plot of Fermentation Time and Urea*

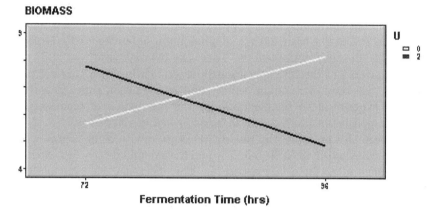

effects, it is possible that the large effect labeled as CH on the normal plot is actually the three-factor interaction. In this case the effect heredity principle may overshadow the hierarchical ordering principle.

A three-factor interaction means that the effect of one factor depends upon the combination of levels of two other factors. A series of interaction plots (like those in the example in Section 3.6) are useful for explaining or interpreting a three-factor interaction. Figure 6.12, produced by SAS ADX, shows the effect of ammonium sulfate upon biomass at the four combinations of rice bran and sodium phosphate.

Figure 6.12 *Interaction Plots To Interpret Three-Factor Ammonium Sulfate by Rice Bran by Sodium Phosphate Interaction*

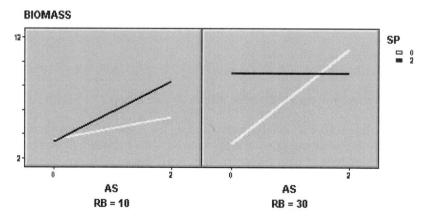

If this three-factor interaction is assumed to be important, the two-factor interaction plot in Figure 6.10 doesn't tell the whole story. In Figure 6.12 it can be seen that adding 2.0 g/L of ammonium sulfate to the growth medium increases the biomass produced in general. However, this effect is greatest when there is 30 g/L of rice bran and no sodium phosphate in the growth medium. The optimum result appears to be with a fermentation time of 72 hours, 30 g/L of rice bran, 2.0 g/L of ammonium sulfate and 0.0 g/L of sodium phosphate in the growth medium. There the biomass yield was 11.57 g/L.

The authors of the article continued experimenting with three of the four factors (rice bran, ammonium sulfate and fermentation time) in a response surface experiment to further optimize the production of biomass protein. These experiments and results will be described in Chapter 10.

To recap this example, the researchers started with eight factors with two levels each as shown in Table 6.5. It would have required $2^8 = 256$ experiments to complete a full factorial experiment. The effect sparsity principle and the hierarchical ordering principle suggest that this many experiments is probably not needed. With eight factors, it is doubtful that all main effects and a large number of interactions would be significant. A series of two fractional factorial experiments, employing a total of 24 runs, reveal that four of the main effects are significant and at most three interactions. Interaction graphs helped make a plausible interpretation of the results and identification of factor levels for optimal results. The interpretation of the interactions was critical for identification of optimal results and would never have been discovered if a seat-of-the-pants approach or vary one-factor-at-a-time plan had been utilized.

6.5 Augmenting Fractional Factorials

The last example introduced the idea of augmenting fractional factorial designs with additional experiments in order to de-confound certain main effects or interactions. In this section some formal procedures will be described that allow this to be done. Section 6.5.1 describes procedures that will preserve the optimal orthogonality property of the augmented design. Section 6.5.2 describes a procedure that may require fewer additional experiments, but may not preserve the orthogonality.

6.5.1 Augmenting by Foldover or Mirror Image Designs

In a resolution III fractional factorial design, main effects are confounded with some two-factor interactions. If more than one effect appears to be significant, after the analysis of data from a resolution III design, it is not clear whether all effects are due to main effects or whether some could be two-factor interactions. For example in an eight-run resolution III design in six factors (designated by 2^{6-3}_{III}) with generators $D = AB$, $E = AC$, and $F = BC$, the defining relation is

$$I = ABD = ACE = BCF = DEF = BCDE = ACDF = ABEF.$$

AUGMENTING FRACTIONAL FACTORIALS

If after analysis of data from this design, effects B and D appear to be significant, it could be that the two main effects are the active effects. However, the effect heredity principle tells us there are two alternative explanations. The first alternative is the following. Since D is confounded with AB and EF, it could be that main effect B and the AB interaction are the active effects and that D only appears large because it is confounded with AB. The second alternative is that main effect D and the AD interaction are the active effects and B only appears large because it is confounded with AD and CF. Without additional experiments, there is no way to determine which of the three explanations of the data is correct.

One way to break the confounding between main effects and two-factor interactions is to run an additional set of experiments that is the same as the first except that the coded factor levels on one or more factors has been reversed (Box et al., 1978). This is called a foldover fraction. For example, in the 2_{III}^{6-3} described in the last paragraph, if the signs were reversed for factor B, the defining relation would become $I = -ABD = ACE = -BCF = DEF = -BCDE = ACDF = -ABEF$. If the two eight-run designs were combined, as shown in Table 6.6, the resulting sixteen-run design would have defining relation $I = ACE = DEF = ACDF$. Although this is still a resolution III design, main effect B is clear, and main effect D is no longer confounded with the AB interaction.

Table 6.6 2_{III}^{6-3} Design Augmented by 2_{III}^{6-3} Design with Signs Reversed on Factor B

Run	A	B	C	D	E	F
1	-	-	-	+	+	+
2	+	-	-	-	-	+
3	-	+	-	-	+	-
4	+	+	-	+	-	-
5	-	-	+	+	-	-
6	+	-	+	-	+	-
7	-	+	+	-	-	+
8	+	+	+	+	+	+
9	-	+	-	+	+	+
10	+	+	-	-	-	+
11	-	-	-	-	+	-
12	+	-	-	+	-	-
13	-	+	+	+	-	-
14	+	+	+	-	+	-
15	-	-	+	-	-	+
16	+	-	+	+	+	+

In general, augmenting a resolution III design with another resolution III design, with the coded factor levels for one factor reversed, will make that factor and all of its two-factor interactions clear of other two-factor interactions.

Montgomery and Runger (1996) show the defining relation for the combined design from the original plus foldover will contain those effects in the original design that were not sign reversed in the foldover fraction.

If a resolution III design is augmented with another resolution III design, with the coded factor levels reversed for all factors (called the mirror image design), the combined design will be resolution IV and all main effects will be clear of two-factor interactions. The defining relation for the combined design in this situation will only contain even length words from the original defining relation. For example, the 2_{III}^{5-2} design with generators $D = AB$, $E = AC$ has defining relation $I = ABD = ACE = BCDE$. If it is combined with its mirror image fraction where the signs of all the coded factor levels have been reversed, the defining relation of the combined design will be $I = BCDE$. The SAS code below shows how this design can be created in the SAS data step. It is also possible to create foldover and mirror image designs in the SAS ADX tool.

```
data orig;
  do x1 = -1 to 1 by 2;
   do x2 = -1 to 1 by 2;
    do x3 = -1 to 1 by 2;
    A=x3; B=x2; C=x1; D=A*B; E=A*C; Block=1; output;
    end;
   end;
  end;
keep A B C D E Block;
data fold; set orig;
A=-A; B=-B; C=-C; D=-D; E=-E; Block=2;
data comb; set orig fold;
proc print; run;
```

When creating a resolution IV design by combining a resolution III design with its mirror image (i.e., signs reversed on all factors), it is possible to add one additional blocking factor that can account for any difference in the average response between the two sets of experiments. By having an additional factor, the experiments can be performed sequentially. The original resolution III design is completed first and the data analyzed. If there is only one factor that appears to be significant, there may be no need for additional experiments. If, on the other hand, two or more effects appear significant, the mirror image design can be completed to clear all main effects from two-factor interactions. The additional factor can account for any changes in the experimental environment that have occurred since the first set of experiments. Including it in the analysis will prevent any unanticipated changes from biasing the effects of the factors studied.

Augmenting a resolution IV design with its mirror image design will not help to break strings of aliased two-factor interactions since the signs will not change on the two-factor interactions. However, augmenting a resolution IV designs by a foldover fraction can be used to break strings of aliased two-factor interactions, when the signs are changed on only one or two of the factors. For example, the 2_{IV}^{8-4} design with $E = BCD$, $F = ACD$, $G = ABC$, and $H = ABD$ has seven strings of confounded two-factor interactions in groups of 4, i.e.,

$$AB + EF + CG + DH$$
$$AC + DF + BG + EH$$
$$CD + BE + AF + GH$$
$$AD + CF + EG + BH$$
$$AE + BF + DG + CH$$
$$BC + DE + AG + FH$$
$$BD + CE + FG + AH.$$

If this design is augmented with another design where the signs are reversed on factor A, all the two-factor interactions involving factor A will be clear of other two-factor interactions.

In general, Montgomery and Runger (1996) show that the alias sets for the combined design are obtained from a partition of the alias sets in the original fraction by grouping effects that are sign reversed. For example, in the 2_{IV}^{6-2} design with generators $E = ABC$, and $F = BCD$ the aliases for the 16 effects that can be estimated (up to three-factor interactions) are:

$$I + ABCE + BCDF + ADEF$$
$$A + BCE + DEF$$
$$B + ACE + CDF$$
$$C + BDF + ABE$$
$$D + BCF + AEF$$
$$E + ABC + ADF$$
$$F + BCD + ADE$$
$$CE + AB$$
$$AC + BE$$
$$AD + EF$$
$$BC + AE + DF$$
$$DE + AF$$
$$CF + BD$$
$$CD + BF$$
$$ABD + CDE + ACF + BEF$$
$$ACD + BDE + CEF + ABF$$

If this design were augmented by the foldover design with all signs for factor A reversed, 32 effects could be estimated, and the aliases for each of them could be determined from the list above. Every effect containing A changes signs. Group effects with like signs together to get the following 32 groups.

$$
\begin{array}{rl}
ABCE + ADEF, & I + BCDF \\
A, & BCE + DEF \\
ACE, & B + CDF \\
ABE, & C + BDF \\
AEF, & D + BCF \\
ABC + ADF, & E \\
ADE, & F + BCD \\
AB, & CE \\
AC, & BE \\
AD, & EF \\
AE, & BC + DF \\
AF, & DE+ \\
ACDE + ABEF, & CF + BD \\
ABDE + ACEF, & CD + BF \\
ABD + ACF, & CDE + BEF \\
ACD + ABF, & BDE + CEF
\end{array}
$$

If the 2_{IV}^{6-2} design were augmented by the same design with signs reversed on factors A and B, effects containing a single A or a single B will reverse signs. Effects containing both A and B or neither A nor B will not reverse signs, and the aliases for the 32 estimable effects will be different than those shown above. When only a few effects are found significant after analysis of data from a resolution IV design, a foldover that will separate the significant main effects from two-factor interactions (that involve the significant main effects) can usually be found.

Consider two designs used to optimize a drinking water filter to illustrate the augmenting of a fractional factorial to estimate interactions. High concentrations of arsenic are reported in ground water in countries such as Argentina, Bangladesh, Chile, China, India, Japan, Mexico, Mongolia, Nepal, Poland, Taiwan, Vietnam, and some parts of the United States, and studies have shown people exposed to high levels of arsenic are prone to develop various forms of cancer. Iron oxide coated sand (IOCS) has been used to remove arsenic from ground water in simple household filtration systems in Bangladesh. Ramakrishna et al. (2006) conducted a study with the objective of systematically studying the effects of several factors on the arsenic removal rate for (IOCS).

AUGMENTING FRACTIONAL FACTORIALS

The experiments consisted of making a coating solution composed of ferric nitrate and sodium hydroxide, with NAOH added to adjust the pH. This solution was aged and then poured over clean dry sand and mixed. The mixture was dried for 12 hours, and then used to filter water solutions spiked with a known concentration of arsenic. The response was the percentage of arsenic removed, and the factors varied in the study are shown in the table below.

Table 6.7 Factors and Levels for Arsenic Removal Experiment

Label	Factors	Levels −	Levels +
A	coating pH	2.0	12.0
B	drying temperature	110°	800°
C	Fe concentration in coating	0.1 M	2 M
D	number of coatings	1	2
E	aging of coating	4 hrs	12 days
F	pH of spiked water	5.0	8.0
G	mass of adsorbent	0.1 g	1 g

A 2_{III}^{7-4} fraction factorial design, with generators $D = AB$, $E = AC$, $F = BC$, and $G = ABC$, was used for the first set of experiments. The results are shown in the top half of Table 6.8. These experiments are shown in the

Table 6.8 2_{III}^{7-4} Design Augmented by Foldover 2_{III}^{7-4} with Signs Reversed on all Factors

Run	Block	A	B	C	D	E	F	G	%removal of As
1	1	−	−	−	+	+	+	−	69.95
2	1	+	−	−	−	−	+	+	58.65
3	1	−	+	−	−	+	−	+	56.25
4	1	+	+	−	+	−	−	−	53.25
5	1	−	−	+	+	−	−	+	94.40
6	1	+	−	+	−	+	−	−	73.45
7	1	−	+	+	−	−	+	−	10.00
8	1	+	+	+	+	+	+	+	2.11
9	2	+	+	+	−	−	−	+	16.20
10	2	−	+	+	+	+	−	−	52.85
11	2	+	−	+	+	−	+	−	9.05
12	2	−	−	+	−	+	+	+	31.1
13	2	+	+	−	−	+	+	−	7.40
14	2	−	+	−	+	−	+	+	9.90
15	2	+	−	−	+	+	−	+	10.85
16	2	−	−	−	−	−	−	−	48.75

standard order, not the random order in which they were run.

A half-normal plot of the effects calculated from the first eight runs (left as an exercise) revealed that factors B (temperature of drying the coated sand) and F (pH of the arsenic spiked water) appeared to be significant. Both factors had negative effects and the high level of each resulted in a lower percentage of arsenic removal. This implies that percent arsenic removed is only influenced by the temperature of drying the coated sand and by the pH of the spiked water solution. However, interpreting the results to mean that only the two main effects are active could be misleading. B is confounded with three two-factor interactions $AD + CF + EG$ and F is confounded with $BC + DE + AG$. According to the effect heredity principle, the model including B, and BC (which imply arsenic removal is influenced by the temperature of drying the coated sand and by the Fe concentration in the coating), and the model including F and CF (which imply arsenic removal is influenced by the pH of the spiked water and by the Fe concentration in the coating) are also plausible explanations of the data. From the eight experiments in the 2_{III}^{7-3} it is impossible to tell which of the three plausible models is appropriate since the effects are completely confounded.

The original eight experiments were augmented with the mirror image design, shown in the bottom half of Table 6.8, in order to de-confound the main effects from strings of two-factor interactions. An additional blocking factor is included in the combined design to account for any difference in the two sets of experiments. A half-normal plot of the effects calculated from the combined experiments (left as an exercise) revealed that the blocking factor representing the difference in the two groups of experiments, main effect F (pH of the arsenic spiked water), main effect B (temperature of drying the coated sand), main effect A (pH of the coating), and the effect representing the confounded string of interactions $AD + CF + EG$ all appeared to be significant. The effect heredity principle would suggest two plausible models to explain the data (ignoring the block term). The first model is (F, B, A, AD), or:

$$\% \ removal = 37.76 - 12.99 \left(\frac{pHs - 7.0}{2.0}\right) - 11.76 \left(\frac{temp - 455°}{345°}\right)$$
$$- 8.89 \left(\frac{pHc - 7.0}{5.0}\right) - 10.09 \left(\frac{pHs - 7.0}{2.0}\right)\left(\frac{number\ coats - .75}{.5}\right).$$

This model suggests that increasing the pH of the ferric nitrate sodium hydroxide solution used to coat the sand has little effect on arsenic removal when the sand is only coated once, but that it substantially decreases the arsenic removal when the sand is coated twice. This can be visualized in the left side of Figure 6.13. If this model is correct, it would imply that maximum arsenic removal in a sand filtration system can be achieved when the coated sand is dried at the low temperature ($B = -$), the coating solution has a low pH ($A = -$), and the sand is coated twice ($D = +$).

AUGMENTING FRACTIONAL FACTORIALS

The second plausible model is (F, B, A, CF), or:

$$\% \ removal = 37.76 - 12.99 \left(\frac{pHs - 7.0}{2.0} \right) - 11.76 \left(\frac{temp - 455°}{345°} \right)$$
$$- 8.89 \left(\frac{pHc - 7.0}{5.0} \right) - 10.09 \left(\frac{Fe - 1.05M}{0.95M} \right) \left(\frac{pHs - 7.0}{2.0} \right).$$

This model suggests that increasing the Fe concentration in the coating solution, increases the arsenic removal when the pH of the water being filtered is low, but decreases it when the pH of the water being filtered is high. This can be visualized in the right side of Figure 6.13. If this model is correct, it would imply that maximum arsenic removal in a sand filtration system can be achieved when the coated sand is dried at the low temperature $(B = -)$, and when the Fe concentration in the coating solution is high if the pH of the water to be filtered is low, or when the Fe concentration in the coating solution is low if the pH of the water to be filtered is high.

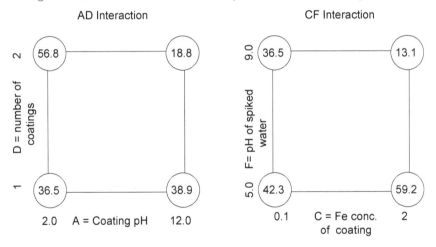

Figure 6.13 *Plausible Interaction Plots for Arsenic Removal Experiment*

The two models are contradictory. The first model implies that the pH of the coating solution and the number of times the sand is coated influence the percent of arsenic the sand filter can remove. However, the second model implies the pH of the coating solution and the number of times the sand is coated does not affect arsenic removal; rather, the Fe concentration of the coating solution does matter. The optimum level of this concentration must be determined by the pH of the water to be filtered.

In sixteen experiments the researchers were able to narrow down the list of factors that influence arsenic removal from the original seven down to five. But, because the two interactions AD and CF are completely confounded, there is no way from the results of these two sets of experiments to determine which model is correct and what is the best way to coat sand for a filtration system. If another foldover fraction of 16 experiments were run reversing the signs on one of the factors A, D, C, or F, the combined set of 32 experiments would allow for estimating all the terms (F, B, A, AD, and CF), and a reliable model could be established. The next section will describe another way of augmenting the data to estimate all the terms with fewer additional experiments.

6.5.2 Augmenting by Optimal Design

In matrix notation the model with the terms (F, B, A, AD, and CF), from the last section, and the sixteen run design shown in Table 6.8 can be written as:

$$y = X\beta + \epsilon,$$

where

$$y = \begin{pmatrix} 69.95 \\ 58.65 \\ 56.25 \\ 53.25 \\ 94.40 \\ 73.45 \\ 10.00 \\ 2.11 \\ 16.20 \\ 52.85 \\ 9.05 \\ 31.10 \\ 7.40 \\ 9.90 \\ 10.85 \\ 48.75 \end{pmatrix}, \quad X = \begin{pmatrix} 1 & -1 & -1 & -1 & 1 & -1 & -1 \\ 1 & -1 & 1 & -1 & 1 & -1 & -1 \\ 1 & -1 & -1 & 1 & -1 & 1 & 1 \\ 1 & -1 & 1 & 1 & -1 & 1 & 1 \\ 1 & -1 & -1 & -1 & -1 & -1 & -1 \\ 1 & -1 & 1 & -1 & -1 & -1 & -1 \\ 1 & -1 & -1 & 1 & 1 & 1 & 1 \\ 1 & -1 & 1 & 1 & 1 & 1 & 1 \\ 1 & 1 & -1 & -1 & -1 & -1 & -1 \\ 1 & 1 & 1 & -1 & -1 & -1 & -1 \\ 1 & 1 & -1 & 1 & 1 & 1 & 1 \\ 1 & 1 & 1 & 1 & 1 & 1 & 1 \\ 1 & 1 & -1 & -1 & 1 & -1 & -1 \\ 1 & 1 & -1 & -1 & 1 & -1 & -1 \\ 1 & 1 & 1 & 1 & -1 & 1 & 1 \\ 1 & 1 & 1 & 1 & -1 & 1 & 1 \end{pmatrix}, \quad \beta = \begin{pmatrix} \beta_0 \\ \beta_{bl} \\ \beta_A \\ \beta_B \\ \beta_F \\ \beta_{AD} \\ \beta_{CF} \end{pmatrix}.$$

This model cannot be fit to the data since the last two columns in the X matrix are identical, and the least squares normal equations are singular. However, it is not necessary to double the number of runs, which would be required if augmented by a foldover, in order to estimate all of the parameters in the model. Doubling the number of runs could preserve the orthogonality property so that all estimated effects would be uncorrelated. However, any additional runs that would make the $X'X$ matrix nonsingular will allow estimation of all the effects of interest.

One criterion that can be checked, to insure the $X'X$ matrix nonsingular, is $|X'X|$, and Dykstra (1971) has shown that choosing a set of additional runs that will maximize $|X'X|$ for the combined set of runs is called the *D-Optimal* design. The motivation for D-Optimal designs comes from the fact that the variance covariance matrix of the regression coefficients in the linear model is $\sigma^2(X'X)^{-1}$ and the reciprocal of the volume of a confidence ellipsoid for

AUGMENTING FRACTIONAL FACTORIALS

$\hat{\beta}$ is proportional to the determinant of $X'X$. Orthogonal designs are always D-Optimal and have the maximum $|X'X|$ for their run size, but when the experimenter is willing to forgo uncorrelated effect estimates in order to reduce the number of runs required, a D-optimal design for a reduced run size will be the best choice.

Another criterion that can be used to choose additional runs is the $\text{tr}(X'X)^{-1}$, since the variances of the regression coefficients are the diagonal elements of $\sigma^2(X'X)^{-1}$. A design that minimizes $\text{tr}(X'X)^{-1}$ is said to be *A-optimal*, since it minimizes the average variance of the regression coefficients.

SAS proc optex can find both D-optimal and A-optimal designs when given the model and a list of candidate design points. This procedure uses an algorithm to find the subset of the candidate design points that will maximize $|X'X|$ or minimize $\text{tr}(X'X^{-1})$. To illustrate how this proc can be used to augment an existing design, consider augmenting the design for the arsenic removal experiments so that all the terms in the model can be determined. The design shown in Table 6.8 was created with a modification of the code used earlier to produce a combined file (original plus foldover), and was stored in the SAS file comb. The SAS code below creates a list of 128 candidate points forming a full factorial in the file cand. Next proc optex is called to augment the file comb by selecting eight additional runs from cand to result in a total of 24 runs in three blocks. The option criterion=D on the generate statement specifies a D-optimal design. This is the default for proc optex if this option is omitted. To generate an A-optimal design change the option to criterion=A. The eight new runs in the third block are printed and shown in

```
proc factex;
  factors A B C D E F G;
  size design=128;
  model estimate=(A|B|C|D|E|F|G);
  output out=cand;
  examine confounding;
run;
data cand; set cand;
  Block=3;
proc optex data=cand seed=37034;
  class Block;
  model Block A B F A*D C*F;
  generate n=24 method=detmax criterion=D augment=comb;
  output out=optim;
run;
data newruns; set optim; if Block=3;
proc print ; run;
```

the SAS output at the top of the next page. After completing the experiments for this block, the model including main effects A, B, and F, and the AD and CF interactions could be fit to the data using a regression program such as proc glm or proc reg.

Obs	A	B	F	D	C	Block
1	-1	-1	-1	1	-1	3
2	-1	-1	-1	-1	1	3
3	-1	1	1	1	1	3
4	-1	1	1	-1	-1	3
5	1	-1	1	-1	1	3
6	1	-1	1	1	-1	3
7	1	1	-1	-1	-1	3
8	1	1	-1	1	1	3

6.6 Plackett-Burman (PB) Screening Designs

Resolution III 2^{k-p} fractional factorial designs are often used for screening experiments where the objective is to determine which factors (from a list assembled by brainstorming) are important enough to be studied in more detail in follow-up experiments. However, the number of runs in an 2^{k-p} fractional factorial design is always a power of 2, i.e., 8, 16, 32 etc., and these limited choices for run size can be restrictive in screening experiments. For example, to examine 8 factors requires at least 16 experiments, and to examine 16 factors requires 32 experiments when using a fractional factorial design.

Resolution III Plackett and Burman (1946) designs, or PB designs, are available in run sizes that are multiples of 4, i.e., 8, 12, 16, 20, etc. These designs were originally discovered by two British statisticians during World War II while studying the effect of a number of factors on the performance of anti-aircraft proximity fuse prototypes. Like 2^{k-p} fractional factorial designs, these designs have two levels for each factor; and for run sizes that are powers of 2, they are the same as a 2^{k-p} design. For other run sizes, they retain the desirable orthogonality property of 2^{k-p} designs, but they do not have generators or a defining relation. The designs for run sizes of 12, 20 and 24 can be created by cyclically rotating the factor levels for the first run. Table 6.9 shows the factor levels for the first run in these designs. Table 6.10 shows

Table 6.9 *Factor Levels for First Run of Plackett-Burman Design*

Run size	Factor Levels
12	+ − + − − − + + + − +
20	+ − + + − − − − + − + − + + + + − − +
24	+ − − − − + − + − − + + − − + + − + − + + + +

how these factor levels are used to create a 12-run Plackett-Burman Design in 11 factors. The factor levels for the first run are copied directly from the first line of Table 6.9. In the second run, the level for the first factor (A) is the level of the eleventh factor (L) in the first run, and the factor levels for factors B through L for the second run are the factor levels for factors A through K from the first run. This pattern is continued, cyclically rotating the factor levels for each run to create the factor levels for the next run until the last run. For run number 12, all the factors are set to their low (−) level. To create

PLACKETT-BURMAN (PB) SCREENING DESIGNS

Table 6.10 *12 Run Plackett-Burman Design*

run	A	B	C	D	E	F	G	H	J	K	L
1	+	-	+	-	-	-	+	+	+	-	+
2	+	+	-	+	-	-	-	+	+	+	-
3	-	+	+	-	+	-	-	-	+	+	+
4	+	-	+	+	-	+	-	-	-	+	+
5	+	+	-	+	+	-	+	-	-	-	+
6	+	+	+	-	+	+	-	+	-	-	-
7	-	+	+	+	-	+	+	-	+	-	-
8	-	-	+	+	+	-	+	+	-	+	-
9	-	-	-	+	+	+	-	+	+	-	+
10	+	-	-	-	+	+	+	-	+	+	-
11	-	+	-	-	-	+	+	+	-	+	+
12	-	-	-	-	-	-	-	-	-	-	-

a 12-run design in k factors, where $k < 11$, proceed exactly as above, but use only k columns in the design to define factor setting. To create a 20 or 24-run Plackett-Burman Design proceed exactly as before, but start with the row in Table 6.9 with the correct run size.

These designs can be created easily in the SAS data step. The code at the top of the next page shows how to create a 12-run design like that shown in Table 6.10. The first data step reads in the factor levels for run 1 that were copied from Table 6.9. The second data step creates the factor levels for runs 1 through 11 by cyclically rotating the levels for the first run. The factor levels for the final run are all set to -1.

There are SAS macros included in the AUTOCALL library that can also be used to create Plackett-Burman Designs. These macros can be included in your SAS code by simply typing the file name preceded by a %. The ADXPBD macro creates a Plackett-Burman design with the statement %adxpbd(*file,nf*), where *file* is the name of the SAS file that will contain the design and *nf* is the number of factors. For example, to create a 28-run Plackett-Burman design with 25 factors, simply enter the commands

```
%adxgen
%adxff
%adxinit
%adxpbd(pb28,25)
proc print data=pb28; run;
```

These macros are the basis for the SAS ADX graphical user interface, and it also can be used to create Plackett-Burman designs. In practice, the order of the runs in a Plackett-Burman Design should be randomized to reduce the chance of bias from unknown factors before conducting the experiments just as they would be in any other design.

```
* Create 12-run Plackett-Burman Design in Data Step;
* first reads factor levels for first line from    ;
* Table 6.9                                        ;
data lis;
  input f1-f11;
  datalines;
1 -1 1 -1 -1 -1 1 1 1 -1 1
data pb11; set lis;
* Next cyclically rotates factor levels;
  array f{11} f1-f11;
  array t{11} t1-t11;
    do i=1 to 11;
      do j= 1 to 11;
        if j<i then k=11-i+1+j;
        else k=j-i+1;
        t{j}=f{k};
      end;
      output;
    end;
  do i=1 to 11;
    t{i}=-1;
  end;
  output;
keep t1-t11;
proc print; run;
```

As an example of the use of a Plackett-Burman design, consider the data in Table 6.11, which comes from a study by Hunter *et al.* (1982) of the fatigue life of weld-repaired castings. Seven factors were assigned to the last seven columns of the Plackett-Burman design matrix shown in Table 6.10, and the first four columns were not assigned to factors in defining the experiments. The purpose of the experiments was to identify the factors that affect the fatigue life of the weld repairs. The response, y, is the log of the lifetime of each weld repair. We will use this data to illustrate the analysis of data from a Plackett-Burman Design. Analysis of this data has been presented in the literature by several others.

Although there is no defining relation in a Plackett-Burman design, it is a resolution III design and the unassigned columns in Table 6.11 represent confounded strings of interactions. A first step in the analysis would be to calculate an effect for each of the 11 columns in Table 6.11 and make a half-normal plot.

PLACKETT-BURMAN (PB) SCREENING DESIGNS

Table 6.11 *Design Matrix and Lifetime Data for Cast Fatigue Experiment*

run	c8	c9	c10	c11	G	F	E	D	C	B	A	
1	+	−	+	−	−	−	+	+	+	−	+	4.733
2	+	+	−	+	−	−	−	+	+	+	−	4.625
3	−	+	+	−	+	−	−	−	+	+	+	5.899
4	+	−	+	+	−	+	−	−	−	+	+	7.000
5	+	+	−	+	+	−	+	−	−	−	+	5.752
6	+	+	+	−	+	+	−	+	−	−	−	5.682
7	−	+	+	+	−	+	+	−	+	−	−	6.607
8	−	−	+	+	+	−	+	+	−	+	−	5.818
9	−	−	−	+	+	+	−	+	+	−	+	5.917
10	+	−	−	−	+	+	+	−	+	+	−	5.863
11	−	+	−	−	−	+	+	+	−	+	+	6.058
12	−	−	−	−	−	−	−	−	−	−	−	4.809

The SAS commands to change the column names for the 12-run Plackett-Burman design created earlier and merge it with the response data from Table 6.11 is shown below.

```
data cast; set pb11;
  c8=t1; c9=t2; c10=t3; c11=t4; G=t5; F=t6; E=t7; D=t8; C=t9;
  B=t10; A=t11; keep c8 c9 c10 c11 G F E D C B A;
data y; input y @@;
datalines;
4.733 4.625 5.899 7.000 5.752 5.682 6.607 5.818 5.917
5.863 6.058 4.809
data castf; merge cast y;
```

Next, a model is fit to the data including all eleven columns in Table 6.11. The SAS commands to do this are shown below.

```
proc glm data=castf;
  model y=A B C D E F G C8 C9 C10 C11/solution;
  ods output ParameterEstimates=sol;
run;
```

A half-normal plot of the absolute regression coefficients was created by copying the code from Section 3.7.5, and the plot is shown in Figure 6.14.

Figure 6.14 *Half-Normal Plot of Absolute Regression Coefficients from Cast Fatigue Experiment*

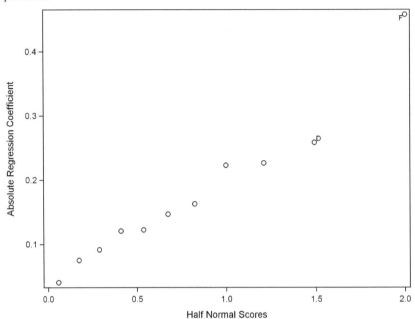

The plot shows that main effect F is clearly significant and that the next largest effect is main effect D. The authors of the original article concluded that the model including these two factors was the correct model, and they fit it to the data by regression obtaining an $R^2 = .5867$. However, the same caution should be exercised that was illustrated in the water filtration example presented in Section 6.5. Because this Plackett-Burman design has resolution III, two-factor interactions are confounded with main effects and there may be other plausible models for the data. A mirror image design could be added to the Plackett-Burman design in order to clear main effects of two-factor interactions; however, this is not necessary due to the complex aliasing for this type design.

By complex aliasing we mean that each interaction is partially confounded with many main effects rather than being completely confounded with one main effect as it would be in a 2_{III}^{k-p} design. Lin and Draper (1992) and Wang and Wu (1995) showed that designs with complex aliasing have a hidden projection property. The hidden projection property allows some interactions to be estimated even though the design is resolution III. For Plackett-Burman designs where only a subset of the factors appear to be important, Wang and Wu have shown that a model can be fit to the data by regression which includes the main effects and all two-factor interactions involving the important main effects. When interaction terms are included in the model, the design becomes non-orthogonal and effects must be estimated by regression, but Wang and Wu have shown that the D-efficiency remains quite high. Usually some type of regression subset selection procedure is used in order to identify the most appropriate model. For the data in Table 6.11, the SAS commands below, create interaction columns and use `proc reg` to perform an all-subsets regression.

```
data castf; set castf;
*Note Defines Interactions;
  AB=A*B; AC=A*C; AD=A*D; AE=A*E; AF=A*F; AG=A*G;
          BC=B*C; BD=B*D; BE=B*E; BF=B*F; BG=B*G;
                  CD=C*D; CE=C*E; CF=C*F; CG=C*G;
                          DE=D*E; DF=D*F; DG=D*G;
                                  EF=E*F; EG=E*G;
                                          FG=F*G;
proc reg data=castf;
  model y=A B C D E F G AB AC AD AE AF AG BC BD BE BF BG CD CE
  CF CG DE DF DG EF EG FG/selection=rsquare best=4 stop=4;
run;
```

The results shown below indicate that when two variables are considered in the model, the model which includes main effect F and the interaction FG fits best and has an $R^2 = 0.8925$, compared to the $R^2 = 0.5867$ for the model that the authors fit to the data (which included main effects D and F). The best model with three variables in the equation only raises R^2 from 0.8925 to 0.9526 and the third variable is probably not needed. The model including F and FG is the model found in the other papers which have reanalyzed this data by various procedures.

```
Number in
  Model     R-Square    Variables in Model

    1       0.4474      FG
    1       0.4451      F
    1       0.4377      AE
    1       0.1812      BC
    ---------------------------------------------
    2       0.8925      F  FG
    2       0.6638      AE FG
    2       0.6621      F  AE
    2       0.6239      BD FG
    ---------------------------------------------
    3       0.9526      F  AE FG
    3       0.9313      F  BD FG
    3       0.9190      D  F  FG
    3       0.9116      F  DG FG
    ---------------------------------------------
    4       0.9686      F  AE EF FG
    4       0.9660      F  AE CD FG
    4       0.9643      F  AD AE FG
    4       0.9642      F  BD CF FG
```

Table 6.12 shows a summary of the data classified by the levels of factors F and G. There it can be seen that the consistently longest log fatigue life for the weld repaired castings occurs when factor F is at its high level and factor G is at its low level. In this example seven factors were under study. After just 12 experiments it was found that only two of the factors (F and G) were important, and that the conditions for maximum life (among the combinations studied) were discovered.

Table 6.12 *Summary of Data from Cast Fatigue Experiment*

Factor F	Factor G	
	−	+
−	4.733	5.899
	4.625	5.752
	4.809	5.818
+	6.058	5.682
	7.000	5.917
	6.607	5.863

In Plackett-Burman designs with a large number of factors, the computational effort required to do an all-subsets regression to search for the most appropriate model may be prohibitive. Methods have been proposed to limit the number of candidate terms for a model search. Hamada and Wu (1992) proposed an iterative forward stepwise regression approach guided by the principle of effect heredity. Box and Meyer (1993) proposed a Bayesian approach for identifying an appropriate model, and Chipman *et al.* (1997) proposed using the Bayesian stochastic search algorithm that incorporates the effect heredity principle through heredity priors that capture the relation between the importance of an interaction term and the main effects from which it is formed. Lawson (2002) proposed limiting the interaction candidates for an all-subsets regression based on the alias structure of the design. There is a SAS macro available to perform this procedure.

Sometime there might be several competing models that are found to fit the data well (from a Plackett-Burman Design) after performing a subset selection to find an appropriate model. If this is the case, additional follow-up experiments may be required to provide data that will allow accurate discrimination between the alternate models. SAS `proc optex` can be used as shown in Section 6.2.2 to find an additional block of data that is D-optimal for the model composed of all the terms from the competing models.

6.7 Mixed Level Factorials and Orthogonal Arrays (OA)

In the preliminary stage of experimentation, where the objective may be to determine which factors are important from a long list of candidates, two-level fractional factorial designs or Plackett-Burman designs are often appropriate. If a factor has quantitative levels, the two levels are denoted symbolically by $(-)$ and $(+)$, where $(-)$ represents the lowest level the experimenter would consider, and $(+)$ represents the highest level the experimenter would consider. The high and low are usually spread out as far as feasibly possible in order to accentuate the signal or difference in response between the two levels. If a factor has qualitative levels, the $(-)$ and $(+)$ designations are arbitrary, but the two levels chosen normally would be two that the experimenter believes should result in the maximum difference in response.

Sometimes, however, two-levels for each factor may not be adequate. In cases where the experimenter would like to consider non-linear effects of quantitative factors or qualitative factors with more than two alternatives, two-level fractional designs will not be suitable. For example, Fannin et al. (1981) report an experiment investigating the effects of four three-level factors and two two-level factors upon the rate of bacterial degradation of phenol for the purpose of evaluating the fate of chemicals in aquatic ecosystems. A full factorial would require $3^4 \times 2^2 = 324$ experiments; however, the study was completed using only a fraction of these runs by utilizing a mixed level fractional factorial design based on an orthogonal array. Taguchi (1986) describes an experiment to determine the factors that affect the durability of an auto clutch spring. The factors and levels are shown in the table below. The levels of factors A,

Factor	Description	Levels
A	Shape	3 alternatives
B	Hole ratio	2 possibilities
C	Coining	2 possibilities
D	Stress σ_t	90 65 40
E	Stress σ_c	200 170 140
F	Shop peening	3 alternatives
G	Outer Perimeter planing	3 alternatives

C, F and G represented discrete alternatives that were of interest. Factors B, D and E were continuous factors and three levels were included in order to determine whether there was a curvilinear relation between these factor levels and durability of the clutch springs. There was also interest in the interaction between factors D and F and the interaction between factors D and G. A full factorial would require $3^5 \times 2^2 = 972$ experiments, but Taguchi was able to get valid data with a fractional design that included only 27 experiments.

An orthogonal array $OA(N, s_1^{m_1}, \ldots, s_\gamma^{m_\gamma}, 2)$ of strength 2 is an $N \times m$ matrix, $m = m_1 + \cdots + m_\gamma$ where column m_i has $s_i (\geq 2)$ symbols (or levels) such that for any subset of two columns all possible combinations of the symbols occur equally often in the matrix. These designs are orthogonal in the main effects and are of resolution III. However, in some cases like the Plackett-Burman designs, they can be used to estimate a limited number of interactions as well.

A necessary condition for the existence of an orthogonal array is that number of runs N be divisible by the product of each possible pair of factor levels. For example, in the experiment described by Taguchi, with five three-level factors and two two-level factors, $3^5 \times 2^2$, no orthogonal array fractional factorial exists with $N = 54$ runs because 54 is not divisible by $2 \times 2 = 4$. In addition, no orthogonal array exists with $N = 24$ runs because 24 is not divisible by $3 \times 3 = 9$. However an orthogonal array with $N = 36$ happens to exist since 36 is divisible by 4, 6, and 9.

Entire books have been written on methods for obtaining particular orthogonal array designs; however, SAS macros exist that will do the work. Specifically macro %mktRuns finds the number of runs required by an orthogonal array for a given number of factors and levels and displays alternative arrays from a vast catalog of designs that are stored or can be created by the software. Macro %mktEx actually finds the array given the number of runs in addition to the number of factors and levels. These macros have the prefix Mkt because orthogonal array designs are commonly used in market research studies such as conjoint analysis and stated choice experiments. These macros are described by Kuhfeld (2009) and the algorithms used by these macros are described by Kuhfeld and Tobias (2005). The call of the macro %mktRuns below illustrates the use of these macros to find a fractional factorial array design that includes five three-level factors and two two-level factors.

```
%mktruns(3**5 2**2)
```

Part of the resulting output is shown below.

```
        Saturated       = 13
        Full Factorial  = 972

        Some Reasonable                    Cannot Be
          Design Sizes      Violations     Divided By
               36 *             0
               72 *             0
               18               1          4
               54               1          4
               24              10          9
               48              10          9
               60              10          9
               30              11          4 9
               42              11          4 9
               66              11          4 9
               13 S            28          2 3 4 6 9

 * - 100% Efficient design can be made with the MktEx macro.
 S - Saturated Design - The smallest design that can be made.
     Note that the saturated design is not one of the
     recommended designs for this problem.  It is shown
     to provide some context for the recommended sizes.
```

MIXED LEVEL FACTORIALS AND ORTHOGONAL ARRAYS (OA)

The output shows that there are a total of $13 = 1 + 5 \times (3-1) + 2 \times (2-1)$ degrees of freedom required to estimate the mean and the six main effects, and that the number of runs required for a full factorial would be 972. Next, the results show that the orthogonal array with the minimum number of runs for studying five three-level factors and two two-level factors is 36, and that another orthogonal array with 72 runs exists.

Another part of the output, which is not shown here, lists all of the known 36-run and 72-run orthogonal array designs. Once the user decides how many runs to include in the orthogonal array, the %mktEx macro can be used to actually create the design as shown below.

```
%mktex(3**5 2**2, n=36)
proc print data=randomized; run;
```

The output of %mktEx indicates that a 100% D-efficient (or orthogonal design) for this run size was found. The design is not automatically printed out, but is stored in the SAS data set **design** and in a random order in the SAS data set **randomized**. The **proc print** statement above prints the contents of this randomized data file.

Sometimes when the purpose of experimentation is to screen the important factors in a preliminary experiment, the number of runs required for an orthogonal array may be more that the experimenter is willing to run. For example, in the $3^5 \times 2^2$ experiment, only 13 degrees of freedom are required to estimate all the main effects. If the experimenter, relying on the effect sparsity principle, believes that only one or two factors and their interaction will be important (like the last example presented), then 36 runs may be more than necessary. The output of the %mktRuns macro listed one 18-run design that had only one violation to the requirement for an orthogonal array. This means that an 18-run design can be constructed, but it will not be an orthogonal array. Sometimes designs with fewer runs than required for an orthogonal array, can still allow for uncorrelated estimates of main effects. These designs are called orthogonal main effect plans.

Although there are methods of constructing orthogonal main effect plans manually, a simple alternative is to use the %mktEx macro. It uses an algorithm to find an 18-run design that is as close to orthogonal as possible by using several methods including **proc optex**, selecting the method that works best for the problem, and then performing more iterations with the selected method. The macro call to produce this design is:

```
%mktex(3**5 2**2,n=18, maxiter=100, seed=104)
```

The output of %mktEx shown below indicates that the macro found a design that is 99.9045% D-efficient.

Design Number	D-Efficiency	A-Efficiency	G-Efficiency	Average Prediction Standard Error
1	99.9045	99.8081	99.0521	0.8498

D-efficiency is defined as:

$$100 \times \frac{1}{N|(\boldsymbol{X'X})^{-1}|^{\frac{1}{p}}},$$

where N is the number of runs in the design and p is the number of degrees of freedom for the mean and main effects. $|(\boldsymbol{X'X})^{-1}| = \frac{1}{N^p}$ for an orthogonal design, thus orthogonal designs are 100% D-efficient. For this example the 18-run design is not orthogonal, but is close. The design can be printed with the command:

```
proc print data=design; run;
```

or

```
proc print data=randomized; run;
```

The %mktEx macro can also search for a fractional design that will allow estimation of some interactions. For example to find a 27-run design for the $3^5 \times 2^2$ design that will allow estimation of the DE and DF interactions, similar to the one used by Taguchi, issue the macro call:

```
%mktex(3 2**2 3**4, interact=x4*x5 x4*x6, n=27)
proc print data=randomized; run;
```

As an example of the use of a design created with the Mkt macros, consider a conjoint analysis described by Wang et al. (2004). These types of studies are commonly done in market research to determine the relative importance of different product attributes to a customer's preference for the product. The purpose of this particular study was to assess the market potential for using low-grade hardwood lumber (with knots and other character marks) to produce fine furniture since the low-grade hardwood is under-utilized. The

MIXED LEVEL FACTORIALS AND ORTHOGONAL ARRAYS (OA)

products studied were hardwood chairs, and the product attributes (or factors) and levels are shown in Table 6.13.

A questionnaire was developed where potential customers were asked to rate product alternatives, composed of combinations of the levels shown in Table 6.13, on a seven-point scale (1 = least preferred to 7 = most preferred). The surveys were conducted in the Vermont state fair where an exhibit booth was set up where respondents were able to see the design and character mark density of the chairs and ask questions about them. A label was attached to each chair indicating the price and guarantee policy. Potential respondents were enticed into filling out the survey by giving them a chance to enter a raffle to win the chair of their choice.

Table 6.13 *Product Attributes and Levels for Conjoint Study*

Factor	Product Attribute	Levels
A	Design	1 = Rocking Chair
		2 = Arm Chair
		3 = Office Chair I
		4 = Office Chair II
B	Price	1 = $350.00
		2 = $425.00
		3 = $500.00
		4 = $575.00
C	Density of Marks	1 = Clear
		2 = Medium
		3 = Heavy
D	Guarantee policy	1 = 1-year
		2 = Unconditional

There are $4^2 \times 3 \times 2 = 96$ possible combinations of levels of the factors shown in Table 6.13, which would be too many to ask a respondent to rate. Twelve profiles or product alternatives were created from these levels using a mixed-level fractional factorial design.

The following macro calls were issued to look for an orthogonal array and create a design. The smallest orthogonal array for examining a $4^2 \times 3 \times 2$ design was 48, which is still too many to ask a respondent to rate. A 12-run design was created using the %mktEx macro, and was found to be 94.8% D-efficient.

```
%mktruns(4 4 3 2)
%mktex(4 4 3 2, n=12, seed=2013)
```

Table 6.14 shows the design in actual factor levels. These 12 combinations

244   FRACTIONAL FACTORIAL DESIGNS

are the profiles each respondent was asked to rate. The average rating in the last column is the average of 122 Vermont respondents. The raw (summary) data for the surveys consisted of a count of the number of respondents rating each profile with a 1, 2,...,7. This is similar to the data for the teaching experiment described in Section 2.5.3.

Table 6.14 *Mixed Level Fractional Factorial for Conjoint Study*

Profile	Design	Price ($)	Density of Marks	Guarantee policy	Average Rating
1	Rocking Chair	350	Clear	1-year	5.32273
2	Rocking Chair	425	Heavy	1-year	5.27871
3	Rocking Chair	575	Medium	Unconditional	5.35539
4	Arm Chair	425	Medium	Unconditional	4.73211
5	Arm Chair	500	Clear	Unconditional	4.75073
6	Arm Chair	575	Heavy	1-year	4.24606
7	Office Chair I	350	Heavy	Unconditional	4.62892
8	Office Chair I	500	Medium	1-year	3.94293
9	Office Chair I	575	Clear	1-year	3.85872
10	Office Chair II	350	Medium	1-year	4.39812
11	Office Chair II	425	Clear	Unconditional	4.71872
12	Office Chair II	500	Heavy	Unconditional	4.51137

The authors of the article used SAS `proc catmod` to fit a model to the data, which accounted for the discrete nature of the data, similar to the way `proc genmod` was used in Section 2.5.3. Since the raw summary data was not published in the article, model fitting will be illustrated using regression (`proc reg`) on the average ratings shown in Table 6.14.

The SAS commands shown below were used to fit the model. The author's

```
Data Recre;
input x1 x2 x3 x4 R;
  if x1=1 then do; D1=1;  D2=0;  D3=0;  end;
  if x1=2 then do; D1=0;  D2=1;  D3=0;  end;
  if x1=3 then do; D1=0;  D2=0;  D3=1;  end;
  if x1=4 then do; D1=-1; D2=-1; D3=-1; end;
G=(x4-1.5)/.5;
Price=275+75*x2;
Density=x3;
datalines;
1       1       1       1       5.32273
...
proc reg;
  model R=Price Density G D1 D2 D3;
run;
```

model used dummy variables (see Faraway, 2004) D1 to D3 to represent the differences in chair designs. D1 represents the difference between rocking chair and office chair II. D2 represents the difference between arm chair and office chair II, and D3 represents the difference between office chair I and office chair II. A dummy variable G was used to represent the difference between a 1-year and unconditional guarantee. Linear functions of both price and a ranking of the density of character marks (1 = clear, 2 = medium, and 3 = heavy) were also included in the author's model.

The SAS output of **proc reg** shown below is similar to what the authors obtained.

Variable	DF	Parameter Estimate	Standard Error	t Value	Pr > |t|
Intercept	1	5.36301	0.12194	43.98	<.0001
Price	1	-0.00156	0.00024024	-6.49	0.0013
Density	1	0.00177	0.02326	0.08	0.9423
G	1	0.18688	0.02014	9.28	0.0002
D1	1	0.71637	0.03371	21.25	<.0001
D2	1	-0.07290	0.03476	-2.10	0.0901
D3	1	-0.42007	0.03371	-12.46	<.0001

These results showed that while price, guarantee policy and chair design had a significant effect on customer preference, the density of character marks did not. This meant that there would be a market for hardwood furniture made with low-grade wood. The regression model could be used to predict the preference (demand) for the different chair designs, and the regression coefficient for price indicated the price elasticity.

In general, orthogonal array designs and orthogonal main effect plans are resolution III designs for fractions of mixed level factorials. Like Plackett-Burman designs, they have no defining relation. Except for fractions of $4^k \times 2^p$ designs, the alias structure is complex in that each interaction is partially confounded with many main effects. They also share the hidden projection property with Plackett-Burman Designs, and even though these designs are resolution III, some interactions may be detected using the methods of analysis described by Hamada and Wu (1992), Box and Meyer (1993), Chipman et al. (1997), and Lawson (2002).

6.8 Review of Important Concepts

When more than one factor is under study, factorial designs are preferable to studying each factor in separate designs. Experimentation is more efficient and interactions can be detected, which could be the most important information obtained from the experiments. However, if the list of factors to be studied is long, factorial designs may require too many experiments to be practical even when there are only two levels of each factor and no replicates. Fractional factorial designs (CRFF, PB or OA) require only a fraction of the runs for a full factorial, yet preserve many of the benefits such as orthogonality and ability to detect interactions. Figure 6.15 illustrates when these designs should be used in relation to the designs presented in earlier chapters.

Figure 6.15 *Design Selection Roadmap*

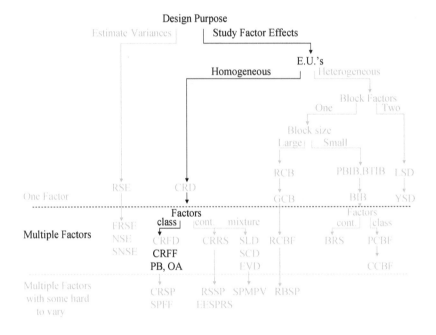

In order to preserve benefits of factorial designs, the runs in a fractional factorial must be carefully chosen. For half-fractions of 2^k designs this is accomplished by choosing a full factorial in $k-1$ factors and assigning the coded factor levels of the kth factor to the product of the coded factor levels of all the other factors. The defining relation is used to determine the alias structure of the design that shows which interaction is confounded or aliased with each estimable effect. No replicates are included in a fractional factorial design and the analysis is conducted by making a normal or half-normal plot of the ef-

REVIEW OF IMPORTANT CONCEPTS

fects estimated by regression. The principles of effect sparsity and hierarchical ordering help in interpreting the results.

The subset of runs for one-quarter and higher fractions of 2^k factorials can also be carefully selected by choosing generators that confound p added factors to interactions among the $k-p$ factors in the base design. From the generators the defining relation and alias structure for the design can be determined. SAS ADX system automatically generates designs that have the highest resolution and minimum aberration among all 2^{k-p} designs with the same number of runs. Designs of resolution R confound main effects with $R-1$ and higher order interactions. These designs also have a projective property. When only $R-1$ factors or less appear significant after running a resolution R design, the insignificant factors can be ignored and a full factorial in $R-1$ factors will remain. The 2^{k-p} fractional factorial projects to a 2^{R-1} full factorial in $R-1$ factors.

After analysis of data from a resolution III fractional factorial, the interpretation may be straightforward if only one factor appears to have a significant effect. When more than one factor appears to have a significant effect, the interpretation may be more difficult due to aliasing of interactions with main effects. Augmenting a resolution III fractional factorial with a foldover or mirror fraction can often eliminate the confounding between apparently important main effects and interactions, and can allow for a simple interpretation of the effects after only a fraction of the runs needed for a full factorial.

When confounded strings of two-factor interactions appear significant when analyzing data from a resolution IV design, augmenting the design with certain foldover fractions or D-Optimal subsets of runs can provide the data to determine which interactions are significant.

Plackett-Burman designs preserve many of the benefits of factorial and fractional factorial designs such as orthogonality among main effects and the projective property. In addition, these designs are available in run sizes that are multiples of 4 rather than powers of 2 as they are for fractional factorial designs. If more than one factor is significant after the analysis of a Plackett-Burman design, significant interactions can sometimes be detected using regression subset selection techniques without requiring additional experiments such as a foldover or D-Optimal fraction.

Finally, when more than two levels are required for some factors, mixed level fractional factorials can be created using orthogonal array and orthogonal main effect plans that are generated with the SAS Mkt macros. These designs are similar to Plackett-Burman designs in that they are orthogonal in the main effects and have no defining relation. Many of these designs also have a projective property, and interactions can be detected without additional experiments using methods of analysis similar to those used for Plackett-Burman designs.

6.9 Exercises

1. Consider a 2^{3-1} factional factorial design,

    (a) Modify the SAS code in Section 6.2 to create a list of the runs in the design.
    (b) Determine the defining relation.
    (c) Determine the complete alias pattern.
    (d) Use SAS ADX to create this design.
    (e) Examine the Alias Structure in SAS ADX.

2. Melo et al. (2007) used a 2^{4-1} factional factorial design with generator $D = ABC$ to study the factors that influence the production of levan by aerobic fermentation using the yeast *Zymomonas mobilis*. Levan is a sugar polymer of the fructan group which has been shown to have anti-tumor activity against sarcoma and Ehrlich carcinoma in Swiss albino mice. The factors varied in the fermentation medium and their levels are shown in the table below.

Label	Factor	Levels −	Levels +
A	Sucrose initial concentration(g.L^{-1})	150	250
B	Temperature(°C)	20	30
C	Yeast extract initial concentration(g.L^{-1})	2.0	5.0
D	Agitation(rpm)	50	100

    (a) What is the defining relation and complete alias structure for this design?
    (b) What is the resolution of this design?
    (c) The fermentations were carried out batchwise in Pyrex flasks. After 72 hours of fermentation, the levan produced was extracted from the fermented medium and its dry weight was determined. The results (in g.L^{-1}) for the 8 experiments (in standard order) were: 4.70, 14.67, 1.71, 3.73, 9.47, 7.61, 0.15, 4.78. From this data, calculate the 7 effects and make a normal probability plot to determine what is significant.
    (d) Delete the smallest three effects and fit the model again. Are the four effects left in the model significant?
    (e) Based on the effect heredity principle, what do you think the significant string of aliased two-factor interactions could represent?
    (f) Can the design be collapsed to a full factorial by ignoring insignificant factors?
    (g) Based on what you said in (e) write an interpretation of the significant effects and interaction. Based on your model determine the factor levels that will produce the maximum dry weight of levan.

3. Use SAS ADX to create a quarter fraction 2^{5-2} design.

EXERCISES 249

(a) Examime the Confounding rules used by SAS ADX to create the design.
(b) Modify the SAS code for exercise 1 to create the same design.
(c) Determine the defining relation.
(d) Verify that the alias structure in SAS ADX (for this design) can be obtained from the defining relation you found in (c).
(e) Suppose that after analysis of data from this experiment that two main effects A and D appear to be significant. Because of the confounding between main effects and two-factor interactions, interpreting the results to mean that these two main effects are the only significant effects could be misleading. Using the effect heredity principle, suggest two alternate plausible models that could explain the data.
(f) List the experiments in a foldover fraction that you could use to augment the original eight experiments in order to remove the aliasing between main effects and two-factor interactions.

4. Erhardt (2007) ran a 2^{8-4} fractional factorial experiment to determine the factors that affect flight time of paper helicopters. The standard design is shown in the figure below with all dimensions in centimeters. The factors

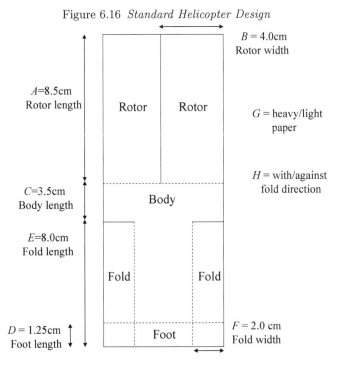

Figure 6.16 *Standard Helicopter Design*

and levels (in centimeters) he studied are shown in the table below. The low level on paper weight was phone book white pages paper, and the high level was standard copy paper. The fold direction was against (opposite to) or with the direction of rotation.

Label	Factor	Levels −	Levels +
A	Rotor length	5.5	11.5
B	Rotor width	3.0	5.0
C	Body length	1.5	5.5
D	Foot length	0.0	2.5
E	Fold Length	5.0	11.0
F	Fold width	1.5	2.5
G	Paper weight	telephone book page	copy paper
H	Fold direction	against	with

(a) What would you recommend for the generators of a 2^{8-4} design?

(b) Create the design you recommend in (a) and randomize the order of the runs. What is the experimental unit?

(c) List the aliases (up to three-factor interactions) for the 15 effects that can be estimated.

(d) Actually run the experiments by making the helicopters and timing them with a stopwatch.

(e) Calculate the effects (or regression coefficients) and make a half-normal plot of the effects.

(f) Using the principle of effect heredity, try to determine what any strings of aliased two-factor interactions, that appear significant on your plot, represent.

(g) Which factors would you include in a follow-up design to determine the combination of factor levels that produces the optimum flight time.

5. Create a 2^{7-3}_{IV} design in the SAS data step with generators $E = ABC$, $F = ABD$, $G = ACD$.

(a) Determine the defining relation for this design.

(b) Determine the confounded strings of two-factor interactions that can be estimated from this design.

(c) Suppose that after the analysis of data from this experiment, the significant effects appear to be B, $BC + AE + FG$ and $BE + AC + DG$. What foldover fraction would you recommend in order to determine which specific interactions are causing the two strings of confounded interactions to appear significant.

(d) List the experiments in the foldover fraction you recommend in part (c).

6. Consider the experiment to study factors that affect the percent removal of arsenic from drinking water using a iron coated sand filter.

EXERCISES

(a) Modify the SAS code in Section 6.5.1 to produce the design shown in Table 6.8

(b) Determine the defining relation for the experiments in the first block and show the confounding of main effects with strings of two-factor interactions.

(c) Calculate the effects and make a half-normal plot of them using only the data from block 1 in Table 6.8.

(d) Determine the defining relation for the combined experiments.

(e) Calculate the effects, including the block effect, and make a half-normal plot of them using all the data in Table 6.8. What interactions are confounded with the block effect?

7. Prince (2007) performed a 2^{5-2} fractional factorial in the process of optimizing a clinical assay to detect *Streptococcus pyogenes* with real-time PCR. Optimization of this process would allow hospitals to detect *Strep* infections in less than 30 minutes with 99% accuracy. The factors he studied were A=number of *S. pyogenes* colonies (1 or 4), B=Boiling time (1 min. or 3 min.), C=Centrifuge time (0 min. or 2 min.), D=cycling temperature (60° or 61°), E=Cycling time (5/10 sec. or 8/13 sec.). The generators for the design were $D = AC$ and $E = BC$.

(a) The response data (in standard order) from the eight experiments were: 1.31091, 1.43201, 1.29951, 1.37199, 1.33566, 1.46820, 1.39023, 1.41531. Calculate the effects and make a half-normal plot of the effects. Identify any effects you believe may be significant.

(b) Determine the confounding pattern or alias structure for the design.

(c) What are the aliases of the largest effect in absolute value?

(d) Prince performed 8 more experiments according to a foldover (similar to that shown in Table 6.8) and the resulting data were: 1.31702, 1.38881, 1.32222, 1.36248, 1.33826, 1.32654, 1.32654, 1.34635. Combining this data with the data from the original 8 runs, calculate the 15 effects including a block effect for the differences in the two groups of experiments.

(e) What is the defining relationship and alias structure for the complete set of 16 experiments?

(f) Make a half-normal plot of the 15 effects and determine what appears to be significant from the combined set of data.

(g) Provide an interpretation for all effects and interactions you determine are significant, and discuss how the conclusions and interpretation after 16 experiments differs from what could have been concluded after the first 8 experiments.

8. Modifying the SAS commands shown in the chapter or using SAS ADX create a Plackett-Burman design to study 15 factors in 20 runs.

9. Consider the data from the chemical reaction experiment from Box and Meyer (1993) shown in the table below. The first five columns defined the factor settings for the experiment, and the remaining unassigned columns represent aliased interactions.

Table 6.15 Design Matrix Response for Reactor Experiment

run	A	B	C	D	E	6	7	8	9	10	11	%Reacted
1	+	-	+	-	-	-	+	+	+	-	+	56
2	+	+	-	+	-	-	-	+	+	+	-	93
3	-	+	+	-	+	-	-	-	+	+	+	67
4	+	-	+	+	-	+	-	-	-	+	+	60
5	+	+	-	+	+	-	+	-	-	-	+	77
6	+	+	+	-	+	+	-	+	-	-	-	65
7	-	+	+	+	-	+	+	-	+	-	-	95
8	-	-	+	+	+	-	+	+	-	+	-	49
9	-	-	-	+	+	+	-	+	+	-	+	44
10	+	-	-	-	+	+	+	-	+	+	-	63
11	-	+	-	-	-	+	+	+	-	+	+	63
12	-	-	-	-	-	-	-	-	-	-	-	61

(a) Use `proc glm` to calculate the effects for all eleven columns and make a half-normal plot.
(b) Use `proc reg` to perform an all subsets regression using all main effects and two-factor interactions as candidate variables.
(c) What model do you feel is appropriate for the data?
(d) What factor settings would you recommend to maximize the %Reacted.

10. Curtis et al. (1982) proposed to conduct an experiment to study the effect of four factors in the software development process upon the software life cycle cost and the number of residual errors. The factors and levels are shown in the table below:

		Levels	
Label	Factor	−	+
A	Design methodology	functionally structured	data structured
B	Programming language	Ada	CMS-2
C	Source language debugger	No	Yes
D	Testing strategy	Functional	Structural

(a) A 2^{4-1} fractional factorial design was proposed to study these factors since the cost of building and testing a multi-module system can be quite high. The author proposed to use a design with generator $C = AB$. Suggest a better design generator and explain why you think it would be better.

EXERCISES

(b) If the number of levels for factor A=Design methodologies and B=Programing languages were three rather than two, what is the minimum number of runs required for an orthogonal array design for studying two factors at three levels and two factors at two levels (i.e., $3^2 \times 2^2$)?

(c) Use the SAS Mkt macros to create a 12-run fractional factorial design for a $3^2 \times 2^2$.

11. Kamohara *et al.* (2001) used an 18 run orthogonal fractional factorial to investigate the effect of 8 factors ($3^7 \times 2$) upon the Vt variance of MOSFETs (metaloxidesemiconductor field-effect transistor). The list of factors and number of levels is shown below. The list of experiments and the response,

Label	Factor	Number of Levels
A	SD Energy	2
B	SDE Energy and Dose	3
C	Halo Energy	3
D	Halo Dose	3
E	Halo Tilt Angle	3
F	Vt-Impala. Dose	3
G	Vt-Impala. Energy	3
H	SD Dose	3

S/N, is shown in the table below. The purpose of the experiments was to find factor level settings to maximize S/N.

Run	A	B	C	D	E	F	G	H	S/N
1	1	1	1	1	1	1	1	1	-0.66
2	1	1	2	2	2	2	2	2	-0.66
3	1	1	3	3	3	3	3	3	-0.69
4	1	2	1	1	2	2	3	3	-0.67
5	1	2	2	2	3	3	1	1	-0.82
6	1	2	3	3	1	1	2	2	-0.69
7	1	3	1	2	1	3	2	3	-0.75
8	1	3	2	3	2	1	3	1	-0.65
9	1	3	3	1	3	2	1	2	-0.66
10	2	1	1	3	3	2	2	1	-0.80
11	2	1	2	1	1	3	3	2	-0.67
12	2	1	3	2	2	1	1	3	-0.65
13	2	2	1	2	3	1	3	2	-1.04
14	2	2	2	3	1	2	1	3	-0.69
15	2	2	3	1	2	3	2	1	-0.67
16	2	3	1	3	2	3	1	2	-0.67
17	2	3	2	1	3	1	2	3	-0.68
18	2	3	3	2	1	2	3	1	-0.71

(a) Read the data into SAS and create coded factor levels for the two-level factor A and linear and quadratic contrasts variables for the three-level factors in the SAS data step (i.e., `if B=1 then do; Bl=-.707107; Bq=.4082483; end; if B=2 then do; Bl=0; Bq=-.816497; end;` etc.) using the contrast coefficients as shown in Section 2.8.1.

(b) Fit a regression model to the data using the coded factor levels for the two-level factor and the linear and quadratic contrast variables for each three-level factor in the design.

(c) Fit the model including only the five independent variables with the highest absolute t-statistics that you found in (b).

(d) Fit the model that includes the coded factor levels for A, the linear contrast on C, the quadratic contrast on D, the interaction or product of the quadratic contrasts for B and D, the interaction or product of the linear contrast of F and the quadratic contrast on H, and experiment with a few other models of your choice.

(e) Do you think a plausible model can be found for the data after these 18 experiments? Why or why not?

CHAPTER 7

Incomplete and Confounded Block Designs

7.1 Introduction

One of the two purposes for randomized block designs, described in Chapter 4, was to group heterogeneous experimental units together in homogeneous subgroups called blocks. This increases the power or precision for detecting differences in treatment groups. The overall F-test for comparing treatment means, in a randomized block design, is a ratio of the variability among treatment means to the variability of experimental units within the homogeneous blocks.

One restriction on randomized block designs is that the number of experimental units in the block must be greater than or equal to the number of levels of the treatment factor. For the RCB design the number of experimental units per block, t, is equal to the number of levels of the treatment factor, and for the GCB design the number of experimental units per block is equal to the number of levels of the treatment factor times the number of replicates per block, tr.

When the number of levels of the treatment factor is large in a randomized block design, the corresponding number of experimental units per block must also be large. This could cause a problem. For example, frequently in agricultural field tests of new hybrid crops (called varietal trials) the number of hybrids could be very large (50-100). Since the experimental units are plots of land, the larger the number of plots included in a block the more diverse these plots will likely be since they cover a wider physical area. When experimental units are small animals, it is possible to group them into small blocks of litter mates that are genetically more homogeneous. However, when larger block sizes are required, animals from different litters will be included within a block, making the groups less homogeneous. Less homogeneity within the blocks results in more variability among the experimental units within a block and defeats one of the purposes for using a randomized block design.

One solution to the problem is to construct block designs where each block only contains a subset of the possible levels of the treatment factor. In that way the number of experimental units per block, or block size, can be kept small. Designs that do this are called incomplete block designs. There are two basic types of incomplete block designs. The first type is called a balanced incomplete block design or BIB, and the other type is called a partially balanced incomplete block design or PBIB. Incomplete block designs are also useful when the physical requirements of experimentation make it impossible to test all levels of the treatment factor within each block of experimental units. Incomplete block designs are commonly used in agricultural experimentation, animal science, educational testing and food science.

As an example of a situation where an incomplete block design would be useful, consider the following situation. In food science, taste panels are often used to test palatability of new recipes for new food products. In these panels, subjects are asked to taste and rate different recipes. The different recipes represent the levels of the treatment factor. A random sample of subjects (or the experimental units) is included to represent the potential market. Since there is wide variability in taste among subjects it is better to use a blocked design where each subject tastes and rates each recipe. However, if there are many recipes to be tested, subjects will lose their ability to discriminate and it will be difficult to detect any differences. A solution is to have each subject (block) taste only a subset of the recipes.

7.2 Balanced Incomplete Block (BIB) Designs

Care must be taken when choosing the subset of treatment combinations tested in each block of an incomplete block design. If one treatment factor level is left out of every block, it cannot be compared to the other treatment levels. If different treatment levels are unequally represented, some comparisons of factor levels will have more precision than others.

The optimal way to create an incomplete block design is to have each treatment level equally replicated and appearing within a block with every other treatment level an equal number of times. This is called a balanced incomplete block design or BIB. By doing this, all pairwise differences of least squares treatment means will have the same standard error, and the power for detecting a difference in any two means will be the same.

The simplest way to construct a BIB for the case where there are t levels of the treatment factor and $k < t$ experimental units per block is to form all possible subsets of k treatment levels chosen from t. For example, in the taste panel described above, if there were $t = 6$ recipes to be tested, and it was determined that each subject could taste at most $k = 3$ recipes without losing discriminatory power, $\binom{6}{3} = 20$ subjects would be required. All possible subsets are listed below.

$$\begin{array}{ccccc}
(1\ 2\ 3), & (1\ 2\ 4), & (1\ 2\ 5), & (1\ 2\ 6), & (1\ 3\ 4) \\
(1\ 3\ 5), & (1\ 3\ 6), & (1\ 4\ 5), & (1\ 4\ 6), & (1\ 5\ 6) \\
(2\ 3\ 4), & (2\ 3\ 5), & (2\ 3\ 6), & (2\ 4\ 5), & (2\ 4\ 6) \\
(2\ 5\ 6), & (3\ 4\ 5), & (3\ 4\ 6), & (3\ 5\ 6), & (4\ 5\ 6)
\end{array}$$

Thus, subject one would taste recipes 1, 2 and 3 in a random order; subject 2 would taste recipes 1, 2 and 4, etc. This plan is completely balanced in that each treatment level or recipe is replicated $r = 10$ times (or tasted by 10 subjects) and each pair of treatment levels occurs together in the same block $\lambda = 4$ times. For example, treatment levels 1 and 2 occur together only in the

BALANCED INCOMPLETE BLOCK (BIB) DESIGNS

first four blocks on the first line above. By inspection, it can be seen that all pairs of treatment levels occur together in only four blocks.

Although taking all possible subsets of size k chosen from t is the simplest way to form a balanced incomplete block design, there may be other balanced incomplete block designs that require fewer blocks. If the precision of the design does not require as many as $\binom{b}{k}$ blocks, there would be no need to use that many. For example, if a practical size difference in recipes could be detected with less than 20 subjects and 10 replicates of each treatment level, in the taste test panel, perhaps a BIB design could be found with less than 20 blocks. If r is the number of times a treatment level is replicated in an incomplete block design, λ is the number of times each treatment level occurs with every other treatment level in the same block, t is the number of levels of the treatment factor, k is the number of experimental units in a block, and b is the number of blocks, then the following requirements must be met in order for that design to be a BIB design.

$$b \geq t \tag{7.1}$$

$$tr = bk \tag{7.2}$$

$$\lambda(t-1) = r(k-1) \tag{7.3}$$

These relations can be used to find the minimum number of blocks required for a BIB design. Since r and λ must be integers, by Equation (7.3) we see that $\lambda(t-1)$ must be divisible by $k-1$. If $t = 6$ and $k = 3$, as in the taste test panel, 5λ must be divisible by 2. The smallest integer λ for which this is satisfied is $\lambda = 2$. Therefore, it may be possible to find a BIB with $\lambda = 2$, $r = 10/2 = 5$, and $b = (6 \times 5)/3 = 10$. In versions of SAS 9.2 and later the macro %Mktbsize provides a quick way of finding values of λ and r to satisfy Equations (7.1) to (7.3) when given the number of levels of the treatment factor t and the blocksize k. For example the command: tries values of b from

```
%mktbsize(t=6, k=3, b=6 to 100)
```

6 to 100 until a solution is found. The results are shown below.

t Number of Treatments	k Block Size	b Number of Blocks	r Treatment Frequency	Lambda Pairwise Frequencies	n Total Sample Size
6	3	10	5	2	30

Fisher (1940) showed that even though Equations (7.1) - (7.3) are satisfied for some combination of t, b, r, λ and k, a corresponding BIB may not exist. If a BIB does exist, Kiefer (1958) and Kshirsager (1958) have shown that it is D-optimal, thus it can be found using SAS **proc optex** that was described in

258   INCOMPLETE AND CONFOUNDED BLOCK DESIGNS

Section 6.5.2. In versions of SAS 9.2 and later the %MktBIBD macro provides a front end and back end to proc optex in finding BIB designs. For example, the macro call below searches for a BIB with $b = 10$ blocks of $k = 3$ experimental units per block and $t = 6$ levels of the treatment factor.

```
%mktbibd(t=6, b=10, k=3)
```

A portion of the output is shown below. A note in the SAS log indicates that a balanced incomplete block design has been found, and this is verified since the blocking design D-efficiency is 100%. According to this plan the three

```
Blocking Design Efficiency Criterion      100.0000
Number of Treatments, t                          6
Block Size, k                                    3
Number of Blocks, b                             10
Treatment Frequency                              5
Pairwise Frequency                               2
Total Sample Size                               30
Positional Frequencies Optimized?              Yes

        Balanced Incomplete Blocks Design

              x1      x2      x3
              6       1       4
              5       2       1
              6       5       4
              4       1       2
              1       5       3
              4       3       2
              2       6       5
              5       4       3
              3       6       1
              3       2       6
```

experimental units in the first block would receive treatment levels 6, 1 and 4, the experimental units in the second block would receive treatment levels 5, 2 and 1, etc. These results are automatically stored in a file called bibd and a sorted version is stored in the file sorted. In older versions of SAS the BIB design can also be created directly with proc optex as shown in the SAS code on the Web site for this book.

Once a BIB design is found, the levels of the treatment factor within each block should be randomized to the experimental units within that block, as illustrated for the RCB design in Section 4.2. The file bibd that is produced by the %mktbibd macro is not randomized but may be optimized for positional frequencies. Therefore, if the experimental units in each block are numbered in a way that any systematic differences are correlated with the experimental unit number, the ordering of treatments in the file bibd may reduce biases due to those systematic differences. Another approach to follow in this situation is to use the Youden Square design described in Section 7.5.

7.3 Analysis of Incomplete Block Designs

The model for an incomplete block design is

$$y_{ij} = \mu + b_i + \tau_j + \epsilon_{ij} \tag{7.4}$$

which is identical to the model for a randomized complete block design given in Section 4.3. However, the analysis is slightly different due to the missing observations.

7.3.1 An Example

Consider the data from a taste panel experiment reported by Moskowitz (1988), shown in Table 7.1. This experiment is a BIB with t=4 levels of the treatment factor or recipe, and blocksize k=2. Thus each panelist tastes only two of the four recipes in a random order, and assigns a category scale score. Category scales are commonly used in assessing food likes or dislikes and consist of numbers 1 to 10 that represent descriptive categories. Only $\binom{4}{2} = 6$ blocks or panelists are required for a BIB, but in this experiment that number was doubled in order to increase the power for detecting differences. Thus the first six panelists and the last six are a repeat of the same BIB design. Subjects participating in the taste panel were randomly assigned to panelist numbers and the order of the two recipes tasted by each panelist was randomized.

Table 7.1 *Data from BIB Taste Test*

	Recipe			
Panelist	A	B	C	D
1	5	5	-	-
2	7	-	6	-
3	5	-	-	4
4	-	6	7	-
5	-	6	-	4
6	-	-	8	6
7	6	7	-	-
8	5	-	8	-
9	4	-	-	5
10	-	7	7	-
11	-	6	-	5
12	-	-	7	4

When analyzing the data from an incomplete block design, the marginal treatment means are not unbiased estimators of the estimable effects $\mu + \tau_i$. For example, in Table 7.1 the marginal mean for recipe A could be biased low by the fact that it was not tasted by panelists 4, 5, 6, 10, 11, and 12 who seem to rate recipes higher. Likewise, the non-centrality factor for the type I sums of squares for treatments (or recipes) will contain block effects as well as treatment effects. The solution to these problems is the same as shown in Section 3.5.3 for analyzing data from factorial designs with an unequal number of replicates per cell. That is, always use the type III sums of squares in SAS `proc glm` and the least squares means rather than marginal means. The commands to analyze the data in Table 7.1 are shown below.

```
data taste;
input panelist recipe $ score;
datalines;
1 A 5
. . .
proc glm;
   class panelist recipe;
   model score=recipe panelist;
   lsmeans recipe;
run;
```

The type III $F_{3,9}$ value for testing recipes was 3.98, which is significant at the $\alpha = 0.05$ significance level. The least squares means produced by `proc glm` are shown below. In order to detect which recipes differ, the estimate

```
            The GLM Procedure
           Least Squares Means

         recipe      score LSMEAN

           A          5.45833333
           B          6.20833333
           C          6.83333333
           D          4.83333333
```

statements shown below could be added to the `proc glm` commands above.

```
estimate 'A-B' recipe  1 -1;
estimate 'A-C' recipe  1  0 -1;
estimate 'A-D' recipe  1  0  0 -1;
estimate 'B-C' recipe  0  1 -1;
estimate 'B-D' recipe  0  1  0 -1;
estimate 'C-D' recipe  0  0  1 -1;
```

This results in the output shown on the top of the next page. There it can be seen that the category scale score for recipe C is significantly higher than the score for recipe D at the $\alpha = 0.05$ level, but that no other differences are significant. Another thing that should be noticed in the output is the fact that the standard error of the differences in means (shown in the third column) is

PBIB-BTIB DESIGNS

the same for all pairs of means. This is a result of the fact that a BIB design was used.

Parameter	Estimate	Standard Error	t Value	Pr > |t|
A-B	-0.75000000	0.61801654	-1.21	0.2558
A-C	-1.37500000	0.61801654	-2.22	0.0531
A-D	0.62500000	0.61801654	1.01	0.3383
B-C	-0.62500000	0.61801654	-1.01	0.3383
B-D	1.37500000	0.61801654	2.22	0.0531
C-D	2.00000000	0.61801654	3.24	0.0102

7.3.2 Determining the Number of Replicates

A rough estimate of the number of replicates of each treatment level, $r = \frac{bk}{t}$, or the number of blocks, $b = \frac{tr}{k}$, required for adequate power for detecting a practical difference in treatment means in a BIB design can be determined using the following strategy. If an estimate of σ^2, the variance of heterogeneous experimental units, and Δ, the size of a practical difference in two treatment means are available, then use the method of Section 3.5.2 to determine the number of replicates, r_{crd}, required for a completely randomized design.

If blocking the heterogeneous experimental units into small blocks of size k is expected to reduce the variance of experimental units within the blocks by a percentage equal to $100 \times (1 - \frac{1}{RE})$, then following Section 4.5, the number of replicates required for the blocked design would be $r = r_{crd}/RE$ and the number of blocks of size k in an incomplete block would be $b = tr/k$.

7.4 PBIB-BTIB Designs

Sometimes a BIB design requires more blocks or experimental units than are available or needed to obtain adequate power for detecting practical differences in treatment means. In this situation, the number of blocks and experimental units can be reduced by relaxing the requirements that: (1) each treatment level be equally replicated and (2) that it appears within a block with every other treatment level the same number of times. By relaxing these requirements, each treatment level will occur more frequently in a block with some treatment levels than it will with other treatment levels. Therefore some pairwise comparisons of treatment means will have smaller standard errors than others. When this is the case, the design is called a partially balanced incomplete block design or PBIB. PBIB designs are also useful in situations where the physical constraints of experimentation prevent some treatment levels from being tested together within the same block. For example, an experiment was conducted to compare the reading of a portable home use blood pressure monitor to other automatic blood pressure monitors in supermarket pharmacies in order to determine if readings from the portable monitor were biased high. The experimental unit was the pressure in a subject's veins at the time it was measured, and the treatment factor levels were the monitors used to make the measurement.

Blood pressure is known to vary widely from person to person and within a person at different times throughout a day. Blood pressure is most consistent in one person within a short period of time. Therefore the experimental units were blocked into homogeneous groups by subject and time. The portable blood pressure monitor could be carried into a store and a subject's blood pressure could be checked within a short period of time by both the automatic monitor within the store and the portable monitor. However, the monitors from two stores could not be compared within a block, or short period of time, because the stores were physically separate. Driving between stores could completely change a subject's blood pressure, so the partially balanced design shown in Table 7.2 was utilized. The response (diastolic blood) pressure is shown in the table. Here it can be seen that treatment level 1 (portable

Table 7.2 *PBIB Design with Blood Pressure Monitors*

Block	Treatment			
	Portable monitor	Store A monitor	Store B monitor	Store C monitor
1=(subject 1, time 1)	85	77	-	-
2=(subject 2, time 1)	80	75	-	-
3=(subject 1, time 2)	89	-	73	-
4=(subject 2, time 2)	80	-	70	-
5=(subject 1, time 3)	78	-	-	76
6=(subject 2, time 3)	80	-	-	70

monitor) appears in a block with every other treatment level, but the other treatment levels never appear together in a block. The analysis of the data using commands similar to those used for the BIB example results in the following table of estimates. Here it can be seen that the standard errors of the differences between the portable monitor mean and the means for the other monitors are smaller than the standard errors of the other comparisons.

```
                                Standard
Parameter             Estimate     Error    t Value   Pr > |t|

Portable-Store A    13.0000000   3.01385689    4.31     0.0229
Portable-Store B     6.0000000   3.01385689    1.99     0.1406
Portable-Store C     6.5000000   3.01385689    2.16     0.1199
Store A-Store B     -4.0000000   3.80058475   -1.05     0.3331
Store A-Store C     -2.0000000   3.80058475   -0.53     0.6176
Store B-Store C      2.0000000   3.80058475    0.53     0.6176
```

The design for the blood pressure monitor experiment is a special case of a PBIB design that Bechhofer and Tamhane (1981) have called a BTIB (balanced with respect to test treatments). In these designs one treatment level is designated as the control level and there is more interest in comparing each of the other treatment levels with the control than there is in comparing the other treatment levels. In a BTIB design each treatment must appear the same number of times (λ_0) in a block with the control treatment, and each test treatment must occur the same number of times (λ_1) in a block with every other test treatment. This results in a design that is more efficient in comparing each treatment with the control but less efficient in comparisons among the other treatment levels. One way to form a PTIB design with t levels of the treatment factor and block size k is to combine a control level to each block of a BIB design with $t-1$ levels of the treatment factor and block size $k-1$.

In general there are multiple ways of creating PBIB designs. Bose et al. (1954) have published tables of some of the most useful plans. Jarrett and Hall (1978) have described a class of PBIB designs called generalized cyclic incomplete block designs which have good statistical properties and are easy to create in SAS. Generalized cyclic incomplete block designs with block size k and $b=t$ blocks or $b=n$ blocks, where t factors into $t = m \times n$, and $m < n$ can be created following the steps listed below.

1. To form a generalized cyclic design with $b = t$,

    (a) Start with a subset of k treatment factor levels as the initial block.

    (b) Add 1 (modulo t) to each treatment level in the initial block to form the next block.

    (c) Continue adding blocks until you have t blocks.

2. To form a generalized cyclic design with $b = n$, where $t = m \times n$, and $m < n$

    (a) Start with a subset of k treatment factor levels as the initial block.

    (b) Add m (modulo t) to each treatment level in the initial block to form the next block.

    (c) Continue adding blocks until you have n blocks.

To illustrate this consider creating an incomplete block design with $t = 6$ and $k = 3$. The BIB design with the smallest number of blocks for this combination is found (solving Equations (7.1) to (7.3) to be $b = 10$). A generalized cyclical incomplete block design for testing $t = 6$ levels of the treatment factor can be found with $b = t = 6$ blocks or $b = n = 3$ blocks since $6 = 2 \times 3$. To find a design with 6 blocks, following the steps above, start with a subset of $k = 3$ levels of the treatment factor to be tested in the initial block, i.e.,

(1 2 4)

264   INCOMPLETE AND CONFOUNDED BLOCK DESIGNS

Next add one to each treatment level to get the treatment levels in the next block, i.e.,

$$(2\ 3\ 5)$$

Continue this modulo 6 (i.e., 7 modulo 6 = 1, etc.) to form the following blocks.

Block	Treatments		
1	1	2	4
2	2	3	5
3	3	4	6
4	4	5	1
5	5	6	2
6	6	1	3

The treatment levels in each block would be randomized to the three experimental units in each block. SAS **proc plan** can create a generalized cyclic design and print the randomization. For example, the code below shows the commands to create a generalized cyclic design with $t = 6$, $b = 3$. This code can be modified to create the design above with $t = 6$ and $b = 6$ by changing the specification of m and b.

```
proc plan seed=33373;
*Note (1 2 4) are contents of the initial block, m=2;
  treatments treat=3 of 6 cyclic (1 2 4) 2;
*Note block=3 specifies b=n=3 blocks, EU=3 specifies k=3;
  factors block=3 EU=3;
  output out=PBIB;
proc print data=PBIB; run;
```

The proc plan stores the design in a SAS data file **PBIB** and produces the output shown below.

```
                    Treatment Factors
                                                Initial
                                                Block /
Factor      Select      Levels       Order      Increment
treat          3           6         Cyclic     (1 2 4) / 2

            block        --EU--      -treat-
              3          3 2 1        1 2 4
              1          3 1 2        3 4 6
              2          2 1 3        5 6 2
```

YOUDEN SQUARE DESIGNS (YSD)

The results under -treat- are the treatment factor levels in each block, and the results under -EU- is a randomized list of the experimental units in each block. This can be read as follows: in block 3, the first experimental unit would get treatment level 4, experimental unit 2 would get treatment level 2, and experimental 3 would get treatment level 1, etc. The proc print statement prints the randomized list shown below. This can be modified in a data step to create a data collection worksheet.

Obs	block	EU	treat
1	3	3	1
2	3	2	2
3	3	1	4
4	1	3	3
5	1	1	4
6	1	2	6
7	2	2	5
8	2	1	6
9	2	3	2

The analysis of PBIB designs is the same as the examples shown above. The type III sums of squares for treatment and the least squares means should be used. The model and assumptions are the same as for the RCB design, and the assumptions of normality and homogeneity of experimental error variance can be checked with the residual plots described in Section 2.4.

7.5 Youden Square Designs (YSD)

Latin square designs with two independent blocking factors were described in Chapter 4. These designs could increase the precision in detecting differences among treatments by adjusting for variability in experimental units in two ways. However, the restriction on Latin square designs was that the number of levels of the row blocking factor, the number of levels of the column blocking factor, and the number of levels of the treatment factor all had to be equal. This restriction may be impractical in some situations.

In Chapter 4 an experiment was described where the purpose was to study the effect of shelf facing on the sales of toothpaste in drug stores. In that example, four levels of the treatment factor (shelf facings), four levels of the row blocking factor (stores), and four levels of the column blocking factor (week of sales) were used. If the researchers desired to expand their study to test 6 different shelf facings instead of 4, they could easily increase the number of levels of the row blocking factor and include 6 stores. However, increasing the number of weeks would prolong the study and could be undesirable.

An alternate design called a Youden square design or YSD utilizes a complete block design in the column blocks, but an incomplete block design in the row blocks. It is created by randomly dropping one or more columns from a Latin square design. This type design can also be created and randomized by SAS `proc plan`. The code below is a modification of the example shown in Chapter 4.

```
proc plan seed=37430;
  factors rows=8 ordered cols=4 ordered/noprint;
  treatments tmts=8 cyclic;
  output out=YS
  rows cvals=('Store 1' 'Store 2' 'Store 3' 'Store 4'
  'Store 5' 'Store 6' 'Store 7' 'Store 8') random
  cols cvals=('Week 1' 'Week 2' 'Week 3' 'Week 4') ordered
  tmts nvals=(1 2 3 4 5 6 7 8) random;
run;
```

This creates the design shown in the Output below. The randomized list is in the SAS file YS and can be used to create a data collection worksheet.

	cols			
	Week 1	Week 2	Week 3	Week 4
	tmts	tmts	tmts	tmts
	Sum	Sum	Sum	Sum
rows				
Store 1	4	8	3	1
Store 2	7	2	6	4
Store 3	5	7	2	6
Store 4	2	6	4	8
Store 5	6	4	8	3
Store 6	8	3	1	5
Store 7	1	5	7	2
Store 8	3	1	5	7

The model and analysis of YSDs is identical to the model and analysis of Latin square designs described in Chapter 4, with the exception that type III treatment sums of squares and least squares treatment means should be used due to the fact that the row blocks are incomplete and do not contain all levels of the treatment factor.

7.6 Confounded 2^k and 2^{k-p} Designs

When the experimental units are heterogeneous and can be grouped into blocks, frequently there are too few experimental units per block to accommo-

CONFOUNDED 2^k AND 2^{k-p} DESIGNS

date all the treatment combinations in 2^k factorial or 2^{k-p} fractional factorial design. Therefore, it is impossible to use complete block designs, like those described in Section 4.5. One solution is to use an incomplete block design described in the first part of this chapter with $t = 2^k$. However this usually results in many more blocks than are necessary. The effect sparsity principle and hierarchical ordering principle tell us that it is unlikely that all interactions (especially those above order 3) will be important. If we are willing to sacrifice the ability to estimate some interactions, 2^k factorial designs can be blocked in a minimum number of blocks as long as the number of experimental units in each block is a power of 2.

For example, a 2^3 factorial design has 8 treatment combinations. By confounding the three-factor interaction with blocks, this design can be run in two blocks of size 4 as shown in Table 7.3 below.

Table 7.3 2^3 in Two Blocks of Size 4

A	B	C	Block=ABC	
-	-	-	1	-
+	-	+	1	-
-	+	+	1	-
+	+	-	1	-
-	-	+	2	+
+	-	-	2	+
-	+	-	2	+
+	+	+	2	+

In this design, the ABC interaction effect is completely confounded between the two blocks and cannot be estimated. However, since interactions between block and treatment effects are assumed negligible, all the other main effects and two-factor interactions are not confounded. Detecting the significant effects from an experiment like this can be done using the graphical methods illustrated in Section 3.7.5 and Chapter 6.

7.6.1 Confounding 2^k Designs

In general a 2^k factorial can be run in blocks of size 2^q by choosing $k-q$ interaction contrasts to confound with blocks. These interactions are called *block defining contrasts*. When a 2^k is run in blocks of size 2^q, there will be $2^k/2^q = 2^{k-q}$ blocks. Therfore, there will be $2^{k-q}-1$ degrees of freedom for blocks. These $2^{k-q}-1$ degrees of freedom are accounted for by the $k-q$ defining block contrasts and all their *generalized interactions*.

As an example consider confounding a 2^4 factorial in blocks of size $4=2^2$. Since $k=4$ and $q=2$, $k-q=2$ defining block contrasts must be chosen. If the three-factor interactions ABD and BCD are chosen as the block defining contrasts, their generalized interaction $ABD(BCD) = AC$ will also be confounded with blocks, accounting for the 3 degrees of freedom for blocks. The SAS code below shows an example of creating this design in standard order using the SAS data step.

```
data conf;
do x4=-1 to 1 by 2;
  do x3= -1 to 1 by 2;
    do x2 = -1 to 1 by 2;
      do x1 = -1 to 1 by 2;
      A=x1; B=x2; C=x3; D=x4;
      ABD=A*B*D; BCD=B*C*D;
      if ABD=-1 and BCD=-1 then Block=1;
      if ABD= 1 and BCD=-1 then Block=2;
      if ABD=-1 and BCD= 1 then Block=3;
      if ABD= 1 and BCD= 1 then Block=4;
      output;
      end;
    end;
  end;
end;
keep A B C D Block;
proc sort; by Block;
proc print; run;
```

The output below is the first eight lines produced when running this code.

Obs	A	B	C	D	Block
1	-1	-1	-1	-1	1
2	1	1	1	-1	1
3	-1	1	-1	1	1
4	1	-1	1	1	1
5	1	-1	-1	-1	2
6	-1	1	1	-1	2
7	1	1	-1	1	2
8	-1	-1	1	1	2

This design is called a completely confounded blocked factorial or CCBF. In practice the treatment combinations within each block would be randomized to experimental units in that block.

The analysis of a confounded blocked factorial is similar to the analysis of an unreplicated 2^k design as described in Section 3.7.5. As an example consider the experiment conducted by Apple (2006). He was interested in the effects of the four factors shown in Table 7.4 upon dishwashing.

His experiment consisted of soaking dishes with baked-on (microwaved) spaghetti sauce in dishwater that consisted of all possible combinations of the factor levels shown above. He drew a 10×10 grid on a white plate with a permanent marker, and his response was the number of clean grid squares after soaking the plate with baked-on spaghetti sauce in the dishwater. The

CONFOUNDED 2^k AND 2^{k-p} DESIGNS

Table 7.4 *Factors for Dishwashing Experiment*

Factor	Levels	
	(−)	(+)
A-Water Temperature	60 Deg F	115 Deg F
B-Soap Amount	1 tbs	2 tbs
C-Soaking Time	3 min	5 min
D-Soap Brand	WF	UP

experimental unit was the baked-on spaghetti sauce, and in order to generalize his conclusions, he included four blocks which consisted of combinations of two different brands of spaghetti sauce and two different baking times in the microwave, as shown in the table below. In this experiment, it would have

Table 7.5 *Blocks for Dishwashing Experiment*

Block	Type Sauce	Microwave Time
1	Store Brand	1 min
2	Premium Brand	1 min
3	Store Brand	1:30 min
4	Premium Brand	1:30 min

been possible to run a complete block factorial design where all combinations of the treatment factor levels from Table 7.4 were included in each block. This would have taken a total of 4×16=64 experiments. However, considering the hierarchical ordering principle, some interactions were confounded in order to reduce the total number of experiments. Using the same design shown earlier, ABD and BCD and their generalized interaction $ABD(BCD) = AC$ were confounded with blocks, resulting in 16 experiments or plates grouped into 4 blocks with 4 experimental units in each block. The design and response (number of clean grid squares) is shown in Table 7.6 on the next page.

To analyze the data, the factorial model was fit that included all main effects and interactions, except ABD, BCD and AC that are confounded with blocks. The SAS `proc glm` commands to do this are shown below.

```
proc glm data =comb;
  class Block;
  model y=Block A B C D A*B A*D B*C B*D C*D
   A*B*C A*C*D A*B*C*D/solution;
  ods output ParameterEstimates=sol;
run;
```

Since there were no replicates in the experiment, a half-normal plot of the effects was used to determine the significant effects and interactions. The SAS code below Table 7.6 produces the half-normal plot. It is the same as the code

Table 7.6 *Design and Data for Dishwashing Experiment*

A	B	C	D	Block	y
−	−	−	−	1	0
+	+	+	−	1	14
−	+	−	+	1	0
+	−	+	+	1	12
+	−	−	−	2	33
−	+	+	−	2	2
+	+	−	+	2	24
−	−	+	+	2	10
+	+	−	−	3	11
−	−	+	−	3	1
+	−	−	+	3	1
−	+	+	+	3	0
−	+	−	−	4	5
+	−	+	−	4	41
−	−	−	+	4	3
+	+	+	+	4	70

shown in Section 3.7.5, except the first five observations in the file **sol** were dropped so that block effects would not be included in the plot.

```
data hnplot; set sol;
  estimate=abs(estimate);
  if _n_>5; drop StdErr tValue Probt;
*proc rank calculates ranks for parameter estimates;
proc rank data=hnplot out=hnplots; var estimate; ranks rnk;
*data step calculates half normal scores for estimates;
data hnplots; set hnplots;
  zscore=probit((((rnk-.5)/12)+1)/2);
  if abs(zscore)<=1.2 then parameter='     ';
ods graphics on;
proc print; run;
proc sgplot data=hnplots;
  scatter x=zscore y=estimate/datalabel=parameter;
  xaxis label='Half Normal Scores';
  yaxis label='Absolute Regression Coefficient';
run;
ods graphics off;
```

The half-normal plot of effects is shown in Figure 7.1. The reference line was added manually. Here it can be seen that main effect A, the water temperature and possibly the BD interaction between soap amount and soap brand appear to be significant. The water temperature main effect was easy to interpret. The means for this factor revealed that 23.125 more grid squares were cleaned with hot water (115°F) than with cold water (60°F).

To interpret the BD interaction an interaction plot or table of means is

Figure 7.1 *Half-Normal Plot of Absolute Regression Coefficients from Dishwashing Experiment*

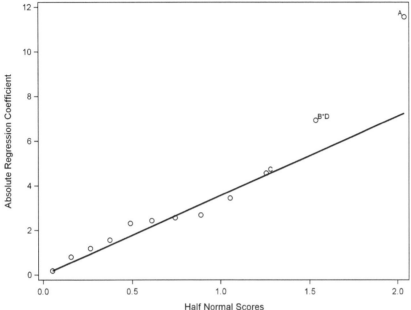

required. Figure 7.2 shows the interaction plot. Here it can be seen that increasing the amount of WF brand soap from 1 tablespoon to 2 tablespoons actually decreases the number of clean grid squares on the plate being soaked, while increasing the amount of UP brand soap from 1 tablespoon to 2 tablespoons has the opposite effect and increases the number of clean grid squares. According to the effect heredity principle, it would be unusual to have an interaction between two factors that don't have significant main effects. This, along with the fact that the BD interaction does not appear to be significant using the graphical techniques of Box and Meyer (1986a), Lawson et al. (1998) or Lenth (1989), would justify ignoring this unusual finding. Therefore, the only thing that seems to have a significant effect on the ease of removing baked-on spaghetti sauce is the temperature of the dishwater.

Blocks (not shown in the normal plot) also accounted for a large proportion of the sums of squares. If the block term were not included in the model, none of the factor effects would be determined significant. If experimental units had been restricted to one brand of spaghetti sauce and one microwave time, the conclusions regarding the dishwater would only apply to that brand of sauce and baking time.

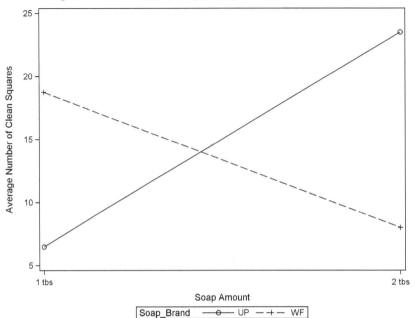

Figure 7.2 *Interaction Plot for Soap Amount × Soap Brand*

7.6.2 Confounding in 2^{k-p} Designs

A 2^{k-p} fractional factorial can also be easily blocked by confounding interactions if the block size is a power of two. For example, consider the 2^{5-2} fractional factorial shown in Section 6.3. In Table 7.7 the eight runs in this design are split into two blocks by using the BC interaction as the block defining contrast. When there is only one block defining contrast in a full factorial

Table 7.7 *Blocked 2^{5-2} Design*

A	B	C	D	E	Block	=BC
−	−	−	+	+	2	+
+	−	−	−	−	2	+
−	+	−	−	+	1	−
+	+	−	+	−	1	−
−	−	+	+	−	1	−
+	−	+	−	+	1	−
−	+	+	−	−	2	+
+	+	+	+	+	2	+

design, only that interaction is confounded with the difference in blocks. However, in a one-fourth fractional factorial, each effect that can be estimated is

CONFOUNDED 2^k AND 2^{k-p} DESIGNS

aliased with three other effects due to the fractionation. In the example above the generators for the quarter fraction were $D = AB$ and $E = AC$, resulting in the defining relation for the fraction $I = ABD = ACE = BCDE$. Multiplying through the defining relation for the fractional factorial by the block defining contrast, we see that $BC = ACD = ABE = DE$ and that four interactions are actually confounded with the blocks.

In general, when a 2^{k-p} fractional factorial is blocked into blocks of size 2^q, there will be $2^{k-p}/2^q = 2^{k-p-q}$ blocks. $k - p - q$ interactions must be chosen as block defining contrasts. The block defining contrasts and all their generalized interactions will account for the $2^{k-p-q} - 1$ degrees of freedom for blocks, and each interaction that is confounded will also be aliased with $2^p - 1$ other interactions due to the fractionization. These designs are called completely confounded blocked fractional factorial or CCBFF.

To illustrate the design and analysis of a confounded 2^{k-p} design, consider the experiment described by Porter and Busch (1978). The purpose was to explore the effects of the eight variables shown in Table 7.8 on survival and growth of neonatal deer mice. Knowledge of survival and growth of any species may aid in understanding aspects of animal distribution. They used a $\frac{1}{16}$th fractional factorial design to identify variables affecting the ability of wild female deer mice to raise young to weaning. This was a screening experiment with the purpose of identifying significant main effects that would be studied later in further experiments.

Table 7.8 *Factors and Levels for Mouse Growth Experiment*

Factor	Description	Levels −	+
A	Sprouts	None	Free Access
B	Frequency of weighing	Once per day	Once per 3 days
C	Nest box	No	Yes
D	Remove young	No	Yes
E	Male presence	Yes	No
F	Exercise wheel	Locked	Free
G	Available food	80 percent	Free access
H	Available water	80 percent	Free access

The generators for the 2^{8-4} fractional design were $E = BCD$, $F = ACD$, $G = ABC$, and $H = ABD$, resulting in the defining relation

$$I = BCDE = ACDF = ABCG = ABDH = ABEF = ADEG = ACEH$$
$$= BDFG = BCFH = CDGH = CEFG = DEFH = BEGH = AFGH$$
$$= ABCDEFGH$$

This was a resolution IV fraction and the eight main effects were confounded with strings of three-factor and higher-order interactions. In addition, there were 7 strings of estimable confounded two-factor interactions listed below.

$$AB + CG + DH + EF$$
$$AC + BG + DF + EH$$
$$BC + AG + DE + FH$$
$$AD + BH + CF + EG$$
$$BD + AH + CE + FG$$
$$CD + AF + BE + GH$$
$$CH + AE + BF + DG$$

The experimenters desired to block the experiments into 8 blocks of $2^1 = 2$ runs each. This would allow them to prevent biases from uncontrolled variables, such as time of year, over the duration of their experiments. Doing this would require choosing $k-p-q = 8-4-1 = 3$ block defining contrasts. These block defining contrasts, along with their generalized interactions, would account for the 8-1=7 degrees of freedom for blocks.

The block defining contrasts that were chosen were $AB + CG + DH + EF$, $AC + BG + DF + EH$, and $AD + BH + CF + EG$. The two-factor generalized interactions are $AB(AC) = BC$ (or $BC + AG + DE + FH$), $AB(AD) = BD$ (or $BD + AH + CE + FG$), and $AC(AD) = CD$ (or $CD + AF + BE + GH$). Finally, the three-factor generalized interaction can be identified by substituting BH for AD, since AD is confounded with BH to get $AB(AC)(AD) = AB(AC)(BH) = CH$, or $CH + AE + BF + DG$. Therefore, using the three block defining contrasts $AB + CG + DH + EF$, $AC+BG+DF+EH$, and $AD+BH+CF+EG$ actually confounds all of the 7 confounded strings of two-factor interactions (shown above) with blocks. The researchers were willing to give up information on all two-factor interactions to prevent biases because this was a screening experiment where they were satisfied to get information about the relative importance of the main effects.

Each run of the design was started on the day a female's young were born and terminated either when they had all died or otherwise at 21 days. The response was the weight at weaning. The design and results are shown in Table 7.9. Mice were grouped into blocks of two sequentially as they gave birth. The two treatment combinations within each block were randomized to the mice in that block.

Table 7.9 *Design and Results for Mouse Growth Experiment*

Block	A	B	C	D	E	F	G	H	weight
1	+	−	−	−	−	+	+	+	9.0
1	−	+	+	+	+	−	−	−	0.0
2	−	−	+	+	−	−	+	+	9.25
2	+	+	−	−	+	+	−	−	4.90
3	+	−	+	−	+	−	−	+	8.80
3	−	+	−	+	−	+	+	−	4.35
4	−	−	−	+	+	+	−	+	0.0
4	+	+	+	−	−	−	+	−	7.43
5	−	+	+	−	−	+	−	+	5.35
5	+	−	−	+	+	−	+	−	9.90
6	+	+	−	+	−	−	−	+	2.60
6	−	−	+	−	+	+	+	−	7.43
7	−	+	−	−	+	−	+	+	6.80
7	+	−	+	+	−	+	−	−	3.93
8	+	+	+	+	+	+	+	+	10.20
8	−	−	−	−	−	−	−	−	4.87

The SAS data step commands to create this blocked fractional factorial are shown on the next page.

INCOMPLETE AND CONFOUNDED BLOCK DESIGNS

```
data design;
do x4 = -1 to 1 by 2;
  do x3 = -1 to 1 by 2;
    do x2 = -1 to 1 by 2;
      do x1 = -1 to 1 by 2;
      A=x1; B=x2; C=x3; D=x4;
* here are the generators for the fraction;
      E=B*C*D; F=A*C*D; G=A*B*C; H=A*B*D;
* here are the block defining contrasts;
      AB=A*B; AC=A*C; AD=A*D;
      if AB=-1 and AC=-1 and AD=-1 then Block=1;
      if AB= 1 and AC=-1 and AD=-1 then Block=2;
      if AB=-1 and AC= 1 and AD=-1 then Block=3;
      if AB= 1 and AC= 1 and AD=-1 then Block=4;
      if AB=-1 and AC=-1 and AD= 1 then Block=5;
      if AB= 1 and AC=-1 and AD= 1 then Block=6;
      if AB=-1 and AC= 1 and AD= 1 then Block=7;
      if AB= 1 and AC= 1 and AD= 1 then Block=8;
      output;
      end;
    end;
  end;
end;
keep A B C D E F G H Block;
proc sort; by Block  descending H;
```

To analyze the data a model involving only the main effects $A - H$ and the block effects can be fit, since they account for all the degrees of freedom. The `proc glm` commands below will fit this model.

```
proc glm data=comb;
  class Block;
  model weight=Block A B C D E F G H/solution;
  ods output ParameterEstimates=sol;
run;
```

Since this was a saturated model with zero degrees of freedom for the error term, the significance of effects could be accessed using the graphical techniques described earlier. In this case the code for the half-normal plot, shown in Section 7.6.1, was modified by changing the third line from if _n_ >5; to if _n_ >9; in order to remove all block effects from the plot. The eighth line was changed from zscore=probit((((rnk-.5)/12)+1)/2); to zscore=probit((((rnk-.5)/8)+1)/2); since there are only eight effects to be plotted, and the fourteenth line was also changed from if abs(zscore)>1.2; to if abs(zscore)>.8; to label the significant points with a smaller number of effects on the plot. This resulted in the half-normal plot of effects shown in Figure 7.3. The reference line was again added manually.

Figure 7.3 *Half-Normal Plot of Absolute Regression Coefficients from Mouse Growth Experiment*

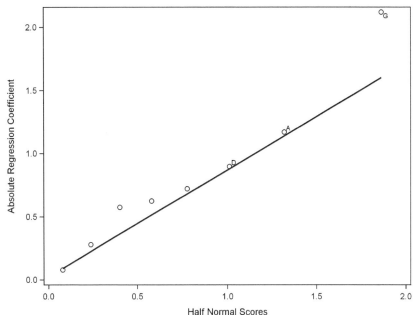

From the half-normal plot it is clear that the only factor having a significant effect on the weaning weight was G - available food. The calculated effect showed that the weaning weight was on the average 4.24g higher for pups of female mice given free access to food compared to pups of mice whose food intake was restricted to 80% of normal. This is the same conclusion that the authors of the article had reached.

7.6.3 Criteria for Choosing Block Defining Contrasts

Care must be exercised when choosing the block defining contrasts for 2^k full factorial or 2^{k-p} fractional factorial designs. To illustrate why, consider the following two examples.

Example 1 consists of blocking a 2^5 factorial into 8 blocks of size $2^2 = 4$. To do this $k - q = 5 - 2 = 3$ block defining contrasts must be chosen. Suppose the interactions ABC, CDE and $ABCDE$ are chosen as the block defining contrasts. Their two-factor generalized interactions $ABC(CDE) = ABDE$, $ABC(ABCDE) = DE$, $CDE(ABCDE) = AB$, and their three-factor generalized interaction $ABC(CDE)(ABCDE) = ABC(AB) = C$ are also confounded, accounting for the 7 degrees of freedom for blocks. The block defining contrasts were all three-factor interactions and above, yet one of their generalized interactions is main effect C that we would not want to confound with blocks. A better choice would be to choose ACE, BCE and $ABCD$ as block defining contrasts. Their generalized interactions include no main effects.

Example 2 consists of blocking a 2^{6-2} fractional factorial in four blocks of size $2^2 = 4$. To do this $k - p - q = 6 - 2 - 2 = 2$ block defining contrasts must be chosen. If the generators for the fractional factorial are $E = ABC$ and $F = ABD$, the defining relation for the fractional factorial is $I = ABCE = ABDF = CDEF$. If the block defining contrasts chosen are BDE and $ACDE$, they and their generalized interaction $BDE(ACDE) = ABC$ will account for the three degrees of freedom for blocks. However, because of the fractionation we see that all of the following aliases also become confounded with blocks.

$$BDE = ACD = AEF = BCF$$
$$ACDE = BD = BCEF = AF$$
$$ABC = E = CDF = ABDEF$$

Again we can see that this choice of generators results in confounding main effect E with blocks. This was an unintended result. A better choice of block defining contrasts would be ACD and AB, which will not result in any main effects being confounded with blocks.

There are many choices for block defining contrasts and fractional factorial generators, and each will result in different set of interactions being confounded with blocks and a different alias structure for the design. Some choices will be better than others and will result in fewer low-order interactions being confounded with blocks and main effects. However, to find the best generators and block defining contrasts for a particular design problem is not a simple task. Fortunately, statisticians have provided tables that show choices that are optimal in certain respects.

Box et al. (1978) provide tables for block defining contrasts that will result in a minimal number of low-order interactions being confounded with blocks in a blocked 2^k design. Wu and Hamada (2000) provide a more extensive table that was generated by the algorithm described by Chen et al. (1993). In some cases, such as the 2^6 blocked into 16 blocks of size 4, Wu and Hamada's tables provide a set of block defining contrasts that are better (in the sense that fewer two-factor interactions are confounded with blocks) than the contrasts shown in Box, Hunter and Hunter's tables.

It is more difficult to find fractional factorial generators and block defining contrasts for 2^{k-p} fractional factorial designs, because of the combination of confounding due to the block effects and fractionization. However, Wu and Hamada's book also has tables for the generators for fractional factorial 2^{k-p} designs along with the corresponding block defining contrasts that will result in highest *estimability order*. They say that a design with estimability of order e is one in which all factorial effects of order e or less are estimable; that is they are not confounded with blocks or factorial effects of order less than $e+1$. These plans were again obtained using the algorithm of Chen et al. (1993).

SAS ADX automatically chooses block defining contrast for blocked 2^k de-

CONFOUNDED 2^k AND 2^{k-p} DESIGNS

signs based on Box, Hunter and Hunter's table. For 2^{k-p} designs ADX also chooses a reasonable set of block defining contrasts and design generators which are in many cases equivalent to the optimal selection given in Wu and Hamada's tables, but in other cases are slightly worse. For example to create a 2^{6-1} design, blocked into 4 blocks of 8 experimental units each, simply define 6 factors in the new two-level menu. Then, when selecting the design, click the show blocked designs option, as shown in Figure 7.4, and select the 32 run half-fraction in 4 blocks.

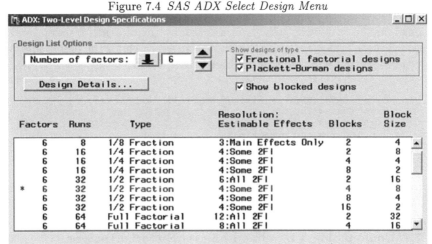

Figure 7.4 *SAS ADX Select Design Menu*

SAS ADX then generates a 2^{6-1} design using the generator $F = ABCDE$ and the block defining contrasts $BCDE$ and $ACDE$. This is a resolution VI fractional factorial, and if this were run in a completely randomized design, all main effects and two-factor interactions would be estimable assuming four-factor and higher-order interactions were negligible. A design like this might be used if the experimenter was interested in estimating two-factor interactions. However if experimental units are heterogeneous, more power or precision for detecting factorial effects and interactions could be achieved by sacrificing some interactions and confounding them with blocks. The block defining contrasts chosen by SAS ADX, are both four-factor interactions, and at first glance, it appears that no two-factor interactions will be sacrificed. However, the generalized interaction $BCDE(ACDE) = AB$ accounts for the third degree of freedom for blocks and due to the fractionization each interaction confounded with blocks has an alias that is also confounded with blocks. In this case the defining relation for the fractional factorial is $I = ABCDEF$ and we see that $BCDE = AF$, $ACDE = BF$ and $AB = CDEF$. So in reality three two-factor interactions AB, AF and BF will be confounded with blocks and lost using this design.

On the other hand, Wu and Hamada's table gives $F = ABCDE$ as the generator of the half-fraction and ACD, BCD as the block defining contrasts for the first 2^{6-1} design in 4 blocks listed in their table. This plan only confounds one two-factor interaction with blocks and is therefore slightly better than the design created by SAS ADX. Knowing the fraction generator and the block defining contrasts, this design could be created in the SAS data step like the mouse growth experiment shown in Section 7.6.2.

Given information about the effects the experimenter would like to be able to estimate, SAS `proc factex` uses an algorithm similar to the one proposed by Franklin (1985) to search for an acceptable set of block defining contrasts and design generators, and then to produce the resulting design. For example, the code below finds the same design produced by SAS ADX. The `size design=32;` statement tells `proc factex` to create a half-fraction, and the `model resolution=max;` statement tells it to search for the design and block generators that will result in the highest resolution.

```
*this code creates 2^(6-1) in 4 blocks of 8;
proc factex;
   factors A B C D E F;
   model resolution=max;
   size design=32
   blocks nblocks=4;
   output out=bd;
   examine aliasing;
```

7.7 Confounding 3 Level and p Level Factorial Designs

If every factor in a factorial has 3 levels we call the design a symmetric 3^k design. These designs can be run in blocks (CCBF) using the same strategy as shown in Section 7.6.1 by confounding portions of interactions with blocks. Confounded 3^k designs can only be run in blocks of size 3^q where $q < k$. To illustrate the method of confounding, consider a 3^2 design with factors A and B, each with 3 levels. Since $k = 2$, this design can only be blocked in blocks of size 3, resulting in 3 blocks. In order to prevent main effect A and B from being completely confounded with blocks, the two-factor interaction AB should be confounded with blocks. However, there are $(3 - 1) \times (3 - 1) = 4$ degrees of freedom for the two-factor interaction while there are only $3 - 1 = 2$ degrees of freedom for blocks. Therefore, only part of the interaction need be confounded with blocks. The four degrees of freedom for AB can be partitioned into two 2 degrees of freedom contrasts $A + B$ and $A + 2B$, and one of these can be used to define the blocks.

If the levels of the factors A and B are represented symbolically by 0, 1 and 2. The levels of the two contrasts are given by $A + B$ modulo 3 and $A + 2B$ modulo 3, as shown in the table on the top of the next page. The levels of these two contrasts can each be seen to have 2 degrees of freedom because they divide the treatment combinations into three groups similar to the way the levels of main effects A and B divide the treatments combinations into three

A	B	A+B	A+2B
0	0	0	0
0	1	1	2
0	2	2	1
1	0	1	0
1	1	2	0
1	2	0	2
2	0	2	2
2	1	0	1
2	2	1	0

groups. Both of these contrasts are orthogonal to main effects A and B since all three levels of each main effect are represented within each level of both contrasts. Therefore if either of these contrasts is used to define blocks neither main effect will be confounded with blocks. The SAS code below shows how a blocked factorial could be created in the data step confounding blocks with the $A+B$ contrast.

```
data blockdes;
do A = 0 to 2;
  do B = 0 to 2;
    Block=mod(A+B,3);
    output;
  end;
end;
```

The model, represented in SAS `proc glm` notation for this design, would be y = Block A B;. The AB interaction, that is confounded with blocks, is left out of the model just as it would have been in the confounded 2^k and 2^{k-p} designs. Main effect A would have 2 degrees of freedom, main effect B would have 2 degrees of freedom, blocks would have 2 degrees of freedom, and the error would have 2 degrees of freedom. For this design to be useful the interaction AB would be assumed negligible.

In general when confounding 3^k experiments in blocks of size 3^q, $k-p$ block defining contrasts must be chosen. They and all of their generalized interactions will also be confounded with blocks. It would be difficult to choose a set of block defining contrasts by trial and error that will result in the fewest low-order interactions being confounded with blocks. Wu and Hamada (2000) give tables of block defining contrasts and design generators for 3^k and 3^{k-p} designs blocked into blocks of size 3^q that will result in the maximum number of estimable effects. These tables were determined using the algorithm described by Chen et al. (1993). SAS `proc factex` uses a different algorithm to search for an acceptable set of block defining contrasts and design generators and then to produce the resulting design. For example, the code below creates a 3^4 design in 9 blocks of 9.

The `factors` statement defines the factor names and the `nlev=3` option

```
*this code creates 3^4 in 9 blocks of 9;
proc factex;
  factors A B C D/nlev=3;
  model estimate=(A B C D A*B A*C A*D B*C B*D C*D);
  blocks nblocks=9;
  output out=tf;
  examine aliasing;
```

declares all factors to have three levels. In **proc factex** the number of levels of the factors must always be a prime number (i.e., 2, 3, 5 etc.). The **model** statement declares which effects the experimenter would like to be able to estimate. In other words, these should not be confounded with block effects or other main effects. In this case, a design is sought for that allows all main effects and two-factor interactions to be estimated. There are three ways to define the model. One is to list the effects to be estimated as shown above. The other two, more relevant for fractional factorial designs, are to declare the resolution desired or the fact that a minimum abberation design is requested. The **blocks** statement declares the number blocks in the design to be created. This will always be a power of the number of levels of the factors. The **examine aliasing;** option prints out a table that shows the block defining contrasts that **proc factex** used to come up with the design. In this example they were $A + 2B + C$, $A + B + 2D$, $A + 2C + D$, and $B + C + D$, which are all contrasts from three-factor interactions. Since each of these contrasts has 2 degrees of freedom, together they account for the 8 degrees of freedom for blocks. Therefore, the model (represented in SAS **proc glm** notation) that could be fit to the data collected using this design would be y = Block A B C D A*B A*C A*D B*C B*D C*D; or y = Block A B C D A*B A*C A*D B*C B*D C*D A*B*C*D; since neither two-factor interactions nor the four-factor interaction are confounded with blocks.

proc factex uses an algorithm to search for an appropriate set of design generators and block defining contrasts. If it cannot find a set that will satisfy the requirements specified in the model statement, a message will be printed in the SAS log stating that no design was found.

Since **proc factex** allows for designs where all factors have p levels, where p is a prime number, the code above could be changed to block a 5^4 factorial in 25 blocks by changing the following options: (1) change the factors option from **nlev=3** to **nlev=5**, and (2) change the number of blocks specified from **nblocks=9** to **nblocks=25**.

In practice, 3^k or 3^{k-p} designs and p^k or p^{k-s} designs are rarely used because it is unusual for all factors in a factorial design to have the same number of levels, unless the number of levels is two. 2^k and 2^{k-p} designs are commonly used in practice because experimenters reduce the number of levels of each factor to two (by choosing the two levels they feel will exhibit the maximum difference) in order to reduce the total number of experiments. In cases where the number of levels of some factors cannot be reduced to 2, because they represent a discrete set of alternatives such as machine settings, mixed level or asymmetric factorials result. The principles shown in this section, to block 3^k or p^k experiments by confounding portions of interactions can be extended to the mixed level factorials as shown in the next section.

7.8 Blocking Mixed-Level Factorials and OAs

A mixed level or asymmetric factorial can be represented as $s_1^{m_1} \times s_2^{m_2} \times \cdots s_\gamma^{m_\gamma}$ involving $n = \sum_{i=1}^{\gamma} m_i$ factors where m_i factors each has s_i levels. For example, a $2^3 \times 3^2 \times 4^1 \times 6^1$ is a factorial with three factors with two levels, two factors with three levels, one factor with four levels, and one factor with six levels. The number of levels of every factor in a mixed-level factorial is either a prime number or a product of prime numbers. If the number of the ith factor $s_i = \prod_{l=1}^{m} p_l$ where p_l are prime numbers, then the levels of the factor s_i can be represented by the combination of levels of m pseudo factors each with a prime number of levels. For example, in the $2^3 \times 3^2 \times 4^1 \times 6^1$ factorial the first three factors (A B and C) have two levels, where two is a prime number. The fourth and fifth factors (D and E) have three levels, where three is a prime number. The sixth factor (F) has four levels, and $4 = 2 \times 2$ is the product of two prime numbers. Finally, the seventh factor (G) has six levels, and $6 = 2 \times 3$ is also a product of prime numbers. The levels of the four- and six-level factors F and G can be represented as combinations of the levels of two- and three-level pseudo factors f_1, f_2, g_1 and g_2 as shown below.

F	f_1	f_2		G	g_1	g_2
0	0	0		0	0	0
1	1	0		1	1	1
2	0	1		2	0	2
3	1	1		3	1	0
				4	0	1
				5	1	2

7.8.1 Blocking Mixed-Level Factorials

Since the levels of each factor in a mixed-level factorial can be represented by a combination of the levels of pseudo factors with a prime number of levels, the entire factorial can be represented by a $p_1^{n_1} \times p_2^{n_2} \times \cdots p_g^{n_g}$ factorial, where all the factors have a prime number of levels. For example the $2^3 \times 3^2 \times 4^1 \times 6^1$ factorial can be represented by a $2^6 \times 3^3$ factorial. The $p_i^{n_i}$ are called sub-experiments, and the classical method of confounding a mixed-level or asymmetric factorial is to apply the method described in the last section to each sub-experiment and then combine the blocks from each sub-experiment to form the blocks of the entire completely confounded blocked factorial (CCBF). This method will result in the greatest efficiency for estimating the effects and interactions that are not confounded with blocks in that they will be completely orthogonal to blocks.

Asymmetric factorials of the form $s_1^{m_1} \times s_2^{m_2} \times \cdots s_\gamma^{n_\gamma}$ can be blocked into b blocks of size k. However, to avoid confounding or partially confounding any main effect, the block size k must be divisible by all the prime numbers p_1 - p_r that represent the number of levels of the factors or number of levels of the pseudo factors used to represent a factor.

To illustrate the classical method of blocking a mixed-level factorial, consider blocking the 36 treatment combinations of a $2^2 \times 3^2$ factorial. This factorial can be represented as a product of two sub-experiments: a 2^2 comprised of two factors A and B each at 2 levels; and a 3^2 comprised of two factors C and D each at 3 levels. Since the levels of all factors are prime, the block size must be divisible by both 2 and 3. The blocks can be found by confounding within each sub-experiment. Since there are only two factors in each sub-experiment, and it would be undesirable to confound main effects with blocks, there is only one interaction that can be confounded in each sub-experiment.

Confounding AB in the 2^2 sub-experiment results in two blocks of size 2. Confounding the $C+D$ contrast in the 3^2 sub-experiment results in three blocks of size 3. The combination of each block from the first sub-experiment with each block of the second sub-experiment results in blocks of size $k = 6$, which is divisible by both 2 and 3. The interactions confounded in the complete factorial will be AB, two degrees of freedom from CD and two degrees of freedom from the product $ABCD$. The main effects A, B, C, D and the AC, AD, BC, BD two-factor interactions and the ABC, ABD, ACD and BCD three-factor interactions are not confounded with blocks and are estimable. The code at the top of the next page shows how to create this design in SAS `proc factex`.

The first use of `proc factex` creates a blocked design in two three-level factors C and D and stores it in the SAS file `d1`. Specification of `resolution=max` on the model statement and the `nblocks=3` on the blocks statement causes `proc factex` to select the $C+D$ contrast to confound this sub-experiment into three blocks. This can be determined by examining the table printed by the `examine aliasing;` statement. The second use of `proc factex` creates a blocked experiment in two two-level factors A and B (the default in `proc factex`). The option `pointrep=d1` in the output statement tells `proc factex` to combine every block from the design `d1` with every block in the design being created and then store the combined result in the SAS file `d2`. This results in a design with block size 6. The block divisions from the first use of `proc factex` are in the variable `block`, while the block divisions from the second use of `proc factex` are in the variable `block2`. The final data step creates the six-level blocking factor from the original two. The final result from the file `df` is shown at the top of the next page. It can be the basis for creating a randomized data collection worksheet.

BLOCKING MIXED-LEVEL FACTORIALS AND OAS

```
proc factex;
  factors C D/nlev=3;
  blocks nblocks=3;
  model resolution=max;
  examine aliasing; output out=d1 ;
proc factex;
  factors A B;
  blocks nblocks=2;
  model resolution =max;
  examine aliasing; output out=d2 pointrep=d1;
data df; set d2;
  blk=(block-1)*3+block2;
  keep blk A B C D;
proc sort; by blk;
proc print; run;
```

Obs	A	B	C	D	blk
1	-1	1	-1	-1	1
2	-1	1	0	1	1
3	-1	1	1	0	1
4	1	-1	-1	-1	1
5	1	-1	0	1	1
6	1	-1	1	0	1
7	-1	1	-1	1	2
8	-1	1	0	0	2
9	-1	1	1	-1	2
10	1	-1	-1	1	2
11	1	-1	0	0	2
12	1	-1	1	-1	2
13	-1	1	-1	0	3
14	-1	1	0	-1	3
15	-1	1	1	1	3
16	1	-1	-1	0	3
17	1	-1	0	-1	3
18	1	-1	1	1	3
19	-1	-1	-1	-1	4
20	-1	-1	0	1	4
21	-1	-1	1	0	4
22	1	1	-1	-1	4
23	1	1	0	1	4
24	1	1	1	0	4
25	-1	-1	-1	1	5
26	-1	-1	0	0	5
27	-1	-1	1	-1	5
28	1	1	-1	1	5
29	1	1	0	0	5
30	1	1	1	-1	5
31	-1	-1	-1	0	6
32	-1	-1	0	-1	6
33	-1	-1	1	1	6
34	1	1	-1	0	6
35	1	1	0	-1	6
36	1	1	1	1	6

This design will allow estimation of all the terms in the model y = A B C D A*C A*D B*C B*D, but the interactions A*B and C*D are confounded with blocks and cannot be estimated.

When the levels of one or more factors in a mixed-level factorial are a product of prime powers, and can be represented by the combination of levels of pseudo factors, no interactions among pseudo factors that represent the same factor can be confounded in any sub-experiment. If any interaction among pseudo factors that represents a factor is confounded, then that main effect will also be confounded. For example, consider blocking the 72 combinations of factor levels in a $3 \times 4 \times 6$ factorial. Factor A has three levels, factor B has four levels, and can be represented by all combinations of two two-level pseudo factors b_1 and b_2, and factor C has six levels that can be represented by all combinations of a two-level pseudo factor c_1 and a three-level pseudo factor c_2. Using the pseudo factors, the $3 \times 4 \times 6$ factorial can be represented by a $2^3 \times 3^2$ factorial in prime level factors and prime level pseudo factors. The block size must be divisible by the prime numbers 2 and 3 to avoid confounding a main effect, therefore blocks of size 6 or 12 may be possible.

The first sub-experiment is a 2^3 composed of two-level pseudo factors b_1, b_2, c_1 and the second sub-experiment is a 3^2 composed of factor A and pseudo factor c_2. The first sub-experiment can only be blocked into 2 blocks of 4 in order to avoid confounding the $b_1 \times b_2$ interaction and therefore the B main effect. Thus the three-factor interaction $b_1 \times b_2 \times c_1$ must be confounded with blocks in the first sub-experiment. In the second sub-experiment the $A + c_2$ contrast of the AC interaction must be confounded to create 3 blocks of 3. The combination of each block from the first sub-experiment combined with each block from the second sub-experiment results in six blocks of 12 treatment combinations, and this is the only block size that is possible without confounding a main effect.

The interactions confounded with the five degrees of freedom for blocks in the combined factorial will be BC (with one degree of freedom) from the first sub-experiment, two degrees of freedom from AC from the second sub-experiment, and two degrees of freedom from the product ABC (i.e., $b_1 \times b_2 \times c_1 \times (A + c_2)$). The SAS code on the next page illustrates how this design can be created using **proc factex** and is similar to the previous example.

By default **proc factex** assigns the levels $(-1, 1)$ to two-level factors and $(-1, 0, \text{and } 1)$ to three-level factors, etc. The first three assignment statements in the data step that creates the SAS file **df** assign the levels $(0, 1, 2 \ldots)$ to the factors A, B, and C based on the default levels or combination of levels of the pseudo factors b_1 and b_2, and c_1 and c_2 that were created by **proc factex**. The final assignment statement combines the block indicators from the first file **d1** and the second file **d2** to get block indicators for the combined file.

```
*This confounds the A*c2 interaction in the 3^2 sub-experiment;
proc factex;
  factors A c2/nlev=3;
  blocks nblocks=3;
  model resolution=max;
  examine aliasing;
  output out=d1 ;
*This confounds b1*b2*c1 interaction in the 2^3 sub-experiment;
proc factex;
  factors b1 b2 c1;
  blocks nblocks=2;
  model estimate=(b1|b2|c1@2);
  examine aliasing;
  output out=d2 pointrep=d1;
data df; set d2;
  A=A+1;
  B=(b2+1)+((b1+1)/2);
  C=2*(c2+1)+((c1+1)/2);
  block=3*(block-1)+block2;
  keep block A B C;
proc sort; by block;
proc print; run;
```

The model that can be fit to the data resulting from this experiment is y = A B C A*B; since none of the terms in this model are confounded with blocks.

Even though this design is optimal for estimating the parameters in the model, it is undesirable since the two-factor interactions AC and BC cannot be estimated. Since there are only five degrees of freedom for blocks, it would appear that there should be a way to confound part of the $2 \times 3 \times 5 = 30$ degrees of freedom for the three-factor interaction ABC with blocks and leave all the two-factor interactions estimable. In fact, if you are willing to sacrifice some of the efficiency in estimating the terms in the model, a better blocked design can be found using a D-optimal search than can be found using the classical method.

Before the days of modern computers and software packages like SAS, it was necessary to completely confound some interactions with blocks, using the classical method, so that other interactions would be left orthogonal to blocks. In that way, the analysis of the data could be completed by hand using the ANOVA sums of squares formulas for balanced data. However, with availability of programs like SAS proc glm sums of squares are computed using matrix operations as shown in Chapters 2 and 3, and it is no longer necessary for each term in the model to be completely orthogonal to blocks in order to compute the ANOVA and F-tests.

The model for a blocked factorial experiment can be written in matrix notation as

$$y = X\tau + Z\beta + \epsilon \qquad (7.5)$$

where y is the $n \times 1$ vector of responses, τ is the vector of estimable treatment effects and interactions, and β is the vector of block effects. One optimality criterion that has been proposed for blocked designs is the D_s criteria (see Atkinson et al., 2007). A D_s optimal design is one that minimizes the covariance matrix of the least squares estimator for $\boldsymbol{\tau}$ or equivalently maximizes the determinant of

$$\boldsymbol{X'QX} \qquad (7.6)$$

where

$$\boldsymbol{Q} = \boldsymbol{I} - \boldsymbol{Z}(\boldsymbol{Z'Z})^{-1}\boldsymbol{Z'}. \qquad (7.7)$$

Designs where blocks are orthogonal to treatment effects, or $\boldsymbol{X'Z} = \boldsymbol{0}$, are 100% D_s efficient.

Applying the classical method separately to symmetric sub-experiments results in designs that have known confounding patterns and are 100% D_s efficient for estimating the effects and interactions, $\boldsymbol{\tau}$, that are not confounded with the block differences (since they will be orthogonal to blocks). However, in practical situations, use of the classical approach does not provide much flexibility in the choice of block size or in the choice of interactions to be confounded with block differences. Since the sub-experiments are often defined in terms of pseudo-factors, interactions of interest often become confounded with blocks.

Cook and Nachtsheim (1989) describe a more general computer algorithm for creating blocked designs by beginning with a non-singular starting design, then sequentially exchanging treatment combinations assigned to one block with those assigned to other blocks in order to maximize $|\boldsymbol{X'QX}|$. The designs resulting from this algorithm may not be 100% D_s efficient for estimating $\boldsymbol{\tau}$, but greater choices of block sizes and estimable interactions are possible.

SAS `proc optex` is capable of creating blocked designs using the algorithm of Cook and Nachtsheim (1989). The next example illustrates the use of this procedure. In the data step shown on the next page, the commands create a candidate set composed of a 3 × 4 × 6 factorial. The model statement in the `proc optex` call specifies that all two-factor interactions be estimable. The blocks statement requests six blocks of size 12.

Since there is more than one treatment factor in the model, `proc optex` does not print the block design D-efficiency, but rather the D_s-efficiency. The result (which is not shown here) indicates that `proc optex` found a design that was 87.35% D_s-efficient for the treatments or terms in the model. Although not 100% efficient (implying all terms in the model are completely orthogonal to blocks), like the design found using the classical method, at least all of the two-factor interactions are estimable. If the code is modified by changing the option on the blocks statement from `structure=(6)12;` to `structure=(12)6;`, `proc optex` finds a 3 × 4 × 6 factorial blocked in 12 blocks of six that still allows estimation of all two-factor interactions. The D_s-efficiency for the treatments in this design was 78.94%, but this is far better than could be accomplished with the classical method. The only way the classical method could block the 3 × 4 × 6 factorial into blocks of size 6 would be to confound part of the main

BLOCKING MIXED-LEVEL FACTORIALS AND OAS 289

effect for factor B. If reducing the block size to 6 reduces the variance of the experimental units within a block, the sacrifice in efficiency will be worthwhile.

```
data cdesign;
do C = 0 to 5;
  do B = 0 to 3;
    do A =0 to 3 ;
    output;
    end;
  end;
end;
*creates a design with 6 blocks of size 12 i.e., structure=(b)k;
*Although not orthogonal to blocks two-factor interactions
 are estimable ;
proc optex data=cdesign coding=orthcan seed=73565;
  class A B C;
  model A B C A*B A*C B*C;
  blocks structure=(6)12;
  generate n=72;
  output out=bdesign blockname=blk;
proc print; run;
```

When using the D_s-optimal approach to finding a confounded block design for a mixed-level factorial, it is no longer required that the block size be divisible by all the prime numbers that represent factor levels in the sub-experiments. If the block size that is most convenient for reducing variability of experimental units within a block is not divisible by the number of levels of all factors, it may prevent finding a design where all factors and interactions of interest are orthogonal to blocks. However, a design found using the D_s-optimal search will usually not have all effects in the model orthogonal to blocks anyway. As long as the treatment D_s-efficiency is not zero, the terms specified in the model will be estimable. Due to the non-orthogonality of these designs, the type III sums of squares and least squares means should be used when analyzing data.

7.8.2 Blocking Orthogonal Arrays, Orthogonal Main Effect Plans and Nearly Orthogonal Main Effect Plans

Orthogonal array designs, orthogonal main effect plans and nearly orthogonal plans discussed in Section 6.7 can also be blocked into CCBFF designs in order to reduce the variation of experimental units within blocks or to allow the list of experiments to be completed on different days or with different batches of raw materials, etc. One way to accomplish this is to include the block factor as one of the factors when initially creating a design using the SAS %mktEx macro.

For example, to create a blocked fraction of $4^1 \times 6^1 \times 2^3 \times 3^2$ experiment, first use the macro call %mktruns(4 6 2**3 3**2) to find that there is a reasonable design, that is close to orthogonal with only 24 runs, that will allow estimation of all 15 degrees of freedom for main effects. If you want to block this experiment into 3 blocks of 8 treatment combinations (or experimental units), add another three-level factor to the request and use the macro call %mktruns(4 6 2**3 3**3). Again, this reveals there is a reasonable 24-run plan that will allow for estimation of all 15 degrees of freedom for main effects and 2 degrees of freedom for blocks. The first macro calls below creates the design, and the second macro call labels the last three-level variable as Block. The %mktlab macro can be used to change the default names x1 - x8 in the SAS data set `randomized` that is produced by the %mktex macro. The SAS log shows that the data set with the newly labeled variables are stored in the SAS data set `final`. Finally the sort and print commands shown below print the results sorted by blocks.

```
%mktex(4 6 2**3 3**3, n=24)
%mktlab(vars=A B C D E F G Block)
proc sort data=final; by Block;
proc print data=final; run;
```

Another way to block a design that has already been created by the %mktex macro is to use the %mktblock macro. This macro tries to create a blocking variable that has the least correlation with each factor in the design. For example if the 24-run fraction of the $4^1 \times 6^1 \times 2^3 \times 3^2$ was created using the first macro call below, the second macro call will block the design into three

```
%mktex(4 6 2**3 3**2, n=24)
%mktblock(data=randomized, nblocks=3)
```

blocks of 8. The result is printed and automatically stored in the SAS file `blocked`. Since this nearly orthogonal main effect plan allows estimation of the block effect and all main effects, the data resulting from this experiment can be analyzed using the main effects model in `proc glm` as shown below.

```
proc glm;
  class Block x1-x7;
  model y = Block x1-x7;
```

7.9 Partially Confounded Blocked Factorial (PCBF)

One of the main purposes for running a factorial experiment with few factors is to estimate all interactions. Yet if the design is confounded into blocks, the experimenter will lose the ability to estimate some interactions. However, including a few additional blocks in the design will allow estimation of all main effects and interactions using the method of partial confounding. This method consists of confounding one or more effects or interactions in one set of blocks, and confounding different effects or interactions in additional sets of blocks

PARTIALLY CONFOUNDED BLOCKED FACTORIAL (PCBF)

(or replicates). By combining all the replicates, all effects and interactions will be estimable, although not orthogonal to blocks.

For example, if an experiment were to be performed to study the effect of two levels of A=calcium supplements and two levels of B=potassium supplements upon the blood pressure of hypertensive subjects, a 2^2 factorial experiment would be performed in order to estimate the two main effects and the interaction. However, if the experiments were blocked into 2 blocks of 2 experimental units (e.g., two pairs of identical twins) by confounding AB, this interaction would be lost. One way to remedy the problem would be to include four additional blocks, confounding main effect A with the difference in two additional blocks, and main effect B with the difference in two more blocks. Combining the six blocks, both main effects and their interaction would be estimable. This design is shown in the Table 7.10, and this general class of designs is called partially confounded blocked factorials or PCBF.

Table 7.10 *Partially Confounded 2^2 in 6 Blocks of 2*

Block	A	B	
1	−	−	
1	+	+	Rep 1
			confounds
2	−	+	AB
2	+	−	
3	−	−	
3	−	+	Rep 2
			confounds
4	+	−	A
4	+	+	
5	−	−	
5	+	−	Rep 3
			confounds
6	−	+	B
6	+	+	

The model `y = Block A B A*B` could then be fit to data from the combined set of blocks. The type III sums of squares and least squares means would be used to analyze the data since the effects are not completely orthogonal to blocks.

If each effect and interaction in the model is confounded an equal number of times in different sets of blocks, as in the example shown above, Butler (2006) shows the design will have favorable properties and the maximum treatment D-efficiency for estimating the factorial effects. Therefore, a design like this could be created using Cook and Nachtsheim's algorithm. The example below shows how this could be done using `proc optex`.

For mixed-level factorial plans Das (1960) has provided a method for con-

```
data cdesign;
do A = 0 to 1;
 do B = 0 to 1;
  output;
 end;
end;
*creates a design with 6 blocks of size 2 i.e., structure=(b)k;
proc optex data=cdesign coding=orthcan seed=73462;
   class A B;
   model A|B;
   blocks structure=(6)2;
   generate n=12;
   output out=bdesign blockname=block;
proc sort; by block;
proc print data=bdesign; run;
```

structing balanced confounded designs where (1) the information recovered for each degree of freedom for any partially confounded interaction is the same, and (2) any contrast of a partially confounded interaction is estimable independently of any contrast of another partially confounded interaction. The information recovered for the ith degree of freedom in the model (7.5) is calculated as c_{ii}/c'_{ii}. c_{ii} is the diagonal of $\boldsymbol{X'X}^{-1}$ matrix corresponding to particular single degree of freedom, and $\sigma^2 c_{ii}$ is the variance of $\hat{\tau}_i$ in a design where the treatment effects are orthogonal to blocks. c'_{ii} is the diagonal $\boldsymbol{X'QX}^{-1}$, and $\sigma'^2 c'_{ii}$ is the variance of $\hat{\tau}_i$ in the partially confounded design where $\boldsymbol{Q}$ is defined in Equation (7.7). In partially confounded designs $c'_{ii} > c_{ii}$, but σ'^2 should be much smaller than σ^2 due to the fact that the experimental units are more homogeneous within the blocks of reduced size.

Constructing a design using Das's method consists of converting the asymmetrical factorial into a fraction of a symmetrical factorial. The partial confounding is performed in replicates of the symmetrical factorial, and then each replicate is reduced by fractionation. Lawson et al. (2009) show how Das's method can be implemented with **proc plan** and the SAS data step. They also provide a SAS macro for generating Das's balanced confounded designs. Creating designs by this method will allow all interactions in the model to be estimated and will result in a design with equal precision for each single degree of freedom of partially confounded effects and interactions. One disadvantage for using this method is that the total number of treatment combinations, $N = s_1^{m_1} \times s_2^{m_2} \times \cdots s_\gamma^{n_\gamma}$ must always be divisible by the block size. This can require a large number of blocks in some cases.

Cook and Nachtsheim's algorithm can also be used to find a partially confounded mixed-level factorial that will allow estimation of all interactions, and there is no restriction on the block size. Using this method can sometimes produce a balanced confounded design like Das's method, and in other cases it will find an approximately balanced design with more choices for the block size and total number of runs. Lawson et al. (2009) illustrate how partially confounded designs can be created with Cook and Nachtsheim's algorithm using

PARTIALLY CONFOUNDED BLOCKED FACTORIAL (PCBF) 293

SAS `proc plan` and `proc optex`, and they compare the properties of designs created by this method to designs created by Das's method. The example below shows the use of `proc plan` and `proc optex` to construct a partially confounded 3×2^2 factorial blocked in 12 blocks of 4.

```
* Create Candidate set with proc plan ;
proc plan;
  factors A=2 B=2 C=3;
  output out=cdesign A nvals=(0 1)
                     B nvals=(0 1)
                     C nvals=(0 1 2);
*creates a balanced confounded design with 12 blocks;
* of size 4: structure=(b)k using prior option;
proc optex data=cdesign coding=orthcan seed=73462;
  class A B C;
  model A, B C B*C, A*B A*C A*B*C/prior=0, 12, 36;
  blocks structure=(12)4 noexchange niter=1000 keep=10;
  generate n=48;
  output out=bdesign blockname=blk; run;
proc print; run;
```

The `prior=0, 12, 36;` option on the model statement prompts `proc optex` to search for a balanced confounded design, which may exist for certain block sizes and number of runs. If the `prior` option is removed, `proc optex` will find an approximately balanced confounded design where the information recovered will be approximately equal for all degrees of freedom in the model.

As an example of the design and analysis of a partially confounded factorial, consider an experiment performed by Dossett et al. (2007). They were investigating methods of storing apple slices in brown bag lunches. The problem was that the apple slices in a sack lunch turn brown and look unappetizing before lunchtime. They wanted to compare the effects of dipping the slices in different treatment solutions prior to storage and to compare the effects of storing them in different storage containers to see if they could find conditions to reduce the amount of browning. The combination of levels of the two treatment factors shown in Table 7.11 comprise a 4×3 factorial. They thought that dif-

Table 7.11 Levels of Factors for Apple Slice Browning Experiment

Factor Level	A=pre treatment solution	B=storage container
0	none	none - open air
1	weak lemon juice	Ziploc bag
2	salt water	Tupperware
3	baking soda water	—

ferent varieties of apples might brown at different rates and therefore wanted to block by apple. Their apple slicer cut the apples into six slices, therefore all

294   INCOMPLETE AND CONFOUNDED BLOCK DESIGNS

$4 \times 3 = 12$ treatment combinations could not be tested within each block or apple. Since they were interested in the possible interaction of their treatment factors, they ran a partially confounded design. The SAS code below Table 7.11 was used to create a 4×3 factorial design in 4 blocks of size 6 using proc optex. The blocks represented four different varieties of apples namely Fuji, Braeburn, Red Delicious and Granny Smith. The experimental units were the six slices from each apple, and these were each randomly assigned to one of the treatment combinations designated for that block by proc optex.

```
* Create candidate set using the data step;
data cdesign;
do B = 0 to 2;
  do A = 0 to 3;
  output;
  end;
end;
* Create approximately balanced design with 4 blocks;
* of 6 E.U.'s per block                                ;
proc optex data=cdesign coding=orthcan seed=73462;
  class A B;
  model A|B;
  blocks structure=(4)6 noexchange niter=1000 keep=10;
  generate n=24;
  output out=bdesign blockname=block;
proc print; run;
```

After storing their treated apple slices for an hour and forty-five minutes, each slice was compared to photographs of an apple slice at various stages of browning and assigned a rating between 1 and 11. The lowest rating was for the least amount of browning and the highest was for the most. All three team members independently rated the slices and the response was the average rating. The results of the experiment are shown in Table 7.12.

The data were analyzed using the model rating = block A B A*B in proc glm, and the resulting type III ANOVA table is shown below.

Source	DF	Type III SS	Mean Square	F Value	Pr >
BLOCK	3	3.0069875	1.0023292	0.14	0.933
A	3	145.9635560	48.6545187	6.77	0.011
B	2	2.2054486	1.1027243	0.15	0.859
A*B	6	7.7347144	1.2891191	0.18	0.975

It can be seen that the only thing significant was the treatment solution. Table 7.13 shows the least squares means for the treatment solution.

The Tukey's HSD method was used to make the pairwise comparisons of the differences in means, and the results are summarized by the underlines in the table. It can be seen that dipping the apple slices in salt water reduces browning the most, but the amount of browning for slices dipped in lemon

Table 7.12 *Blocked 4 × 3 Factorial Design and Results for Apple Slice Browning*

Block 1			Block 2			Block 3			Block 4		
A	B	rating	A	B	rating	A	B	rating	A	B	rating
0	0	7.33	2	0	1.00	3	2	10.33	2	2	1.00
2	1	1.67	1	0	3.33	1	0	2.00	3	0	8.33
0	2	6.67	2	2	1.00	2	1	2.33	1	1	4.33
1	1	1.33	0	1	8.67	0	2	7.00	1	2	1.33
2	0	1.67	0	0	8.33	3	1	3.67	0	1	3.33
3	0	8.00	3	2	4.00	1	2	6.00	3	1	9.33

Table 7.13 *Least Squares Means for Factor A*
(means underlined by the same line are not significantly different)

salt water	lemon juice	none	baking soda
1.53	2.97	6.97	7.19

juice was not significantly worse. The experimenters recommended further studies varying the concentration of lemon juice to see if they could improve the results and eliminate the aftertaste left by salt water.

7.10 Review of Important Concepts

When experimental units are heterogeneous and can be grouped into smaller blocks of more homogeneous experimental units, a blocked design should be used. When the number of experimental units per block or block size is smaller than the number of levels of the treatment factor or combinations of levels of treatment factors in a factorial design, an incomplete block design should be used.

When there is only one treatment factor, there is a choice between two types of incomplete block designs. Balanced incomplete block (BIB) designs require that every treatment level occurs in a block an equal number of times with every other treatment level. BIB designs can be created with SAS proc optex. The advantage of these designs is that the precision (or standard error) of the difference in every possible pair of treatment means will be equal. The disadvantage is that many blocks and experimental units may be required to achieve the balance. The other alternative design for one treatment factor is the partially balanced incomplete block (PBIB) designs.

The advantage of PBIB designs is that the total number of blocks and experimental units required can be reduced. The disadvantage is that the standard error of differences in pairs of treatment means will not be constant. There are many different methods of creating PBIB designs, and some of the more useful designs have been tabled. One type of PBIB called a BTIB (balanced with respect to test treatments) can be easily created from a BIB design. This design is useful when there is more interest in comparing one treatment level (such as a control) to all other treatment levels than there is in comparisons among the other treatment levels. Another class of PBIB designs that can be easily created using SAS `proc plan` is called generalized cyclic designs.

Latin square designs introduced in Chapter 4 have two orthogonal blocking factors, and contain a complete block design in both the row blocks and column blocks. If an incomplete block design is required in either the row or column blocks, a Youden square design (YSD) can be utilized. A YSD can be constructed by deleting one or more columns from a Latin square design.

When experimenting with multiple factors and the block size is not large enough to accommodate all possible treatment combinations, there are two alternative methods for creating an incomplete block design. The first method is to completely confound some interactions with blocks in a completely confounded blocked factorial (CCBF) design, or a completely confounded blocked fractional factorial (CCBFF) design. The advantage of these designs is that the total number of blocks and experimental units can be minimized. The disadvantage is that some interactions will be completely confounded with blocks and will be inestimable. The other method is to use a partially confounded blocked factorial (PCBF) design. Figure 7.5 illustrates when these designs should be used in relation to the designs presented in other chapters.

When each factor has only two or three levels, there is a classical method for creating completely confounded factorial designs, and tables are available for block defining contrasts that will minimize the number of low order interactions confounded. For factorials where all factors have p (a prime number) of levels, or where the experiment can be broken down into sub-experiments where all factors have the same prime number of levels, the classical method for creating a completely confounded design can also be used. SAS `proc factex` can be used to find block defining contrasts, if they exist, that will allow estimation of the effects and interactions of interest. A D-optimal search can also be used to create a confounded design to estimate all effects and interactions of interest and confound others with blocks. The advantage of this approach is that it has a higher possibility of finding a design capable of estimating a specified list of effects and interactions, and it is more flexible in terms of block sizes available. The disadvantage of designs created with a D-optimal search is that treatment levels are not orthogonal to blocks and they are estimated less efficiently than they would be in a design created with the classical method.

Partially confounded blocked factorials are used in factorials where there are only a few factors, like two or three, and there is a need to estimate all

REVIEW OF IMPORTANT CONCEPTS

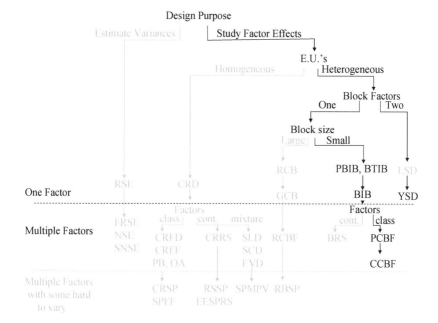

Figure 7.5 *Design Selection Roadmap*

interactions. These designs require more blocks and experimental units than completely confounded designs, but do allow estimation of all interactions. These designs can be created by confounding all estimable effects an equal number of times in different replicates of blocks or by using a D-optimal search.

The analysis of incomplete block designs and partially confounded factorial designs is similar to the analysis of complete block designs, except care must be taken to always use the type III sums of squares and least squares means. For completely confounded factorial designs, the interactions that are confounded with blocks must be left out of the model, and the significant effects can be identified by graphical techniques such as a half-normal plot of effects.

7.11 Exercises

1. An experimenter wanted to determine how the gear selected on a 10 speed bike would affect the time it takes a rider to pedal up a long, gradual hill. The experimental unit will be the trial or point in time when a rider will ride up the hill. Treatment will be the gear selected on bicycle. Since the experimenter would like his conclusions to apply to all riders, he would like to run a blocked experiment where several riders (blocks) test each gear. However, if each rider who participates in his study must test all gears in separate trials, he or she will become fatigued and variability in time to climb the hill will increase. Therefore, the experimenter decided to do an incomplete block design wherein each rider participating in his study will test only a subset of 4 of the 10 gears.

    (a) If the experimenter were to use a balanced incomplete block (BIB) design by choosing all possible subsets of 4 selected from 10 as his blocks, how many riders or blocks would he need to recruit to his study?

    (b) Using Equations (7.1) to (7.3), determine the minimum possible number of blocks that would be required for BIB design.

    (c) Modify the SAS code in Section 7.2 to create a BIB with the number of blocks you found in (b) using `proc optex`.

2. Consider an incomplete block design with $t=10$ levels of the treatment factor.

    (a) If a balanced incomplete block design (BIB) was constructed using the simplest method of forming blocks as all possible combinations of $\binom{10}{k}$, how many blocks would be required for the cases where $k=3$, 4, 5, or 6?

    (b) Using Equations (7.1) to (7.3), determine the minimum number of blocks that could possibly be required for a BIB with $t=10$, and $k=3$, 4, 5, or 6?

    (c) Is it possible to find a BIB with $t=10$, and $k=3$, 4, 5, or 6, by modifying the SAS code in Section 7.2?

    (d) Can one or more generalized cyclic incomplete block design with $t=10$ and $k=3$ be found with less blocks than required for a BIB design? What number of blocks would be required for these designs? Using SAS proc plan as illustrated in Section 7.4, find these designs.

3. Consider blocking a 2^5 factorial into 8 blocks with 4 experimental units per block using the block defining contrasts ACE, BCE and $ABCD$ (shown in Section 7.6.2). Show that the generalized interactions of these contrasts, which would be confounded with blocks, contain no main effects.

4. Consider blocking a 2^{6-2} fractional factorial, with generators $E = ABC$, and $F = ABD$, into 4 blocks of size 4. If the block defining contrasts are ACD and AB (shown in Section 7.6.1), show all interactions and their aliases that will be confounded with blocks.

EXERCISES

5. Choice based conjoint studies (see Section 6.7 for an example and description of conjoint analysis) are frequently used in marketing research. In choice based conjoint studies, potential customers are asked to choose among several alternatives. The alternatives are combinations of levels of several product characteristics. The response for each combination of product characteristics is a count of the number of potential customers who choose that alternative. Analysis of the data helps marketing researchers determine what product characteristics have the most influence on customer choices and what mix of alternatives they should offer in order to satisfy the needs of the market. Consider the following scenario described by Street and Burgess (2007). Researchers want to study how the attributes of a take-out pizza outlet affect consumer choice of where they buy their pizza. The table below lists the product characteristics (i.e., factors in experimental design terminology) and their levels.

	Levels	
Attribute	−	+
Pizza type	Traditional	Gourmet
Type of crust	Thin	Thick
Ingredients	Some canned	All fresh
Size of pizza	One size only	Three sizes
Price	$13	$17
Delivery time	30 minutes	45 minutes

The alternatives that customers must choose from in a survey are all 2^6 combinations of attribute levels.

(a) Since it would be impractical to ask a potential customer to choose from a list of $2^6 = 64$ combinations, it would be better to have each customer surveyed make a choice from 4 alternatives. Create a blocked 2^6 design with block size 4, so that each customer surveyed must only choose from among the 4 combinations of attribute levels in one block.

(b) Since higher-order interactions are unlikely to be important, construct a 16-run 2^{6-2} design blocked into 4 blocks of size 4. Are all main effects estimable from your design?

(c) Block a 16-run 2^{6-2} design blocked into 8 blocks of size 2, so that each customer surveyed only has to choose between two alternatives. Are all main effects estimable from this design?

(d) If the number of alternatives of the type of crust and the size of pizza were increased from 2 to 3, the factorial combination of all attribute levels would change from a 2^6 to a $2^4 \times 3^2$ factorial. Using the %mkt macros, create a 24-run fraction of a $2^4 \times 3^2$ factorial, blocked in 8 blocks of 3, for this problem.

(e) What model in SAS proc glm would you use for the analysis of data arising from the design you created in (d)?

6. For a 2^{6-1} experiment with generator $F = ABCDE$ that is blocked into four blocks with 8 experimental units per block using the block defining contrasts ACD and BCD,

   (a) Determine the defining relation for the fractional factorial.
   (b) Determine the complete alias structure for the fractional factorial.
   (c) Determine the generalized interaction of the block defining contrasts.
   (d) Show what interactions are confounded with blocks.

7. Consider blocking a $2 \times 3^2 \times 4$ factorial into six blocks of 12 treatment combinations each.

   (a) Naming the factors A, B, C, and D, which factors must be represented by pseudo factors in order to block using the classical method?
   (b) What are the two sub-experiments composed of factorials with prime number of levels?
   (c) What interactions of factors and pseudo factors would you confound in each sub-experiment in order to allow estimation of the main effects? What effects and interactions will be unconfounded with blocks and estimable?
   (d) Create the design using SAS `proc factex`.
   (e) Can you create a design using `proc optex` that has six blocks of 12 and allows estimation of all two-factor interactions?

8. Show why it is impossible to block a $3 \times 4 \times 6$ into blocks of size 6 using the classical method without confounding a main effect.

9. Create a partially confounded 2^3 design in 14 blocks of size 4. Generate random data, and use `proc glm` to illustrate that all main effects and interactions can be estimated from the design you created.

10. Using the `proc optex` create a partially confounded 2×3^2 factorial in 6 blocks of size 6.

11. Consider the apple slice browning experiment shown in Section 7.9.

    (a) If the apple slicer made eight slices of each apple instead of six, find a partially confounded 4×3 factorial in three blocks of eight (instead of 4 blocks of 6) by modifying the `proc optex` code shown in that example.
    (b) Analyze the data from the experiment shown in the text and verify the results shown.
    (c) Check the equal variance and normality assumptions. Is a transformation warranted? If so, what transformation would you use and does it change the conclusions?

CHAPTER 8

Split-Plot Designs

8.1 Introduction

A factorial design is a powerful tool for aiding a researcher. Use of a factorial design in experimentation can dramatically increase the power of detecting factor main effects through hidden replication, while additionally affording a researcher the ability to detect interactions or joint effects of factors. When there is an interaction between two factors, the effect of one factor will depend on the level of the other factor, and determination of optimal factor combinations must take that into account. Interactions occur frequently in the real world and they can only be quantified through factorial designs.

In the factorial and fractional factorial designs discussed in Chapters 3, 4, 6 and 7, it was assumed that combinations of factor levels in the design could be randomly assigned to the experimental units within a block for a blocked factorial design, or to the entire group of experimental units for a completely randomized factorial design. Randomization of treatment combinations to experimental units guarantees the validity of the conclusions reached from the analysis of data. However, sometimes the levels of one or more factors in the design are more difficult to change or require more experimental material to evaluate. In this situation complete randomization of factor-level combinations to experimental units could make the experiment much more time consuming or perhaps impossible to conduct. This is the case frequently in process improvement studies where the process consists of several steps, and some factors in an experiment relate to one process step while others relate to a later process step.

When treatment combinations cannot be completely randomized to experimental units due to one or more hard-to-vary factors, the experimental unit is different for the hard-to-vary factors than it is for the easy-to-vary factors. In order to ensure validity, the design of the experiment and the analysis of the data must take into account these different experimental units. The designs that do this are called split-plot experiments. The name split-plot comes from the original application in agricultural experiments, where the levels of some factors (called whole-plot factors) could only be varied between plots of land, while the levels of other factors could be varied within a plot (see Yates, 1937).

A compromise is made in split-plot experiments. The randomization is such that the hard-to-vary factors are not changed as frequently as easy-to-vary factors. This makes it more convenient to conduct the list of experiments, but there is less efficiency or power for detecting effects of the hard-to-vary

8.2 Split-Plot Experiments with CRD in Whole Plots CRSP

To understand the concept of a split-plot experiment, consider the following example problem. Recipes for chocolate and orange cookies include exactly the same ingredients up to the point where the syrup was added to the batch. However, after the cookies were baked, the chocolate cookies had an appealing round and plump appearance, while the orange cookies spread during the baking process and became thin, flat and unappealing. A factorial experiment was devised to determine if there was a way to change the process of making the orange cookies that would reduce the spreading during baking. The factors that were chosen to be varied were A: the amount of shortening in the dough batch (80% of what the recipe called for or 100%), B: the baking temperature (below, at or above the temperature called for by the recipe), and C: the temperature of the cookie sheet upon which the cookies were placed to be baked (hot out of the oven, or cooled to room temperature). A response that could quantify the objective of the experiment was the diameter of the baked cookies.

Table 8.1 shows a completely randomized list for this 3×2^2 factorial experiment. The experimental unit is a batch of cookies and, this plan calls for baking 12 batches. There are no replicate batches, and therefore no estimate of the variance experimental error from replicates in this plan.

Table 8.1 *Completely Randomized Cookie Baking Experiment*

	Factor A	Factor B	Factor C
Batch	Shortening	Bake Temp.	Tray Temp.
1	100%	low	RoomT
2	100%	low	Hot
3	100%	norm	RoomT
4	100%	high	RoomT
5	80%	high	RoomT
6	80%	norm	RoomT
7	100%	high	Hot
8	100%	norm	Hot
9	80%	low	RoomT
10	80%	high	Hot
11	80%	norm	Hot
12	80%	low	Hot

However, the cookie-making process consists of the two steps as shown in Figure 8.1. Factor A was related to process step 1, while factors B and C were related to process step 2. The amount of shortening was a hard-to-vary factor because each time it was changed it required making a new batch of cookie dough, while the baking temperature and and tray temperature were easy to vary. Once a batch of dough was made, there was enough to make six trays of cookies, and all six combinations of baking temperature and tray temperature could be tested within each dough batch, greatly reducing the total number of batches required.

Figure 8.1 *Two-Step Cookie Making Process*

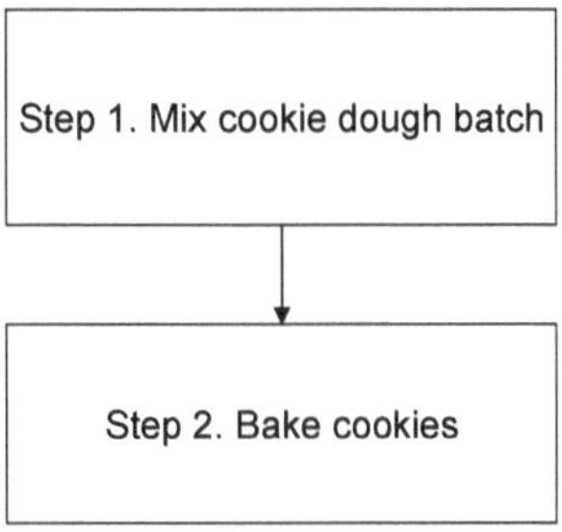

```
┌─────────────────────────────────┐
│ Step 1. Mix cookie dough batch  │
└─────────────────────────────────┘
                │
                ▼
┌─────────────────────────────────┐
│ Step 2. Bake cookies            │
└─────────────────────────────────┘
```

Therefore, an alternate way of designing the experiment would be to follow a two-step plan. First, plan to make four batches of cookie dough and randomly assign two batches to use 80% of the recipe recommended amount of shortening and two batches to receive the full amount of shortening recommended by the recipe. This represents a completely randomized design in one factor and the experimental unit is a batch of cookie dough. This first step is called the whole-plot design, and the whole-plot factor is A the amount of shortening. Next, bake six trays of cookies from each batch of dough and completely randomize the six combinations of bake temperature and tray temperature to the six trays of cookies within each batch. This is a randomized block 3×2 factorial within each batch of cookie dough and is called the sub-plot design. The sub-plot block is the batch of cookies, the sub-plot experimental unit is a tray of cookies and the sub-plot factors are B the bake temperature and C the tray temperature. The combination of the whole-plot and sub-plot design is called a split-plot design with CRD in the whole plots or CRSP.

By designing the experiment in this way only four batches of cookies are required instead of 12, and there are replicate whole-plot and sub-plot experimental units from which to estimate the variance of the two experimental error terms. There is less power for testing the shortening effect than there would be with the 12 batch completely randomized design shown in Table 8.1, but due to the blocking by batch, there is actually more power for detecting the baking temperature, tray temperature and interaction effects. For this reason split-plot experiments are often called super efficient.

8.2.1 Creating a Split-Plot Design with CRD in Whole Plots with SAS

`proc factex` that was described in Chapter 7 can be used to create a randomized plan for a split-plot experiment. The commands to do this are shown below.

```
*Create sub-plot design with factex;
proc factex;
  factors trayT;
  output out=s1    trayT cvals=('RoomT' 'Hot');
run;
  factors bakeT /nlev=3;
  output out=s2 pointrep=s1   bakeT cvals=('low' 'norm' 'high');
run;
*Add the whole-plot design and randomize;
  factors short;
  size design=4;
  output out=sp pointrep=s2 randomize(1012)  short cvals=('80%' '
      100%');
run;
```

The first two sequences of the `factors` statement followed by an `output` statement create a file s2 that contains a full factorial in the three-level factor B and the two-level factor C in a sequential order with the levels of factor C changing fastest. The last sequence of the `factors` statement followed by an `output` statement creates a file sp that consists of a random ordering of the levels of the whole-plot factor A augmented by each line of the file s2 in a random order to create the split-plot design. The option `randomize(seed)` on the output statement creates the random ordering. Using the same seed will result in the same random ordering each time the code is run.

The commands below create a file of sequential batch and tray numbers within a batch and merge it with the randomized design created by `proc factex`.

```
data batches;
do batch=1 to 4;
 do tray = 1 to 6;
 output;
 end;
end;
data sp; merge batches sp;
proc print data=sp; run;
```

The resulting output appears on the next page. There it can be seen that batch 1 was randomly assigned to be made with 80% of the recipe-recommended shortening and the first tray of cookies from batch 1 was to be baked at the recipe-recommended temperature on a tray hot out of the oven, etc.

SPLIT-PLOT EXPERIMENTS WITH CRD IN WHOLE PLOTS

Obs	batch	tray	short	bakeT	trayT
1	1	1	100%	high	Hot
2	1	2	100%	high	RoomT
3	1	3	100%	norm	Hot
4	1	4	100%	low	Hot
5	1	5	100%	low	RoomT
6	1	6	100%	norm	RoomT
7	2	1	80%	low	RoomT
8	2	2	80%	low	Hot
9	2	3	80%	high	Hot
10	2	4	80%	norm	RoomT
11	2	5	80%	high	RoomT
12	2	6	80%	norm	Hot
13	3	1	80%	norm	Hot
14	3	2	80%	low	RoomT
15	3	3	80%	high	Hot
16	3	4	80%	high	RoomT
17	3	5	80%	low	Hot
18	3	6	80%	norm	RoomT
19	4	1	100%	norm	Hot
20	4	2	100%	high	Hot
21	4	3	100%	low	Hot
22	4	4	100%	high	RoomT
23	4	5	100%	low	RoomT
24	4	6	100%	norm	RoomT

In version 9.2 and later SAS `proc factex` has a `uniteffect` statement, that when used in conjunction with the `blocks` statement, can create split-plot designs with CRD in the whole plots as long as all the factors have the same number of levels. For example, in the cookie-baking experiment if only two levels of baking temperature were used rather than three, all factors would have two levels and the code below will create the split-plot design.

```
proc factex;
  factors shortening TrayTemp BakeTemp;
  size design=16;
  blocks units=(batch=4);
  model estimate=(shortening|TrayTemp|BakeTemp);
  uniteffects batch / whole=(Shortening)
                      sub=(TrayTemp|BakeTemp);
  output out=sptwo shortening cvals=('80%' '100%')
                   TrayTemp cvals=('RoomT' 'Hot')
                   BakeTemp cvals=('norm' 'high');
run;
proc print;
  var batch shortening TrayTemp BakeTemp;
run;
```

In this code `proc factex` uses the default of two levels on all factors since the `nlev=` option was left of the factors statement. The `blocks` statement specifies 4 levels of 'batch'. The `uniteffects` statement specifies batch as the whole-plot experimental unit. The option `whole = (Shortening) sub=(TrayTemp|BakeTemp);` specifies that the factor Shortening will be aliased with differences in the whole-plot experimental units, and that TrayTemp, BakeTemp and their interaction will be sub-plot factors that are independent (or varied within levels) of the whole-plot experimental units.

8.2.2 Model for a Split-Plot Experiment with CRD in Whole Plots

It is easy to write the model for a split-plot design with a completely randomized design in the whole plots by following the same two-step procedure used to create the design. The first step was to create a completely randomized design in the whole plots by assigning the levels of the whole-plot factor at random to the whole plots. The model for this step can be represented as

$$y_{ij} = \mu + \alpha_i + w_{(i)j} \tag{8.1}$$

where y_{ij} represents the average response for the ith level of the whole-plot factor and the jth whole plot, α_i represents the fixed effect of the whole-plot factor A, $w_{(i)j}$ represents the effect of the nested random effect of the jth whole plot assigned to the ith level of the whole-plot factor. This is the same model used for the completely randomized design in Chapter 2.

The second step in creating the design was to split the whole plots and assign each sub-plot at random to one of the levels of the sub-plot factor (or combination of levels of the sub-plot factors if there are more than one). Added to the model at this step is the effect of the sub-plot factor and the interaction between the whole-plot and sub-plot factor which are included after the random block term $w_{(i)j}$ to arrive at the final model shown below

$$y_{ijk} = \mu + \alpha_i + w_{(i)j} + \beta_k + \alpha\beta_{ik} + \epsilon_{ijk} \tag{8.2}$$

where y_{ijk} is the measured response for the ith level of the whole-plot factor and the kth level of the split-plot factor within the jth whole plot. β_k is the effect of the fixed split-plot factor, $\alpha\beta_{ik}$ is the fixed interaction effect, and ϵ_{ijk} is the random effect of the kth split plot within the jth plot. If there are more than one sub-plot factors, as in the cookie-baking experiment described above, the model would be:

$$y_{ijkl} = \mu + \alpha_i + w_{(i)j} + \beta_k + \gamma_l + \beta\gamma_{kl} + \alpha\beta_{ik} + \alpha\gamma_{il} + \alpha\beta\gamma_{ikl} + \epsilon_{ijkl} \tag{8.3}$$

and if there are more than one whole-plot factors the model would be written as:

$$y_{ijkl} = \mu + \alpha_i + \beta_j + \alpha\beta_{ij} + w_{(ij)k} + \gamma_l + \alpha\gamma_{il} + \beta\gamma_{jl} + \alpha\beta\gamma_{ijl} + \epsilon_{ijkl} \tag{8.4}$$

and likewise for other combinations of numbers of whole-plot and sub-plot factors.

8.2.3 Analysis of A Split-Plot Design

The data from the cookie-baking experiment are shown below in the SAS commands to read the data and merge it with the factor levels created above in the file called `sp`. The response was the average cookie diameter per tray of cookies.

```
data cook;
  input y @@ ;
  datalines;
2.37 2.46 2.45 2.28 2.11 2.08 1.18 1.77 1.74 1.33 1.33 2.09
1.72 0.94 1.72 1.28 1.34 1.23 2.26 2.34 2.14 2.19 2.01 2.07
data comb;
merge sp cook;
```

Since the model for a split-plot experiment includes both fixed and random factors, the data can be analyzed using either `proc glm` or `proc mixed` as described in Section 5.8 by utilizing the `random` statement to declare the batch effect as a random factor. `proc glm` produces the familiar ANOVA table and the table of expected mean squares that are used to determine the proper F-statistics, while `proc mixed` provides the correct F-tests and the correct standard errors of least squares means and estimated contrasts, but does not produce the ANOVA table.

The expected mean squares produced by `proc glm` is shown in Table 8.2. σ_w^2 in the table represents the variance of the whole-plot experimental units (batches of cookie dough in this experiment), and σ_s^2 represents the variance of the split-plot experimental units which are trays of cookies in this example. From this table of EMSs it can be seen that the proper denominator for the F test of the whole-plot factor (shortening) is the mean square for batch, while the proper denominator for all other effects is the error mean square.

Table 8.2 *Table of EMS for Split-Plot Cookie Experiment*

Source	df	EMS
A=Shortening	1	$\sigma_s^2 + 6\sigma_w^2 + Q(A, AB, AC, ABC)$
batch	2	$\sigma_s^2 + 6\sigma_w^2$
B=Bake Temp	2	$\sigma_s^2 + Q(B, AB, BC, ABC)$
C=Tray Temp	1	$\sigma_s^2 + Q(C, AC, BC, ABC)$
AB	2	$\sigma_s^2 + Q(AB, ABC)$
AC	1	$\sigma_s^2 + Q(AC, ABC)$
BC	2	$\sigma_s^2 + Q(BC, ABC)$
ABC	2	$\sigma_s^2 + Q(ABC)$
Error	10	σ_s^2

The SAS commands to analyze the data using proc mixed are shown below.

```
proc mixed  data= comb;
  class   short bakeT trayT batch;
  model y =short bakeT trayT short*bakeT
          short*trayT bakeT*trayT short*bakeT*trayT;
  random batch(short);
  estimate 'Shortening Effect' short -1 1;
  lsmeans bakeT short*trayT;
run;
```

The first part of the output from **proc mixed**, shown below, is the estimated variances of the random factors shown below. Here we see that $\hat{\sigma}_w^2 = 0.01236$ that represents the variability of batches of dough, and $\sigma_s^2 = 0.009634$ that represents the variability from tray to tray within a batch.

```
            The Mixed Procedure

           Covariance Parameter
                 Estimates
         Cov Parm          Estimate
         batch(short)       0.01233
         Residual           0.009831
```

Since the expected value of the denominator for the F-test of the whole-plot factor, $\sigma_s^2 + 6\sigma_w^2$, is larger than σ_s^2, and the analysis of the data ignoring the random whole plots uses $\hat{\sigma}_s^2$ as the denominator for all F-tests, the analysis of data without consideration of the random whole-plot effect can be completely misleading. Generally the variability among the whole-plot experimental units is larger than the variability among the sub-plot units in a split-plot experiment. The **proc mixed** output below shows the correct F-tests for the fixed effects in the model. Here we see that shortening, baking temperature, tray temperature and the interaction between shortening and tray temperature all had significant effects.

```
              Type 3 Tests of Fixed Effects

                         Num      Den
     Effect              DF       DF      F Value    Pr > F

     short                1        2       41.07     0.0235
     bakeT                2       10       10.44     0.0036
     trayT                1       10       68.15     <.0001
     short*bakeT          2       10        2.78     0.1095
     short*trayT          1       10       19.96     0.0012
     bakeT*trayT          1       10        2.60     0.1236
     short*bakeT*trayT    4       10        0.04     0.9587
```

`proc mixed` calculates the correct standard errors of estimable functions. Shown in the top two lines of output below is the estimate of the average change in cookie diameter as the amount shortening increases from 80% to 100%. Its standard error is a function of both σ_s^2 and σ_w^2. `proc mixed` also computes the correct standard errors of the least squares means. The last part of the output shows the least squares means for the factor bake temperature. There, it can be seen that reducing the bake temperature reduces the cookie diameter.

```
                              Estimates

                              Standard
Label                Estimate  Error   DF  t Value  Pr > |t|
Shortening Effect    -0.7575   0.1182   2   -6.41    0.0235


                         Least Squares Means

              bake              Standard
Effect        T       Estimate   Error    DF  t Value  Pr > |t|
bakeT         high    1.9287    0.06566   10   29.37   <.0001
bakeT         low     1.7212    0.06566   10   26.21   <.0001
bakeT         norm    1.9037    0.06566   10   28.99   <.0001
```

Since there is a significant interaction between shortening and tray temperature, the effect of shortening depends upon the tray temperature. The best way of depicting this is through the interaction plot shown in Figure 8.2.

There it can be seen that reducing the amount of shortening from 100% to 80% of that called for in the recipe reduces the cookie diameter more when using room temperature trays than when using trays hot out of the oven. It also can be seen that the smallest diameter cookies are the result of using 80% of the recommended shortening and placing the cookie dough on a room temperature tray before baking.

8.3 RCB in Whole Plots RBSP

A second way to run a split-plot experiment is to utilize a randomized complete block design in the whole plots. To illustrate this design consider another simple example. A fisherman performed an experiment to determine the effect

Figure 8.2 *Interaction of Shortening and Tray Temperature in Cookie Experiment*

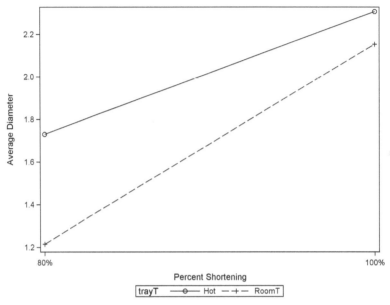

of A: line weight, B: pole stiffness, and C: lure weight upon the distance he could cast. He had three spools of fishing line. One with 6-lb. monofilament test line, one with 10-lb. test line and one with 20-lb. test line. He had a medium light weight pole and a medium weight pole. He had a lightweight lure and a heavier lure, but he only had one fishing reel.

A single experiment or run would consist of selecting the line, the pole, and the lure, then making a cast on a grass field and measuring the distance from the fisherman's feet to the spot where the lure landed. The experimental unit consisted of the conditions such as the wind speed and direction and the caster's ability at the time a certain trial or combination of treatment factors was employed.

This was a 3×2^2 factorial experiment. The experimenter could easily clip his fishing reel onto one fishing pole or the other between casts, and he could easily change the lure between casts, but to change the line weight he had to take all the line off his reel, wind it up on a spindle, and then fill the reel with a different weight line. Thus, the line weight factor was hard to vary. This took several minutes, and rather than completely randomizing the order of the $3 \times 2^2 = 12$ treatment combinations to be tested, it was much easier to perform the experiment in the order shown in Table 8.3.

RCB IN WHOLE PLOTS RBSP

Table 8.3 *Casting Experiment*

Trial	Line Wgt	Pole Stiffness	Lure Wgt
1	10lb.	medium light	heavy
2	10lb.	medium light	light
3	10lb.	light	light
4	10lb.	light	heavy
5	20lb.	light	light
6	20lb.	medium light	light
7	20lb.	medium light	heavy
8	20lb.	light	heavy
9	6lb.	light	light
10	6lb.	medium light	light
11	6lb.	light	heavy
12	6lb.	medium light	heavy

Here the 10-lb. weight line was randomly chosen to be used for the first four casts. The order of the four combinations of pole stiffness and lure weight were randomized in the first four trials. Next the 20-lb. weight line was randomly selected to be used for the next four trials and again the order of the four combinations of pole stiffness and lure weight were randomized in these trials. Finally, the last four casts were made using the 6-lb. weight line. By conducting the experiments in this way the fisherman had inadvertently created a split-plot experiment. The whole-plot experimental unit is a collection of four individual trials or casts and the sub-plot experimental unit is an individual trial or cast.

The problem is that there is no replication of the whole-plot units. If the wind speed or direction changed during the course of the trials, the line weight effect could be biased, and there is no estimate of whole-plot variability from replicates. There is no bias of the pole stiffness and lure weight effects, however, since the four combinations of pole stiffness and lure weight are repeated in a random order within each of the three blocks. To get unbiased estimates of the whole-plot factor effect and gain an estimate of whole-plot experimental error, the fisherman could repeat the whole experiment again as a second block. This was done and the second set of casting trials were completed by a second fisherman so that the results could be generalized to more than one fisherman.

By repeating the experiments with a second fisherman doing the casting, a randomized complete block experiment design is created in the whole plots. The Fisherman represents the block factor, and each line weight is tested by each fisherman in a random order. All four combinations of the sub-plot factors of pole stiffness and lure weight are tested in a random order within each combination of fisherman and line weight.

The model for this experiment is shown in Equation (8.5).

$$y_{ijkl} = \mu + b_i + \alpha_j + b\alpha_{ij} + \beta_k + \gamma_l + \alpha\beta_{jk} + \alpha\gamma_{jl} + \beta\gamma_{kl} + \alpha\beta\gamma_{jkl} + \epsilon_{ijkl} \quad (8.5)$$

Here b_i is the random block or fisherman effect, α_j is the fishing line weight effect, $b\alpha_{ij}$ is the random block by treatment interaction term which represents the whole-plot error term. The first three terms of the model are the same as the model for a RCB design shown in Section 4.3. β is the pole stiffness effect and γ_l is the lure weight effect. These are sub-plot factors along with all the interactions. The error term for the sub-plot factors and interactions is ϵ_{ijkl}.

8.3.1 Creating a Split-Plot Design with RCB in Whole Plots with SAS

The code below creates the split-plot design with an RCB design in the whole plots or a RBSP design. Each block of the whole-plot design is created with a separate use of **proc factex**, and the blocks are concatenated in the data step. The different seeds (in parentheses following the **randomize** option of the **output** statement) creates a different randomization for each block. Each fisherman performing the experiment would be randomly assigned to one of the two blocks.

```
*Create Block 1 of Whole Plot;
proc factex;
*Create sub-plot design;
  factors PoleStiff LureWgt;
  output out=s1  PoleStiff cvals=('light' 'medlgt')
              LureWgt cvals=('light' 'heavy');
run;
*Create whole-plot design for block 1;
  factors LineWgt /nlev=3;
  size design=3;
  output out=b1 pointrep=s1 randomize(101)
  LineWgt cvals=('6lb.' '10lb.' '20lb.');
run;
*Create Block 2 of whole-plot;
proc factex;
*Create sub-plot design;
  factors PoleStiff LureWgt;
  output out=s2  PoleStiff cvals=('light' 'medlgt')
              LureWgt cvals=('light' 'heavy');
run;
*Create whole-plot design for block 2;
  factors LineWgt /nlev=3;
  size design=3;
  output out=b2 pointrep=s2 randomize(201)
  LineWgt cvals=('6lb.' '10lb.' '20lb.');
run;
data spdesign;
set b1(in=in1) b2(in=in2);
if in1 then Block_Fisherman=1;
if in2 then Block_Fisherman=2;
proc print; run;
```

The commands for analyzing the data using `proc glm` from this experiment would be:

```
class Fisherman LineWgt PoleStiff LureWgt;
model y=Fisherman LineWgt Fisherman*LineWgt
PoleStiff LureWgt LineWgt*PoleStiff LineWgt*LureWgt
PoleStiff*LureWgt LineWgt*PoleStiff*LureWgt;
random Fisherman Fisherman*LineWgt/test;
```

These commands could be modified to use `proc mixed` by leaving the random terms `Fisherman Fisherman*LineWgt` out of the model statement and removing the option `test` from the `random` statement. The correct error term for testing the whole-plot (line weight) effect is the interaction of the whole-plot factor with the block term fisherman. This is the same error term shown for the analysis of the generalized complete block design in Section 5.8. The sub-plot effects and interactions are tested with the residual error term.

8.3.2 Example of a Split-Plot Experiment with RCB in Whole Plots

As an example of the analysis of a split-plot experiment with an RCB design in the whole plots, consider an example from industry. Sausage casings are made from collagen in a two-step process. In the first step, collagen must be broken down and reconstituted as a gel with consistent and predictable traits. In the second step, a sophisticated process is used to extrude the gel into a tube that is strong enough to hold the sausage, but tender enough for the final customer. Sausages can be cooked in many ways from steaming to deep-fat frying, and the casing must be able to handle the stress and temperature changes without bursting. Experiments were run to determine how the combination of levels of two factors A and B in the gel-making process, and the combination of levels of two factors C and D in the gel-extrusion step affected the bursting strength of the final casing. A combination of factor levels that would result in achieving a bursting strength at or above a threshold value was sought. The actual factor names, levels and units of the response are not provided in this example for proprietary reasons.

Two levels were chosen for each of the four factors. A different gel batch had to be made for each combination of levels of factors A and B in the gel-making process. The collagen used to make a gel batch was the experimental unit for factors A and B. However, several combinations of the levels of the factors C and D, in the extrusion process, could be tested with the gel from one gel batch. For this reason, the experiment was run as a split-plot experiment and the experimental unit for factors C and D was a portion of the gel from each batch. If the experiments had been run as a completely randomized factorial design (CRFD), it would have been necessary to produce 16 gel batches, one for each of the 2^4 treatment combinations. By using a split-plot experiment fewer gel batches were required.

Table 8.4 shows the results of experimentation with 4 gel batches. The left side of the table shows a standard 2^2 factorial plan in factors A and B that were involved in the gel-making process. One gel batch was made for each of the four combinations of levels of these two factors. Once a gel batch was made, there was enough gel to test all four combinations of levels of the factors C and D that were involved in the extrusion process. In the table, this is represented by showing the combinations of levels of factors C and D horizontally across the top of the table. The responses (bursting strength) shown on the first line of the table were all measured from the same gel batch with differing conditions in the extrusion process. In performing the experiments, the four combinations of levels of A and B were randomized to different samples of collagen, and the order of the four combinations of factors C and D were randomized during the extrusion of each gel batch.

Table 8.4 *First Four Batches for Sausage-Casing Experiment*

Gel Batch	A	B	C D	- -	+ -	- +	+ +
1	-	-		2.07	2.07	2.10	2.12
2	+	-		2.02	1.98	2.00	1.95
3	-	+		2.09	2.05	2.08	2.05
4	+	+		1.98	1.96	1.97	1.97

The problem with the design as shown in Table 8.4 is that there is no replication of the combinations of levels of the whole-plot factors. With only two factors and four runs, there is no adequate way to separate the effects of factors A, B, and their interaction from the differences in the whole-plot experimental units. For this reason the whole experiment was repeated a week later making four more gel batches from a different shipment of collagen. The results for these experiments are shown in Table 8.5. Table 8.4 now represents the first block of whole plots, and Table 8.5 represents the second block. The blocks correspond to two different shipments of collagen.

Even with the addition of a second block, only eight gel batches were re-

Table 8.5 *Second Block of Four Batches for Sausage-Casing Experiment*

Gel Batch	A	B	C D	- -	+ -	- +	+ +
1	-	-		2.08	2.05	2.07	2.05
2	+	-		2.03	1.97	1.99	1.97
3	-	+		2.05	2.02	2.02	2.01
4	+	+		2.01	2.01	1.99	1.97

quired in total for the split-plot experiment. Sixteen gel batches would be required for an unreplicated completely randomized factorial design. The model for the whole-plot part of the experiment can be represented as:

$$y_{ijk} = \mu + b_i + \alpha_j + \beta_k + \alpha\beta_{jk} + w_{ijk} \qquad (8.6)$$

where b_i is the random block or collagen shipment effect, α_j is the fixed effect of factor A, β_k is the fixed effect of factor B, $\alpha\beta_{jk}$ is the fixed interaction effect and w_{ijk} is the random whole-plot error term. The model terms b_i, α_j, β_k, and $\alpha\beta_{jk}$ each has one degree of freedom. The whole-plot error term w_{ijk} is a combination of the random terms $b \times \alpha_{ij}$, $b \times \beta_{ik}$, and $b \times \alpha\beta_{ijk}$ and thus has three degrees of freedom. This is the same model that was used for the randomized complete block factorial (RCBF) shown in Section 4.6.

The model for the complete split-plot experiment is obtained by adding the split-plot factors C and D and all their interactions with the other factors as shown in Equation (8.7).

$$\begin{aligned}
y_{ijklm} = &\mu + b_i + \alpha_j + \beta_k + \alpha\beta_{jk} + w_{ijk} \\
&+ \gamma_l + \delta_m + \gamma\delta_{lm} + \alpha\gamma_{jl} + \alpha\delta_{jm} \\
&+ \beta\gamma_{kl} + \beta\delta_{km} + \alpha\beta\gamma_{jkl} + \alpha\beta\delta_{jkm} \\
&+ \alpha\gamma\delta_{jkl} + \beta\gamma\delta_{klm} + \alpha\beta\gamma\delta_{jklm} + \epsilon_{ijklm}
\end{aligned} \qquad (8.7)$$

The SAS commands to read the data and complete the analysis with proc glm are shown below.

```
data sausage;
input Block    A     B    C    D    ys;
datalines;
1        -1    -1    -1   1    2.10
1        -1    -1    1    1    2.12
         .
         .
         .
proc glm data=sausage;
  class A B Block C D;
  model ys=Block A B A*B Block(A*B) C D C*D A*C A*D
        B*C B*D A*B*C A*B*D A*C*D B*C*D A*B*C*D;
  random Block Block(A*B)/test;
run;
```

The random interactions $b\alpha_{ij}$, $b\beta_{ik}$, and $b\alpha\beta_{ijk}$ were combined to create the whole-plot error term by using the term Block(A*B) in the model. Table 8.6, on the next page, shows the expected mean squares for the ANOVA. From this we see that the denominator for the F tests on the whole-plot effects α_j, β_k, and $\alpha\beta_{jk}$ is the whole-plot error mean square. The denominator for the F

tests on all other effects and interactions in the model is the mean square for the split-plot error term.

Table 8.6 *Expected Mean Squares for Sausage Experiment*

Source	df	Type III Expected Mean Square
Block	1	$\sigma^2 + 4\sigma_W^2 + 16\sigma_B^2$
A	1	$\sigma^2 + 4\sigma_W^2 + Q(A, AB, AC, AD, ABC, ABD, ACD, ABCD)$
B	1	$\sigma^2 + 4\sigma_W^2 + Q(B, AB, BC, BD, ABC, ABD, BCD, ABCD)$
AB	1	$\sigma^2 + 4\sigma_W^2 + Q(AB, ABC, ABD, ABCD)$
Whole Plot Error	3	$\sigma^2 + 4\sigma_W^2$
C	1	$\sigma^2 + Q(C, CD, AC, BC, ABC, ACD, BCD, ABCD)$
D	1	$\sigma^2 + Q(D, CD, AD, BD, ABD, ACD, BCD, ABCD)$
CD	1	$\sigma^2 + Q(CD, ACD, BCD, ABCD)$
AC	1	$\sigma^2 + Q(AC, ABC, ACD, ABCD)$
AD	1	$\sigma^2 + Q(AD, ABD, ACD, ABCD)$
BC	1	$\sigma^2 + Q(BC, ABC, BCD, ABCD)$
BD	1	$\sigma^2 + Q(BD, ABD, BCD, ABCD)$
ABC	1	$\sigma^2 + Q(ABC, ABCD)$
ABD	1	$\sigma^2 + Q(ABD, ABCD)$
ACD	1	$\sigma^2 + Q(ACD, ABCD)$
BCD	1	$\sigma^2 + Q(BCD, ABCD)$
ABCD	1	$\sigma^2 + Q(ABCD)$
Split-Plot Error	12	σ^2

The commands for the analysis using **proc mixed** are shown below.

```
proc mixed data=sausage;
  class A B Block C D;
  model ys=A B A*B C D C*D A*C A*D B*C B*D A*B*C A*B*D
          A*C*D B*C*D A*B*C*D;
  random Block Block(A*B);
  lsmeans A;
  lsmeans C;
  lsmeans A*B*C;
  run;
```

The output table below produced by `proc mixed` shows the estimates of the variance components σ_B^2, σ_W^2 and σ^2.

```
              The Mixed Procedure

           Convergence criteria met.
         Covariance Parameter Estimates

              Cov Parm          Estimate
              Block                    0
              Block(A*B)        0.000340
              Residual          0.000239
```

`proc mixed` performs the correct F tests on the fixed effects, and as shown below, main effects A, C and the three-way interaction ABC were found to be significant at the $\alpha = 0.05$ level.

```
           Type 3 Tests of Fixed Effects

                  Num      Den
    Effect         DF       DF    F Value    Pr > F
    A               1        3      28.65    0.0128
    B               1        3       1.65    0.2897
    A*B             1        3       0.71    0.4623
    C               1       12      16.05    0.0017
    D               1       12       2.21    0.1626
    C*D             1       12       1.06    0.3233
    A*C             1       12       0.64    0.4386
    A*D             1       12       3.79    0.0755
    B*C             1       12       0.33    0.5777
    B*D             1       12       1.06    0.3233
    A*B*C           1       12       5.78    0.0333
    A*B*D           1       12       2.95    0.1117
    A*C*D           1       12       0.12    0.7373
    B*C*D           1       12       0.12    0.7373
    A*B*C*D         1       12       0.12    0.7373
```

The least squares means for the two main effects produced by `proc mixed` are shown on the next page. Here it can be seen that main effect A had the most influential effect on the bursting strength. However, the presence of the three-factor interaction means that the effect of factor A depends upon the levels of both factors B and C. This can be visualized in the interaction plot shown in Figure 8.3. There it can be seen that increasing factor A from its low to high level has caused the greatest decrease in bursting strength when factors B is at its low level and C is at its high level. If the linear relation between the bursting strength and the factor levels is reasonable, there appears to be many combinations of factor levels that would result in bursting strengths greater than the threshold value of 2.0.

Effect	A	C	Estimate	Standard Error	DF	t Value	Pr > |t|
A	-1		2.0613	0.009990	3	206.33	<.0001
A	1		1.9856	0.009990	3	198.76	<.0001
C		-1	2.0344	0.007573	12	268.62	<.0001
C		1	2.0125	0.007573	12	265.73	<.0001

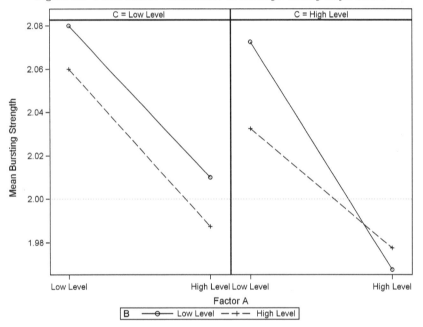

Figure 8.3 *Three-Factor Interaction in Sausage-Casing Experiment*

Further measurements on the casing produced by this series of experiments revealed factor combinations that met the threshold bursting strength requirement and also produced casing tender enough for the final customer.

8.4 Analysis Unreplicated 2^k Split-Plot Designs

When there are no replicates in a split-plot experiment, the significant effects cannot be determined with an Analysis of Variance because there will be zero degrees of freedom for the whole plots. For example, if only the first block of experiments for the sausage-casing experiments (shown in Table 8.4) had been run, it would represent a full 2^4 factorial with no replicates. One way to determine the significant effects in an unreplicated completely randomized factorial design (CRFD) with two levels for each factor is to make a normal or half-normal plot of the effects as described in Section 3.7.5. In this way the standard error of the effects can be estimated graphically as the slope of the straight line through the small effects in the center of the normal plot, and the significant effects will appear above to the right or below to the left of the straight line. However, this approach will not work for a split-plot experiment because there are two standard errors, one for the whole-plot effects and one for the split-plot effects. For the sausage-casing experiments the whole-plot effects were A, B and their interaction AB, while the split-plot effects were C, D, CD, AC, AD, BC, BD, ABC, ABD, ACD, BCD, and $ABCD$. One solution is to make separate normal plots of the whole-plot and split-plot effects.

Bisgaard *et al.* (1996) describe an unreplicated split-plot experiment to study the plasma treatment of paper to make it more susceptible to ink. The data from the experiment, written in the split-plot form, is shown in Table 8.7. When energy is supplied to molecules in a gaseous state, molecules start breaking up and form a mixture of unstable particles called a plasma. The highly charged electrons and protons in a plasma can be used to modify the surface characteristics of materials. At the Engineering Research Center for Plasma Aided Manufacturing at the University of Wisconsin-Madison, plasma was created in a low vacuum chamber reactor between electrodes, and paper samples were placed in the chamber on a grounded electrode. There was room for two paper samples in the reactor.

Table 8.7 *Plasma Experiment Factor Levels and Response*

A	B	C	D	E	
				−	+
−	−	−	−	48.6	57.0
+	−	−	−	41.2	38.2
−	+	−	−	55.8	62.9
+	+	−	−	53.5	51.3
−	−	+	−	37.6	43.5
+	−	+	−	47.2	44.8
−	+	+	−	47.2	54.6
+	+	+	−	48.7	44.4
−	−	−	+	5.0	18.1
+	−	−	+	56.8	56.2
−	+	−	+	25.6	33.0
+	+	−	+	41.8	37.8
−	−	+	+	13.3	23.7
+	−	+	+	47.5	43.2
−	+	+	+	11.3	23.9
+	+	+	+	49.5	48.2

The factors under study in the experiment were A: low and high pressure, B: low and high power, C: low and high gas flow rate, D: type of gas oxygen or SiCl$_4$, and E: paper type. The response, wettability of the paper, was measured by placing a droplet of water on the treated paper and measuring the contact angle between the droplet and the paper surface with a special microscope. Because the plasma was created in a vacuum, it takes up to a half an hour of pumping to get the reactor down to the appropriate vacuum level each time the chamber is opened to insert new paper samples. Therefore, rather than completing a completely randomized 2^5 factorial, the experiments were completed in the following way. The experimenter first randomized the order of the combinations of factors A to D to 16 reactor runs. During each run two samples of paper were placed in the reactor (one of each type), and a coin was tossed to determine which paper type was placed on the right or left side of the reactor. By doing this a split-plot experiment was created where the experimental unit for factors A to D was a reactor run, while the experimental unit for factor E, the paper type, was the position inside the reactor for a particular run.

The main effects A, B, C, D, and all their interactions are whole-plot factors, and their significance should be tested with the whole-plot error term. Factor E and its interactions with A, B, C, D, and all their interactions, are split-plot factors, and their significance should be tested with the split-plot error term. The SAS code below illustrates how to make separate normal plots of the whole-plot and split-plot effects. The first commands fit the saturated model exactly as was done in Section 3.7.5.

```
data plasma;
input A B C D E y;
datalines;
-1    -1    -1    -1    -1    48.6
 1    -1    -1    -1    -1    41.2
           . . .
proc glm;
*fit the full factorial model;
   model y=A|B|C|D|E;
*this ouputs the parameter estimates to a file;
   ods output ParameterEstimates=sol;
run;
```

The next section of code, shown at the top of the following page, selects the whole-plot calculated effects A, B, AB, C, AC, BC, ABC, D, AD, BD, ABD, CD, ACD, BCD, $ABCD$, which are the second through sixteenth estimates on the output, and makes a normal plot of them. Other than the statement if 1<_n_<17 used to select the whole-plot effects, this code is the same as that shown in Section 3.7.5. The resulting plot, with a manually added reference line, is shown in Figure 8.4. It can be seen that the whole-plot main effects A: pressure, D: type of gas, and their interaction appear significant.

ANALYSIS UNREPLICATED 2^k SPLIT-PLOT DESIGNS

```
*this creates the normal plot of whole plot effects;
ods graphics on;
data nplot; set sol;
  estimate=estimate;
  if 1<_n_<17;
  drop StdErr tValue Probt;
* proc rank calculates normal scores for parameter estimates ;
proc rank data=nplot out=nplots normal=blom; var estimate;
  ranks zscore;
data nplots; set nplots;
  if abs(zscore)<=1.2 then parameter='      ';
proc sgplot data=nplots;
scatter x=zscore y=estimate/datalabel=parameter;
xaxis label='Normal Scores';
run;
ods graphics off;
```

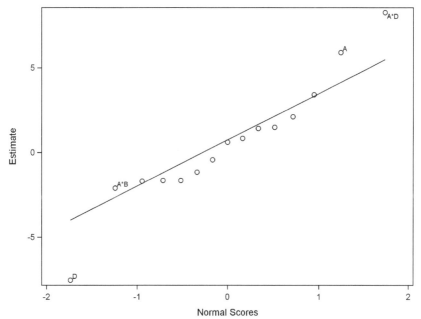

Figure 8.4 *Normal Plot of Whole-Plot Effects - Plasma Experiment*

The normal plot of the split-plot effects E, AE, BE, ABE, CE, ACE, BCE, $ABCE$, DE, ADE, BDE, $ABDE$, CDE, $ACDE$, $BCDE$, $ABCDE$ can be made by changing the statement `if 1<_n_<17;` to `if 16<_n;`. In this plot it can be seen that main effect E: paper type and its interaction with A also appear to be significant. Notice the difference between the vertical scales in Figures 8.4 and 8.5. The standard error of the whole-plot effects which is the slope of the added straight line in Figure 8.4 is much larger than the

Figure 8.5 *Normal Plot of Sub-Plot Effects - Plasma Experiment*

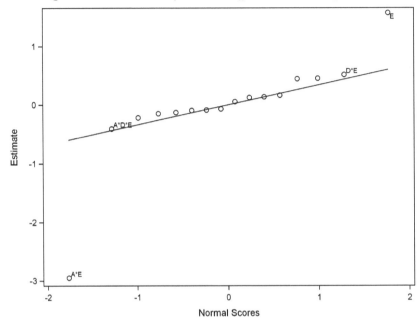

standard error of the split-plot effects, which is the slope of the straight line in Figure 8.5.

If an unreplicated split-plot experiment like that shown in Table 8.7 was mistakenly analyzed as a completely randomized design by making a normal plot of all the calculated effects on the same graph, the mixture of the two error distributions causes the line through the insignificant points to follow a sigmoid curve like that shown in Figure 8.6, rather than a straight line as shown in Figures 8.4 and 8.5. In Figure 8.6 it can be seen that whole-plot main effects A, D and their interaction appear to be significant. However, the split-plot main effect E and the interaction AE that were clearly identified to be significant on the normal plot of split-plot effects now appear to be buried among the whole-plot error effects. Whenever a normal plot of effects from a factorial experiment takes this appearance, Daniel and later Bisgaard *et al.* (1996) warned that the treatment combinations may not have been completely randomized, and that the results actually represent a split-plot experiment.

In this example there were enough whole-plot and split-plot effects to make separate normal plots. In other unreplicated 2^k split-plot experiments this may not be the case. For example, if only the first block of the sausage-casing experiment (that was described in Section 8.3.2) had been run, there would be enough split-plot effects (C, D, CD, AC, AD, BC, BD, ABC, ABD, ACD, BCD, and $ABCD$) to identify significant ones on a normal plot of

ANALYSIS UNREPLICATED 2^k SPLIT-PLOT DESIGNS

Figure 8.6 *Normal Plot of All Effects - Plasma Experiment*

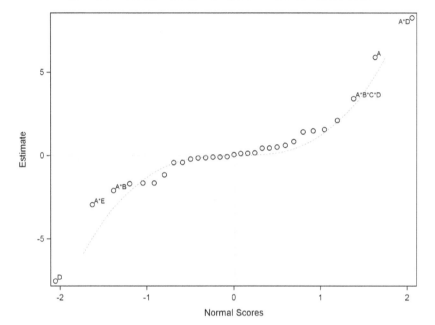

effects, however, there were not enough whole-plot effects $(A, B,$ and $AB)$ to effectively identify the significant ones on a plot.

One strategy to use in this situation is the following. If any interactions between whole-plot factors and split-plot factors appear significant in the normal plot of split-plot effects, the *effect heredity* principle might imply that the involved whole-plot main effect may also be significant. If the suspected whole-plot main effect has a relatively large effect this hypothesis is further supported, and it may be fruitful to include that main effect in the model before calculating the least squares interaction cell means and interpreting the significant interaction. Any other whole-plot effects that are larger in magnitude than the whole-plot main effect involved in a significant interaction may also be considered to be significant.

Another strategy can be used in this situation if there is some prior knowledge about variability to be expected in whole-plot units. For example, if the experiment is performed in a two-step manufacturing process, and there is history (such as control charts) maintained on the output of the first process step, then Gilmour and Goos (2006) propose using a Bayesian analysis with an informative prior on the whole-plot error random effects.

8.5 2^{k-p} Fractional Factorials in Split Plots (FFSP)

The 2^{k-p} fractional factorial experiments described in Chapter 6 assumed that treatment combinations could be assigned to experimental units completely at random. However, sometimes this is impractical for the same reasons cited in Section 8.1. In this case a fractional factorial split-plot (FFSP) design can be utilized. There are many ways to create a fractional factorial split-plot design with the same number of factors and runs. One method is shown in Table 8.8 which is similar in format to Tables 8.4 and 8.5. Here the whole-plot factors are labeled A, B, and C, while the sub-plot factors are labeled P, Q, and R. A 2^{3-1} fractional factorial is created with the whole-plot factors using the defining relation $I = ABC$. The four whole-plot treatment combinations would then be assigned at random to whole-plot experimental units. Each whole plot is divided into four sub-plots which are assigned at random to receive one of the four treatment combinations in the 2^{3-1} fractional factorial created with the sub-plot factors using defining relation $I = PQR$.

Table 8.8 *Cartesian Product Fractional Factorial Split-Plot Design*

				\multicolumn{4}{c}{$(I = PQR)$}
			P	−	+	−	+
$(I = ABC)$			Q	−	−	+	+
A	B	C	R	+	−	−	−
−	−	+		x	x	x	x
+	−	−		x	x	x	x
−	+	−		x	x	x	x
+	+	+		x	x	x	x

Since the same four combinations of sub-plot treatment factors are used in each whole plot, this design is called a *cartesian product design* and the "x"s in Table 8.8 indicate which treatment combinations are used. With six factors and a total of sixteen treatment combinations, this design is a $\frac{1}{4}$ fraction. The overall defining relation can be obtained by multiplying the defining relation for the whole-plot fraction by the defining relation for the sub-plot fraction, i.e., $(I + ABC) \times (I + PQR) = I + ABC + PQR + ABCPQR$. Thus, the design has resolution III and all main effects are confounded with one two-factor interaction, one four-factor interaction and one-five factor interaction.

A better confounding pattern can be obtained by using whole-plot factors in the sub-plot generators as shown in Table 8.9. Here a different fraction of the combinations of sub-plot factors is used in each whole plot. This is

2^{k-p} FRACTIONAL FACTORIALS IN SPLIT PLOTS (FFSP)

similar to the way a fractional factorial was augmented with a mirror image design in Section 6.5. By using the fraction of sub-plot factors obtained with the generator $P = -QR$ when whole-plot factor A is at its low level, and by using the fraction obtained with the generator $P = +QR$ when the whole-plot factor A is at its high level, it is equivalent to using the generator $P = AQR$ or the defining relation $I + APQR$ for the fractional factorial in the sub-plots. This method is called *split-plot confounding* by Bisgaard (2000).

Table 8.9 *Split-Plot Confounding in Fractional Factorial Split-Plot Design*

($I = ABC$)			
A	B	C	
−	−	+	$I = -PQR$
+	−	−	$I = +PQR$
−	+	−	$I = -PQR$
+	+	+	$I = +PQR$

This design can be randomized similar to the way the design shown in Table 8.8 can be randomized. The four whole-plot treatment combinations should be assigned at random to whole-plot experimental units. Each whole-plot is divided into four sub-plots which are assigned at random to receive one of the four sub-plot treatment combinations in the 2^{3-1} fractional factorial created using the generator $P = +QR$ if the whole-plot received the high level of factor A, or to receive one of the four sub-plot treatment combinations in the 2^{3-1} fractional factorial created with generator $P = -QR$ if the whole-plot received the low level of factor A. This design can be created easily in the SAS data step using the commands shown at the top of the next page.

```
*Create Plot numbers for randomized list;
data labels;
 do WholePlot=1 to 4;
  do SubPlot=1 to 4;
   output;
  end;
 end;
data wp;
*Create whole-plot design using I=ABC;
do A = -1 to 1 by 2;
 do B = -1 to 1 by 2;
  C= A*B; ranwp=ranuni(0);
*Create sub-plot design for case where A = -;
  if A=-1 then do;
   do Q=-1 to 1 by 2;
    do R=-1 to 1 by 2;
     P=-Q*R;  ransp=ranuni(0); output;
    end;
   end;
  end;
*Create sub-plot design for case where A = +;
  if A=1 then do;
   do Q=-1 to 1 by 2;
    do R=-1 to 1 by 2;
     P=Q*R; ransp=ranuni(0); output;
    end;
   end;
  end;
 end;
end;
*Randomize whole-plot and sub-plot treatment combinations;
proc sort data=wp; by ranwp ransp;
data splitplot;
merge labels wp; drop ranwp ransp;
proc print; run;
```

This produces the randomized list shown on the next page, which can be printed out and used for a data collection worksheet. The SAS data set `splitplot` can be saved as a permanent data set that can be merged with the response data after the experiment is run.

The overall defining relation for this design can again be determined by multiplying the defining relation for the whole-plot fraction by the defining relation for the sub-plot fraction, i.e., $(I + ABC) \times (I + APQR) = I + ABC + APQR + BCPQR$. This design is also resolution III, but by using the split-plot confounding it has less abberation (see Section 6.4) than the design shown in Table 8.8 since it has only one three-letter word in the defining relation. In this design, whole-plot main effects are confounded with two-factor interactions but sub-plot factors are clear of two-factor interactions. Therefore, the subset of whole-plot factors has resolution III, while the subset of sub-plot factors has resolution IV. In general, using split-plot confounding by including whole-plot factors in the sub-plot generators will result in higher resolution among the sub-plot factors similar to the way resolution III frac-

2^{k-p} FRACTIONAL FACTORIALS IN SPLIT PLOTS (FFSP)

Obs	Whole Plot	Sub Plot	A	B	C	Q	R	P
1	1	1	-1	-1	1	1	-1	1
2	1	2	-1	-1	1	-1	1	1
3	1	3	-1	-1	1	1	1	-1
4	1	4	-1	-1	1	-1	-1	-1
5	2	1	1	1	1	1	1	1
6	2	2	1	1	1	-1	-1	1
7	2	3	1	1	1	-1	1	-1
8	2	4	1	1	1	1	-1	-1
9	3	1	1	-1	-1	1	-1	-1
10	3	2	1	-1	-1	-1	-1	1
11	3	3	1	-1	-1	-1	1	-1
12	3	4	1	-1	-1	1	1	1
13	4	1	-1	1	-1	1	-1	1
14	4	2	-1	1	-1	1	1	-1
15	4	3	-1	1	-1	-1	1	1
16	4	4	-1	1	-1	-1	-1	-1

tional factorials augmented by a mirror image result in a resolution IV design. However, as Bingham and Sitter (2001) state, the converse is not true. The resolution of the whole-plot factors cannot be increased by including sub-plot factors in the whole-plot generators; in fact, it would destroy the split-plot nature of the design.

When there are several whole-plot and split-plot factors, there may be several ways of defining the split-plot confounding. For example, consider an experiment with four whole-plot factors, labeled A, B, C and D, and four split-plot factors labeled P, Q, R, and S. One way of creating a 2^{8-3} fraction by taking a $\frac{1}{2}$ fraction of the whole-plot treatment combinations would be to use the sub-plot generators $APQR$, $BQRS$ and their generalized interaction $ABPS$. This would result in the resolution IV fractional factorial split-plot design with defining relation $I = ABCD = APQR = BQRS = ABPS = BCDPQR = ACDQRS = CDPS$. Another way of creating a 2^{8-3} fraction, taking a $\frac{1}{2}$ fraction of the whole-plot treatment combinations, would be to use the sub-plot generators $ABPR$, $BCPQS$ and their generalized interaction $ACQRS$. This results in the resolution IV design with defining relation $I = ABCD = ABPR = BCPQS = ACQRS = CDPR = ADPQS = BDQRS$. Although both designs are resolution IV, the second design has less abberation since it has only three four-letter words in the defining relation while the first design has five four-letter words in the defining relation. Recall from Section 6.4 that designs with less aberration have less confounding of main effects with lower order interactions.

8.5.1 Minimum Aberration Fractional Factorial Split Plots

For FFSP designs of the same resolution and level of whole-plot and sub-plot fractionation, there is always a minimum aberration design. Huang *et al.*

(1998) and Bingham and Sitter (1999) have given tables of the generators for minimum abberation split-plot designs for a small to moderate size number of factors. SAS ADX automatically generates the minimum aberration split-plot designs using the generators found in these tables. SAS `proc factex`, in SAS versions 9.2 and later, can sometimes find the same generators using an algorithm (see Section 6.3).

Once the number of whole-plot factors and the number of sub-plot factors has been decided upon, there are other things to consider when generating a FFSP design with SAS ADX. One is the degree of fractionation of the whole-plot factors and the number of whole plots. Another is the degree of fractionation of the sub-plot factors and the number of sub-plots per whole plot. For example, if we define four whole-plot factors and four sub-plot factors by choosing the Create New Split-Plot Design menu and filling out the template tabs for Whole Plot Factor and Sub-plot Factor as shown in Figure 8.7.

Figure 8.7 *SAS ADX Define Variables Template for Split-Plot Design*

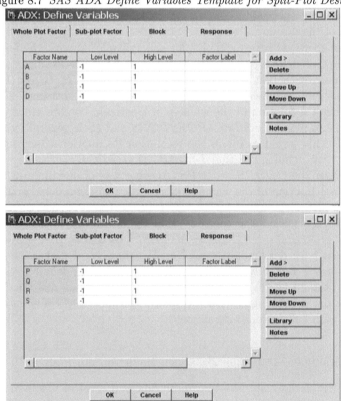

2^{k-p} FRACTIONAL FACTORIALS IN SPLIT PLOTS (FFSP)

SAS ADX lists four alternate minimum aberration designs as shown in Figure 8.8.

Figure 8.8 *SAS ADX Split-Plot Design Specification Template*

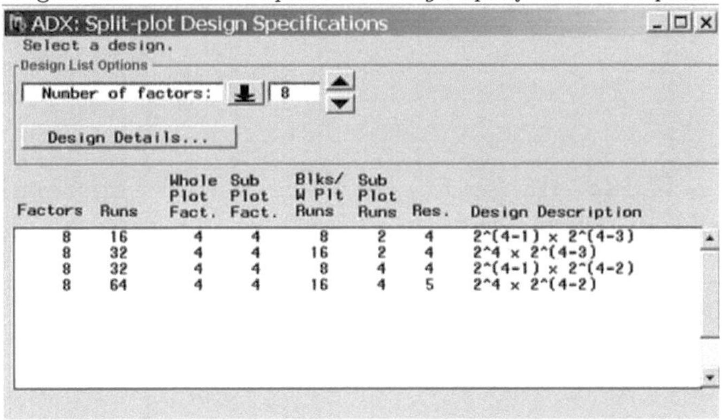

The first design uses a one-half fraction of the combinations of whole-plot treatment levels with defining relation $I = ABCD$, resulting in eight whole plots each split into two sub-plots. By clicking on this design and then the Design Details ... button the information shown in Figure 8.9 is shown on the design properties tab. Here it can be seen that this design uses the split-plot

Figure 8.9 *SAS ADX Design Details Results for $2^{4-1} \times 2^{4-3}$ Design*

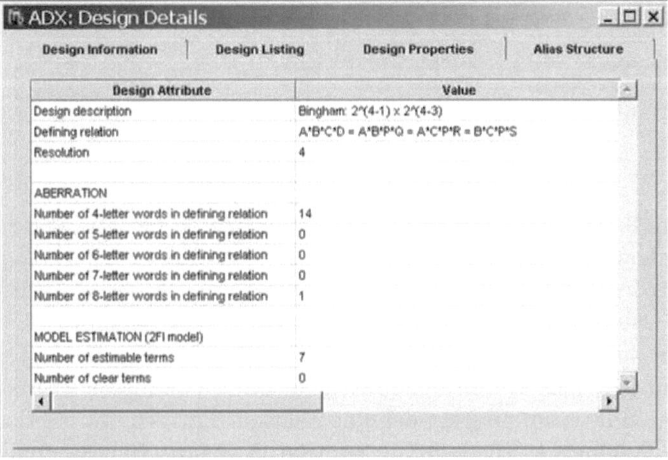

confounding generators found in Bingham and Sitter's tables. The design is resolution IV but no two factor-interactions are estimable clear of other two-factor interactions. Clicking on the Alias Structure tab, the aliases among the two-factor interactions are shown in Figure 8.10.

Figure 8.10 *SAS ADX Alias Structure for* $2^{4-1} \times 2^{4-3}$ *Design*

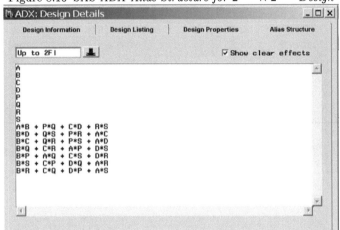

SAS `proc factex` finds the same generators and the same design when using the commands below.

```
proc factex;
  factors A B C D P Q R S;
  size design=16;
  blocks units=(Block=8);
  model resolution=4/minabs;
  uniteffect Block / whole=(A B C D)
                    sub=(P Q R S);
  examine aliasing(2) confounding;
  output out=sp;
proc print; run;
```

Figure 8.11 shows the results of checking the design details for the second design shown in Figure 8.8. It utilizes a full factorial in the whole-plot factors resulting in 16 whole plots each divided into two sub-plots. This design was created using the generators in the tables of Huang *et al.* (1998).

It can be seen that the design is also resolution IV, but 13 of the 28 two-factor interactions are estimable clear of other two-factor interactions. This can be seen by examining the defining relation. This is a $\frac{1}{8}th$ fraction and what SAS ADX labels as the defining relation in Figure 8.11 is actually a list of the three generators for the design. The complete defining relation, including all generalized interactions is: $I = ABPQ = ACPR = BCDPS = BCQR = ACDQS = ABDRS = DPQRS$. The estimable effects (up to two-factor interactions except for the case where no effect of order two or less is aliased with the given effect) are shown below Figure 8.11.

Figure 8.11 *SAS ADX Design Details Results* $2^4 \times 2^{4-3}$ *Design*

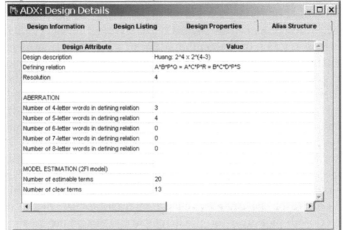

Effects Tested with Whole-Plot Error	Effects Tested with Sub-Plot Error
A	P
B	Q
C	R
D	S
$AB + PQ$	$AP + BQ + CR$
$AC + PR$	$AQ + BP$
AD	$AR + CP$
$BC + QR$	AS
BD	$BR + CQ$
CD	BS
$ABC + CPQ + BPR + AQR$	CS
$ABD + RS$	DP
$ACD + QS$	DQ
$BCD + PS$	DR
$ABCD$	DS
	$BDQ + ADP + CDR$

The estimable effects (shown above) are split into the two columns for the following reason. A full 2^4 factorial is conducted with factors A, B, C, and D in the 16 whole plots, therefore the factorial effects A, B, C, D, AB, AC, AD, BC, BD, CD, ABC, ABD, ACD, BCD, $ABCD$ and all their aliases are whole-plot effects and should be tested with the whole-plot error term. All the other estimable effects are sub-plot effects and are tested with the sub-plot error term. If three factor and higher order interactions are assumed negligible, it can be seen that all main effects and the 13 two-factor interactions AD, BD,

CD, RS, QS, PS, AS, BS, CS, DP, DQ, DR, and DS can be estimated clear of main effects or two-factor interactions. This is what is referred to by the number of clear terms in Figure 8.11.

The third design shown in Figure 8.8 is also resolution IV, but by utilizing a one-half fraction of the combinations of whole-plot treatment levels and then splitting each whole-plot into four sub-plots, only eight whole plots are required. The generators for this design are $ABCD$, $ABPR$, and $BCPQS$. In this design seven effects, namely main effects A, B, C, and D and confounded strings of interactions $AB + CD + PR$, $AC + BD$, and $AD + BC$, are estimated less precisely and should be compared to whole-plot variability for testing significance. This design also allows estimation of 13 of the 28 two-factor interactions clear of main effects and other two-factor interactions, so it appears to be similar to the second design listed in Figure 8.8. However, because of the reduced number of whole plots and the increased number of sub-plots all 13 of the clear two-factor interactions can be tested with the more precise sub-plot error term, whereas in the second design six of the clear two-factor interactions (namely AD, BD, CD, RS, QS, and PS) must be tested with the less precise whole-plot error term.

The fourth and last design shown in Figure 8.8 consists of a full factorial in the four whole-plot factors A, B, C and D in 16 whole plots each divided into 4 sub-plots. The defining relation for this design is $I = ABCPR = ADPQS = BCDRQS$. This design is resolution V, and all main effects and two-factor interactions can be estimated clear of other main effects and two-factor interactions.

The choice between these four alternative designs would depend upon the relative cost of whole plot and sub-plot experimental units and the precision desired for the various estimable two-factor interactions. For more discussion on this topic, see Kulahci *et al.* (2006), Bisgaard (2000) and Bingham and Sitter (2001).

When using `proc factex`, the user does not have a choice about the degree of fractionation in the whole-plot and sub-plot design. `proc factex` only allows choice of the overall degree of fractionation. For example, the code below creates a 2^{8-3} fractional factorial with four whole-plot factors and four sub-plot factors. The search algorithm in `proc factex` only finds a $2^{4-1} \times 2^{4-2}$

```
proc factex;
  factors A B C D P Q R S;
  size design=32;
  blocks units=(Block=16);
  model resolution=4/minabs;
  uniteffect Block / whole=(A B C D)
                     sub=(P Q R S);
  examine aliasing(2) confounding;
  output out=sp;
proc print; run;
```

2^{k-p} FRACTIONAL FACTORIALS IN SPLIT PLOTS (FFSP)

design with generators $ABCD$, $ABPQ$, and $ACRS$. The complete defining relation for this design is $I = ABCD = CDPQ = BDRS = ADPQRS = ABPQ = ACRS = BCPQRS$ which is not the minimum aberration design. For a 2^{8-3} split-plot fractional factorial, SAS ADX gave a choice of a $2^4 \times 2^{4-3}$ split-plot design or a $2^{4-1} \times 2^{4-2}$ split-plot design. Both of these designs are minimum aberration. The generators for the $2^{4-1} \times 2^{4-2}$ created by ADX come from Huang et al. (1998)'s tables. It has only three four-letter words in the defining relation and is better than the design found by `proc factex`.

8.5.2 Analysis of a Fractional Factorial Split-Plot

Bingham and Sitter (2001) describe an experiment to study effects of five factors upon the geometric distortion of drive gears. The factors were A: furnace track, B: tooth size, C: part positioning, P: carbon potential and Q: operating mode. It was more efficient to perform the experimental trials by keeping levels of factors A, B and C constant in blocks and varying factors P and Q within the blocks. A schematic of the $2^3 \times 2^{2-1}$ design is shown in Table 8.10.

Table 8.10 *Fractional Factorial Split-Plot Design for Gear Distortion*

			P	−	+	−	+
A	B	C	Q	−	−	+	+
−	−	−		x	x		
+	−	−		x			x
−	+	−		x			x
+	+	−			x	x	
−	−	+		x			x
+	−	+			x	x	
−	+	+			x	x	
+	+	+		x			x

This is a split-plot design due to the fact that factors A, B and C are constant in blocks of runs. It is also evident that split-plot confounding has been utilized by the fact that a different subset of the combinations sub-plot factor levels is used in each whole plot. The defining relation for the design is $I = ABCPQ$, and the response was the dishing of the gears. This design is one of the minimum aberration designs in Bingham and Sitter (2001)'s tables and can be easily generated using SAS ADX, or `proc factex`.

SAS ADX cannot perform an analysis of the data of a FFSP design when there are no replicates of the whole plots. Therefore if the design shown in Table 8.10 is created in ADX, it must be exported to a SAS data file for analysis. The ADX examples on the lecture slides for this book illustrate how to export SAS ADX designs. Alternatively, the design can be created directly in a SAS data file using `proc factex`. The commands to do that are shown below. In this case, the algorithm used by `proc factex` comes up with the same minimum aberration design shown in Bingham and Sitter (2001)'s tables. A randomized data collection form can be created from the data file `sp` by modifying the last three lines of the SAS code in Section 8.5.

```
proc factex;
  factors A B C P Q;
  size design=16;
  blocks units=(Block=8);
  model resolution=5/minabs;
  uniteffect Block / whole=(A B C)
                     sub=(P Q);
  examine aliasing(3) confounding;
  output out=sp;
run;
proc print; var Block A B C P Q; run;
```

The design is a half-fraction with defining relation $I = ABCPQ$, and the estimable whole-plot and sub-plot effects are shown in Table 8.11.

Table 8.11 *Estimable Effects for Gear Distortion Experiment*

Whole-Plot Effects	Sub-Plot Effects
$A + BCPQ$	$P + ABCQ$
$B + ACPQ$	$Q + ABCP$
$C + ABPQ$	$AP + BCQ$
$AB + CPQ$	$AQ + BCP$
$AC + BPQ$	$BP + ACQ$
$BC + APQ$	$BQ + ACP$
$ABC + PQ$	$CP + ABQ$
	$CQ + ABP$

The model for the experiment written in the split-plot form is shown in Equation (8.8).

$$y_{ijklm} = \mu + \alpha_i + \beta_j + \gamma_k + \alpha\beta_{ij} + \alpha\gamma_{ik} + \beta\gamma_{jk} + \alpha\beta\gamma_{ijk} + w_{ijk}$$
$$+ \rho_l + \phi_m + \alpha\rho_{jl} + \alpha\phi_{im} + \beta\rho_{jl} + \beta\phi_{jm} + \gamma\rho_{kl} + \gamma\phi_{km} + \rho\phi_{lm} + \epsilon_{ijklm}$$
(8.8)

Since there are no replicates of the whole plots, w_{ijk} cannot be estimated, and

2^{k-p} FRACTIONAL FACTORIALS IN SPLIT PLOTS (FFSP)

since each factor has only two levels, the model can be rewritten in the form shown in Section 3.7.2 as:

$$\begin{aligned}y = {}& \beta_0 + \beta_A X_A + \beta_B X_B + \beta_C X_C + \beta_{AB} X_{AB} + \beta_{AC} X_{AC} \\ & + \beta_{BC} X_{BC} + \beta_{ABC} X_{ABC} + \beta_P X_P + \beta_Q X_Q + \beta_{AP} X_{AP} \\ & + \beta_{AQ} X_{AQ} + \beta_{BP} X_{BP} + \beta_{BQ} X_{BQ} + \beta_{CP} X_{CP} + \beta_{PQ} X_{PQ} \\ & + \beta_{CQ} X_{CQ} + \epsilon\end{aligned} \quad (8.9)$$

where the βs represent the regression coefficients or half of the effects, and the Xs represent the coded factor levels.

After completing the experiments and merging the response data with the design, the SAS data file containing the coded factor levels and response would appear as shown below.

Obs	Block	A	B	C	P	Q	y
1	1	-1	1	1	1	-1	22.0
2	1	-1	1	1	-1	1	19.0
3	2	1	-1	-1	-1	-1	13.0
4	2	1	-1	-1	1	1	-4.5
5	3	1	-1	1	-1	1	17.5
6	3	1	-1	1	1	-1	14.5
7	4	-1	1	-1	1	1	15.0
8	4	-1	1	-1	-1	-1	22.5
9	5	1	1	-1	-1	1	0.5
10	5	1	1	-1	1	-1	5.5
11	6	-1	-1	1	1	1	17.0
12	6	-1	-1	1	-1	-1	27.5
13	7	-1	-1	-1	1	-1	21.5
14	7	-1	-1	-1	-1	1	18.0
15	8	1	1	1	-1	-1	24.0
16	8	1	1	1	1	1	13.5

The SAS commands to fit the model to the data using **proc glm** is shown below. Since there are only 16 observations, the model is saturated with zero degrees of freedom for error.

```
proc glm data=spexp;
  model y=A|B|C P Q A*P A*Q B*P B*Q C*P C*Q;
  ods ouput ParameterEstimates=sol;
run;
```

The fitted coefficients produced by **proc glm** are shown on the next page. To determine which effects are significant, the SAS code, shown in Section 8.4 for producing separate normal plots of the whole-plot effects and sub-plot effects, was modified to produce Figures 8.12 and 8.13.

```
Parameter            Estimate

Intercept         15.40625000
A                 -4.90625000
B                 -0.15625000
A*B                0.53125000
C                  3.96875000
A*C                2.90625000
B*C                0.40625000
A*B*C              0.59375000
P                 -2.34375000
Q                 -3.40625000
A*P               -0.90625000
A*Q               -0.34375000
B*P                1.09375000
B*Q                0.15625000
C*P               -0.28125000
C*Q                0.78125000
```

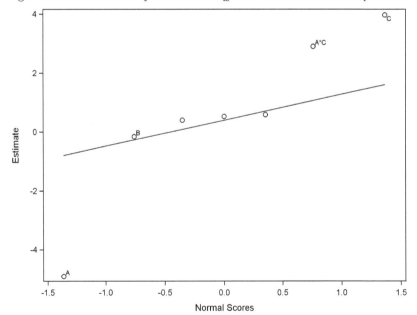

Figure 8.12 *Normal Plot of Whole-Plot Effects - Gear Distortion Experiment*

2^{k-p} FRACTIONAL FACTORIALS IN SPLIT PLOTS (FFSP)

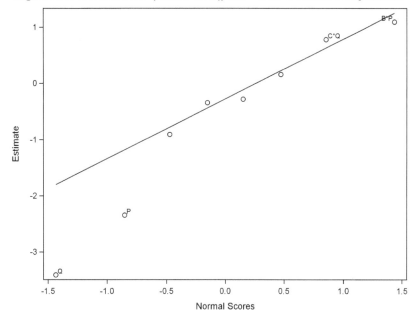

Figure 8.13 *Normal Plot of Sub-Plot Effects - Gear Distortion Experiment*

Here it can be seen that whole-plot effects A and C along with their interaction appear to be significant along with sub-plot factors P and Q.

The interpretation of the significant effects can be explained as follows. Factor P: carbon potential has a negative regression coefficient in the `proc glm` output, so increasing the carbon potential decreases the response gear dishing. Factor Q: operating mode also had a negative coefficient so using the operating mode represented by the $+1$ level of this factor also reduces gear dishing. Since there is an interaction between factors A and C, we must consider the interaction plot shown in Figure 8.14 to interpret their effects. In that figure, it can be seen that changing from the low to high levels of either A:furnace track or C:part positioning will decrease gear dishing, but if the high level of both these factors is chosen there will be a disproportionate decrease in gear dishing. Therefore, to decrease gear dishing the high levels of factors A:furnace track, P: carbon potential, and Q: operating mode should be selected, and the low level of factor C:part positioning.

Figure 8.14 *Interaction of A:Furnace Track and C:Part Position on Gear Dishing*

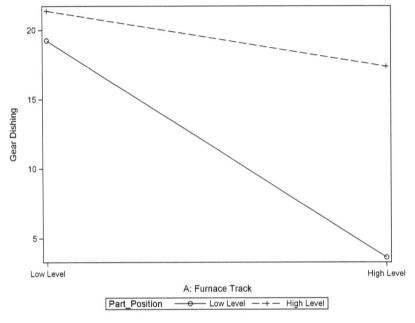

8.6 Sample Size and Power Issues for Split-Plot Designs

To determine the power for detecting whole-plot factor effects (or the sample size required to obtain a specified power level) in a split-plot design with with a CRD in the whole plots (CRSP), the sub-plot factors can be ignored and the formulas and SAS examples presented in Section 2.6 for completely randomized designs with one factor (CRD), or Section 3.5.2 for completely randomized factorial designs (CRFD) can be used by letting σ^2 represent the expected variance of whole-plot experimental units. If each factor in a CRSP factorial design has only two levels, then the simplified formula in Section 3.7.2 can be used to determine the required number of whole plots to achieve power of approximately 0.95.

For split-plot experiments with a RCB in the whole plots, the formulas and SAS code shown in Section 4.5 can be used to determine the power or required sample size for detecting whole-plot factor effects.

Since the sub-plot treatments in any split-plot design are randomized to sub-plots within each whole plot, the sub-plot treatment design is a randomized block design. Therefore, to determine the power or number of whole plots required to detect sub-plot effects, the formulas and SAS code shown in Section 4.5 can again be used, this time letting σ^2 represent the expected variance of sub-plot experimental units within a whole plot.

8.7 Review of Important Concepts

When the levels of one or more treatment factors are difficult to change or require more experimental material for evaluation than other treatment factors, it may be impractical to completely randomize the order of experiments in a factorial or fractional factorial design. This is frequently the case in process improvement studies where some treatment factors belong to one process step and others belong to a later process step. Split-plot designs allow for restricted randomization where some factor levels are varied more frequently than others, and the models for split-plot designs will allow accurate assessment of the significance of both hard and easy-to-vary factors despite the restricted randomization.

In split-plot designs the experimental unit are larger for the hard-to-vary factors and are called whole plots (based on the agricultural origins of the designs). The levels of the easy-to-vary factors can be varied within parts of whole plots (called sub-plots) which are the experimental units for these factors. Whole plots (like blocks of homogeneous experimental units) are generally more variable than sub-plots within whole plots, and therefore there is less power for detecting differences in the hard-to-vary (or whole-plot factors) than there is for the sub-plot factors in split-plot designs.

If the whole-plot experimental units are homogeneous and can be assigned at random to the levels or combinations of levels of the whole-plot factors the design that should be used is split-plot with CRD in the whole plots or a CRSP design as shown in Figure 8.15. The model for a CRSP design with one whole-plot factor and one sub-plot factor is written as:

$$y_{ijk} = \mu + \alpha_i + w_{(i)j} + \beta_k + \alpha\beta_{ik} + \epsilon_{ijk} \tag{8.10}$$

where the first three terms in the model are the same as the model for a CRD given in Chapter 2. In the analysis of data using this model, then mean square for the random whole-plot error term $w_{(i)j}$ is the denominator for the F-test for α_i, while the means square for the sub-plot error term ϵ_{ijk} is the denominator for F-test of the sub-plot factor β_k.

If whole-plot experimental units are not homogeneous and can be grouped into more uniform blocks, such as groups in time, the whole plots should be randomized to levels or combinations of levels of whole-plot factors within each block. This results in a split-plot design with a RCB in the whole plots or a RBSP design as shown in Figure 8.15. The model for a RBSP design with one whole-plot factor, α, and one sub-plot factor β is:

$$y_{ijk} = \mu + b_i + \alpha_j + b\alpha_{ij} + \beta_k + \alpha\beta_{ik} + \epsilon_{ijk} \tag{8.11}$$

where the random block effect is b_i and the random block by treatment interaction, $b\alpha_{ij}$ is the error term for testing the significance of the whole-plot effect α.

When there are no replicates of the levels or combination of levels of whole-plot treatments the analysis of split-plot experiments is accomplished by making separate normal plots of the whole-plot effects and the sub-plot effects.

When there are many whole-plot and sub-plot factors, the number of whole plots and sub-plots required for experimentation can be reduced by utilizing a split-plot fractional factorial design or FFSP as shown in Figure 8.15. FFSP designs can be created as a cartesian product of all possible combinations of treatment levels of a fractional factorial design in the whole-plot factors with combinations of treatment levels in a fractional factorial in the sub-plot factors. However, higher resolution in the subset of sub-plot factors can be achieved by using split-plot confounding where one or more whole-plot factors are included among the generators for the sub-plot fractional factorial.

Tables of generators for minimum aberration fractional factorial split-plot designs have been developed and used when creating a FFSP design using SAS ADX. Analysis of FFSP designs is similar to the analysis of unreplicated split-plot designs and involves judging significant effects using two separate normal plots.

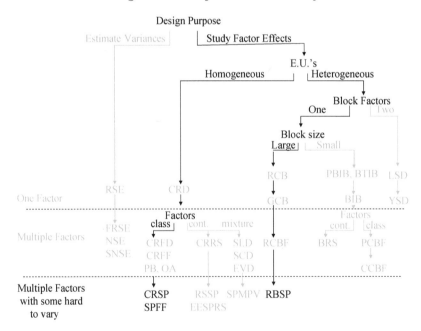

Figure 8.15 *Design Selection Roadmap*

EXERCISES

8.8 Exercises

1. Why wouldn't you treat the diameter of each cookie as the response in the cookie-baking experiment described in Section 8.2? What would you call the individual measurements on each cookie on each tray?

2. Modify the SAS code in Section 8.2.1 to create

   (a) A randomized list for a split-plot experiment with a completely randomized whole plots where there is one whole-plot factor A with 3 levels and two replicate whole plots for each level, and one split-plot factor B with three levels. Label the levels with the nvals=(1 2 3) rather than the cvals=( ) shown in Section 8.2.1.

   (b) Write the model for the design you created in (a).

   (c) A randomized list for a split-plot experiment with a completely randomized whole plots where there are two whole-plot factors A and B each with two levels and two replicate whole plots per treatment combination and two split-plot treatments C and D each with three levels.

   (d) Write the model for the design you created in (c).

3. Kuehl (2000) reports the results of an experiment conducted at a large seafood company to investigate the effect of storage temperature and type of seafood upon bacterial growth on oysters and mussels. Three storage temperatures were studied (0°C, 5°C, and 10°C). Three cold storage units were randomly assigned to be operated at each temperature. Within each storage unit, oysters and mussels were randomly assigned to be stored on one of the two shelves. The seafood was stored for two weeks at the assigned temperature, and at the end of the time the bacterial count was obtained from a sample on each shelf. The resulting data (log bacterial count) is shown below.

Storage Unit	Temp.	Seafood Type Oysters	Mussels
1	0	3.6882	0.3565
2	0	1.8275	1.7023
3	0	5.2327	4.5780
4	5	7.1950	5.0169
5	5	9.3224	7.9519
6	5	7.4195	6.3861
7	10	9.7842	10.1352
8	10	6.4703	5.0482
9	10	9.4442	11.0329

   (a) What is the experimental unit for temperature?

   (b) Why was it necessary to include 9 storage units instead of 3?

   (c) What is the experimental unit for seafood type?

(d) Write the model for the data.

(e) Analyze the data to determine what temperature and type of seafood have significant effects.

(f) Interpret any significant effects.

(g) Check the assumptions of the model you used for analysis.

4. Modify the SAS code in Section 8.3.1 to create the design for the sausage-casing experiment with two whole-plot factors, two blocks in the whole plots and two split-plot factors. Each factor has only two levels. Label the factor levels with cvals=('-' '+').

5. Ramirez and Tobias (2007) present the data shown below from an experiment to compare performance, based on a quality characteristic, of a new and old film type and three manufacturing pressures. One film type was

Time Block	Film Type	P1	P2	P3
1	Old	15.39	15.49	15.39
1	New	15.40	15.62	15.14
2	Old	15.97	15.79	14.99
2	New	15.25	15.37	15.55
3	Old	15.88	15.91	15.48
3	New	15.92	15.26	15.43
4	Old	15.36	15.51	15.47
4	New	15.30	15.53	15.66
5	Old	15.86	15.19	14.93
5	New	15.42	15.03	15.26
6	Old	15.53	15.61	15.49
6	New	15.32	15.55	15.50
7	Old	15.91	16.06	15.53
7	New	15.75	15.54	15.68
8	Old	16.03	15.55	15.49
8	New	15.75	15.31	15.62

selected at random and it was run through the manufacturing process randomly assigning each pressure to $\frac{1}{3}rd$ of the roll. Quality measurements were made on each third of the roll. Next, the other film type was run through the process again randomly assigning each pressure to $\frac{1}{3}rd$ of the roll. This two-step experimental process was repeated eight different times.

(a) Was this a completely randomized or randomized block design in the whole plots?

(b) What is the model for the data?

(c) Analyze the data to determine if type or pressure have any effect on the quality characteristic measured.

EXERCISES

(d) Describe any significant differences you find and interpret what these differences mean by referring to tables or graphs of means or multiple comparison tests.

(e) Check the assumptions of equal variance and normality of the whole-plot and split-plot error terms as described in Section 5.9.

6. Czitrom and Spagon (1997) describe a split-plot experiment conducted in a semiconductor manufacturing facility. Oxidation of silicon in a well-controlled environment is a critical step in fabrication of modern integrated circuits. A nine-nanometer thick oxide layer is grown on 200-mm silicon wafers in a vertical furnace. A quartz boat that rests on the floor of the furnace has room for 160 silicon wafers facing up. When furnace is sealed the temperature is slowly ramped up to $900°$, then oxygen gas is introduced which oxidizes the silicon on the surface of the wafer. When the thickness of the oxide reaches nine nanometers, the furnace run ends. The purpose of the experiments was to determine if the oxide thickness was affected by the position in the furnace or the site on the wafer. Oxide thickness was measured at four positions chosen in the furnace and nine specific sites or locations on each wafer in each of the four furnace positions. These were the two factors considered in the experiments. A schematic of the furnace is shown in Figure 8.16. The experimental unit for the furnace position effect was a wafer, while the experimental units for the site on a wafer were individual sites within a wafer. After one furnace run, the effect of furnace location was completely confounded with differences in wafers. Therefore, eight replicate furnace runs or blocks were made randomizing the wafers placed in each of the four positions measured on each run. The data resulting the eight separate furnace runs is shown in Table 8.12 in a format similar to Tables 8.4 and 8.5.

Figure 8.16 *Diagram of Oxide Growth Furnace*

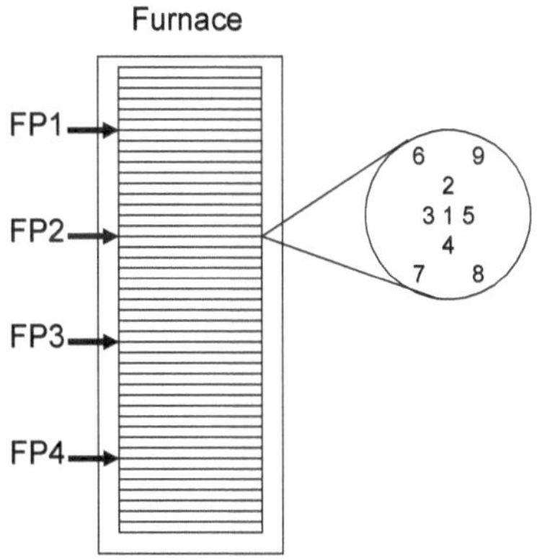

Table 8.12 *Data from Oxide Growth Experiment*

Run	FP	St 1	St 2	St 3	St 4	St 5	St 6	St 7	St 8	St 9
1	1	90.1	90.1	92.8	87.8	88.2	88.2	90.4	92.1	91.8
1	2	91.9	93.3	94.1	89.1	91.4	92.2	87.5	91.2	91.1
1	3	88.1	90.8	91.5	88.2	90.5	92.3	87.4	92.6	92.4
1	4	90.0	93.1	92.7	91.6	89.2	92.6	87.0	93.2	95.2
2	1	90.7	90.8	90.3	92.7	88.4	89.0	89.1	92.6	92.8
2	2	88.6	89.1	91.5	89.5	86.6	93.4	89.9	91.8	92.3
2	3	90.2	90.4	90.9	94.7	91.3	91.3	90.0	91.6	92.0
2	4	90.8	92.6	92.6	88.4	92.4	89.9	89.9	91.9	94.1
3	1	89.4	90.0	93.0	90.4	90.4	89.9	91.6	92.6	93.0
3	2	89.7	90.1	92.1	88.6	90.0	92.6	89.2	92.5	93.0
3	3	86.6	94.9	91.0	89.0	90.9	92.3	90.5	93.6	93.6
3	4	93.2	93.9	91.7	90.3	90.5	93.0	89.7	92.5	94.6
4	1	87.8	93.2	91.7	85.6	90.3	87.9	89.1	93.2	90.9
4	2	86.6	92.4	90.9	90.9	91.4	90.4	89.7	92.6	92.7
4	3	91.9	93.5	97.9	90.1	87.7	92.1	89.0	92.0	93.4
4	4	89.1	92.1	94.6	92.0	89.6	92.4	92.9	96.2	96.1
5	1	91.8	90.4	91.7	91.8	89.0	90.0	88.9	93.8	92.3
5	2	89.3	94.5	94.6	95.8	93.0	91.7	89.2	93.3	95.2
5	3	90.0	92.0	95.0	92.7	88.5	91.3	90.0	92.1	93.9
5	4	90.2	90.4	93.4	92.4	88.8	91.7	89.4	96.7	92.5
6	1	90.3	91.1	93.3	93.5	87.2	88.1	90.1	91.9	94.5
6	2	91.1	89.8	91.5	91.5	90.6	93.1	88.9	92.5	92.4
6	3	92.4	91.7	91.6	91.1	88.0	92.4	88.7	92.9	92.6
6	4	94.1	91.5	95.3	92.8	93.4	92.2	89.4	94.5	95.4
7	1	90.3	91.2	93.0	89.7	88.1	91.0	89.7	95.0	95.4
7	2	92.7	89.3	90.9	90.2	88.8	92.5	89.9	94.2	93.6
7	3	87.0	94.0	95.8	91.7	89.7	88.7	90.7	94.9	91.4
7	4	91.8	91.8	91.6	94.7	92.7	92.5	90.1	94.9	92.8
8	1	89.0	89.8	89.0	90.5	90.1	88.6	90.5	91.3	93.3
8	2	89.9	90.6	90.4	91.8	88.3	93.1	88.4	92.1	93.1

(a) What is the model for the data?
(b) Analyze the data using `proc glm` or `proc mixed`.
(c) Describe any significant differences you find and interpret what these differences mean by referring to tables or graphs of means or multiple comparison tests.
(d) Check the assumptions of equal variance and normality of the whole-plot and split-plot error terms as described in Section 5.9.

EXERCISES

7. In SAS ADX create a new split-plot design defining four whole-plot factors A, B, C, and D, and four sub-plot factors P, Q, R, and S as shown in Figure 8.7.

   (a) Click on the Select Design button to see the four alternate designs shown in Figure 8.10.
   (b) What are the generators shown on the Design Properties tab for the third design ($2^{4-1} \times 2^{4-2}$) shown in Figure 8.8?
   (c) From the generators find the complete defining relation including all generalized interactions, and compare this to the defining relation for the $2^{4-1} \times 2^{4-2}$ design created by `proc factex` in Section 8.5.1.
   (d) List the seven effects and all their aliases that would be tested with the whole-plot error.
   (e) List the remaining 24 effects that would be tested with the sub-plot error.

8. In SAS ADX create a new 16-run split-plot design defining three whole-plot factors A, B, and C, and four sub-plot factors P, Q, R, and S.

   (a) What alternate split-plot fractional factorial designs are listed?
   (b) Which one would you prefer, and why?
   (c) From the design you recommend, list the effects that must be tested with whole-plot error and those that must be tested with sub-plot error.
   (d) Can you create the same design using `proc factex`? If so, show the commands to do it.
   (e) If whole-plot experimental units cost four times more than that of sub-plot experimental units, and only information on whole-plot main effects is desired, which ADX design would you recommend?
   (f) Can you create the same design using `proc factex`? If so, show the commands to do it.
   (g) Show the complete defining relation for the design created by ADX.

9. Bisgaard and Kulahci (2001) describe an experiment conducted by a manufacturer of boxed cake mixes. The manufacturer wanted to find a way to make the recipe more robust so that if customers bake at a slightly different temperature or time from that recommended on the box, the cake would still taste good. The manufacturer varied F:the amount of flour, S:the amount of shortening, and E:the amount of egg powder in the recipe. These factors were varied according to a 2^3 design illustrated in the left-hand side of the table above. For each of the eight recipes, a large batch of dough was made. From each dough batch two cakes were baked, one at each of the T:temperature and t:time conditions listed across the top of the table. Since a different combination of levels of T:temperature and t:time were used depending on the sign of FSE, it can be seen that split-plot confounding was utilized. There were 16 cakes baked in total representing a 2^{5-1} fraction. The responses shown in the table below are averages of

	Design Factors				T:	−	+	−	+
Recipe	F	S	E	FSE	t:	−	−	+	+
1	−	−	−	−		−	1.4	1.0	−
2	+	−	−	+		1.8	−	−	6.1
3	−	+	−	+		1.7	−	−	2.1
4	+	+	−	−		−	3.7	4.0	−
5	−	−	+	+		1.9	−	−	4.7
6	+	−	+	−		−	6.4	6.2	−
7	−	+	+	−		−	2.1	2.3	−
8	+	+	+	+		4.9	−	−	5.7

scores between 1 and 7 obtained from a taste panel. The higher the score the better.

(a) What is the generator and defining relation for this design?
(b) List the estimable effects and their aliases.
(c) Separate the list of estimable effects into two subgroups: first the group that would be tested with whole-plot error and second the group that would be tested with sub-plot error.
(d) Use SAS `proc glm` to calculate the 15 estimable effects.
(e) Modify the SAS code in Section 8.4 to produce separate normal plots of the whole-plot effects and sub-plot effects. Does anything appear to be significant?
(f) Interpret any significant main effects and interactions using an interaction plot if needed.

CHAPTER 9

Crossover and Repeated Measures Designs

9.1 Introduction

Crossover and repeated measures designs are usually used in situations where runs are blocked by human subjects or large animals. The purpose of crossover designs (COD) is to increase the precision of treatment comparisons by comparing them within each subject or animal. In a crossover design, each subject or animal will receive all treatments in a different sequence, but the primary aim is to compare the effects of the treatments and not the sequences. Crossover designs are used frequently in pharmaceutical research, sensory evaluation of food products, animal feeding trials, and psychological research.

The primary purpose of repeated measures designs, on the other hand, is to compare trends in the response over time rather than to look at a snapshot at a particular point in time. Each subject or animal receives the same treatment throughout the experiment, and repeated measures are taken on each subject over time. Repeated measures experiments are similar to split-plot experiments in that there are two sources of error, treatments are compared to the less precise subject to subject error, and the comparison of trends over time between treatments will be compared to the more precise within subject experimental error.

9.2 Crossover Designs (COD)

Crossover designs (CODs) are useful for comparing a limited number of treatments, usually from two to six. Since each subject (that will be referred to as the block or whole-plot experimental unit) will receive each treatment sequentially in time, the number of levels of the treatment factor must remain small, otherwise drop-outs over time will cause problems in the analysis of the data. For this reason CODs are usually not used for factorial treatment plans other than simple 2^2 factorials.

Since the treatments are applied sequentially over time to each block or whole-plot experimental unit, CODs are only useful for comparing temporary treatments of a chronic condition. For example, if a clinical trial was performed to compare pain relievers for the treatment of headaches, there would be no symptom to treat in the second period if the treatment given in the first period cured the headache.

The aspect of crossover trials that requires special attention is the fact that the treatment applied to the block experimental unit in one period may affect the response during that period and during a later period. The latter effect is called a carryover effect. Special designs and models for estimating and testing the significance of carryover effects will be described.

9.3 Simple AB, BA Crossover Designs for Two Treatments

The typical layout for a two-period crossover design is illustrated in Figure 9.1. In this representation two treatments A and B are being compared. $n = n_1 + n_2$ subjects are randomly assigned to either group 1 or group 2, and the number of subjects in group 1, n_1, is not necessarily equal to the number of subjects in group 2, n_2. Group 1 receives treatment A in the first period followed by treatment B in the second period. A washout period where no treatment is given may be included between the two treatment periods. Group 2 receives the treatments in the opposite order.

Figure 9.1 *Layout for Two-Period Crossover*

	Period 1	Washout	Period 2
Group I	A	—	B
Group II	B	—	A

Defining π_i as the period effect, τ_j as the treatment effect, and μ to be the grand average, the expected response in each period for the two groups can be represented as shown below.

	Period 1	Period 2
Group 1 (AB)	$\mu + \pi_1 + \tau_1$	$\mu + \pi_2 + \tau_2$
Group 2 (BA)	$\mu + \pi_1 + \tau_2$	$\mu + \pi_2 + \tau_1$

The design is a 2×2 Latin Square, where groups represent rows and periods represent columns. As in all Latin Square Designs, it is usually assumed that there is no treatment $\times$ period interaction, therefore the model for the data can be written as:

$$y_{ijk} = \mu + s_i + \pi_j + \tau_k + \epsilon_{ijk} \tag{9.1}$$

where s_i is a random subject effect (the block factor) and ϵ_{ijk} is the within subject experimental error. As in a RCB design, the period and treatment effects are compared to the within block (subject) error. The comparison of group 1 to group 2 is just a comparison of two groups of subjects that have been randomly grouped together, and this is not considered here.

As an example of a simple AB, BA crossover study, consider the data reported by Wallenstein and Fisher (1977) from a clinical trial comparing the

SIMPLE AB, BA CROSSOVER DESIGNS FOR TWO TREATMENTS

plasma levels of two formulations of an antifungal agent. Seventeen normal male volunteers were randomly assigned to two groups. The first group received formulation A for 13 days, followed by a 13-day washout period, and then formulation B for 13 days. Group two received the treatments in the B,A order. The response shown was the area under the 12-hour plasma curve taken on the last day of each period.

Table 9.1 *Data from Two-Period Crossover Study of Antifungal Agent*

Group	Subject	Period 1	Period 2
1	2	12.8	8.2
1	3	16.5	13.1
1	6	18.7	15.9
1	8	11.6	14.2
1	11	13.6	12.8
1	12	9.8	15.3
1	16	12.8	14.0
1	18	12.1	12.0
2	1	10.9	12.3
2	4	13.5	11.5
2	5	13.7	16.0
2	7	12.2	14.8
2	9	12.6	16.2
2	10	13.0	17.5
2	14	10.7	7.5
2	15	14.2	12.4
2	17	12.2	12.8

The following commands can be used to read the data from the format in Table 9.1.

```
data antifungal;
input group subject p1 p2;
 if group=1 then do;
  period=1; pl=p1; treat='A'; output;
  period=2; pl=p2; treat='B'; output;
 end;
 if group=2 then do;
  period=1; pl=p1; treat='B'; output;
  period=2; pl=p2; treat='A'; output;
 end;
keep group subject period treat pl;
datalines;
1 2 12.8 8.2
1 3 16.5 13.1
 . . .
```

The commands to perform the analysis using SAS `proc glm` are shown below,

```
proc glm;
  class subject period treat;
  model pl=subject period treat;
  lsmeans treat/pdiff;
  estimate 'B-A' treat -1 1;
run;
```

and following is the resulting ANOVA table. In this case the `proc glm` type III sums of squares table is used because there is an unequal number of subjects in the two groups.

Source	DF	Type III SS	Mean Square	F Value	Pr > F
subject	16	114.6423529	7.1651471	1.57	0.1942
period	1	0.7343791	0.7343791	0.16	0.6939
treat	1	2.9932026	2.9932026	0.66	0.4306

Here it can be seen that there is no significant treatment effect, meaning there is no difference in the amount of antifungal agent reaching the blood between the two formulations. In addition, the period effect after adjusting for treatments is not significant. The `lsmeans` statement produces the following table of least squares means for the two treatments.

Least Squares Means

treat	pl LSMEAN	H0:LSMean1=LSMean2 Pr > |t|
A	13.4560458	0.4306
B	12.8616013	

The estimate statement produces the following table showing the difference of the treatment means, their standard error, and a t-test which is equivalent to the F-test shown in the ANOVA table.

Parameter	Estimate	Standard Error	t Value	Pr > |t|
B-A	-0.59444444	0.73395036	-0.81	0.4306

The usual assumptions of equal variance and normality of within subjects experimental error can be checked using residual plots as shown in Chapter 2.

9.3.1 Sample Size Estimation

Since the test for difference in treatment effects can be reduced to a t-test in the two-period crossover design, the number of subjects required to attain a given power for detecting a difference, Δ, in treatment means can be determined using the non-central t-distribution. For example, the SAS code below, modified from Jones and Kenward (2003), creates a table of the power of the test for treatments as a function of $n = n_1 + n_2$, the significance level (α), the expected difference in treatment means $\Delta = \bar{\mu}_{..1} - \bar{\mu}_{..2}$, and the within patient variance. In the example shown below, $\alpha = 0.05$, $\Delta = 10$, and $\sigma^2 = 326$. The table that results from running this code shows that a power of 0.8 for detecting a treatment difference of 10 can be achieved with a sample size of $n = 54$, and a power of 0.9 can be achieved with a sample size of $n = 72$.

```
*Sample Size Estimation;
data sampsize;
* alpha (two sided);
alpha=0.05;
* within-patient variance;
sigma2=326;
* size of a difference to detect;
delta=10;
* sample size (total number of patients);
do n=40 to 80 by 2;
* standard error of a difference;
stderrdiff=sqrt(2*sigma2/n);
* degrees of freedom for t-test;
df=n-2;
* critical value of the t-distribution;
* (two-sided alpha level);
t1=tinv(1-alpha/2,df);
* noncentrality parameter;
gamma=delta/(stderrdiff);
* power;
power=1-probt(t1,df,gamma);
output;
end;
run;
proc print data=sampsize;
var alpha n delta power;
run;
```

These results can also be reproduced using **proc power** or the SAS Power and Sample Size Application.

9.3.2 Carryover Effects

In cases where the effect of the treatment given in the first period persists and affects the response in the second period, a simple model for the expected response for the two groups in the two periods can be written as shown in the table on the next page,

	Period	
	1	2
Group 1 (AB)	$\mu_{11} = \mu + \pi_1 + \tau_1$	$\mu_{22} = \mu + \pi_2 + \tau_2 + \lambda_1$
Group 2 (BA)	$\mu_{12} = \mu + \pi_1 + \tau_2$	$\mu_{21} = \mu + \pi_2 + \tau_1 + \lambda_2$

where μ_{ij} refers to the expected value for the ith period and the jth treatment, τ_1 and τ_2 are the direct effects of the treatments on the response in the current period, π_1 and π_2 are the period effects, and λ_1 and λ_2 are defined as the carryover effects of treatment A and B. In this case, the expected group difference $(\mu_{12} + \mu_{21})/2 - (\mu_{11} + \mu_{22})/2$ is an estimate of the difference in the carryover effects, $\lambda_2 - \lambda_1$, assuming the expected response for the subjects in the first and second groups are the same. However, the difference in the carryover effects biases the estimate of the treatment difference

$$(\mu_{12} + \mu_{22})/2 - (\mu_{11} + \mu_{21})/2 = \tau_2 - \tau_1 + (\lambda_1 - \lambda_2)/2.$$

Therefore, no test of treatment difference is possible without assuming the carryover effects are equal (i.e. $\lambda_1 = \lambda_2$). Assuming the expected response for the subjects in the first and second groups is the same, a test of the significance of the differences in carryover effects can be made by testing the group effect, ψ_i, in the model

$$y_{ijkl} = \mu + \psi_i + s_{ij} + \pi_k + \tau_l + \epsilon_{ijkl} \qquad (9.2)$$

where y_{ijkl} is the response for the jth subject in the ith group during period k receiving treatment l; ψ_i is the group effect or carryover effect; s_{ij} is the random effect of the jth subject in the ith group; and the π_k and τ_l are the period and direct treatment effects as defined before. This model resembles a split-plot model where there are two error terms s_{ij}, and ϵ_{ijkl}. Jones and Kenward (2003) show that the appropriate error term for testing the difference in carryover effects, ψ_i, is the between subjects error s_{ij}, while the error term for the other effects in the model is the more precise within subjects error term ϵ_{ijkl}. The analysis using this model can be made using either SAS proc glm or proc mixed. Below are the commands for analyzing the data in Table 9.1 using proc mixed.

```
proc mixed data=antifungal;
  class group subject period treat;
  model pl=group period treat;
  random subject(group);
  estimate 'B-A' treat -1 1;
  estimate 'carryover diff' group 1 -1;
run;
```

This code produces the results shown on the next page. They show that there is no significant group effect or carryover difference, and no significant

SIMPLE AB, BA CROSSOVER DESIGNS FOR TWO TREATMENTS

direct treatment effect. The between subjects variance component, σ_s^2, is estimated to be 1.5078, while the within subject variance component, σ^2, is estimated to be 4.563. The estimate statement produces results that are equivalent to the REML F-tests.

```
                  The Mixed Procedure
                  Covariance Parameter
                        Estimates

              Cov Parm                Estimate
              subject(group)           1.5078
              Residual                 4.5630

              Type 3 Tests of Fixed Effects
                       Num      Den
         Effect         DF       DF     F Value    Pr > F
         group           1       15       0.13     0.7262
         period          1       15       0.16     0.6939
         treat           1       15       0.66     0.4306

                          Estimates

                        Standard
Label           Estimate   Error     DF   t Value   Pr > |t|
B-A              -0.5944   0.7340    15    -0.81     0.4306
carryover diff    0.3375   0.9459    15     0.36     0.7262
```

Notice the standard error of the carryover difference effect is larger than the standard error for the direct treatment effects since it includes both variance components. Therefore, the group or carryover effect is like a whole-plot effect that is estimated less precisely. The direct effects are estimated more precisely with the within subject variability.

Grizzle (1965) proposed a preliminary test of the carryover difference at an inflated significance level, like $\alpha = 0.10$, followed by a different test of the direct treatment effects depending on the outcome of the preliminary test of carryover difference. However, because the of the low power for the preliminary test of carryover differences, Freeman (1989) showed that this two-stage procedure inflates the type I error and produces biased estimates of the direct treatment effects, and Senn (2002) recommends against using it. Therefore, the simple AB, BA crossover trial should only be used when the carryover effects can be assumed to be equal.

9.3.3 Modified Designs When Carryover Is Possible

The disadvantage of the simple two-period crossover design presented in the last section is that the direct treatment effects are biased by any difference in carryover effects, and that the test for carryover effects lacks power because

CROSSOVER AND REPEATED MEASURES DESIGNS

it is tested with the between subjects error term. Jones and Kenward (2003) show higher order designs which include additional sequence groups or periods. These designs allow for within-subject estimators of both the carryover and direct treatment effects. Some designs that are optimal in terms of providing minimum variance unbiased least squares estimates of the difference in direct and carryover treatment effects are shown below.

Table 9.2 *Design 1*

Sequence	Period 1	Period 2
1	A	A
2	B	B
3	A	B
4	B	A

Table 9.3 *Design 2*

Sequence	Period 1	Period 2	Period 3
1	A	B	B
2	B	A	A

Table 9.4 *Design 3*

Sequence	Period 1	Period 2	Period 3
1	A	B	B
2	B	A	A
3	A	A	B
4	B	B	A

Design 3 allows for an independent estimate of the direct treatment by period interaction in addition to the direct and carryover treatment effects. Following Grizzle (1965), the model for higher order two-treatment designs can be written in the form

$$y_{ijk} = \mu + \psi_i + s_{(i)k} + \pi_j + \tau_{d(i,j)} + \lambda_{d(i,j-1)} + \epsilon_{ijk} \tag{9.3}$$

where y_{ijk} is the response of the kth subject in the ith group and jth period. ψ_i is the sequence group effect. s_{ij} is the subject within group or block effect which can be treated as a fixed or random factor. π_j is the period effect, $d(i,j)$

SIMPLE AB, BA CROSSOVER DESIGNS FOR TWO TREATMENTS

is the treatment assigned to period j for sequence group i, τ_l is the effect of the lth treatment, and λ_l is the first-order carryover effect of the lth treatment.

As an example of the use of this model, consider the data in Table 9.5 that comes from a bioequivalence study presented by Chi (1994), where half of the 36 subjects were randomly assigned to the ABB sequence and the other half to the BAA sequence in design 2.

In order to fit model 9.3 to the data the treatment, $d(i, j)$, given in the jth period and the ith treatment group, is created in the SAS data step as the data is read in. The carryover effect, $d(i, j - 1)$ in the jth period and the ith treatment group is simply the treatment given in the $j - 1$st period (if one was given) or none. The SAS code below shows how to read the data and create the treatment and carryover indicators.

```
data bioequiv;
input group subject y1 y2 y3;
  if group =1 then do;
    period=1; y=y1; treat='A'; carry='none'; output;
    period=2; y=y2; treat='B'; carry='A';    output;
    period=3; y=y3; treat='B'; carry='B';    output;
  end;
  if group = 2 then do;
    period=1; y=y1; treat='B'; carry='none'; output;
    period=2; y=y2; treat='A'; carry='B';    output;
    period=3; y=y3; treat='A'; carry='A';    output;
  end;
drop y1 y2 y3;
datalines;
1 2 112.25 106.36 88.59
1 3 153.71 150.13 151.31
 . . .
proc print; run;
```

A portion of the proc print results are shown below, and we can see the carryover indicator is always labeled **none** in the first period. In the second period the carryover indicator is the treatment given in the first period, and in the third period the carryover indicator is the treatment given in the second period.

Obs	group	subject	period	y	treat	carr
1	1	2	1	112.25	A	none
2	1	2	2	106.36	B	A
3	1	2	3	88.59	B	B
4	1	3	1	153.71	A	none
5	1	3	2	150.13	B	A
6	1	3	3	151.31	B	B

Table 9.5 *Data from Bioequivalence Study Where Subjects in Sequence Group 1 were Given ABB and Subjects in Sequence Group 2 were Given BAA*

Group	Subject	Period 1	Period 2	Period 3
1	2	112.25	106.36	88.59
1	3	153.71	150.13	151.31
1	6	278.65	293.27	295.35
1	8	30.25	35.81	34.66
1	10	65.51	52.48	47.48
1	12	35.68	41.79	42.79
1	13	96.03	75.87	82.81
1	14	111.57	115.92	118.14
1	18	72.98	70.69	74.20
1	19	148.98	157.70	178.62
1	21	140.22	119.83	139.48
1	23	60.44	44.09	35.53
1	26	136.10	161.76	163.57
1	28	111.19	101.83	101.70
1	31	85.87	99.60	107.48
1	34	111.25	114.90	135.94
1	36	58.79	96.42	122.60
1	129	299.50	303.45	385.34
2	1	52.66	47.65	13.91
2	4	128.44	173.22	140.44
2	5	233.18	88.18	31.93
2	7	53.87	89.18	70.08
2	9	62.07	54.99	73.39
2	11	183.30	153.88	122.41
2	15	51.91	73.01	23.10
2	16	90.75	89.70	111.94
2	17	59.51	56.26	48.87
2	24	83.01	73.85	71.30
2	25	85.37	86.67	92.06
2	27	84.85	75.41	79.45
2	30	70.33	40.80	46.91
2	32	110.04	102.61	113.18
2	33	93.58	87.31	87.58
2	35	66.54	43.29	84.07
2	120	59.67	56.86	69.49
2	122	49.02	50.29	51.71

SIMPLE AB, BA CROSSOVER DESIGNS FOR TWO TREATMENTS

Once the data is read in and the indicators for treatment and carryover effects are created, model 9.3 can be fit using the following commands.

```
proc glm data=bioequiv;
  class group subject period treat carry;
  model y= subject period treat carry;
  lsmeans treat/pdiff;
run;
```

The results below show that the treatments are not equivalent (since there is a significant difference in treatments at the $\alpha = 0.05$ significance level) and that the carryover effects are nonsignificant. The type III ANOVA sums of squares are used for the test since inclusion of carryover effects makes the design unbalanced. The results of the lsmeans statement shows that treatment B has a higher average response.

Source	DF	Type III SS	Mean Square	F Value	Pr > F
subject	35	403586.2044	11531.0344	21.92	<.0001
period	1	38.0628	38.0628	0.07	0.7888
treat	1	2209.0889	2209.0889	4.20	0.0443
carry	1	1050.5764	1050.5764	2.00	0.1622

Least Squares Means

treat	y LSMEAN	H0:LSMean1=LSMean2 Pr > |t|
A	96.587616	0.0443
B	106.181644	

When comparing two treatments in a crossover design, the simple AB, BA two-period design should only be used when it can safely be assumed that there are no carryover effects. When carryover effects are possible, the choice between designs 1, 2 and 3 shown in Tables 9.2-9.4, will depend upon the cost of extra periods or extra subjects. If only two periods are possible, the only choice is design 1. However, this design is much less efficient for estimating carryover effects than the three period designs. If subjects are expensive to recruit, it would be wise to obtain three observations from each subject rather than two. Design 2 is most efficient for estimating direct treatment effects and carryover effects, but it does not allow estimation of a treatment by period interaction. If a treatment by period interaction is a possibility, it would be better to randomize the total number of subjects available to four groups rather than two and use design 3.

9.4 Crossover Designs for Multiple Treatments

Many designs, with more than two treatments, have been proposed in the literature for crossover studies. Which design is best depends upon the assumptions made. For the two-treatment crossover design presented in the last section, only first-order carryover effects were assumed. That is to say, the treatment in a given period may affect the response in that period or the next period, but will not persist beyond that. Other more complicated models may assume second- or- third-order carryover effects, or interactions between the carryover effect and the treatment given in the current period. These assumptions will result in more complicated designs. In this section designs will be described for the simple first-order carryover effects. For guidance on designs for more complicated models refer to Jones and Kenward (2003).

One desirable characteristic of a crossover design is variance balance where the variance of a difference in any pair of direct (or carryover) treatment effects is the same regardless of the pair. This balance can be achieved if every treatment is preceded by every treatment an equal number of times. For example, the simple AB, BA two-period crossover design is not balanced, because treatment A is only preceded by treatment B, and treatment B is only preceded by A. By adding the third period as in design 2, treatment A is preceded by B once and A once in the second sequence group, and treatment B is preceded by A once and B once in the first sequence group. Therefore design 2 is balanced.

9.4.1 Williams' Latin Square Crossover Designs

Balanced designs in t treatments and $p = t$ periods can be created from Latin square designs. If there are no carryover effects, then a variance balanced design for direct treatment effects can be achieved by choosing any arbitrary $t \times t$ Latin square design. If carryover effects are possible, then variance balance can only be achieved when each treatment is preceded by every other treatment an equal number of times. Williams (1949) showed that this property could be achieved with only one particular Latin Square if t is an even number, and two particular Latin squares if t is an odd number. An easy algorithm for constructing these particular squares was described by Sheehe and Bross (1961) and is described below.

1. Number the treatments from 1 to t.

2. Start with a cyclic Latin square design where the treatments in the ith row are $i, i+1, \ldots, t, 1, \ldots, i-1$.

3. Create another Latin square whose rows are the mirror images of the rows in the first Latin square; e.g., the mirror image of (1,2,3,4) is (4,3,2,1).

4. Interlace each row of the first Latin Square with the corresponding row of the second Latin square to produce a $t \times 2t$ rectangle. For example, interlacing (1, 2, 3, 4) with (4, 3, 2, 1) results in (1, 4, 2, 3, 3, 2, 4, 1).

CROSSOVER DESIGNS FOR MULTIPLE TREATMENTS

5. Split the $t \times 2t$ rectangle down the middle to form two $t \times t$ Latin squares.
6. If t is even use both Latin squares, if t is odd choose either one.

Using this algorithm, the rows of the resulting Latin square(s) are indicators of the sequence groups, and the columns are indicators of the periods. Randomization is accomplished by randomizing the treatments to the level indicators $1, 2, \ldots, t$, and then randomizing one or more subjects to each sequence group.

Sheehe and Bross's algorithm can be easily implemented using SAS proc plan to create the cyclic and mirror image Latin squares, and proc sort along with the SAS data step can be used to do the interleaving and splitting. For example, the code below will produce the two Latin squares needed for a three-treatment design.

```
*Create Williams Balanced Crossover Design for 3 periods;
*Create cyclic 3x3 latin square;
proc plan seed=12345;
  factors group=3 ordered  period=3 ordered;
  treatments treat=3 cyclic;
  output out=lsd treat nvals=(1 2 3)ordered;
run;
data lsd; set lsd;
  square=1;
*Create mirror image of 3x3 latin square;
proc plan seed=12345;
  factors group=3 ordered  period=3 ordered;
  treatments treat=3 cyclic 2;
  output out=lsdm treat nvals=(3 2 1)ordered;
run;
data lsdm; set lsdm;
  square=2;
*Interlace each row with mirror image;
data comb; set lsd lsdm;
proc sort; by group period square;
*make new period square designations;
data d;
  do group=1 to 3;
    do square =1 to 2;
      do period=1 to 3; output;
      end;
    end;
  end;
data design; merge comb d; by group;
proc print data=design;
run;
```

To create the mirror image of the cyclic Latin square design, the initial levels of the treatment were reversed from (1,2,3) to (3,2,1) in the treat nvals=(3 2 1)ordered option on the output statement. Next, the treatment level indicators were cycled in the reverse order with the `cyclic 2` option on the treatment statement in place of the `cyclic` option that was used in creating the original cyclic design. To modify this code to create a design for a four-treatment design, the reverse cycling would be achieved using the option `cyclic 3`, and for a five-treatment design the option `cyclic 4` would be used, and so forth.

Running the SAS code on the previous page produces a data file `design` that contains the design shown below in Table 9.6, where it can be seen that each treatment is preceded by every other treatment twice.

Table 9.6 *Williams' Design for Three Treatments*

	Square					
	I			II		
	Period			Period		
Group	1	2	3	1	2	3
1	1	3	2	2	3	1
2	2	1	3	3	1	2
3	3	2	1	1	2	3

Wang *et al.* (2009) have created a SAS macro for the random construction of a Williams design and the relevant procedure for randomization. The results can be generated with the same random seed for a reproducible design to satisfy regulatory concerns. The example below shows the call of the macro to construct a plan for a clinical trial crossover design with three treatments (a test drug, an active control and a placebo) with a sample size of 18 patients (usually a multiple of the total number of sequences).

```
%WILLIAMS(
  TmtsNum=3,
  TmtNames=TestDrg ActCtrl Placebo,
  SampleSize=18,
  OutRtf=D:\WilliamsN18.rtf,
  SeedNum=1234);
```

A randomized treatment assignment list is created by the macro and printed as the second portion of the output. It is shown at the top of the next page.

```
Sub_ID              Sequence
 001       Placebo--ActCtrl--TestDrg
 002       ActCtrl--TestDrg--Placebo
 003       Placebo--TestDrg--ActCtrl
 004       TestDrg--Placebo--ActCtrl
 005       TestDrg--ActCtrl--Placebo
 006       ActCtrl--Placebo--TestDrg
 007       ActCtrl--TestDrg--Placebo
 008       Placebo--ActCtrl--TestDrg
 009       Placebo--TestDrg--ActCtrl
 010       TestDrg--Placebo--ActCtrl
 011       TestDrg--ActCtrl--Placebo
 012       ActCtrl--Placebo--TestDrg
 013       ActCtrl--TestDrg--Placebo
 014       ActCtrl--Placebo--TestDrg
 015       Placebo--ActCtrl--TestDrg
 016       TestDrg--Placebo--ActCtrl
 017       TestDrg--ActCtrl--Placebo
 018       Placebo--TestDrg--ActCtrl
```

As an example of the use of Williams' Design for three treatments, consider an experiment conducted by Chipman (2006). The purpose of the experiment was to determine how the surface (grass, cement, or rubberized running track) affected the time to sprint 40 yards. Twelve subjects were recruited for the study, and in order to compare the surface effect within each subject, all subjects ran on all three surfaces. To adjust for the lingering exhaustion effect of each run, the crossover design shown in Table 9.6 was used. Two subjects were randomized to each sequence group in each square. The data resulting from this experiment is the time in seconds for each subject to sprint 40 yards and is shown in Table 9.7. In the following table, treatment level 1 represents cement, treatment level 2 represents the rubberized track, and treatment level 3 represents grass. These surfaces were side by side at the BYU track stadium, which was a convenient location to conduct the experiments.

The SAS commands to read this data are shown below. Notice the carryover effects were manually added to the data by coding them in the same way the carryover effects were coded in the bioequivalence study shown in the last section.

```
data chipman;
input square group subject period treat carry y;
datalines;
1 1 1 1 1 0 5.47
1 1 2 1 1 0 6.03
1 1 1 2 3 1 5
1 1 2 2 3 1 5.42
1 1 1 3 2 3 5.08
1 1 2 3 2 3 5.38
  . . .
```

Table 9.7 *Williams' Design and Data for Sprint Time Experiment*

		Square I			Square II			
Group	Subject	Period	Treat	Time	Subject	Period	Treat	Time
1	1	1	1	5.47	7	1	2	5.68
1	2	1	1	6.03	8	1	2	5.90
1	1	2	3	5.00	7	2	3	5.27
1	2	2	3	5.42	8	2	3	5.70
1	1	3	2	5.08	7	3	1	5.23
1	2	3	2	5.38	8	3	1	5.54
2	3	1	2	7.69	9	1	3	5.97
2	4	1	2	6.32	10	1	3	7.87
2	3	2	1	7.03	9	2	1	5.73
2	4	2	1	5.43	10	2	1	6.97
2	3	3	3	7.57	9	3	2	4.97
2	4	3	3	5.77	10	3	2	6.85
3	5	1	3	8.05	11	1	1	6.19
3	6	1	3	7.51	12	1	1	7.39
3	5	2	2	7.12	11	2	2	5.66
3	6	2	2	6.49	12	2	2	6.55
3	5	3	1	7.18	11	3	3	5.57
3	6	3	1	6.35	12	3	3	7.09

The SAS commands to analyze the data with **proc glm** are shown below.

```
proc format;
value surface 1='cement' 2='track' 3='grass';
run;
proc glm;
class subject period treat carry;
model y=subject period treat carry;
means treat carry;
format treat carry surface.;
run;
```

Running this code produces the following type III ANOVA table shown on the next page. Here it can be seen that the running surface caused a significant difference in running times, and that there were significant carryover effects.

The **means** statement estimates $\hat{\mu} + \hat{\tau}_i$ and $\hat{\mu} + \hat{\lambda}_l$ in model 9.3. The result shown below the ANOVA table on the next page indicates that the rubberized running track produced the fastest running times, and that grass produced the slowest running times.

CROSSOVER DESIGNS FOR MULTIPLE TREATMENTS

Source	DF	Type III SS	Mean Square	F Value	Pr > F
subject	11	24.25471000	2.20497364	86.01	<.0001
period	1	0.00183750	0.00183750	0.07	0.7920
treat	2	0.63916333	0.31958167	12.47	0.0004
carry	2	0.23320556	0.11660278	4.55	0.0252

The GLM Procedure

Level of treat	N	Mean	Std Dev
cement	12	6.21166667	0.76463697
grass	12	6.39916667	1.12486329
track	12	6.14083333	0.83989673

The carryover effects are summarized in the following table of means. Subtracting the grand mean from these estimates, it can be seen that the carryover effects of grass and cement are negative which means that the sprinting times will be slightly faster in the sprint following one on grass or cement. The "0" level of carryover means no prior treatment. This is where the longest sprint times occur.

Level of carry	N	Mean	Std Dev
0	12	6.67250000	0.94744753
cement	8	5.97375000	0.92968409
grass	8	5.94250000	0.80331012
track	8	6.20250000	0.80786756

9.4.2 Designs with $p > t$ or $t > p$

One problem with Williams' designs is that the direct treatment effects and the carryover effects are not orthogonal. The type III ANOVA table makes a test for significance of treatment effects adjusted for carryover and a test for carryover adjusted for treatments; however, SAS `proc glm` is unable to produce least squares (adjusted) means for treatments and carryover due to the imbalance in these factors. In addition the direct treatment effects are estimated more precisely than the carryover effects since each marginal mean is the average of 12 observations for the direct effects but only 8 for the carryover effects.

Lucas (1957) proposed a solution to this problem by simply adding an extra period where all sequence groups receive the same treatment that they received in the tth period. This makes the design completely balanced in the sense that every treatment follows every other treatment an equal number of times, including itself (like design 2 for two treatments presented in Section 9.3.3). This makes the direct treatment and carryover treatment effects orthogonal and increases the efficiency of estimating the carryover effects. However, it does not come without cost. Adding an extra period may increase the cost or difficulty in carrying out the experiments, and by adding the extra period the treatment effects are no longer orthogonal to the subjects since each subject will receive one treatment twice. Therefore, the efficiency of estimating the direct treatment effects actually decreases slightly by adding the extra period. Unless carryover effects are strongly suspected, the extra period design for k treatments may not be necessary.

In experiments like sensory evaluation of food products, it may be desirable for the number of periods p to be less than the number of treatments t. In this way each subject will only have to taste a subset of the food items being compared. When too many items are tasted by a subject, the taste buds become overwhelmed. Designs with $p < t$ are similar to incomplete block design with subjects representing the blocks. The simplest way to create a design with $p < t$ is to delete one or more periods from an orthogonal set of Latin squares, as long as at least three periods remain. See Jones and Kenward (2003) for other methods of constructing designs with $p < t$.

9.5 Repeated Measures Designs

In a repeated measures design, the objective is to compare the trend over time in the response between treatment groups when the response is measured repeatedly on each subject. As an example consider comparing different weight-loss diets. If subjects were randomized to one of t diets that they would then follow for three months, it would probably be of interest to compare how fast subjects lost weight on the different diets in addition to comparing the average weight loss over time between the treatment diets. Repeated measures designs are like the completely randomized or randomized blocks designs discussed in Chapters 2 and 4, where subjects are randomized to one of several treatments. The difference is that the response is measured repeatedly throughout the treatment period instead of once at the end of the treatment period.

As an example of a repeated measures design, consider the data in Table 9.8. This is part of the data from a study presented in Diggle *et al.* (1994) for the purpose of determining how the diet of dairy cows affects the protein in the milk. Seventy-nine Australian cows were randomized to receive one of three diets: barley alone, a mixture of barley and lupins, or lupins alone. The protein was measured in a weekly sample of the milk from each cow. Table 9.8 shows the data for the first five weekly samples from the first 10 cows in each group as an illustration of the type of data that results from a repeated measures design. Figure 9.2 shows the trend in average protein over time. It can be seen that the lupins diet results in lowest protein levels, and that

UNIVARIATE ANALYSIS OF REPEATED MEASURES DESIGN 365

Table 9.8 *Data from Repeated Measures Experiment on Dairy Cow Diets*

Diet	Cow	Week 1	2	3	4	5
Barley	1	3.63	3.57	3.47	3.65	3.89
Barley	2	3.24	3.25	3.29	3.09	3.38
Barley	3	3.98	3.6	3.43	3.30	3.29
Barley	4	3.66	3.5	3.05	2.90	2.72
Barley	5	4.34	3.76	3.68	3.51	3.45
Barley	6	4.36	3.71	3.42	3.95	4.06
Barley	7	4.17	3.6	3.52	3.10	3.78
Barley	8	4.40	3.86	3.56	3.32	3.64
Barley	9	3.40	3.42	3.51	3.39	3.35
Barley	10	3.75	3.89	3.65	3.42	3.32
Mixed	11	3.38	3.38	3.10	3.09	3.15
Mixed	12	3.80	3.51	3.19	3.11	3.35
Mixed	13	4.17	3.71	3.32	3.10	3.07
Mixed	14	4.59	3.86	3.62	3.60	3.65
Mixed	15	4.07	3.45	3.56	3.10	3.92
Mixed	16	4.32	3.37	3.47	3.46	3.31
Mixed	17	3.56	3.14	3.60	3.36	3.37
Mixed	18	3.67	3.33	3.20	2.72	2.95
Mixed	19	4.15	3.55	3.27	3.27	3.65
Mixed	20	3.51	3.9	2.75	3.37	3.51
Lupins	21	3.69	3.38	3.00	3.50	3.09
Lupins	22	4.20	3.35	3.37	3.07	2.82
Lupins	23	3.31	3.04	2.80	3.17	2.92
Lupins	24	3.13	3.34	3.34	3.25	2.79
Lupins	25	3.73	3.61	3.82	3.61	3.45
Lupins	26	4.32	3.7	3.62	3.50	3.57
Lupins	27	3.04	2.89	2.78	2.84	2.61
Lupins	28	3.84	3.51	3.39	2.88	3.41
Lupins	29	3.98	3.3	3.02	2.99	2.84
Lupins	30	4.18	4.12	3.84	3.65	3.65

the mixed diet, while initially similar to the barley diet in protein, quickly decreases to the level of the lupins diet.

9.6 Univariate Analysis of Repeated Measures Design

The repeated measures design is similar to a split-plot design with whole-plots being subjects (cows in the example) and subplots being different observation times on each subject. In the dairy cow example, cows were randomized to treatment diets, so the diets are the whole-plot treatments. The repeated measures over time are measured within each cow, so the difference in weekly

366   CROSSOVER AND REPEATED MEASURES DESIGNS

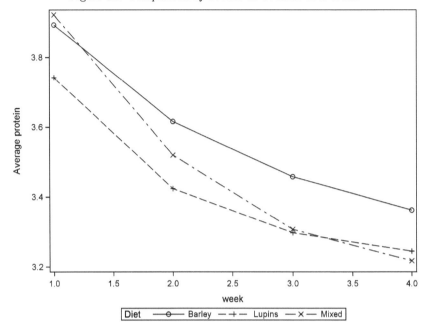

Figure 9.2 *Comparison of Trends in Protein over Time*

measurements and the interaction between diets and weekly measurements are the subplot treatments. A simple univariate analysis of the data could be made using the split-plot model with either **proc glm** or **proc mixed**. For example, the commands to make the analysis in **proc mixed** are shown below.

```
proc mixed data=strung;
  class Diet week ;
  model protein=Diet week Diet*week;
  random Cow(Diet);
  lsmeans Diet week;
  estimate 'Lupins vs Barley' Diet 1 -1 0;
  estimate 'Mixed va Barley' Diet 1 0 -1;
run;
```

In this analysis the diet factor is compared to the cow-to-cow variation and is found to be non-significant. The difference in protein levels over time is compared to the within cow variation and is found to be significant at the $\alpha = 0.01$ level, indicating that the protein level decreases over the first five weeks for all three diets. The comparison of the trends in protein over time,

UNIVARIATE ANALYSIS OF REPEATED MEASURES DESIGN 367

or the interaction that is shown in Figure 9.2, is also compared to the within cow variation and is not significant.

The usual assumptions for analysis of variance are that the experimental error terms should have equal variance and be independent and normally distributed. For split-plot models, the assumptions of normality and equal variance can be checked by making residual plots and plots EBLUPs (as described in Section 5.9.3). However, in past chapters the independence assumption was never checked. For completely randomized designs, the randomization guaranteed independence of experimental error. In split-plot designs, subplot experimental units are usually not independent because observations on pairs of subplots within the same whole plot will be correlated. However, randomization of subplot treatments to subplots in a split-plot design equalizes the correlation between all possible pairs of subplots. This, along with equal variances of subplot responses, creates a condition that is called compound symmetry of the covariance matrix of subplot observations. This condition, like independence of experimental error, justifies the normal analysis by ANOVA or REML as shown in Chapter 8.

With repeated measures designs, however, the subplot treatments are repeated measures over time, and they cannot be randomized. They must be observed in temporal order. Therefore, the compound symmetry condition that justified the normal split-plot analysis may not be satisfied for repeated measures designs.

9.6.1 Test of Huynh-Feldt Condition

Huynh and Feldt (1970) showed the conditions justifying the usual analysis of variance of repeated measures designs are less stringent than compound symmetry. They showed that if repeated measures at different times, y_i and y_j, satisfy the condition

$$\sigma^2_{(y_i - y_j)} = 2\lambda \quad \text{for } i \neq j \tag{9.4}$$

for some $\lambda > 0$, then the covariance matrix of the repeated measures is said to obey the Huynh-Feldt condition. The usual split-plot analysis of repeated measures data is justified if the Huynh-Feldt condition is satisfied.

The Mauchly (1940) sphericity test can be used to test whether the covariance matrix of the repeated measures has the Huynh-Feldt condition. SAS `proc glm` has an option to perform the sphericity test. To do this, read the repeated measures as different variables as in the data step shown on the next page. Next all the repeated measures variables are listed as dependent variables in the `proc glm` model statement, and the repeated statement, with the option `printe`, prints the correlation matrix and performs the sphericity test.

```
data dairy;
input Diet$ Cow p1 p2 p3 p4 p5;
datalines;
Barley 1 3.63 3.57 3.47 3.65 3.89
Barley 2 3.24 3.25 3.29 3.09 3.38
 . . .
proc glm data=dairy;
class Diet;
model p1-p5=Diet/nouni;
repeated time 5 / printe;
run;
```

The portion of the output listing below shows the section that contains the results of the sphericity test. The chi-square value on the "Orthogonal Components" line is the correct test. Since this chi-square value with 9 degrees of freedom is non-significant ($p = 0.35$), it indicates that there is no significant departure from the Huynh-Feldt condition and that it would be justified to use the standard split-plot analysis shown earlier.

```
                       Sphericity Tests

                           Mauchly's
Variables              DF  Criterion   Chi-Square  Pr > ChiSq

Transformed Variates    9  0.2264746   37.746862   <.0001
Orthogonal Components   9  0.6751597    9.9838194   0.3518
```

The commands above also produce a multivariate analysis of variance, or MANOVA, but it will not be discussed here.

9.6.2 Adjusted P-values

When the Huynh-Feldt condition does not hold, there is less information in the estimate of the variance of the within plot experimental error due to the correlation in responses. Greenhouse and Geisser (1959) and Huynh and Feldt (1976) proposed adjustments to the degrees of freedom for the standard subplot F-tests in repeated measures designs. These adjustments were developed for designs where there is only one within subjects treatment factor. Both of these adjustments consist of calculating an ϵ value that is a function of the elements in the covariance matrix of the repeated measurements. Once the ϵ value is computed, the degrees of freedom for the within subplot F-tests are modified by multiplying by ϵ. Huynh and Feldt's adjustment is slightly less conservative than Greenhouse and Geisser's.

Using the repeated statement in SAS `proc glm` causes the procedure to automatically compute both the Greenhouse-Geisser ϵ, the Huynh-Feldt ϵ, and make the adjusted F-tests. For example, the portion of the output that

UNIVARIATE ANALYSIS OF REPEATED MEASURES DESIGN

shows these adjusted F-tests for the dairy cow experiment in Table 9.9 is shown below.

```
              The GLM Procedure
        Repeated Measures Analysis of Variance
Univariate Tests of Hypotheses for Within Subject Effects

                      Type    Mean                      Adj Pr > F
Source          DF    III SS  Square   F     Pr > F    G - G    H - F

time             4    6.562   1.640   30.95  <.0001    <.0001   <.0001
time*Diet        8    0.413   0.051    0.98  0.4594    0.4511   0.4594
Error(time)    108    5.723   0.052

        Greenhouse-Geisser Epsilon     0.8168
        Huynh-Feldt Epsilon            1.0115
```

The Greenhouse-Geisser $\epsilon = 0.8168$ and the Huynh-Feldt $\epsilon = 1.0115$. The P-values for the Greenhouse-Geisser adjusted F-test is listed under the G - G column in the output, and the Huynh-Feldt adjusted F-test is listed under the H - F column in the output. In this example, both ϵ values are close to 1.0, and the P-values for the adjusted F-tests are almost the same as the values for the unadjusted F-tests. This is due to the fact that the covariance matrix of the repeated measures seems to satisfy the Huynh-Feldt condition, which justified the unadjusted F-tests.

9.6.3 Conservative Analysis of Summary Statistics

One simple way to overcome the problem of correlated observations over time within each subject is to summarize the observations over time. This is similar to averaging correlated observational units as discussed in Section 5.8. Consider a simple case. When testing the effectiveness of a treatment, the response may be measured before and after treatment on each subject as illustrated in Table 9.9, the normal two-sample t-statistic

$$t_{2(n-1)} = \frac{\bar{y}_{\cdot 1} - \bar{y}_{\cdot 2}}{s_p \sqrt{2/n}}$$

where

$$s_p^2 = [\sum_{i=1}^{n}(y_{i1} - \bar{y}_{\cdot 1})^2 + \sum_{i=1}^{n}(y_{i2} - \bar{y}_{\cdot 2})^2]/2(n-1)$$

would not be appropriate because the before and after observations on each subject are correlated and the two samples of data are not independent.

The normal solution to this problem is to use the paired t-test statistic

$$t_{n-1} = \bar{d}/s_d,$$

Table 9.9 Before and After Treatment Observations

Subject	Before	After
1	y_{11}	y_{12}
2	y_{21}	y_{22}
$\vdots$	$\vdots$	$\vdots$
n	y_{n1}	y_{n2}

where $d_i = (y_{i1} - y_{i2})$, $\bar{d} = \sum_{i=1}^{n} d_i/n$, $s_d = [\sum_{i=1}^{n}(d_i - \bar{d})^2]/(n-1)$. This eliminates the problem of correlated responses, since there is now only one response (d_i) per subject. The paired t-test is more conservative than adjusted F-tests since it reduces the degrees of freedom from $2(n-1)$ to $n-1$.

A similar conservative analysis can be made of data from a repeated measures design. Consider an experiment conducted at FMC agricultural chemical R&D center in 1985. The purpose was to determine what factors influenced the rate of degradation of a chemical herbicide in soil stored in pots. The factors varied on the pots were storage temperature (10 or 30°C), moisture content (low or high), and soil type (two different types of soil). A 2^3 factorial design with replicates was employed and each of the eight combinations of factor levels was randomly assigned to two pots. Next, the same concentration of herbicide was applied to the soil in all the pots. The pots were stored at the two specified temperatures in separate chambers. Samples of the soil from each pot were taken on days 0, 7, 14, 30 and 60, and the concentration of the herbicide was determined for each sample. Repeated measures data are shown in Table 9.10.

The repeated concentration measurements within the same pot on different days are correlated and the Huynh-Feld condition does not hold for the covariance matrix; however, the objective is to determine whether the factors influence the rate of degradation and not the average concentration in the pot. From first-order kinetics the concentration on day t is given by the equation

$$C = C_0 e^{-kt}$$

where C_0 is the initial concentration, t is the day, and k is the degradation rate constant. Therefore, the data for each pot, can be summarized with the estimate of the rate constant, $\hat{k}$, obtained by regressing $ln(C)$ on t, or with the estimated half-life of the herbicide in the soil which is calculated as

$$ln\left(\frac{C_0}{2}\right) = -ln(2)/-\hat{k}.$$

For example, a representative plot is shown in Figure 9.3 where $\hat{k}$ is estimated to be 0.010462 and the half-life in the soil is estimated to be

$$-ln(2)/-0.010462 = 66.2.$$

Therefore, as can be seen in the graph, the concentration of herbicide for

UNIVARIATE ANALYSIS OF REPEATED MEASURES DESIGN

Table 9.10 *Concentration of Herbicide from Degradation Study*

Temp.	Moisture	Soil	0	7	14	30	60
10	L	C	0.77	0.77	0.76	0.74	0.72
10	L	P	0.78	0.76	0.75	0.72	0.66
10	H	C	0.76	0.74	0.71	0.66	0.57
10	H	P	0.78	0.76	0.74	0.70	0.63
30	L	C	0.77	0.74	0.71	0.65	0.54
30	L	P	0.79	0.73	0.68	0.58	0.42
30	H	C	0.78	0.73	0.69	0.60	0.46
30	H	P	0.78	0.71	0.65	0.53	0.36
10	L	C	0.77	0.76	0.75	0.72	0.66
10	L	P	0.77	0.76	0.74	0.71	0.65
10	H	C	0.78	0.77	0.75	0.72	0.67
10	H	P	0.77	0.75	0.73	0.68	0.60
30	L	C	0.79	0.75	0.72	0.65	0.54
30	L	P	0.78	0.74	0.69	0.60	0.45
30	H	C	0.79	0.72	0.65	0.53	0.35
30	H	P	0.78	0.70	0.63	0.49	0.31

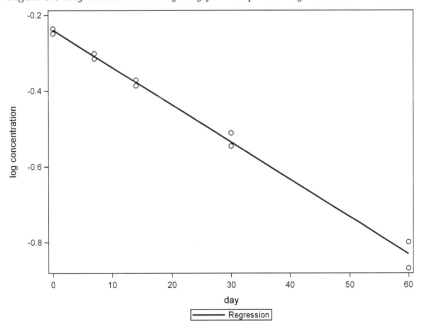

Figure 9.3 *Log Concentration by Day for Temp.=30 Deg. Moisture=L Soil=P*

372 CROSSOVER AND REPEATED MEASURES DESIGNS

this combination of factor levels has degraded to just over half the original concentration by day 60 (the last measurement).

The SAS data step commands below read the data from Table 9.10 and calculate the degradation rate constant and estimated half-life for each pot.

```
*Reads data from Table 9.10 and calculates
 rate constant k and half life;
data residue;
   input temp moisture $ soil$ c1 c2 c3 c4 c5;
*Calculate Log concentration;
   lc1=log(c1); lc2=log(c2); lc3=log(c3); lc4=log(c4); lc5=log(c5);
*Calculate degradation rate constant by least squares;
   sp=7*lc2+14*lc3+30*lc4+60*lc5; *sum of products;
   sm=lc1+lc2+lc3+lc4+lc5; *sum log concentrations;
   num=5*sp-111*sm;
   den=5*4745-(111)**2;
   k=num/den;
   half_life=-log(2)/k;
   logHL=log(half_life);
   T=(Temp-20)/10; if moisture='L' then M=-1; else M=1;
   if soil ='C' then S=-1; else S=1;
datalines;
10 L C 0.77 0.77 0.76 0.74 0.72
10 L P 0.78 0.76 0.75 0.72 0.66
 . . .
proc print;  Var temp moisture soil k half_life;
run;
```

The resulting summary statistics for each pot are shown in the output of proc print shown on the next page.

Either of these summary statistics can be analyzed to determine if the rate constant is influenced by the factors. The following commands can be used to fit the model to the log half-life with proc glm

```
proc glm data=residue;
   class temp moisture soil;
   model logHL=temp|moisture|soil;
   lsmeans temp moisture;
run;
```

This analysis showed that increasing temperature and increasing soil moisture both had significant effects in increasing the rate of degradation. The soil type and all of the interactions had no significant effects on degradation at the $\alpha = 0.05$ level, so the additive model can be used to predict the degradation rate for any combination of temperature and moisture.

UNIVARIATE ANALYSIS OF REPEATED MEASURES DESIGN

Obs	temp	moisture	soil	k	half_life
1	10	L	C	-0.001202	576.845
2	10	L	P	-0.002728	254.065
3	10	H	C	-0.004828	143.572
4	10	H	P	-0.003547	195.402
5	30	L	C	-0.005903	117.415
6	30	L	P	-0.010462	66.255
7	30	H	C	-0.008767	79.066
8	30	H	P	-0.012854	53.927
9	10	L	C	-0.002603	266.325
10	10	L	P	-0.002855	242.764
11	10	H	C	-0.002566	270.115
12	10	H	P	-0.004192	165.340
13	30	L	C	-0.006293	110.151
14	30	L	P	-0.009222	75.165
15	30	H	C	-0.013538	51.202
16	30	H	P	-0.015393	45.029

The approach to fitting a curve to repeated measure over time and then analyzing coefficients from the fitted model is often called the growth curve approach, and it is used so often that it is incorporated as a feature in SAS proc glm. By adding the actual levels of the within subject (time) factor and the polynomial option to the repeated statement as shown in the commands below,

```
proc glm data=residue;
  class temp moisture soil;
  model lc1 lc2 lc3 lc4 lc5 = temp|moisture|soil;
  repeated time 5 (0 7 14 30 60) polynomial/
      printe summary;
run;
```

proc glm automatically computes polynomial contrasts over the repeated measures and analyzes each one separately. For example, two of the tables resulting from this command are shown on the next page.

The table for time_1 shown on the next page is the analysis of the linear contrast in log concentration over days. The significance levels (P-values) are similar to those resulting from the analysis of the log half-life. Both temperature and moisture have significant effects on the linear contrast, and soil (P-value=0.0364) is also significant at the $\alpha = 0.05$ level. In the analysis of log half-life described above the P-value for soil was 0.0526.

The table for time_2 shown on the next page is the analysis of the quadratic contrast in log concentration over days. Here it can be seen that none of the factors affect the quadratic trend. The polynomial option on the repeated statement in proc glm calculates and analyzes up $l-1$ polynomial contrasts for l repeated measures time points.

```
                    The GLM Procedure
             Repeated Measures Analysis of Variance
             Analysis of Variance of Contrast Variables

time_N represents the nth degree polynomial contrast for time

Contrast Variable: time_1

Source              DF   Type III SS    Mean Square   F Value  Pr > F

Mean                 1   1.63058592     1.63058592    297.79   <.0001
temp                 1   0.47804051     0.47804051     87.30   <.0001
moisture             1   0.08498935     0.08498935     15.52   0.0043
temp*moisture        1   0.02381646     0.02381646      4.35   0.0705
soil                 1   0.03448920     0.03448920      6.30   0.0364
temp*soil            1   0.01821844     0.01821844      3.33   0.1056
moisture*soil        1   0.00126468     0.00126468      0.23   0.6437
temp*moisture*soil   1   0.00000176     0.00000176      0.00   0.9861
Error                8   0.04380496     0.00547562

Contrast Variable: time_2

Source              DF   Type III SS    Mean Square   F Value  Pr > F

Mean                 1   0.00004201     0.00004201     1.91    0.2045
temp                 1   0.00000785     0.00000785     0.36    0.5669
moisture             1   0.00004169     0.00004169     1.89    0.2061
temp*moisture        1   0.00000005     0.00000005     0.00    0.9644
soil                 1   0.00000268     0.00000268     0.12    0.7361
temp*soil            1   0.00000068     0.00000068     0.03    0.8647
moisture*soil        1   0.00001497     0.00001497     0.68    0.4335
temp*moisture*soil   1   0.00003402     0.00003402     1.55    0.2490
Error                8   0.00017615     0.00002202
```

9.7 Review of Important Concepts

The crossover and repeated measures designs presented in this chapter are special cases of the randomized complete block (RCB), Latin square design (LSD), and split-plot (CRSP, RBSP) designs presented in Chapters 4 and 8 and highlighted in Figure 9.4. These designs are used in situations where the blocking factor is a human subject or large animal, and the subplot experimental units are successive periods of time.

Crossover designs are most useful when a more precise test of the treatment effects is desired and each subject can be assigned each treatment in successive periods. Crossover experiments should not be used in cases where the experimental subjects are permanently changed by the initial treatment. Otherwise, the result of applying following treatments will not be the same as it would be if they had been applied first. Except for the simple two-period AB, BA case, the levels of the whole-plot factor in crossover experiments are just random groupings of subjects and can usually be ignored in the analysis. Therefore, the model for the analysis of crossover experiments, except for the

REVIEW OF IMPORTANT CONCEPTS

Figure 9.4 *Design Selection Roadmap*

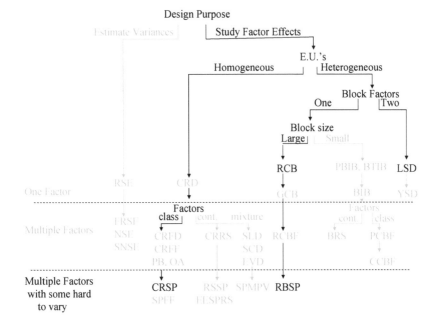

simple case, is the same as it is for Latin square designs. For the simple two-period AB, BA case, the model is the same as a split-plot model, where the whole-plot treatment factor is the carryover effect.

Repeated measures designs are useful when the objective is to compare the trend in the response across repeated measures over time between levels or combinations of levels of treatment factors. The model is the same as a split-plot experiment, where treatment factors are the whole-plot factors and repeated measures over time are the subplot factors. When the covariance matrix of the repeated measures satisfies the Huynh-Feldt condition, the same ANOVA or REML analysis used for split-plot experiments is appropriate. When the Huynh-Feldt condition does not hold, there are two alternative univariate methods of analysis that can be used. The first method is a correction to the usual ANOVA F-tests based on degree of freedom adjustments proposed by Greenhouse and Geisser or Huynh and Feldt. The second univariate analysis method that can be used is the conservative growth curve approach. In this approach a curve is fit to each set of repeated observations over time, and summary statistics from the curve fits, which are independent from one combination of whole-plot factor levels to another, are analyzed as separate responses.

9.8 Exercises

1. Consider the data from a study reported by Varma and Chilton (1974), that was later re-analyzed by Jones (1980).

	Group 1		Group 2	
Subject	Period 1	Period 2	Period 1	Period 2
	Placebo	Test	Test	Placebo
1	0.83	1.83	1.67	0.33
2	1.00	2.17	2.50	0.50
3	0.67	1.67	1.00	-0.17
4	0.50	1.50	1.67	0.50
5	0.50	2.33	1.83	0.50
6	0.83	1.83	0.50	0.33
7	1.00	0.50	1.33	0.67
8	0.67	0.33	1.33	0.00
9	0.67	0.50	0.50	0.17
10	0.33	0.67	2.17	0.83
11	0.00	0.83	1.67	0.33
12	1.17	1.33	1.50	0.00
13	0.00	0.67	1.33	0.50
14	0.50	1.83	1.50	0.50
15	0.33	1.50	1.33	0.00
16	0.33	1.50	0.67	-0.17
17	0.50	1.17	1.67	0.50
18	1.00	1.67	2.50	0.67
19	0.00	1.33	1.83	0.00
20	0.50	1.50	0.83	0.67
21	-0.50	2.83	2.33	0.17
22	0.17	2.33	1.17	0.50
23	1.00	1.33	1.33	0.00
24	1.00	1.67	1.33	0.83
25	1.33	0.67	0.33	1.33
26	0.33	0.83	2.17	1.17
27	2.00	1.00	1.00	0.33
28	4.00	0.17	0.33	1.00
29	0.83	1.67	1.17	0.17
30	0.50	1.33	0.50	0.50
31	0.50	1.50		
32	0.50	1.67		
33	2.17	1.33		
34	0.67	1.17		

This was a simple crossover dental study comparing a test compound to a placebo with regard to their effect on dental hygiene as measured by

EXERCISES

the change in a dental hygiene index. There were thirty-four patients randomized to the first group and thirty patients randomized to the second group.

(a) Use `proc glm` to analyze the data from this experiment.

(b) Is there a difference of the test compound and placebo with respect to their effect on dental hygiene? If so, which produces the larger change in the index?

(c) Is there any evidence of carryover effect of the test compound?

2. Lucas (1957) presented data from an extra-period crossover study on dairy cows. Three diet treatments were being compared and the response was the average daily production of FCM. The data is shown in the table below.

Table 9.11 *Dairy Cow Experiment - Response Is Average Daily FCM Production in Pounds, the Treatment Indicator Is in Parentheses*

		Period			
Group	Cow	1	2	3	4
1	1	(1) 38.66	(2) 37.43	(3) 34.39	(3) 31.30
1	2	(1) 25.72	(2) 26.13	(3) 23.35	(3) 18.69
2	3	(2) 48.85	(3) 46.88	(1) 41.99	(1) 39.61
2	4	(2) 30.80	(3) 29.29	(1) 26.41	(1) 23.16
3	5	(3) 34.64	(1) 32.27	(2) 28.50	(2) 27.13
3	6	(3) 25.35	(1) 26.00	(2) 23.86	(2) 19.92
4	7	(1) 35.19	(3) 33.50	(2) 28.41	(2) 25.12
4	8	(1) 21.80	(3) 23.91	(2) 21.69	(2) 17.55
5	9	(2) 32.90	(1) 33.12	(3) 27.52	(3) 25.10
5	10	(2) 21.37	(1) 21.97	(3) 19.38	(3) 16.57
6	11	(3) 30.40	(2) 29.50	(1) 26.70	(1) 23.09
6	12	(3) 22.84	(2) 20.97	(1) 18.59	(1) 16.10

(a) Does every treatment follow every other treatment, including itself?

(b) Analyze the data using `proc glm`. Are the treatment and carryover effects orthogonal? Check by comparing their type I and type III sums of squares.

(c) Use the `lsmeans` statement to get least squares adjusted means for treatments. Why is this possible when it was not possible for the example in Section 9.4.1 ?

3. Consider planning an experiment to compare the taste of chocolate brownie recipes. The treatments will consist of three brands of boxed brownie mixes and one from-scratch home recipe. A panel of tasters will be used to evaluate the brownies. Each taster will taste each brownie and give it a rating from 1 to 10 (10 being the best possible).

(a) Why do you think it might be appropriate to use a crossover design for this study? What is the alternative?

(b) Modify the SAS code in Section 9.4.1 to create Williams' crossover design for four treatments to use for this experiment.

(c) Decide how many subjects to include in your study and create a randomized list assigning each subject to one of the four sequence groups.

(d) Write the model you would use to analyze this data.

(e) Make a data input form and show how you would code the indicators for the subject, period, treatment and carryover effects.

4. Consider the data from the repeated measures design shown in Table 9.10.

(a) Use `proc glm` to compute the sphericity test and determine if the Huynh-Feldt condition holds.

(b) Use `proc glm` to make the Greenhouse-Geisser and Huynh-Feldt adjusted F-tests of the subplot effects.

(c) Do you find more terms significant using the adjusted F-tests than were found using the growth curve analysis shown in the text? If so, explain why.

5. Consider the results of the repeated measurements experiment from Usage note: Stat-40 (1997) from the University of Texas at Austin. In this experiment the purpose was to determine how diet and exercise affect the pulse of subjects. Eighteen subjects were randomly assigned to one of three exercises (1=aerobic stair climbing, 2=racquetball, 3=weight training) and one of two diets (1=include meat, 2=vegetarian). After a training period where all subjects followed their assigned exercise and diet routine, their pulse was measured three successive times (once after a warm-up, again after jogging and finally after running). The results are shown in the table on the next page.

(a) Make a plot of the average trend in pulse over the three repeat measures for each of the six combinations of diet and exercise.

(b) Do a normal split-plot analysis of the data treating diet and exercise as whole-plot factors and the repeated measurements as a subplot factor.

(c) Use the sphericity test to determine if the normal analysis is justified.

(d) Make the Huynh-Feldt and Greenhouse-Geisser adjusted F-tests for the subplot effects. Are the results different than the unadjusted tests?

(e) Do a conservative growth-curve analysis of the linear contrast in pulse measurements over time. Do the results of this analysis confirm what you found in (b) and (d)? If not, why?

EXERCISES

Exercise	Diet	Pulse 1	Pulse 2	Pulse 3
1	1	112	166	215
1	1	111	166	225
1	1	89	132	189
1	2	95	134	186
1	2	66	109	150
1	2	69	119	177
2	1	125	177	241
2	1	85	117	186
2	1	97	137	185
2	2	93	151	217
2	2	77	122	178
2	2	78	119	173
3	1	81	134	205
3	1	88	133	180
3	1	88	157	224
3	2	58	99	131
3	2	85	132	186
3	2	78	110	164

CHAPTER 10

Response Surface Designs

10.1 Introduction

In a response surface experiment, the independent variables or factors can be varied over a continuous range. The goal is to determine the factor settings that produce a maximum or minimum response or to map the relationship between the response and the factor settings over this contiguous factor space. As a practical matter, if we want to know a lot about the relationship between the factors and the response, it will require many experiments. For that reason, response surface designs are rarely conducted with more than six factors.

Response surface experiments are normally used at the last stage of experimentation. The important factors have already been determined in earlier experiments, and at this stage of experimentation the purpose is to describe in detail the relationship between the factors and the response. It is usually known or assumed that a simple linear model, even with interactions, is not good enough to represent that relationship. In order to locate maximums or minimums in the response as a function of the factor settings, at least three levels of each factor should be utilized.

Response surface methods generally refer to a complete package of statistical design and analysis tools that are used for the following three steps.

1. Design and collection of data to fit an equation to approximate the relationship between the factors and response.
2. Regression analysis to fit a model to describe the data.
3. Examination of the fitted relationship through graphical and numerical techniques.

Response surface methods have found considerable use in industry especially in chemical processes where the reaction yield or cost of production can be optimized as a function of the settings of controllable process factors. Since their origin these designs have also found successful applications in food science, engineering, biology, psychology, textiles, education and many other areas.

10.2 Fundamentals of Response Surface Methodology

10.2.1 Empirical Quadratic Model

In response surface methods, the relationship between the response (y) and the factor settings (x's) is assumed to be a nonlinear equation given by $\mathbf{y} = f(\mathbf{x}) + \epsilon$. For two factors or independent variables we could write this equation as

$$y = f(x_1, x_2) + \epsilon \tag{10.1}$$

where f is a nonlinear function and ϵ is the random effect of experimental error. When f is unknown it can be approximated near the point (x_{10}, x_{20}) using the two-term Taylor series approximation, i.e.,

$$\begin{aligned}
f(x_1, x_2) \approx{} & f(x_{10}, x_{20}) + (x_1 - x_{10}) \left.\frac{\partial f(x_1, x_2)}{\partial x_1}\right|_{x_1=x_{10}, x_2=x_{20}} \\
& + (x_2 - x_{20}) \left.\frac{\partial f(x_1, x_2)}{\partial x_2}\right|_{x_1=x_{10}, x_2=x_{20}} \\
& + \frac{(x_1 - x_{10})^2}{2} \left.\frac{\partial^2 f(x_1, x_2)}{\partial x_1^2}\right|_{x_1=x_{10}, x_2=x_{20}} \\
& + \frac{(x_2 - x_{20})^2}{2} \left.\frac{\partial^2 f(x_1, x_2)}{\partial x_2^2}\right|_{x_1=x_{10}, x_2=x_{20}} \\
& + \frac{(x_1 - x_{10})(x_2 - x_{20})}{2} \left.\frac{\partial^2 f(x_1, x_2)}{\partial x_1 \partial x_2}\right|_{x_1=x_{10}, x_2=x_{20}},
\end{aligned}$$

$$\tag{10.2}$$

which leads to a general quadratic equation of the form

$$y = \beta_0 + \beta_1 x_1 + \beta_2 x_2 + \beta_{11} x_1^2 + \beta_{22} x_2^2 + \beta_{12} x_1 x_2 + \epsilon \tag{10.3}$$

where $\beta_1 = \left.\frac{\partial f(x_1, x_2)}{\partial x_1}\right|_{x_1=x_{10}, x_2=x_{20}}$, etc. If the region of interest is of moderate size, this general quadratic equation will provide a good approximation to f and can be used for interpolation within this region. The general quadratic equation is quite flexible and with appropriate coefficients it can describe a wide variety of surfaces such as hilltops, valleys, rising or falling ridges, or saddle points as shown with contour plots in Figure 10.1.

With k factors or independent variables the general quadratic equation can be written in the form

$$y = \beta_0 + \sum_{i=1}^{k} \beta_i x_i + \sum_{i=1}^{k} \beta_{ii} x_i^2 + \sum_{i<j}^{k}\sum^{k} \beta_{ij} x_i x_j + \epsilon, \tag{10.4}$$

and unless the function f is known, this equation forms the basis of response surface methods. Response surface designs were created to provide data to approximate this equation, and mathematical tools were created to explore the fitted surface represented by this equation.

Figure 10.1 *Surfaces That Can Be Described by General Quadratic Equation*

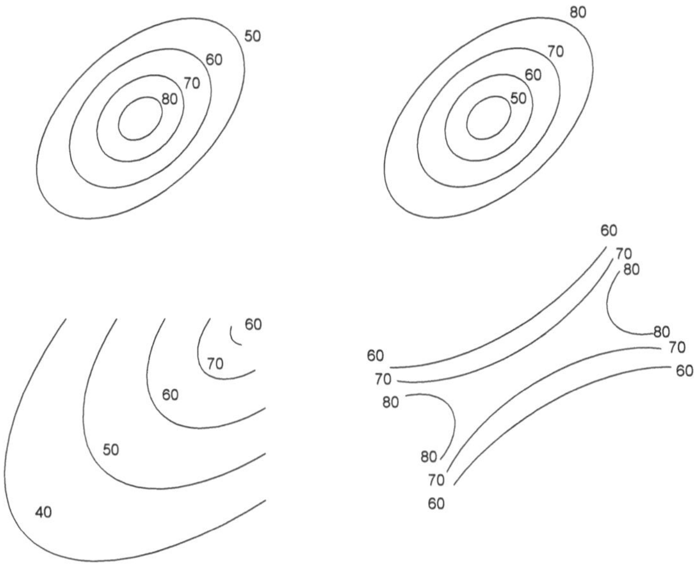

10.2.2 Design Considerations

The first requirement of a response surface design is that it should provide data that will allow estimation of the coefficients in model (10.4). This model has $1+2k+k(k-1)/2$ coefficients, thus any response surface design must have at least three levels for each factor (to allow estimation of quadratic terms) and at least $1+2k+k(k-1)/2$ total runs. 3^k factorial designs have three levels for every factor. However, Box and Wilson (1951) showed that they were less satisfactory as a response surface design than an alternative design they called the central composite design. The central composite will be the first standard response surface design we will discuss in the next section.

Normally the factors or independent variables (x_i) in the general quadratic Equation (10.4) are coded and scaled as shown in Sections 3.7.2 and 3.7.3 so that the experimental region is a hyper-cube or hyper-sphere with radius R in the coded space. This general quadratic model can be written in matrix terms as

$$\mathbf{y} = \mathbf{x}\mathbf{b} + \mathbf{x}'\mathbf{B}\mathbf{x} + \epsilon \qquad (10.5)$$

where $\mathbf{x}' = (1, x_1, x_2, \ldots, x_k)$, $\mathbf{b}' = (\beta_0, \beta_1, \ldots, \beta_k)$ and the symmetric matrix

$$\mathbf{B} = \begin{pmatrix} \beta_{11} & \beta_{12}/2 & \cdots & \beta_{1k}/2 \\ & \beta_{22} & \cdots & \beta_{2k}/2 \\ & & \ddots & \\ & & & \beta_{kk} \end{pmatrix}.$$

When fitting a linear regression model of the form $\mathbf{y} = \mathbf{xb}$, the design points are chosen to minimize the variance of the fitted coefficients $\hat{\mathbf{b}} = (\mathbf{X'X})^{-1}\mathbf{X'y}$. Since the variance covariance matrix of estimated regression coefficients is $\sigma^2(\mathbf{X'X})^{-1}$, this means the design points should be chosen such that the $(\mathbf{X'X})$ matrix is diagonal, or that the design is orthogonal like the 2^k designs discussed in Chapter 3 and the 2^{k-p} designs discussed in Chapter 6. When the $(\mathbf{X'X})$ matrix is diagonal, the diagonal elements of $(\mathbf{X'X})^{-1}$ will be minimized.

On the other hand, when fitting the general quadratic model, the primary purpose is not to understand the mechanism of the underlying relationship between the response and the factors. Therefore, the specific coefficients in the general quadratic model are of less importance. What is more important in a response surface study is to develop a prediction equation with the eventual goal of determining the optimum operating conditions. Thus, the variance of a predicted value at ,$\mathbf{x}$, that is given by the equation $Var[\hat{y}(\mathbf{x})] = \sigma^2 \mathbf{x'}(\mathbf{X'X})^{-1}\mathbf{x}$ is of more importance than the variance of the fitted coefficients.

Since it is not known before the experiment is conducted where the optimum will lie in the design region, one desirable property of a response surface design is to have the variance of a predicted value nearly the same everywhere in the design region. In this way the precision in predicting the optimum will not depend upon its unknown position in the design region. The first step in equalizing the variance of prediction over the design region is to find a design that is *rotatable* (Myers and Montgomery, 2002). A rotatable design is one in which the variance of the predicted value at the point $\mathbf{x}$ is only a function of the distance from the design origin to $\mathbf{x}$. Box and Wilson's central composite design and other standard response surface designs have this property.

Box and Hunter (1957) showed that certain rotatable designs could be modified to have *uniform precision*, which means the variance of a predicted value is the same at the origin and at the radius of one in the coded design region. A design with uniform precision is close to having the variance of the response equal throughout the design region. Many standard response surface designs have this property or are close to having this property. When creating response surface designs using a computer algorithm, the G-optimality criterion is similar to uniform precision. A G-optimal design is one that minimizes the maximum variance of a predicted value in the design region. Computer code for creating G-optimal designs is not normally used, but in the limit G-optimality and D-optimality are equivalent. Programs, such as SAS `proc optex`, are widely used for creating D-optimal designs, and when a standard response surface design does not meet the needs of an experimenter, the D-optimality criterion can be used in a computer search to create a design. Giovannitti-Jensen and

STANDARD DESIGNS FOR SECOND ORDER MODELS

Myers (1989) and Myers *et al.* (1992) describe a variance dispersion plot which can be used to visually evaluate whether any response surface design is close to having the properties of rotatability and uniform precision.

In addition to providing a good distribution of the variance of a predicted value, two other desirable properties of response surface designs are (1) to provide an estimate of the "pure" experimental error so the adequacy of the general quadratic model can be checked, and (2) to allow for blocking so that an experimenter can begin with a linear design and add a second block to estimate curvature if necessary. To estimate pure error, at least one design point must be replicated. When blocking, the first block would consist of a design such as a 2^k or 2^{k-p} augmented with center points that allow for checking the adequacy of the linear model. If the linear model is adequate, no further experimentation is required. If the linear model is not adequate, a second block of experiments can be added which will allow estimation of the quadratic coefficients in the general quadratic model.

10.3 Standard Designs for Second Order Models

This section presents some of the more popular completely randomized response surface designs (CRRS).

10.3.1 Central Composite Design (CCD)

Box and Wilson (1951)'s central composite design (CCD) consists of a 2^k factorial or 2^{k-p} fractional factorial design augmented with center points and axial points as shown in Figure 10.2. The 2^k or resolution V 2^{k-p} design allows

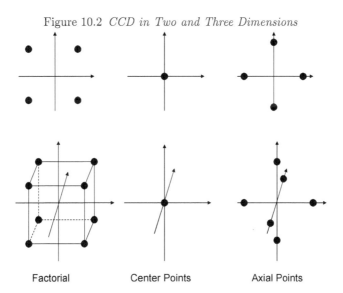

Figure 10.2 *CCD in Two and Three Dimensions*

Factorial   Center Points   Axial Points

estimation of the linear and linear-by-linear terms in the general quadratic model. The addition of the center points and axial points allow estimation of the quadratic terms. By choosing the distance from the origin to the axial points (α in coded units) equal to $\sqrt[4]{F}$ where F is the number of points in the factorial portion of the design, a CCD will be rotatable. By choosing the correct number of center points, the CCD will have the uniform precision property.

As an example of a CCD, consider the results of the experiment described by Kamon et al. (1999) in Table 10.1. The experiment was conducted to find the optimal formulation of a two-component admixture to improve the workabilty of cement grouts. The factors were the water-to-cement ratio, the percent of black liquor added, and the percent of SNF added. The left side of the table shows the coded factors x_1-x_3. The axial points were chosen at $\pm 1.68 = \sqrt[4]{8}$ in coded units to make the design rotatable. The coded levels are found as $(actual\ level - center\ value)/(half - range)$. For example, $x_2 = (Black Liq. - 0.150)/0.03$. The actual factor levels on the right side of the table can be obtained from the coded values on the left side of the table by solving $actual\ level = (half - range) \times x_i + center\ value$.

Table 10.1 CCD in Coded and Actual Units for Cement Workability Experiment

Run	x_1	x_2	x_3	Water/Cement	Black Liq.	SNF	y
1	-1	-1	-1	0.330	0.120	0.080	109.5
2	1	-1	-1	0.350	0.120	0.080	120.0
3	-1	1	-1	0.330	0.180	0.080	110.5
4	1	1	-1	0.350	0.180	0.080	124.5
5	-1	-1	1	0.330	0.120	0.120	117.0
6	1	-1	1	0.350	0.120	0.120	130.0
7	-1	1	1	0.330	0.180	0.120	121.0
8	1	1	1	0.350	0.180	0.120	132.0
9	0	0	0	0.340	0.150	0.100	117.0
10	0	0	0	0.340	0.150	0.100	117.0
11	0	0	0	0.340	0.150	0.100	115.0
12	-1.68	0	0	0.323	0.150	0.100	109.5
13	1.68	0	0	0.357	0.150	0.100	132.0
14	0	-1.68	0	0.340	0.100	0.100	120.0
15	0	1.68	0	0.340	0.200	0.100	121.0
16	0	0	-1.68	0.340	0.150	0.066	115.0
17	0	0	1.68	0.340	0.150	0.134	127.0
18	0	0	0	0.340	0.150	0.100	116.0
19	0	0	0	0.340	0.150	0.100	117.0
20	0	0	0	0.340	0.150	0.100	117.0

By including six center points, the design has uniform precision and the variance of a predicted value will be the same at the origin $(0, 0, 0)$, in coded factor levels, and at any point at radius 1 in coded units. Box and Hunter

STANDARD DESIGNS FOR SECOND ORDER MODELS

(1957) have tabulated the number of center points required to make a central composite design uniform precision for various values of k =number of factors. These tables are not necessary when using SAS to create designs since the tools that will be shown in Section 10.3.5 automatically determine the number of center points required.

Figure 10.3 shows Myers *et al.* (1992)'s variance dispersion graph for this design, from SAS ADX. The variance dispersion graph plots the maximum, minimum and average scaled variance of a predicted value $(NVar(\hat{y}(\mathbf{x}))/\sigma^2)$ as a function of the distance from the origin of the design. When the design is not rotatable, there will be three distinct lines on the graph. The closer together the lines, the closer the design is to being rotatable. In Figure 10.3 the lines overlap showing that the design is rotatable. It can also be seen that the value of the scaled variance of a predicted value at the origin is nearly equal to the variance at radius 1.0, indicating the design has uniform precision.

Figure 10.3 *Variance Dispersion Graph for Uniform Precision CCD in Three Factors*

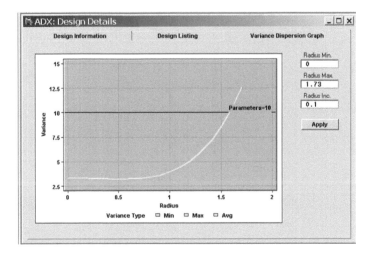

The runs in Table 10.1 should be run in a random order, but they are listed in a non-random order to illustrate another useful aspect of central composite designs. This design can be run as a sequence of two blocks. In some applications this is a desirable approach. The first 8 runs in the table represent a full 2^3 factorial. Runs 9-11 are center points at the mid-level of each factor. If these 11 experiments were completed in the first block, the data could be analyzed to see if a linear model

$$y = \beta_0 + \beta_1 x_1 + \beta_2 x_2 + \beta_3 x_3 + \beta_{12} x_1 x_2 + \beta_{13} x_1 x_3 + \beta_{23} x_2 x_3 \qquad (10.6)$$

is adequate to represent the data. If so, runs number 12-20 need not be completed. If the linear model is not adequate to represent the data, the second set of experiments can be completed and a block variable added to the model to account for any changes in background variables that occurs between the first and second set of experiments.

10.3.2 Box-Behnken Design

Whereas the CCD requires five levels $(-\alpha, -1, 0, +1, +\alpha)$ for each factor, Box and Behnken (1960) developed some three level designs that will allow estimation of the general quadratic model. These designs consist of 2^2 factorials in each pair of factors with all other factors held constant at their mid-level plus a few center points. No Box-Behnken design exists for only two factors. An example of a Box-Behnken design in three coded factors is shown in Table 10.2.

Table 10.2 *Box-Behnken Design in Three Factors*

run	x_1	x_2	x_3
1	-1	-1	0
2	1	-1	0
3	-1	1	0
4	1	1	0
5	-1	0	-1
6	1	0	-1
7	-1	0	1
8	1	0	1
9	0	-1	-1
10	0	1	-1
11	0	-1	1
12	0	1	1
13	0	0	0
14	0	0	0
15	0	0	0

Box-Behnken designs have two advantages over CCDs. The first advantage is that they only require that factors be varied over three levels. This may make experimentation less costly if actual prototypes are being constructed in the experimentation. The second advantage is that they usually (except for the five-factor case) require less total runs than the central composite design. For example, the three-factor CCD required twenty runs whereas the three-factor Box-Behnken design only required 15 runs.

A disadvantage of a Box-Behnken design compared to a central composite design is that it cannot be built up in two steps beginning with a 2^k design. This can be visualized in Figure 10.4, which diagrams a three-factor Box-Behnken design. A 2^3 design consists of runs at all corners of the cube as shown in Section 3.7.1, but as can be seen in Figure 10.4 the Box-Behnken design does not include those points. Therefore a Box-Behnken design should

STANDARD DESIGNS FOR SECOND ORDER MODELS

be used if the experimenter is reasonably sure that a linear model will not adequately represent the relationship between the factors and the response, and he or she wants to save a few runs. Another possible disadvantage is that the factors have only three levels. While this may be less costly when building prototypes, having three levels leaves nothing to check the adequacy of the quadratic model

Figure 10.4 *Graphical Representation of Three-Factor Box-Behnken Design*

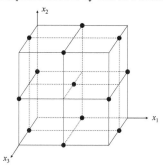

Figure 10.5 shows the variance dispersion graph for the three-factor Box-Behnken Design.

Figure 10.5 *Variance Dispersion Graph for Box-Behnken Design in Three Factors*

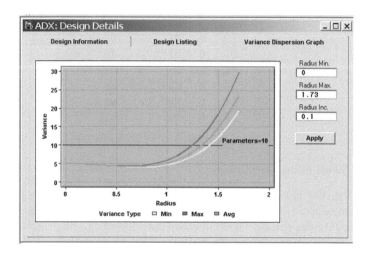

Since the lines for the minimum, maximum, and average scaled variance of a predicted value do not overlap as in Figure 10.3, the Box-Behnken design is not rotatable. But, since the lines are very close together within the coded

experimental region to a radius of 1.0, the Box-Behnken design is close enough to being rotatable and having uniform precision for practical use. This is true for Box-Behnken designs will $k = 3$ to 6.

As an example of a Box-Behnken design, consider the experiment performed by Anderson (2003). He experimented with a scale model trebuchet (medieval missile launcher) that was issued to all engineering students at the South Dakota School of Mines and Technology for hands-on experimentation. He wanted to determine the settings that would allow him to toss an object a given distance to fend off attackers or door-to-door salespeople. He varied three factors on the trebuchet. A, the arm length 4 to 8 inches from the counterweight end to the point where the weights were hung; B, counterweight 10 to 20 pounds; and C, missile weight 2 to 3 ounces. The Box-Behnken design in actual factor levels is shown in Table 10.3 along with the response (the distance the missile flew).

Table 10.3 *Box-Behnken Design Results for Trebuchet Experiments*

run	A	B	C	y
1	4	10	2.5	33
2	8	10	2.5	85
3	4	20	2.5	86
4	8	20	2.5	113
5	4	15	2	75
6	8	15	2	105
7	4	15	3	40
8	8	15	3	89
9	6	10	2	83
10	6	20	2	108
11	6	10	3	49
12	6	20	3	101
13	6	15	2.5	88
14	6	15	2.5	91
15	6	15	2.5	91

10.3.3 Small Composite Design

When the cost of each experiment is high, and an experimenter is willing to compromise some of the desirable properties of a response surface design in order to reduce the run size, a small composite design can be utilized. In a small composite design the 2^k or resolution V 2^{k-p} part of the central composite design is replaced with a resolution III 2^{k-p}. In the central composite design, the 2^k or resolution V 2^{k-p} was included to allow estimation of all linear main effect terms and linear by linear two-factor interactions. Hartley (1959) has shown that when center points and axial points are included, two-factor interactions can be estimated with a resolution III fractional factorial. Westlake

STANDARD DESIGNS FOR SECOND ORDER MODELS

(1965) obtained additional small composite designs by substituting irregular fractions for the factorial portion of the design, and Draper (1985) and Draper and Lin (1990) substituted Plackett-Burman Designs for the factorial portion to come up with even more small composite designs.

Figure 10.6 shows a comparison of the central composite design for $k = 2$ and the small composite design for $k = 2$. The central composite design has five center points for a total of thirteen runs, while the small composite design has three center points with a total of nine runs.

Figure 10.6 *Graphical Comparison of CCD and Small Composite (with $I = AB$) for $k=2$*

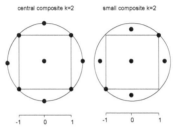

Figure 10.7 compares the variance dispersion graph for the central composite and small composite designs with $k = 2$ factors. Here it can be seen the central composite is rotatable and has uniform precision while beyond a radius of 0.75 in coded units, the scaled variance of a predicted value depends on the location for a small composite design. The coefficients for the general quadratic model can still be estimated when a small composite design is used, but the variance of a predicted value is not as uniform as it is for a central composite design. This is the sacrifice for decreasing the number of runs.

10.3.4 Hybrid Design

Roquemore (1976) developed hybrid designs that require even fewer runs than the small composite designs. These designs were constructed by making a central composite design in $k - 1$ factors and adding a kth factor so that the **X'X** has certain properties and the design is near rotatable. Table 10.4 shows an example of the Roquemore 310 design. It is labeled 310 because it has 3 factors and 10 runs. It can be seen that there is a central composite design in columns 1 and 2. This design is near rotatable as shown in Figure 10.8, but since it only has 10 runs, there are zero degrees of freedom for estimating the experimental error. Some of the Roquemore hybrid designs leave one degree of freedom for estimating experimental error, but none of them have pure replicates that are required for testing the adequacy of the general quadratic model. Therefore these designs should only be used when the experimenter is confident that the general quadratic model will represent the data well, and there is an independent estimate of experimental error from previous data.

Figure 10.7 *Comparison of Variance Dispersion Graph for CCD and Small Composite Design with k=2*

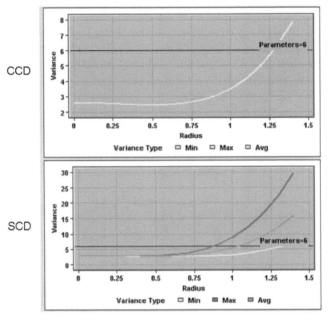

Table 10.4 *Roquemore 310 Design*

Run	x_1	x_2	x_3
1	0	0	1.2906
2	0	0	-0.1360
3	-1	-1	0.6386
4	1	-1	0.6386
5	-1	1	0.6386
6	1	1	0.6386
7	1.736	0	-0.9273
8	-1.736	0	-0.9273
9	0	1.736	-0.9273
10	0	-1.736	-0.9273

10.4 Creating Standard Designs in SAS

Central Composite Designs (if you know the axial distance and number of center points) and Box-Behnken designs can be easily created in the SAS data step by following the descriptions in the last section. There are also SAS macros in the AUTOCALL library that can produce central composite designs. For example, the commands below will produce the central composite design in uncoded units shown in Table 10.1.

Figure 10.8 *Variance Dispersion Graph for Roquemore 310 Design*

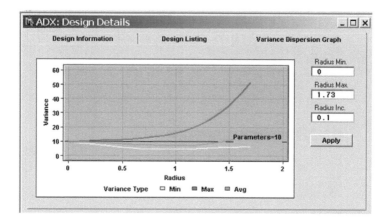

```
%adxgen
%adxff
%adxcc
%adxinit
%adxccd(rs1,3,8,6,1.68,1)
```

The first argument in the call statement, %adxccd(rs1,3,8,6,1.68,1), is rs1: the file where the design will be stored in non-random order, the second argument 3: is the number of factors, the third argument 8: is the number of points in the factorial portion of the design, the fourth argument 6: is the number of center points in the design, the fifth argument 1.68: is the value of the axial extreme, and the last argument 1: is the number of blocks in the design.

However, the easiest and most versatile method of creating response surface designs in SAS is to use the automated design of experiments tool SAS ADX. For example, to create a three-factor central composite design, click on Solutions ▶ Analysis ▶ Design of Experiments to open the ADX system. Next, click on the icon for New Response Surface Design, or click File ▶ Create New Design ▶ Response Surface... To create a design for three factors, click the Define Variables... button. On the Factors tab, click on Add > and then define the variables. To create the uncoded design shown in Table 10.1, fill out the template as shown in Figure 10.9. The Low Levels and High Levels shown in the template are the extremes of the factorial portion of the design for a central composite design.

Figure 10.9 *Add Factors Template from SAS ADX*

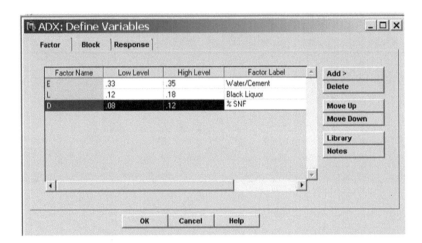

When finished with the Add Factors template, click the Ok button and click the Yes button in response to the question, "Do you want to save changes?" Next, click on the Select Design button, and the template shown in Figure 10.10 will appear. This template shows all the completely randomized response surface designs available for three factors.

The central composite-uniform precision is the design shown in Table 10.1. The other option for a central composite design (central composite-orthogonal) will create a central composite design that has a diagonal $\mathbf{X'X}$ when the correct axial value is chosen and the factors, interactions and squared factor levels are properly scaled. Before the widespread use of computer programs to fit the general quadratic equation, orthogonality was important because it simplified the calculations required to obtain least squares estimates of the coefficients in the model. Using SAS, orthogonality is not important and the uniform precisions design should be chosen if a central composite is to be used. It can also be seen that the Box-Behnken, Small Composite and Roquemore designs discussed in the last section are available choices.

To choose a design on the Select Design template, single left click on the line for the design you want. You can see a description of the design, a listing of the runs in the design or the variance dispersion graph for the design by clicking the Design Details... button. When you have decided what design you would like to use, click on the line listing the design and then click on the "x"

NON-STANDARD RESPONSE SURFACE DESIGNS

in the upper right-hand corner to close the Design Selection template. ADX will query "Do you want to use the selected design?"; respond by clicking the Yes button.

Figure 10.10 *Select Design Template from SAS ADX*

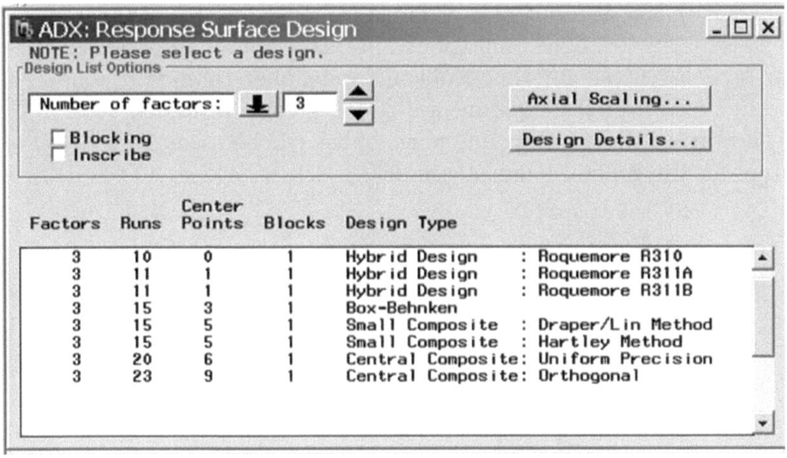

```
ADX: Response Surface Design
NOTE: Please select a design.
Design List Options

  Number of factors:  3              Axial Scaling...
    Blocking                         Design Details...
    Inscribe

                Center
  Factors  Runs Points Blocks  Design Type
     3      10    0      1     Hybrid Design    : Roquemore R310
     3      11    1      1     Hybrid Design    : Roquemore R311A
     3      11    1      1     Hybrid Design    : Roquemore R311B
     3      15    3      1     Box-Behnken
     3      15    5      1     Small Composite  : Draper/Lin Method
     3      15    5      1     Small Composite  : Hartley Method
     3      20    6      1     Central Composite: Uniform Precision
     3      23    9      1     Central Composite: Orthogonal
```

SAS ADX produces the design chosen in a non-random order. You can export the design to a SAS data set by clicking File ▶ Export. You can choose to export the coded or uncoded design. The default name for the file exported is work._adx$i$, where the number i will be shown to you before the file is exported. You can use the SAS data step to create a randomized data collection worksheet with the exported design, and you can augment the design with the response data after the experiments are complete so that you can fit the general quadratic model using the SAS procedure described in Section 10.6. Alternatively you can input the response data to the ADX tool and use the point and click menus to analyze the data.

10.5 Non-Standard Response Surface Designs

Some design situations do not lend themselves to the use of standard response surface designs for reasons such as (1) the region of experimentation is irregularly shaped, (2) not all combinations of the factor levels are feasible, (3) there is a nonstandard linear or a nonlinear model. In these situations, standard response surface designs will not be appropriate. One way to construct a response surface design in these cases is to use a computer algorithm to construct a D-optimal design as described in Section 6.5.2.

For an example of the first situation where the region of experimentation is irregularly shaped, Atkinson et al. (2007) describe an experiment to investigate the performance of an internal combustion engine. Two factors under study were the spark advance and the torque. Both are independently variable so the coded design region is a square. However, for factor combinations outside the pentagonal region shown in Figure 10.11, it is likely the engine would not run or would be seriously damaged.

Therefore the experimental region is irregular and the standard two-factor response surface designs would not fit in this region. A response surface design can be constructed for this problem by defining candidate points that fall within a grid over the experimental region (shown in Figure 10.11). This can be done in a SAS data step, and `proc optex` can be used to select a D-optimal subset for the general quadratic model.

Figure 10.11 *Experimental Region for Engine Experiment*

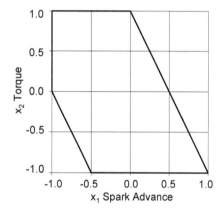

As an example of the second situation where not all possible combinations of the factor are feasible, consider a typical study in QSAR (Quantitative Structure Activity Relation) used in drug design. In this type of study, when a lead compound is discovered with desirable bio-activity, many derivatives of this compound are considered in an attempt to find one that will increase the desired effect. For example, Figure 10.12 shows the general structure of hydroxyphenylureas which have been shown to be active as anti-oxidants.

The compound has four receptor sites (R, R', R'' and R''') (shown in the figure) where different substituent atoms or molecules can be attached to modify the basic structure. With many possible substituent molecules that can be added at each receptor site, there is a large library of candidates that could be synthesized and tested for activity. Table 10.5 is an illustrative list that shows 36 different possible hydroxyphenylureas from a longer list considered by Deeb et al. (2008).

For each possible variant compound in the library there are several physical chemistry descriptors that can be calculated. In Table 10.5 three descrip-

Figure 10.12 *General Structure of Hydroxyphenylureas*

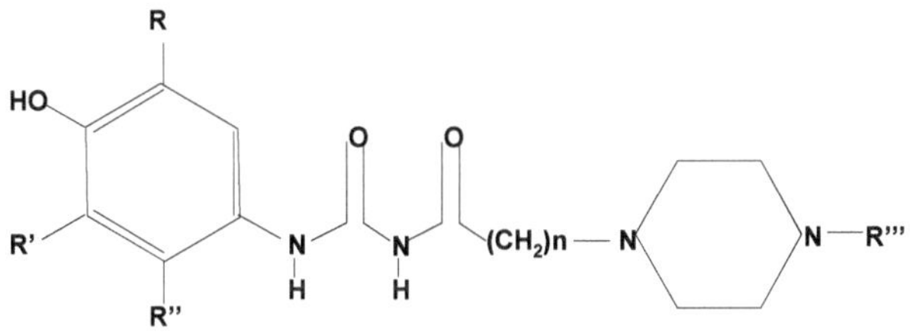

tors are shown: Hydration Energy (HE), Molecular Dipole Moment at the z-direction (DM_z), and Symmetry Index (S0K). The objective of a QSAR study would be to (1) synthesize a subset of the molecules from the large library and test them to determine their ability to scavenge oxygen FRs as measured by the binding constant $logK_{app}$, and (2) fit a model relating the response $logK_{app}$ to several physical chemistry descriptors (in this case, the variables HE, DM_z, and S0K). Once a model relating the response as a function of the independent variables has been found, the combination of the independent variables that is predicted to give the maximum response can be determined. Compounds in the large library that have values of the physical chemistry descriptors (independent variables) closest to those predicted to result in a maximum response can then be synthesized and tested for activity. In this way variant compounds that have increased activity are discovered.

In QSAR studies a standard response surface designs cannot be used because not all potential combinations of the independent variables are possible. Only a subset of the combinations of independent variables that exist in the large library can be chosen in a subset or experimental design to be tested. Again `proc optex` can be utilized to create a D-optimal subset of the larger library. In this case the list of candidates will be all the compounds in the library.

Table 10.5 *Library of Substituted Hydroxyphenylurea Compounds*

Compound	R	R'	R''	R'''	HE	DMz	S0K
1	H	H	H	CH_3	-12.221	-0.162	64.138
2	H	H	H	CH_2Ph	-14.015	-0.068	88.547
3	H	H	H	Ph	-14.502	0.372	85.567
4	H	H	H	$2CH_3OC_6H_4$	-14.893	1.035	96.053
5	H	OCH_3	H	CH_3	-12.855	1.091	74.124
6	H	OCH_3	H	CH_2Ph	-14.628	1.115	99.002
7	H	OCH_3	H	Ph	-15.123	1.554	96.053
8	H	OCH_3	H	$2CH_3OC_6H_4$	-15.492	2.221	106.607
9	H	OC_2H_5	H	CH_3	-11.813	1.219	77.02
10	H	OC_2H_5	H	CH_2Ph	-13.593	1.188	101.978
11	H	OC_2H_5	H	Ph	-14.088	1.621	99.002
12	CH_3	OC_2H_5	H	$2CH_3OC_6H_4$	-14.46	2.266	109.535
13	CH_3	H	CH_3	CH_3	-8.519	-0.56	71.949
14	CH_3	H	CH_3	CH_2Ph	-10.287	-0.675	96.6
15	CH_3	H	CH_3	Ph	-10.798	-0.134	96.62
16	CH_3	H	CH_3	$2CH_3OC_6H_4$	-11.167	0.418	104.047
17	H	H	H	CH_3	-12.245	-0.609	67.054
18	H	H	H	CH_2Ph	-13.98	-0.518	91.546
19	H	H	H	Ph	-14.491	-0.561	88.547
20	H	H	H	$2CH_3OC_6H_4$	-14.888	-1.478	99.002
21	H	OCH_3	H	CH_3	-11.414	-1.888	77.02
22	H	OCH_3	H	CH_2Ph	-13.121	-1.692	101.978
23	H	OCH_3	H	Ph	-13.66	-1.893	99.002
24	H	OCH_3	H	$2CH_3OC_6H_4$	-14.012	-2.714	109.535
25	H	OC_2H_5	H	CH_3	-10.029	-1.891	79.942
26	H	OC_2H_5	H	CH_2Ph	-11.74	-1.652	104.977
27	H	OC_2H_5	H	Ph	-12.329	-1.902	101.978
28	OCH_3	OC_2H_5	H	$2CH_3OC_6H_4$	-12.637	-2.762	112.492
29	OCH_3	OCH_3	H	CH_3	-12.118	-2.994	81.106
30	OCH_3	OCH_3	H	CH_2Ph	-13.892	-2.845	106.299
31	OCH_3	OCH_3	H	Ph	-14.456	-2.926	103.23
32	OCH_3	OCH_3	H	$2CH_3OC_6H_4$	-14.804	-3.78	113.856
33	CH_3	H	CH_3	CH_3	-9.209	-0.423	74.871
34	CH_3	H	CH_3	CH_2Ph	-10.97	-0.302	99.603
35	CH_3	H	CH_3	Ph	-11.488	-0.453	96.6
36	CH_3	H	CH_3	$2CH_3OC_6H_4$	-11.868	-1.322	107.01

The design must consist of a subset of the compounds in the library. If the general quadratic model will be used to fit the data, `proc optex` can be used as shown on the next page to create a design.

NON-STANDARD RESPONSE SURFACE DESIGNS

```
*Reads in the library of candidates;
data qsar;
input Compound HE DMz SOK ;
datalines;
1 -12.221 -0.162 64.138
   . . .
proc optex data=qsar coding=static; id Compound;
  model HE DMz SOK HE*HE DMz*DMz SOK*SOK HE*DMz HE*SOK DMz*SOK;
  generate  iter= 20 keep=5 ;
  output out =desgn1 ;
proc print data=desgn1;
run;
```

With three factors there are ten coefficients in the general quadratic model so the number of runs in the design must be at least $n = 10$. By default **proc optex** creates a design with $n = p + 10$ candidates where p is the number of coefficients in the model. The option **iter=20 keep=5** on the **generate** statement instructs **proc optex** to make 20 different searches for the D-optimal design with a random starting design and to keep the results of the best 5 searches. The **output** statement instructs **proc optex** to store the best design found in the SAS file **desgn1**. The results of the **proc print** statement are shown below.

Obs	HE	DMz	SOK	Compound
1	-15.492	2.221	106.607	8
2	-15.123	1.554	96.053	7
3	-14.888	-1.478	99.002	20
4	-14.804	-3.780	113.856	32
5	-14.804	-3.780	113.856	32
6	-14.502	0.372	85.567	3
7	-14.460	2.266	109.535	12
8	-13.121	-1.692	101.978	22
9	-12.637	-2.762	112.492	28
10	-12.637	-2.762	112.492	28
11	-12.221	-0.162	64.138	1
12	-12.221	-0.162	64.138	1
13	-12.118	-2.994	81.106	29
14	-12.118	-2.994	81.106	29
15	-11.813	1.219	77.020	9
16	-11.813	1.219	77.020	9
17	-11.167	0.418	104.047	16
18	-11.167	0.418	104.047	16
19	-8.519	-0.560	71.949	13
20	-8.519	-0.560	71.949	13

Here it can be seen that of the twenty design points selected there are only thirteen unique ones, and compounds 32, 28, 1, 29. 9, 16 and 13 are to be tested twice as replicates. If the library was longer than the 36 compounds listed in Table 10.5, there would be more choices of possible design points and therefore fewer replicates. Once the design is created with proc optex, it can be imported to the SAS ADX system using the File ▶ Import design option. The variance dispersion graph of the D-optimal design is shown in Figure 10.13 (created by SAS ADX in the coded design space where coded factor levels range from -1 to 1). There the design is neither rotatable or uniform precision, but it is the best it can be with the limited library of factor combinations available in Table 10.5.

Figure 10.13 *Variance Dispersion Graph for D-optimal QSAR Design*

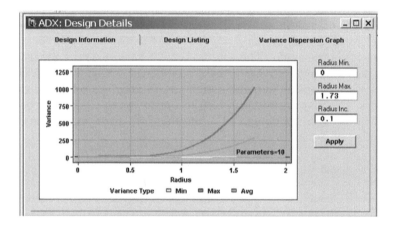

As an example of the third situation where the model is nonstandard or nonlinear, consider modeling the metabolism of tetracycline (a common antibiotic). Figure 10.14 shows a diagram of the two-compartment model useful in pharmacokinetics for representing the metabolism of drugs. When the drug is taken orally, γ_0 represents the initial concentration in the gastrointestinal (GI) tract. $\gamma_1(x)$ represents the concentration in the blood at time x. k_1 represents the rate constant at which the drug moves from the GI tract to the blood and k_2 is the rate constant at which the drug is eliminated from the blood.

Based on chemical kinetics, an equation relating the concentration in the blood at time x can be derived and is shown in Equation (10.7)

$$y = \gamma_1(x) = \gamma_0 [e^{-k_1(x-t_0)} - e^{-k_2(x-t_0)}] \tag{10.7}$$

where t_0 is the dead time. This is an example of a mechanistic model that is derived from physical principles. To determine the parameters in the model (γ_0, k_1, k_2, and t_0) a subject takes an oral dose of the drug, and small blood

Figure 10.14 *Diagram of Two-Compartment Model for Tetracycline Metabolism*

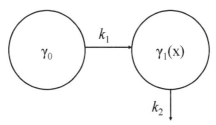

samples are taken at times x_i, $i = 1, \ldots n$. The response, or concentration of drug in the blood $\gamma_1(x_i)$, is determined at each sampling time or factor level and the model 10.7 is fit to the data. Once fit, the mechanistic model can be used to predict response values outside the experimental range of x_i, $i = 1, \ldots n$ since it is based on physical principles. This is not true for the general quadratic model (10.4) since it is just a Taylor's series approximation of the true equation about some point (x).

In this problem there is only one independent variable and one response, and there are four parameters in the model. Since the model is nonlinear in the parameters, a simple design of four equally spaced times (to sample blood) would not be efficient for estimating the parameters. Instead, a D-optimal search can be made to determine the design points.

The first step in creating a D-optimal design for a nonlinear model is to linearize the model about some point as described by Bates and Watts (2007). For example, linearizing $f(x, \gamma_0, k_1, k_2, t_0)$ about the point $(\gamma_0^*, k_1^*, k_2^*, t_0^*)$

$$
\begin{aligned}
f(x, \gamma_0, k_1, k_2, t_0) = f(x, \gamma_0^*, k_1^*, k_2^*, t_0^*) &+ (\gamma_0 - \gamma_0^*) \left(\frac{\partial f}{\partial \gamma_0} \right) \bigg|_{\gamma_0 = \gamma_0^*} \\
&+ (k_1 - k_1^*) \left(\frac{\partial f}{\partial k_1} \right) \bigg|_{k_1 = k_1^*} \\
&+ (k_2 - k_2^*) \left(\frac{\partial f}{\partial k_2} \right) \bigg|_{k_2 = k_2^*} \\
&+ (t_0 - t_0^*) \left(\frac{\partial f}{\partial t_0} \right) \bigg|_{t_0 = t_0^*}
\end{aligned} \quad (10.8)
$$

which is linear in the variables $\frac{\partial f}{\partial \gamma_0}$, $\frac{\partial f}{\partial k_1}$, $\frac{\partial f}{\partial k_2}$, and $\frac{\partial f}{\partial t_0}$ that are all functions of

x. For the compartment model in Equation (10.7)

$$\frac{\partial f}{\partial \gamma_0} = e^{-k_1(x-t_0)} - e^{-k_2(x-t_0)}$$

$$\frac{\partial f}{\partial k_1} = -\gamma_0(x-t_0)e^{-k_1(x-t_0)}$$

$$\frac{\partial f}{\partial k_2} = -\gamma_0(x-t_0)e^{-k_2(x-t_0)}$$

$$\frac{\partial f}{\partial t_0} = \gamma_0 k_1 e^{-k_1(x-t_0)} - \gamma_0 k_2 e^{-k_2(x-t_0)}$$

The strategy is to create a grid of candidates in the independent variable x, calculate the values of the four partial derivatives using initial guesses of the parameter values at each candidate point, and then use **proc optex** to select a D-optimal subset of the grid. In the SAS code below the initial parameter estimates $\gamma_0^0 = 2.65$, $k_1^0 = .15$, $k_2^0 = .72$ and $t_0^0 = 0.41$ were used.

```
*Create a candidate grid in partial derivatives;
data grid;
  k1=.15; k2=.72; gamma0=2.65; t0=.41;
  do x=1 to 25 by 1;
    dfdk1=-gamma0*exp(-k1*(x-t0))*(x-t0);
    dfdk2=gamma0*exp(-k2*(x-t0))*(x-t0);
    dfdgamma0=exp(-k1*(x-t0))-exp(-k2*(x-t0));
    dfdt0=gamma0*exp(-k1*(x-t0))*k1-gamma0*exp(-k2*(x-t0))*k2;
    output;
  end;
*Create D-optimal p point Design;
proc optex data=grid coding=orth;
  model dfdk1 dfdk2 dfdgamma0 dfdt0/noint;
  generate n=4;
  id x;
  output out=desnl; run;
proc print data=desnl; run;
```

Since there are $p = 4$ parameters in the model, **proc optex** will always generate at most $p = 4$ distinct design points, therefore the option **n=4** is used on the **generate** statement to eliminate any replicates in the file **desnl**. An experimenter can then replicate each design point as many times as necessary to get an estimate of experimental error. The code above results in the four distinct times for taking blood samples (1, 2, 5 and 13), and due to the form of the model and initial guesses of the parameters these points are far from equally spaced.

As stated by Bates and Watts (2007), the efficiency of the design for the nonlinear model will depend on the stage at which a researcher is in the investigation. If the form of the equation is known but not the parameter values, the design created by **proc optex** will be called an initial design. After

FITTING THE RESPONSE SURFACE MODEL WITH SAS   403

the data is collected from an initial design and the model is fit, the parameter estimates will be an improvement on the initial estimates. Therefore, the initial guesses of the parameter values can be replaced by the estimates and the initial design can be augmented with $p = 4$ additional design points using the `augment=desnl` option on the `generate` statement similar to the example shown in Section 6.5.2.

10.6 Fitting the Response Surface Model with SAS

This section will describe how to fit a response surface model using SAS `proc rsreg` and a nonlinear mechanistic model using SAS `proc nlin`.

10.6.1 Fitting a Linear Model and Checking for Curvature

If the factorial portion of a central composite design along with center points is completed in an initial block of experiments as described in Section 10.3.1, the linear model (10.6) should be fit to the data and checked for adequacy. If the linear model is adequate, the axial points and additional center points in the central composite design will not be necessary. To check for adequacy the residual sums of squares from model 10.6 should be partitioned into the portion due to pure replication and the portion due to quadratic departure from the model. This can be easily done using SAS `proc rsreg`. For example, the SAS code below reads in the first eleven data points from Table 10.1, and calls `proc rsreg` for the analysis.

```
data grout1;
input run Wat_Cem BlackL SNF Work;
datalines;
    1       0.33        0.12        0.08        109.5
    2       0.33        0.12        0.12        120.0
    3       0.33        0.18        0.08        110.5
    4       0.33        0.18        0.12        124.0
    5       0.35        0.12        0.08        117.0
    6       0.35        0.12        0.12        130.0
    7       0.35        0.18        0.08        121.0
    8       0.35        0.18        0.12        132.0
    9       0.34        0.15        0.10        117.0
   10       0.34        0.15        0.10        117.0
   11       0.34        0.15        0.10        115.0
proc rsreg data=grout1;
  model work=Wat_cem BlackL SNF;
run;
```

Normally the model 10.6 could be fit using SAS `proc glm` with the model statement `model work=Wat_cem BlackL SNF Wat_cem*BlackL Wat_cem*SNF BlackL*SNF;`. This would produce a report where the sums of squares for the corrected total has 10 degrees of freedom, the model has 6 degrees of freedom and the error has 4 degrees of freedom. Since **proc rsreg** was created for fitting the general quadratic model, it is not necessary to define the cross-product terms in the model statement. They are created by default. In addition **proc rsreg** creates quadratic terms. When the axial points are left out of the central composite design, all the quadratic terms become confounded. However, the third table of results produced by **proc rsreg** shows a test of the confounded quadratic terms. The results below show this table that resulted from the **proc rsreg** command above.

Regression	DF	Type I Sum of Squares	R-Square	F Value	Pr > F
Linear	3	465.125000	0.9137	80.31	0.0023
Quadratic	1	37.878788	0.0744	19.62	0.0214
Crossproduct	3	0.250000	0.0005	0.04	0.9859
Total Model	7	503.253788	0.9886	37.24	0.0065

Residual	DF	Sum of Squares	Mean Square
Total Error	3	5.791667	1.930556

The sums of squares for Linear account simultaneously for the three linear terms in the model, and the sums of squares for Crossproduct account simultaneously for the three interaction terms in the model. The sums of squares for Quadratic accounts for the departure from linearity, and the sums of squares for Residual represents the pure error sums of squares due to replicate center points. When fitting model 10.6 with **proc glm** the sums of squares for Quadratic would be combined with the error sums of squares resulting in four degrees of freedom for error. The F-test on the quadratic term is a test of the adequacy of the linear model. Since it is significant at the $\alpha = 0.05$ level in the above table, it indicates that the quadratic departure from the linear model is significant, and thus the linear model (10.6) is not adequate for representing the first eleven data points in Table 10.1. Therefore, if the experiments shown in Table 10.1 were to be run in two blocks, it would be necessary to run the second block that includes the axial points. This test for adequacy of the linear model can also be made using SAS ADX.

10.6.2 Fitting the General Quadratic Model

proc rsreg is most useful for fitting the general quadratic model and analyzing the fitted surface. For example, consider fitting the general quadratic model to the trebuchet data in Table 10.3. The commands to read in the data and fit the model using **proc rsreg** are shown on the next page.

FITTING THE RESPONSE SURFACE MODEL WITH SAS

```
data trebuchet;
input run  A  B  C   y;
datalines;
1       4      10      2.5     33
2       8      10      2.5     85
 . . .
proc rsreg;
  model y=A B C/lackfit;
run;
```

Notice that the uncoded factor levels are read in the statements above. `proc rsreg` automatically codes and scales the factor levels and prints the coefficients for the model in coded and uncoded factor levels. The first table in the output on the next page shows the values that `proc rsreg` used to code and scale the factor levels. The second table in the results shows the summary statistics from the fit. The Response mean is simply the average of the response data and the Coefficient of Variation is $\hat{\sigma}$ = Root MSE divided by the Response Mean.

The third table of results shows the ANOVA partition shown in the last section. In this example it can be seen that the linear quadratic and cross-product terms in the model are all significant. The error sums of squares are also partitioned into the pure error sums of squares and the lack of fit sums of squares as a result of the `lackfit` option on the model statement. By doing this, a simple F-test can be made for the adequacy of the quadratic model. For example, the responses at the three center points in Table 10.3 are 88, 91 and 91. Therefore, the pure sums of squares due to replication are:

$$\sum_{i=1}^{3} y_i^2 - \frac{(\sum_{i=1}^{3} y_i)^2}{3} = 88^2 + 91^2 + 91^2 - \frac{(88+91+91)^2}{3} = 6.0$$

The difference in the total error sums of squares (20.5) and the pure error sums of squares is called the lack of fit sums of squares. An F-test of the lack of fit sums of squares is a test of the adequacy of the quadratic model. In this case it is clearly insignificant, indicating that predictions made from the general quadratic model for this experiment can be considered just as accurate as running additional experiments, as long as no lurking variables change before additional experiments can be run.

The RSREG Procedure
Coding Coefficients for the Independent Variables

Factor	Subtracted off	Divided by
A	6.000000	2.000000
B	15.000000	5.000000
C	2.500000	0.500000

Response Surface for Variable y

Response Mean	82.466667
Root MSE	2.024846
R-Square	0.9975
Coefficient of Variation	2.4554

Regression	DF	Type I Sum of Squares	R-Square	F Value	Pr > F
Linear	3	7299.000000	0.9011	593.41	<.0001
Quadratic	3	351.483333	0.0434	28.58	0.0014
Crossproduct	3	428.750000	0.0529	34.86	0.0009
Total Model	9	8079.233333	0.9975	218.95	<.0001

Residual	DF	Sum of Squares	Mean Square	F Value	Pr > F
Lack of Fit	3	14.500000	4.833333	1.61	0.4051
Pure Error	2	6.000000	3.000000		
Total Error	5	20.500000	4.100000		

Parameter	DF	Estimate	Standard Error	t Value	Pr > |t|	Parameter Estimate from Coded Data
Intercept	1	-35.875000	38.386494	-0.93	0.3929	90.00000
A	1	35.500000	4.339859	8.18	0.0004	19.75000
B	1	2.600000	1.735944	1.50	0.1945	19.75000
C	1	-24.500000	22.803874	-1.07	0.3317	-11.50000
A*A	1	-2.343750	0.263441	-8.90	0.0003	-9.37500
B*A	1	-0.625000	0.101242	-6.17	0.0016	-6.25000
B*B	1	-0.055000	0.042151	-1.30	0.2488	-1.37500
C*A	1	4.750000	1.012423	4.69	0.0054	4.75000
C*B	1	2.700000	0.404969	6.67	0.0011	6.75000
C*C	1	-13.500000	4.215052	-3.20	0.0239	-3.37500

The RSREG Procedure

Factor	DF	Sum of Squares	Mean Square	F Value	Pr > F
A	4	3691.519231	922.879808	225.09	<.0001
B	4	3465.980769	866.495192	211.34	<.0001
C	4	1372.557692	343.139423	83.69	<.0001

When the lack of fit test is significant, it indicates that the general quadratic model is not adequate for prediction. This could be due to the fact that the experimental region is so large that the quadratic model does not provide a good approximation to the true response function over the entire region, or due to outliers, or extreme nonlinearity in certain corners of the experimental region where the approximate function does not fit. These conditions can be detected by making residual plots to check the least squares assumptions or by fitting the model using a "robust" technique such as M-estimators, for details and discussion see Lawson (1982).

The fourth table shows the estimates of individual coefficients in the general quadratic model along with their standard errors, t-values, and P-values. The estimated coefficients for the model fit to the uncoded factor levels are shown in the third column and the coefficients for the model fit to the coded factor levels are shown in the last column. Some statisticians advocate dropping the insignificant terms from the model and refitting, but this is not possible with `proc rsreg`. Other statisticians advocate retaining all terms in the general quadratic model, whether significant or not, since the individual model terms have no physical meaning. A compromise is to eliminate a model factor if all linear, quadratic and interaction terms involving that factor are insignificant. By eliminating a factor the response surface model is reduced from p dimensions to $p-1$ dimensions.

`proc rsreg` automatically performs a test on the factors as shown in the fifth table in the output. Each term has 4 degrees of freedom representing the linear quadratic and interaction term involving each factor. All the factors in this model are seen to be significant. If any factor had been found to be insignificant, it could be eliminated from the `proc rsreg` model statement and the response surface equation would simplify from three dimensions to two.

Block terms can also be included in the `proc rsreg` model statement. If, for example, the experiments on cement grout shown in Table 10.1 were completed in two blocks, where the first block consisted of the factorial plus center points shown in runs 1-11 and the second block consisted of the axial and center points shown in runs 12-20. The commands to read the data and fit the model are shown at the top of the next page.

The block term is entered first in the model, and the `covar=1` option on the model statement tells `proc rsreg` not to create square and interaction terms for the first term in the model. The block factor will then account for any average differences between the first and second group of experiments.

If the response surface design is created in SAS ADX or imported to SAS ADX, the general quadratic model can also be fit using the point and click interface. ADX provides the table of model coefficients and standard errors and by clicking on the Model ▶ Fit Details, after fitting the model ADX automatically creates a table showing the lack of fit test. However, ADX does

408   RESPONSE SURFACE DESIGNS

not provide a table partitioning the sums of squares to test each factor as
proc rsreg does.

```
data grout2;
input run block Wat_Cem BlackL SNF Work;
datalines;
    1     1   0.33       0.12       0.08       109.5
    .  .  .
   11     1   0.34       0.15       0.10       115.0
   12     2   0.32318    0.15000    0.10000    109.5
   13     2   0.35682    0.15000    0.10000    132.0
    .  .  .
   20     2   0.34000    0.15000    0.10000    117.0
proc rsreg data=grout2;
   model work=block Wat_cem BlackL SNF/covar=1;
run;
```

10.6.3 Fitting a Nonlinear Mechanistic Model

The general quadratic model is linear in the coefficients (β's), and the $\mathbf{X'X}$ is full rank so the least square estimates are obtained by solving the normal equations through straightforward matrix inversion. When the model is nonlinear like the two-compartment model shown in Equation (10.7), the least squares estimates are more difficult to obtain and must be found by iterative numerical techniques. However, SAS proc nlin makes this all transparent to the user. For example, to fit the two-compartment model to the data from Wagner (1967) shown in Table 10.6, use the following commands.

```
data Tetracycline;
input x conc;
datalines;
1  .7
 . . .
16 .3
proc nlin data=Tetracycline best=5;
  parms k1 =.05 to .25 by .05
        k2 =.25 to .75 by .05
        gamma0 = 0 to 20 by 5
        t0 = 0 to 1 by .25;
   model conc=gamma0*(exp(-k1*(x-t0))-exp(-k2*(x-t0)));
   output out=results p=pconc;
proc print data=results; run;
```

The parms statement specifies a range of initial guesses for the parameter values. proc nlin begins by making a grid search over this space of guesses

Table 10.6 *Tetracycline Concentration in Plasma Over Time*

Time (hr)	Tetracycline Conc. (μg/ml)
1	0.7
2	1.2
3	1.4
4	1.4
6	1.1
8	0.8
10	0.6
12	0.5
16	0.3

to find the parameter set with the minimum sum of squares for residuals. The `best=5` option prints the best five combinations from the grid search. Next, `proc nlin` uses the best parameter guesses from the grid search as initial estimates and proceeds by default (other options are available) to use the Gauss-Newton method with numerical derivatives to find the least squares estimates. The model statement defines the equation for the response and the output statement can be used to output predicted values and residuals. A portion of the resulting output summarizing the fit is shown on the next page. The first table is an ANOVA breakdown of the model and error sums of squares. The second table gives the parameter estimates and their standard errors. The distribution of the parameter estimates depends on the nonlinear function. Therefore no t-tests are performed, and only approximate confidence intervals are given.

```
                          The NLIN Procedure
                    Sum of       Mean                Approx
Source          DF  Squares      Square     F Value  Pr > F

Model            4  8.3900       2.0975     1044.01  <.0001
Error            5  0.0100       0.00201
Uncorrected Total 9 8.4000


                                           Approx
Parameter   Estimate   Std Error  Approximate 95%
                                  Confidence Limits
k1          0.1488     0.0144     0.1118     0.1858
k2          0.7158     0.1261     0.3917     1.0398
gamma0      2.6496     0.3645     1.7128     3.5865
t0          0.4122     0.0949     0.1682     0.6563
```

10.7 Determining Optimum Operating Conditions

10.7.1 Contour Plots

Once a response surface model has been fit to the data from an experiment, there are several ways to identify the optimum conditions. If there are only two factors, the simplest method is to make a contour plot or a three-dimensional plot of the response surface. If there are more than two factors in the model, several two-dimensional slices through the experimental region can be made by holding some factors constant and making a contour or 3D plots with respect to the remaining two factors at each slice.

`proc rsreg` automatically generates a panel of contour plots if the ODS graphics is on. The commands below show that by simply adding the option `plots=(surface)`, the panel of contour plots shown in Figure 10.15 is created.

```
ods graphics on;
proc rsreg data=trebuchet plots=(surface);
  model y=A B C;
run;
ods graphics off;
```

Figure 10.15 *Contour Plots of Predicted Trebuchet Distance*

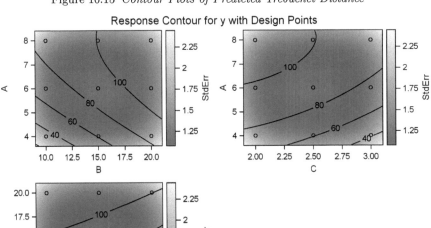

Value when Fixed: C=2.5, B=15, A=6

DETERMINING OPTIMUM OPERATING CONDITIONS

In this panel of plots, the first plot (upper left) is made at a slice with C: missile weight held constant at its mid-value of 2.5, and factors A: arm length and B: counter weight on the axis. The lines represent contours of predicted values from the quadratic model, the small circles show the location of the design points where data was collected, and the shading represents the standard error of the predicted value. The near circular shading illustrates the fact that the Box-Behnken design is near uniform precision. The remaining two plots show factors A and C on the axis with factor B held constant at its mid-value of 15, and factors B and C on the axis with A held constant at its mid-value of 6.

By changing the option from `plots=(surface)` on the `proc rsreg` statement above to `plots=(surface(3D))`, the panel of three-dimensional surface plots shown in Figure 10.16 is created.

Figure 10.16 *Three-Dimensional Surface Plots of Predicted Trebuchet Distance*

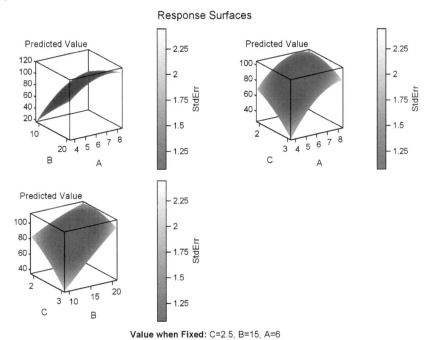

The contour or 3D plots in the panel can be unpacked and the constant values of the factors not shown on the axis can also be specified. The **unpack** option on the plots statement shown in the commands at the top of the next page separates the plots, and the surface option (**at (CounterWeight = 20)** specifies the constant value of B: counter weight.

RESPONSE SURFACE DESIGNS

```
data decrip; set trebuchet;
rename y=Distance A=ArmLength B=CounterWeight C=MissileWeight ;
proc print; run;
ods graphics on/imagefmt=png border=off;
ods listing style=journal2 image_dpi=250;
proc rsreg data=decrip plots(unpack)=(surface (at (
    CounterWeight = 20)));
  model Distance=ArmLength CounterWeight MissileWeight;
run;
ods graphics off;
```

The resulting contour plot at a slice with B: counterweight = 20lbs. is shown in Figure 10.17. In this plot it can be seen that with a counterweight of 20lbs., the predicted maximum distance will be achieved using a missile weight of about 2.375 ounces and an arm length of about 7.5 inches. The predicted distance at these conditions is about 115 feet, with a standard error of about 1.5 feet.

Figure 10.17 *Contour Plot of Predicted Trebuchet Distance with Counterweight=20lbs.*

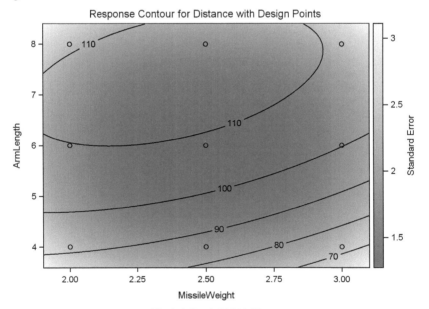

10.7.2 Canonical Analysis

When the response surface is a hilltop or a valley with a distinct maximum or minimum within the experimental region, the exact factor coordinates of the maximum or minimum can be determined by simultaneously setting the derivatives of the fitted equation with respect to each factor equal to zero and solving the simultaneous set of equations. This is useful when there are more than two factors and the maximum or minimum would be difficult to identify with multiple contour plots. The vector solution (of factor settings) to the simultaneous set of homogeneous equations can be expressed in matrix terms as

$$\mathbf{x_0} = -\hat{\mathbf{B}}^{-1}\hat{\mathbf{b}}/2 \tag{10.9}$$

where $\hat{\mathbf{B}}$ and $\hat{\mathbf{b}}$ are the least squares estimates of the matrix and vector of regression coefficients defined in Equation (10.5). This solution is actually called the stationary point because it could be a maximum, a minimum, or a saddle point as shown in Figure 10.1. To determine what the solution is, it is useful to express the response surface Equation (10.5) in a canonical form with the origin translated to $\mathbf{x_0}$ and the axis rotated (as shown in Figure 10.18).

Figure 10.18 *Representation of Canonical System with Translated Origin and Rotated Axis*

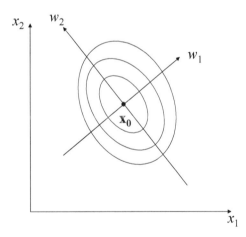

Letting $\mathbf{z} = \mathbf{x} - \mathbf{x_0}$, Equation (10.5) can be written as

$$y = y_0 + \mathbf{z}'\mathbf{Bz}, \tag{10.10}$$

where $y_0 = \mathbf{x_0}\mathbf{b} + \mathbf{x_0'}\mathbf{Bx_0}$. This translates the origin to $\mathbf{x_0}$. Through the orthogonal rotation $\mathbf{z} = \mathbf{Mw}$, the second order part of Equation (10.10) can be reduced to a linear combination of squares of the rotated variables w_i given by

$$\mathbf{z}'\mathbf{Bz} = \mathbf{w}'\mathbf{M}'\mathbf{BMw} = \lambda_1 w_1^2 + \lambda_2 w_2^2 + \cdots + \lambda_k w_k^2 \tag{10.11}$$

If all the coefficients λ_i are negative, it indicates the stationary point is a maximum. If all the coefficients are positive, it indicates the stationary point is a minimum, and if there are mixed signs of the coefficients it indicates the stationary point is a saddle point. The matrix $\mathbf{M}$ is the matrix containing the eigenvectors of $\mathbf{B}$ as the columns, and the coefficients λ_i are the eigenvalues of $\mathbf{B}$.

SAS `proc rsreg` automatically calculates the eigenvalues and eigenvectors of the least squares estimates of $\hat{\mathbf{B}}$. For example, for the cement grout experiment shown in Table 10.1, `proc rsreg` produces the following table of results following the output summarizing the fitted model.

```
Canonical Analysis of Response Surface Based on Coded Data

                          Critical Value
            Factor        Coded           Uncoded

            Wat_Cem       -1.148152       0.320688
            BlackL        -0.108888       0.144512
            SNF           -0.996193       0.066528

      Predicted value at stationary point: 107.031254


                                    Eigenvectors
   Eigenvalues      Wat_Cem          BlackL              SNF

      4.259335      0.193581         0.346182         0.917979
      4.008004      0.892674         0.326026        -0.311193
      3.561788     -0.407014         0.879696        -0.245915

              Stationary point is a minimum.
```

The coordinates of the stationary point are given in both coded and uncoded units, and the predicted value at the stationary point, $\mathbf{y_0}$, is given. Next the eigenvalues and eigenvectors of $\hat{\mathbf{B}}$ are listed. Since all the eigenvalues are positive, the stationary point is a minimum.

For this particular problem, knowing the factor settings to achieve a minimum response is not useful. The goal of experimentation was to find the factor combinations that would maximize the workability of the cement grout. In general, when seeking a maximum or minimum response within the experimental region, the stationary point will only be useful if (1) the stationary point is within the experimental region, and (2) the stationary point is of the type sought (i.e., maximum or minimum). If the stationary response is outside the experimental region, or if it is a saddle point or minimum when seeking the maximum, another method must be used to identify the optimum.

10.7.3 Ridge Analysis

Another method of finding the optimum within the experimental region is to use *ridge analysis* (Hoerl, 1959; Draper, 1963). This method seeks to find the maximum or minimum of $y = xb + x'Bx$ subject to the constraint that $x'x = R^2$, or that the coordinates of the optimum are on a radius R from the origin in coded units. The solution is obtained in a reverse order using Lagrange multipliers. The resulting optimal coordinates are found to be the solution to the equation

$$(B - \mu I_k)x = -b/2. \qquad (10.12)$$

To solve this system, choose a value for μ, insert it into Equation (10.12), and solve for the vector of coded factor settings x. Once the coded factor settings have been found, the radius of the solution is $R = \sqrt{\sum x_i^2}$. Inserting values of μ larger than the largest eigenvalue of the matrix B will result in a solution for the maximum response on the radius R, and inserting values of μ smaller than the smallest eigenvalue of the matrix B will result in a solution for the minimum response on the radius R. Trial and error is required to find the solution on a specific radius. However, the **proc rsreg ridge** statement does all the work.

As an example, consider again finding the maximum distance the trebuchet can toss an object within the experimental ranges defined in Table 10.3. This was done approximately using a contour plot in Section 10.7.1. The stationary point analysis, printed by **proc rsreg** for this data, showed the stationary point was a saddle point that was outside the experimental region. Therefore, the maximum within the experimental region has to be on the extreme boundary of the experimental region. Since the experimental region for a Box-Behnken design (that was used for this experiment) is a sphere with radius $\sqrt{2} = 1.412$ in coded units, the **ridge** statement is added to the commands shown earlier to give rise to the list shown below.

```
ods graphics on;
proc rsreg data=trebuchet plots=ridge;
  model y=A B C;
  ridge max radius =.112 to 1.412 by .1;
run;
ods graphics off;
```

This asks **proc rsreg** to find the maximum on each radius from .112 to 1.412 in increments of .1. The resulting table that **proc rsreg** prints is shown on the next page.

It can be seen that the predicted maximum on a radius R increases as R increases along the path shown in Figure 10.19, and that it finally reaches a maximum of 117.24 feet at the boundary of the experimental region (1.412 in coded units). At that point the factor levels in uncoded units are shown to be exactly A: arm length = 6.8 inches; B: counterweight = 21.8 lbs.; and C: missile weight = 2.5 oz.

The RSREG Procedure

Estimated Ridge of Maximum Response for Variable y

			Uncoded Factor Values		
Coded Radius	Estimated Response	Standard Error	A	B	C
0.112000	93.251100	1.164507	6.142638	15.378158	2.479159
0.212000	95.935691	1.153212	6.262527	15.738522	2.461603
0.312000	98.422609	1.136306	6.373760	16.123890	2.445469
0.412000	100.722064	1.115928	6.474549	16.537574	2.431299
0.512000	102.847076	1.095470	6.562998	16.982072	2.419754
0.612000	104.813794	1.079975	6.637396	17.458066	2.411552
0.712000	106.641367	1.076429	6.696633	17.963426	2.407340
0.812000	108.351128	1.093482	6.740605	18.492884	2.407520
0.912000	109.965128	1.140099	6.770345	19.038880	2.412130
1.012000	111.504443	1.223394	6.787766	19.593315	2.420851
1.112000	112.987838	1.346900	6.795165	20.149231	2.433125
1.212000	114.431106	1.510481	6.794765	20.701640	2.448320
1.312000	115.847040	1.711613	6.788459	21.247518	2.465839
1.412000	117.245771	1.946882	6.777737	21.785363	2.485175

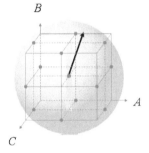

Figure 10.19 *Path of Maximum Ridge Response Through Experimental Region*

When the ODS graphics is on, the `plots=ridge` option on the `proc rsreg` statement also produces a graph of the predicted response and factor coordinates as a function of the distance from the origin (in coded units) as shown in Figure 10.20 at the top of the next page. This is a common way to present and summarize the results of a ridge analysis.

10.7.4 Nonlinear Optimization

Canonical analysis and ridge analysis work well for finding the optimum of a response surface if the model is the general quadratic model and the experimental region is spherical in coded units. If the response surface equation is not a general quadratic model or the experimental region is irregular, the exact optimum can always be found using more general numerical methods. SAS `proc nlp`, which uses a variety of nonlinear programming methods, can

DETERMINING OPTIMUM OPERATING CONDITIONS

be used to quickly find a numerical optimum. This procedure automatically uses the method that is best suited to the problem at hand.

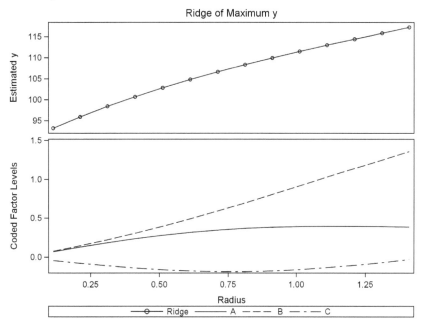

Figure 10.20 *Factor Levels To Achieve Maximum Predicted Value*

As an example of using `proc nlp` to find a response surface optimum, consider finding the maximum of a nonlinear mechanistic response surface. In a chemical reaction the reactant is transformed to the product at rate constant k_1 and the product degrades to decomposition byproducts at rate constant k_2. From first-order kinetics, the concentration of the product, p, at time t is given by the equation

$$p = -R_0[e^{-k_2 t} - e^{-k_2 t}]k_1/(k_1 + k_2) \tag{10.13}$$

where the rate constants are found by the Arrhenius equations in terms of the temperature in Kelvin, T.

$$k_1 = a \times \exp[-b(1/T - 1/400]$$
$$k_2 = c \times \exp[-d(1/T - 1/400]$$

If experiments had been conducted in the region where $t = 0$ to 25 hours and $T = 375$ to 425 degrees Kelvin and the coefficients in the model were estimated to be $\hat{R}_0 = 132.0$, $\hat{a} = 0.523$, $\hat{b} = 9847$, $\hat{c} = 0.20$ and $\hat{d} = 12327$, then the `proc nlp` statements below can be used to find the maximum product concentration. The **parms** statement gives starting values of the factors in the center of the experimental region. The **bounds** statement gives the bounds on the experimental region.

```
proc nlp;
  max p;
  parms time Temp= 12.5 400;
  bounds 0 <= time <=25,
         375 <=Temp <=425;
         k1=.523*exp(-9847*((1/Temp)-(1/400)));
         k2=.2*exp(-12327*((1/Temp)-(1/400)));
         p=-132*(exp(-k1*time)-exp(-k2*time))*k1/(k1-k2);
run;
```

The results show the maximum product concentration, p, within the experimental region is $\hat{p} = 82.88$ at $t = 18.16$ hours and $T = 375$ degrees Kelvin.

Since this problem only has two factors (time and temperature) the results can be visualized in the contour plot shown in Figure 10.21 which displays the maximum at the boundary of the experimental region.

Figure 10.21 *Maximum Product Concentration*

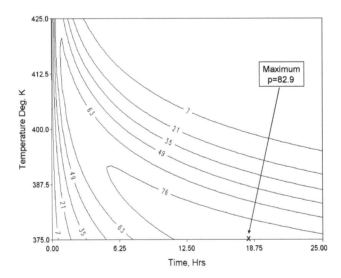

A rough approximation to numerical optimization can be made by doing a grid search. Create a fine grid of the experimental region in a SAS data step and then evaluate the response surface function at each grid point to get

DETERMINING OPTIMUM OPERATING CONDITIONS

predicted values. Finally, sort the resulting data file to locate the optimum predicted response within the experimental region. The numerical optimization option in SAS ADX uses a grid search to locate the optimum.

10.7.5 Multi-Response Optimization

Often the optimum factor settings will depend on more than one response. For example, optimum drug regimens depend on efficacy and side effects, optimum formulation of food products depends on consumer taste and health issues, and optimum manufacturing processes depend on product quality and cost of production. One way to determine optimal factor settings when there is more than one response is to use constrained numerical optimization. As an example, consider a simple problem with only two responses.

Dry bleach products that are safe for colored fabrics consist of sodium perborate in combination with an activator that enhances the bleaching ability of perborate in low-temperature wash water. Response surface experiments were run to determine the optimal ratio of a new activator-to-perborate and the amount of perborate in a dry bleach product. The three factors in the study were the wash temperature, the ratio of activator to perborate, and the amount of perborate (measured in PPM of active oxygen). The range of the three factors studied were $70 \leq$ Temp ≤ 140, $0.5 \leq$ Ratio ≤ 1.5, and $5.0 \leq$ AOPPM ≤ 65.0. The response was the percent tea stain removal.

From the results of the experiment, the general quadratic model (in actual factor levels) relating the response to the factors was estimated to be

$$tsr = -226.38 + 3.375(\text{Temp}) + 86.5(\text{Ratio}) + 2.646(\text{AOPPM})$$
$$- .0128(\text{Temp}^2) - 17.5(\text{Ratio}^2) - .0121(\text{AOPPM}^2)$$
$$- .3857(\text{Ratio})(\text{Temp}) - .0126(\text{AOPPM})(\text{Temp}) - .0333(\text{AOPPM})(\text{Ratio})$$

The cost per wash (in cents) was a known function of the amount of activator and perborate in the product and is given by the equation

$$\text{cost} = .8313 + 1.27(\text{Ratio}) + .37(\text{Ratio})(\text{AOPPM})$$

The goal was to maximize the tea stain removal for a cost of ten cents per wash or less. This optimum can be found by `proc nlp` by including the `nlincon` (non-linear constraint) statement as shown at the top of the next page.

Running this code shows that the maximum percent tea stain removal that can be expected for the activator being tested is 40.27% at Temp=103 degrees, Ratio=0.55 and AOPPM=41.9.

```
proc nlp;
  max tsr;
  parms   Temp Ratio AOPPM= 120 1 35;
  bounds 70 <= Temp  <=140,
         0.5 <=Ratio<=1.5,
      5.0 <=AOPPM<=65.0;
  nlincon cost<=10;
  tsr=-226.38+3.375*Temp+86.5*Ratio+2.646*AOPPM-
       .0128*(Temp**2)-17.5*(Ratio**2)-.0121*(AOPPM**2)
      -.3857*Ratio*Temp-.0126*AOPPM*Temp-.0333*AOPPM*Ratio;
  cost=.8313  +1.27*Ratio+.37*Ratio*AOPPM;
run;
```

When there are multiple (k) responses, another method of optimization that is sometimes useful is to combine the responses into one desirability function proposed by Derringer and Suich (1980). The idea is to convert each response into an individual desirability function $0 \leq d_i(y_i) \leq 1$, where this function is 1 if the response, y_i obtains its most desirable value, and zero when the response is in an unacceptable range. Next, the overall desirability function is defined as

$$(d_1 \times d_2 \times \cdots \times d_k)^{(1/k)} \qquad (10.14)$$

and this function can be maximized using numerical optimization or grid search. If it is desirable to obtain a target value for a response $\hat{y}_i$, for example, the individual desirability function can be defined as

$$d_i = \begin{cases} 0 & \text{if } \hat{y}_i < L \\ \left(\frac{\hat{y}_i - L}{T - L}\right)^r & \text{if } L \leq \hat{y}_i \leq T \\ \left(\frac{\hat{y}_i - U}{T - U}\right)^s & \text{if } T \leq \hat{y}_i \leq U \\ 0 & \text{if } \hat{y}_i > U \end{cases} \qquad (10.15)$$

where $L \leq T \leq U$, T is the target value for the response, and the regions where $\hat{y}_i < L$ or $\hat{y}_i > U$ are undesirable. The powers r and s control how critical the researcher feels it is to be close to the target. If it is desirable for one of the responses to be maximized, modify Equation (10.15) by setting $T = U$ and defining $d_i = 1$ for $\hat{y}_i \geq U$. If it is desirable to minimize a response, modify Equation (10.15) by setting $T = L$ and defining $d_i = 1$ for $\hat{y}_i \leq L$, and $d_i = (U - \hat{y}_i)/(U - L)^s$ if $L \leq y_i \leq U$.

The desirability function could be maximized with proc nlp, but SAS ADX has a tool to automatically maximize the desirability function with a graphical interface. In an input form the user supplies the lower (L), upper (U) and target (T) values for each response and the powers r and s. However, much care should be exercised when choosing these values, otherwise less than desirable conditions can result.

10.8 Blocked Response Surface (BRS) Designs

When experimental units are not homogeneous, it is always advantageous to group them into more homogeneous blocks and use a blocked experimental design. Earlier it was mentioned that the three-factor central composite designs could be run in two blocks; the factorial portion plus three center points in one block, and the axial portion and remaining center points in the other block. This is actually an example of an incomplete block design, like those described in Chapter 7, since not all the treatment combinations are represented in each block. It is possible to create an incomplete block design out of any response surface design so that the coefficients can be estimated more precisely.

Standard response surface designs such as the central composite designs (CCDs) and Box-Behnken designs (BBDs) have been blocked in a way that the blocks are orthogonal to the coded factor levels, squares of coded factor levels, and interactions among coded factor levels. In this way the least squares estimates of the parameters of the general quadratic model are not correlated with the block effects and these standard designs are 100% D_s efficient as described in Section 7.8.1. The example described earlier, where the three-factor central composite design was blocked into two blocks with block sizes 11 and 9, is not 100% D_s efficient. In order to achieve 100% efficiency, some additional restrictions must be made in terms of the number of blocks, the block sizes, and the axial radius. This will affect the distribution of the variance of a predicted value over the design region, and a compromise must be made. Table 10.7 shows the number of blocks and block sizes for orthogonally blocked CCDs and BBDs with $k = 2$ to 5 factors.

Table 10.7 *Number of Blocks and Block Sizes for Orthogonally Blocked CCD and BBD Designs*

Number Factors	CCD No. Blocks	CCD Block Sizes	BBD No. Blocks	BBD Block Sizes
2	2	7,7		
3	3	6, 6, 8		
4	3	10, 10, 10	3	9, 9, 9
5	2	22, 11	2	23, 23

These designs can be created easily with the SAS ADX tool. It will automatically use the optimal block size and axial radius. To create a blocked response surface design, check the Blocking box on the ADX Select Design Template, as shown in Figure 10.22, and you will see the blocked designs available.

Figure 10.22 *Design Selection Template with Blocking Box Checked*

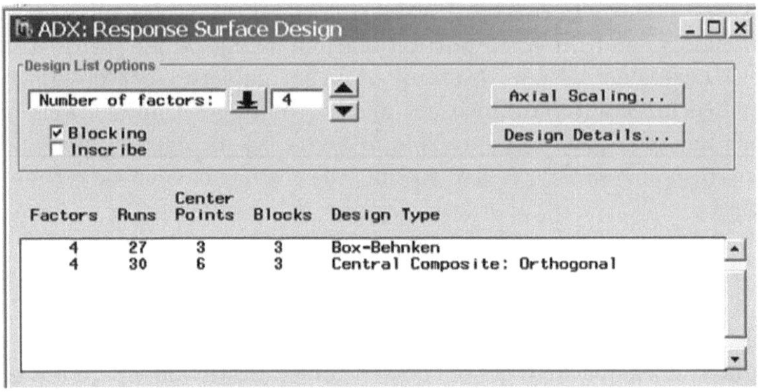

```
ADX: Response Surface Design                              _ □ x
 Design List Options
   Number of factors:  ▼ 4  ▲            Axial Scaling...
      ☑ Blocking                         Design Details...
      ☐ Inscribe

                      Center
   Factors   Runs   Points   Blocks   Design Type
      4       27       3        3     Box-Behnken
      4       30       6        3     Central Composite: Orthogonal
```

The number of blocks and block sizes available for orthogonally blocked CCD and BBD designs may be too restrictive for some applications. For example, Gilmour and Trinca (2000) presented the data from an experiment to determine how the characteristics of pastry dough depended upon parameters of the mixing process. In the experiment three factors were varied, the feed flow rate (FR), the initial moisture content (MC) and the screw speed (SS). One of the responses (y) measured light reflectance of the resulting dough in particular bands of the spectrum in order to see how the color of the dough was affected by the mixing parameters. Only four runs or experiments could be conducted on one day and the experimenters anticipated day to day variation. None of the designs in Table 10.7 have blocks of size four. Thus the experimenters ran the modified design shown in Table 10.8, with the coded factor values $x_1 = (FR - 37.5)/7.5$, $x_2 = (MC - 21)/3$, and $x_3 = (SS - 350)/50$.

This is an example of a face-centered cube design that is a special case of a central composite design with the axial points (in coded units) pulled in to the face of the cube at ± 1. The 2^3 factorial portion of this design was replicated and there were six center points. The twenty-eight runs of this design were blocked into seven blocks (days) of four runs in a way that the main effects are orthogonal to the blocks. However, as Goos (2002) pointed out, the quadratic and interaction terms of the general quadratic model are not orthogonal to blocks and a better way of blocking the twenty-eight experiments design can be achieved using a D_s optimal design. This can be done using **proc optex**.

First create the face centered cube design using the %adxccd macro as shown in the code below Table 10.8 on the next page. This macro creates all twenty-eight design points shown in Table 10.8 and stores them in the SAS file **fccd** with the factors labeled **t1**, **t2**, and **t3**.

Table 10.8 *Design and Response for Pastry Dough Experiment*

Block	x_1	x_2	x_3	y
1	-1	-1	-1	12.92
1	-1	1	1	13.91
1	1	-1	1	11.66
1	1	1	-1	14.48
2	-1	-1	1	10.76
2	-1	1	-1	14.41
2	1	-1	-1	12.27
2	1	1	1	12.13
3	-1	1	-1	14.22
3	0	-1	0	12.35
3	1	0	0	13.5
3	0	0	1	12.54
4	1	-1	1	10.55
4	-1	0	0	13.33
4	0	1	0	13.84
4	0	0	-1	14.19
5	-1	-1	-1	11.46
5	1	1	1	11.32
5	0	0	0	11.93
5	0	0	0	11.63
6	-1	-1	1	12.2
6	1	1	-1	14.78
6	0	0	0	14.94
6	0	0	0	14.61
7	-1	1	1	12.17
7	1	-1	-1	11.28
7	0	0	0	11.85
7	0	0	0	11.64

```
%adxgen
%adxff
%adxcc
%adxinit
%adxccd(fccd,3,16,6,1.0,1)
```

Next use **proc optex** to group the runs into seven blocks of four runs as shown on the next page. The **noexchange** option on the **blocks** statement tells **proc optex** to keep all the runs in the file **fccd** in the design and arrange them into seven blocks of four in a way to maximize the D_s efficiency.

To create a blocked response surface design that is a subset of the possible runs in a file of candidates, replace the file **fccd** in the commands above

```
proc optex data=fccd coding=orthcan seed=2347;
   model t1 t2 t3 t1*t1 t1*t2 t1*t3 t2*t2 t2*t3 t3*t3;
   blocks structure=(7)4 noexchange niter=1000 keep=10;
   generate n=28 augment=fccd;
   output out=bpastry blockname=block; run;
proc print data=bpastry;
run;
```

by the file of candidates and remove the noexchange option of the blocks statement. For example, to block the experiments in the QSAR example in Section 10.5 so that they could be run in five different labs (that may have differing measurements of bio-activity) modify the code from Section 10.5 as shown below.

```
proc optex data=qsar; id Compound;
   model HE DMz SOK HE*HE DMz*DMz SOK*SOK HE*DMz HE*SOK DMz*SOK;
   blocks structure=(5)4;
   generate  n=20 iter=200 keep=5;
   output out =desgn1 ;
proc print data=desgn1; run;
```

To analyze a blocked design with more than two blocks using `proc rsreg`, create dummy variables for the blocks as shown below for the pastry dough experiment shown in Table 10.8.

```
data pastry;
input Block x1 x2 x3 y;
b1=(Block=1);
b2=(Block=2);
b2=(Block=2);
b3=(Block=3);
b4=(Block=4);
b5=(Block=5);
b6=(Block=6);
b7=(Block=7);
datalines;
1 -1 -1 -1 12.92
 . . .
proc rsreg;
   model y1=b1-b6 x1 x2 x3/covar=6;
run;
```

10.9 Response Surface Split-Plot (RSSP) Designs

When response surface designs contain factors from different process steps or some factors are hard to vary, it may be costly or inconvenient to completely randomize the runs in a standard response surface design. For ex-

RESPONSE SURFACE SPLIT-PLOT (RSSP) DESIGNS

ample, consider a simple response surface design to identify the baking temperature and baking time that would result in the optimal moist cake. The design and data are shown in Table 10.9 where $x_1 = $ (bake temp. $- 350)/25$, $x_2 = $ (bake time $- 31.5)/4$, and y is the reading from a moisture tester.

Table 10.9 *Data for Cake Baking Experiment*

Oven run	x_1	x_2	y
1	-1	-1	2.7
1	-1	1	2.5
1	-1	0	2.7
2	1	-1	2.9
2	1	1	1.3
2	1	0	2.2
3	0	-1	3.7
3	0	1	2.9
4	0	0	2.9
4	0	0	2.8
4	0	0	2.9

The design is a face-centered cube design with three levels for each factor and three center points. The experiments were not performed in a random order. Once the oven temperature was set and the oven was preheated, as many as four cakes could be baked at once. Performing the experiments in a completely random order would have required eleven oven runs, but by baking as many as three cakes at the same time (as shown in the table) only four oven runs were required. The oven runs were performed in a random order with the two or three cakes for each run placed in random positions in the oven. The cakes in each run were withdrawn from the oven at different baking times except for oven run 4 when all cakes were baked for 31.5 minutes.

By running the experiments in this way, a response surface split-plot (RSSP) type design results, where x_1, the baking temperature, is the whole-plot factor and x_2, baking time, is the subplot factor. Since the same subplot treatment combinations are not run in each whole plot, Letsinger *et al.* (1996) show that the least squares estimates of the model parameters, which would be obtained by `proc rsreg`, will be unsatisfactory if the ratio of the whole-plot to subplot error variances is greater than one. They recommend using REML estimators, which can be computed in `proc mixed`.

Combining the coefficients for the linear, quadratic, and interaction terms in the same vector, the general quadratic model for a completely randomized response surface design can be written in matrix form as

$$\mathbf{y} = \mathbf{X}\beta + \epsilon \tag{10.16}$$

When the response surface design is run as a split plot, like the example in

Table 10.9, the general quadratic model can be written in the form

$$\mathbf{y} = \mathbf{X}\beta + \omega + \epsilon \tag{10.17}$$

where β is the vector of regression coefficients for the whole-plot and subplot effects, ω is a vector of random whole-plot errors, and ϵ is a vector of random subplot errors. It is assumed that $\omega + \epsilon$ has zero mean and variance covariance matrix given by $\mathbf{\Sigma} = \sigma^2\mathbf{I} + \sigma_\omega^2\mathbf{J}$, where σ_ω^2 and σ^2 are the variances of the whole-plot and subplot experimental units and

$$\mathbf{J} = \begin{pmatrix} \mathbf{1}_1\mathbf{1}'_1 & 0 & \cdots & 0 \\ 0 & \mathbf{1}_2\mathbf{1}'_2 & \cdots & 0 \\ & & \ddots & \\ 0 & 0 & \cdots & \mathbf{1}_m\mathbf{1}'_m \end{pmatrix}$$

The length of $\mathbf{1}_i$ is n_i the number of subplot runs within the ith whole plot. The least squares estimates of β is $(\mathbf{X}'\mathbf{X})^{-1}\mathbf{X}'\mathbf{y}$, while the best linear unbiased estimate of β is

$$\hat{\beta} = (\mathbf{X}'\mathbf{\Sigma}^{-1}\mathbf{X})^{-1}\mathbf{X}'\mathbf{\Sigma}^{-1}\mathbf{y} \tag{10.18}$$

and its variance covariance matrix is given by

$$Var(\hat{\beta}) = (\mathbf{X}'\mathbf{\Sigma}^{-1}\mathbf{X})^{-1} \tag{10.19}$$

σ^2, σ_ω^2 must be known in order to compute the best linear unbiased estimate of β. However, estimates of σ^2, σ_ω^2, and the large sample best linear unbiased estimates that is found by substituting $\hat{\sigma}^2$, and $\hat{\sigma}_\omega^2$ into Equation (10.18) can be obtained using the REML method that was described in Chapter 5 and is available in **proc mixed**. The commands to fit the general quadratic model to the data in Table 10.9 using **proc mixed** are shown below.

```
data CakeBake;
input wp x1 x2 y;
datalines;
1 -1 -1 2.7
 . . .
proc mixed;
   class wp;
   model y=x1 x2 x1*x1 x2*x2 x1*x2/
      solution;
   random wp;
run;
```

A comparison of the parameter estimates obtained by the least squares method and REML is shown in Table 10.10. Here it can be seen that the estimated linear main effects are the same for both methods, but the standard errors are much smaller for the subplot effect and larger for the whole-plot effect using the more correct REML method. The other estimates differ and again the tests for the subplot effects are more sensitive using REML and the tests for whole-plot effects are less sensitive. The REML estimated whole-plot and subplot error variances are shown at the bottom on the right side of the table. Letsinger et al. (1996) showed that when $\sigma_\omega^2/\sigma^2 < 0.25$ the least squares estimates will be reasonably close to the REML estimates and can be used in practice. However, since σ_ω^2/σ^2 is usually not known, the REML method should be used first.

Table 10.10 *Comparison of Least Squares and REML Estimates for Split-Plot Response Surface Experiment*

	Least Squares (proc rsreg)			REML (proc mixed)		
Factor	$\hat{\beta}$	$s_{\hat{\beta}}$	P-value	$\hat{\beta}$	$s_{\hat{\beta}}$	P-value
intercept	2.979	0.1000	<.001	3.1312	0.2667	0.054
x_1	-0.2500	0.0795	0.026	-0.2500	0.2656	0.399
x_2	-0.4333	0.0795	0.003	-0.4333	0.0204	<.001
x_1^2	-0.6974	0.1223	0.002	-0.6835	0.3758	0.143
x_2^2	0.1526	0.1223	0.016	-0.0965	0.0432	0.089
$x_1 x_2$	-0.3500	0.0973	0.268	-0.3500	0.0250	<.001

$$\hat{\sigma}_\omega^2 = 0.1402, \ \hat{\sigma}^2 = 0.0025$$

From a practical point of view, it would be advantageous to be able to use the least squares estimates, since proc rsreg that uses the least squares method, can also automatically compute the canonical analysis, the ridge analysis, or produce predicted values over a grid for use in contour plotting.

Vining et al. (2005) proved an equivalence theorem that shows the least squares estimates of all of the regression coefficients in a split-plot response surface design will be the same as the REML estimates if (1) the design is balanced (in that each whole plot contains the same number of subplots), (2) the subplot designs are orthogonal (although not necessarily the same), and (3) the axial runs for the subplots are run in a single whole plot. In addition, to get separate estimates of the variances of whole-plot and subplot experimental units at least two whole plots of center points on all factors should be included. Designs that have these properties will be referred to here as *estimation equivalent split-plot response surface designs* or EESPRS. No EESPRS is possible when there is only one subplot factor because property (2) cannot be satisfied.

When there are two or more subplot factors, EESPRS designs can be easily obtained by modifying standard response surface designs. For example, consider the two central composite designs run in a split-plot arrangement shown in Figure 10.23. If there is one whole-plot factor, A, and two subplot factors, P and Q, the standard central composite design with three center points shown on the left can be grouped into five whole plots by sorting on the whole-plot factor A as exhibited in the figure. Since the number of runs in each whole-plot are not equal for the design on the left, it is not balanced, and it is not

428   RESPONSE SURFACE DESIGNS

Figure 10.23 *Comparison of Split-plot CCDs*

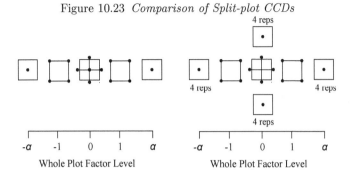

Standard CCD in SP				EESPRS CCD			
Whole Plot	A	P	Q	Whole Plot	A	P	Q
1	$-\alpha$	0	0	1	$-\alpha$	0	0
2	$-$	$-$	$-$	1	$-\alpha$	0	0
2	$-$	$+$	$-$	1	$-\alpha$	0	0
2	$-$	$-$	$+$	1	$-\alpha$	0	0
2	$-$	$+$	$+$	2	$-$	$-$	$-$
3	0	$-\alpha$	0	2	$-$	$+$	$-$
3	0	α	0	2	$-$	$-$	$+$
3	0	0	$-\alpha$	2	$-$	$+$	$+$
3	0	0	α	3	0	0	0
3	0	0	0	3	0	0	0
3	0	0	0	3	0	0	0
3	0	0	0	3	0	0	0
4	$+$	$-$	$-$	4	0	$-\alpha$	0
4	$+$	$+$	$-$	4	0	α	0
4	$+$	$-$	$+$	4	0	0	$-\alpha$
4	$+$	$+$	$+$	4	0	0	α
5	α	0	0	5	0	0	0
				5	0	0	0
				5	0	0	0
				5	0	0	0
				6	$+$	$-$	$-$
				6	$+$	$+$	$-$
				6	$+$	$-$	$+$
				6	$+$	$+$	$+$
				7	α	0	0
				7	α	0	0
				7	α	0	0
				7	α	0	0

RESPONSE SURFACE SPLIT-PLOT (RSSP) DESIGNS

an EESPRS design. By removing the center point from the whole plot that has the axial points for the subplot factors P and Q, adding replicates to the whole plots that contain the axial points for the whole-plot factor A, and adding replicates in the whole plots that contain center points, the design on the right becomes balanced. Since the columns for the two subplot factors P and Q are orthogonal (i.e., $\sum_i p_i q_i = 0$) within each whole plot, and all the axial points for the subplot factors are in the same whole plot, this is an EESPRS design. Duplicate whole plots (3 & 5) containing all center points make it possible to estimate σ_ω^2 and σ^2 using the REML method in `proc mixed`.

While the least squares and REML estimates of the regression coefficients are the same when an EESPRS design is used, the standard errors of the coefficients will not be the same, because the covariance matrix of the least squares estimates, $\sigma^2(\mathbf{X'X})^{-1}$, is not the same as the covariance of the REML estimates $(\mathbf{X'\hat{\Sigma}^{-1}X})^{-1}$. Therefore, if an EESPRS design is used, `proc mixed` should be used to make hypothesis tests about the parameters (in order to determine if the model can be simplified), but `proc rsreg` can be used for doing a canonical analysis, ridge analysis or contour plots.

In a balanced EESPRS design, the number of whole plots is m and the number of subplots within each whole plot is n, making the total number of runs $N = nm$. The number of factors is $k = k_1 + k_2$, where k_1 is the number of whole-plot factors and k_2 is the number of subplot factors. In an estimation equivalence central composite design, the number of runs in each whole plot will be $2 \times k_2$, since one whole plot must contain all the axial runs for the subplot factors. Table 10.11 on the next page shows some EESPRS designs that can be constructed by modifying standard central composite designs. The designation $k_1(k_2)$ in the first column refers to the number of whole and subplot factors.

The design in the first row of Table 10.11 is the design in the right side of Figure 10.22 that has one whole-plot factor and two subplot factors. In this design, the factorial portion (whole plots 2 and 6) has the four treatment combinations of a full 2^2 factorial in the subplot factors, randomized to the four subplots. In other designs shown in Table 10.11, the subplots in the factorial portion of the design may contain a fractional factorial in the subplot factors, or a fractional factorial in the subplot factors augmented by two center points as indicated in the last column in the table.

Using Table 10.11 as a guide, these designs can be created easily in the SAS data step. The example code below Table 10.11 on the next page creates the design in row four of the table which has two whole-plot factors and two subplot factors in a non-random order. In this code each whole plot is created in a separate data step, and all the whole plots are combined in the last data step.

Table 10.11 *EESPRS CCD Designs*

$k_1(k_2)$	k	m	n	N	Factorial Subplot Design
1(2)	3	7	4	28	2^2
1(3)	4	7	6	42	$2^{3-1}(I = \pm PQR) + 2$ cp
1(4)	5	7	8	56	$2^{4-1}(I = \pm PQRS)$
2(2)	4	11	4	44	2^2
2(3)	5	11	6	66	$2^{3-1}(I = \pm PQR) + 2$ cp
2(4)	6	11	8	88	$2^{4-1}(I = \pm PQRS)$
3(2)	5	18	4	72	2^2
3(3)	6	18	6	108	$2^{3-1}(I = \pm PQR) + 2$ cp

```
*Create 2(2) CCD design in SAS;
*Factorial portion;
data wp1;
  Wp=1; A=-1; B=-1;
  do P=-1 to 1 by 2;
   do Q=-1 to 1 by 2;
     output;
    end;
   end;
data wp2;
  Wp=2; A=1; B=-1;
  do P=-1 to 1 by 2;
   do Q=-1 to 1 by 2;
     output;
    end;
   end;
data wp3;
  Wp=3; A=-1; B=1;
  do P=-1 to 1 by 2;
   do Q=-1 to 1 by 2;
      output;
    end;
   end;
data wp4;
  Wp=4; A=1; B=1;
  do P=-1 to 1 by 2;
   do Q=-1 to 1 by 2;
     output;
    end;
   end;
*Subplot Axial portion;
data wp5;
  Wp=5; A=0; B=0;
  input P Q @@;
  datalines;
-2 0 2 0 0 -2 0 2
```

RESPONSE SURFACE SPLIT-PLOT (RSSP) DESIGNS

```
* Whole Plot Axial portion;
data wp6;
  Wp=6; A=-2; B=0;
  do i = 1 to 4;
    P=0; Q=0; output;
  end;
  drop i;
data wp7;
  Wp=7; A=2; B=0;
  do i = 1 to 4;
    P=0; Q=0; output;
  end;
  drop i;
data wp8;
  Wp=8; A=0; B=-2;
  do i = 1 to 4;
    P=0; Q=0; output;
  end;
  drop i;
data wp9;
  Wp=6; A=0; B=2;
  do i = 1 to 4;
    P=0; Q=0; output;
  end;
  drop i;
*Center point portion;
data wp10;
  Wp=10;
  do i = 1 to 4;
    A=0; B=0; P=0; Q=0; output;
  end;
  drop i;
data wp11;
  Wp=11;
  do i = 1 to 4;
    A=0; B=0; P=0; Q=0; output;
  end;
drop i;
data SPDs; set wp1 wp2 wp3 wp4 wp5 wp6 wp7 wp8 wp9 wp10 wp11;
proc sort ; by WP;
proc print data=SPDs; run;
```

Vining et al. (2005) presented an example of the central composite EESPRS design with two whole-plot factors and two subplot factors. In this example, an engineer was studying the effects of two hard-to-change factors (A, zone 1 furnace temperature, and B, zone 2 furnace temperature), and two easy-to-change factors (P, amount of binder in formulation, and Q, grinding speed of the batch) upon the strength of ceramic pipe. In this example, the axial points are at ± 1 (i.e., face-centered cube design). The design in coded levels and resulting strength measurements are shown in Table 10.12. This design used three whole plots that consisted of nothing but center points.

Analysis of this data (left for an exercise) will show the regression coefficients

432   RESPONSE SURFACE DESIGNS

for the general quadratic model obtained by the method of least squares and REML are the same, although the standard errors will differ.

Table 10.12 *Design and Strength Measures for Ceramic Pipe Experiment*

WP	A	B	P	Q	y	WP	A	B	P	Q	y
1	-1	-1	-1	-1	80.40	7	0	-1	0	0	80.07
1	-1	-1	1	-1	71.88	7	0	-1	0	0	80.79
1	-1	-1	-1	1	89.91	7	0	-1	0	0	80.20
1	-1	-1	1	1	76.87	7	0	-1	0	0	79.95
2	1	-1	-1	-1	87.48	8	0	1	0	0	68.98
2	1	-1	1	-1	84.49	8	0	1	0	0	68.64
2	1	-1	-1	1	90.84	8	0	1	0	0	69.24
2	1	-1	1	1	83.61	8	0	1	0	0	69.2
3	-1	1	-1	-1	62.99	9	0	0	-1	0	78.56
3	-1	1	1	-1	49.95	9	0	0	1	0	68.63
3	-1	1	-1	1	79.91	9	0	0	0	-1	74.59
3	-1	1	1	1	63.23	9	0	0	0	1	82.52
4	1	1	-1	-1	73.06	10	0	0	0	0	74.86
4	1	1	1	-1	66.13	10	0	0	0	0	74.22
4	1	1	-1	1	84.45	10	0	0	0	0	74.06
4	1	1	1	1	73.29	10	0	0	0	0	74.82
5	-1	0	0	0	71.87	11	0	0	0	0	73.6
5	-1	0	0	0	71.53	11	0	0	0	0	73.59
5	-1	0	0	0	72.08	11	0	0	0	0	73.34
5	-1	0	0	0	71.58	11	0	0	0	0	73.76
6	1	0	0	0	82.34	12	0	0	0	0	75.52
6	1	0	0	0	82.20	12	0	0	0	0	74.74
6	1	0	0	0	81.85	12	0	0	0	0	75.00
6	1	0	0	0	81.85	12	0	0	0	0	74.90

Box-Behnken designs can also be easily modified to create EESPRS designs. Simply sort the Box-Behnken design by the whole-plot factors and determine the number of subplots in each whole plot to match the block with the maximum number. Add center points with respect to the subplot factors in whole plots where the whole-plot factor levels are not at the center value, and add at least two whole plots consisting of entirely center points. Table 10.13 shows some EESPRS designs that can be constructed by modifying standard Box-Behnken designs. Using Table 10.13 as a guide, these designs can also be created easily in the SAS data step. Example SAS code to create the design in row two of the table which has one whole-plot factor and three subplot factors in a nonrandom order is shown below the table on the next page.

Table 10.13 *EESPRS BBD Designs*

$k_1(k_2)$	k	m	n	N
1(2)	3	5	4	20
1(3)	4	7	6	42
1(4)	5	7	8	56
2(2)	4	11	4	44
2(3)	5	13	6	78

```
*Create Box-Behnken SP 1(3);
*factorial portion of BBD;
data o1;
  do A=-1 to 1 by 2;
   do P=-1 to 1 by 2;
     Q=0; R=0; output;
   end;
  end;
data o2;
  do A=-1 to 1 by 2;
   do Q=-1 to 1 by 2;
     P=0; R=0; output;
   end;
  end;
data o3;
  do A=-1 to 1 by 2;
   do R=-1 to 1 by 2;
     P=0; Q=0; output;
   end;
  end;
data o4;
  do P=-1 to 1 by 2;
   do Q=-1 to 1 by 2;
     A=0; R=0; output;
   end;
  end;
data o5;
  do P=-1 to 1 by 2;
   do R=-1 to 1 by 2;
     A=0; Q=0; output;
   end;
  end;
data o6;
  do Q=-1 to 1 by 2;
   do R=-1 to 1 by 2;
     A=0; P=0; output;
   end;
  end;
*non-zero level of A portion;
data c; set o1 o2 o3;
proc sort; by A;
data c; set c; by A;
  retain WP 0;
  if first.A then WP=WP+1;
```

```
*zero level of A portion;
data ic; set o4(in=in4) o5(in=in5) o6(in=in6);
  if in4 then WP=3;
  if in5 then WP=4;
  if in6 then WP=5;
data au1;
  WP=3;
  do i =1 to 2;
    A=0; P=0; Q=0; R=0; output;
  end;
  drop i;
  WP=4;
  do i =1 to 2;
    A=0; P=0; Q=0; R=0; output;
  end;
  drop i;
  WP=5;
  do i =1 to 2;
    A=0; P=0; Q=0; R=0; output;
  end;
  drop i;
*center point whole plots;
data wp6;
  do i=1 to 6;
    WP=6; A=0; P=0; Q=0; R=0; output;
  end;
  drop i;
data wp7;
  do i=1 to 6;
    WP=7; A=0; P=0; Q=0; R=0; output;
  end;
  drop i;
data sp ; set c ic au1 wp6 wp7;
proc sort ; by WP;
proc print data=sp; run;
```

In cases where the designs shown in Tables 10.11 and 10.13 are not flexible enough in terms of the number of subplots per whole plot, Goos and Vandebroek (2001) and Goos (2002) describe a computer algorithm to create optimal split-plot designs.

10.10 Review of Important Concepts

In response surface methods the factors can be varied over a continuous range, and the primary goal is not to determine the factor effects, but to rather map the relationship between the response and the factor levels so that optimal factor settings can be determined. If the model for the relationship between the factors and response is unknown, a general quadratic model is used as an approximation. Experimental designs are constructed to equalize the variance of a predicted value over the design region, rather than to give maximum power for detecting factor effects. A graphical tool called the variance dispersion graph can be used to determine how well a design meets the objective of equalizing the variance of a predicted value.

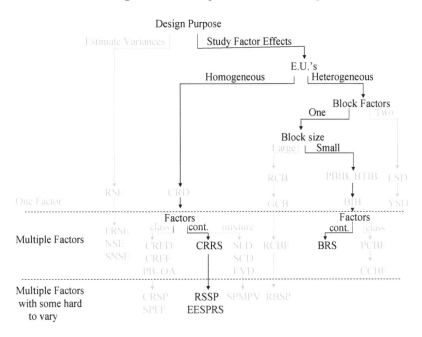

Figure 10.24 *Design Selection Roadmap*

Figure 10.23 illustrates the situations where the response surface designs shown in this chapter should be used. Standard response surface designs (CRRS) such as central composite, Box-Behnken, small composite and hybrid designs are most easily constructed in SAS ADX. The model can be fit and analyzed using SAS ADX or exported to a SAS file where the more powerful `proc rsreg` can be used. This procedure can fit the general quadratic model as well as determine the factor settings that produce the maximum or minimum response using canonical analysis or ridge analysis. Contour plots are also useful for visually identifying optimal factor settings.

Nonstandard response surface designs are used when a non-standard model (such as a nonlinear model) is used or the experimental region is irregular or consists of a discrete set of candidates. SAS `proc optex` can be used to find a design in these situations. Nonlinear models can be fit with the SAS `proc nlin`. Finding the optimal factor settings in irregular design regions or with a nonlinear model can be accomplished using nonlinear programming with SAS `proc nlp`.

When experimental units are not homogeneous, blocked response surface (BRS) designs should be utilized. SAS ADX can be used to create orthogonally blocked response surface designs for central composite and Box-Behnken designs. When the number of blocks and block sizes are not flexible enough for the problem at hand, `proc optex` can be used to create D_s optimal blocked response surface designs.

When there are hard-to-vary factors which make it inconvenient or costly to completely randomize a response surface design, a split-plot response surface (RSSP) design results. In general, the coefficients in the quadratic model for a split-plot design should be determined using the REML method of estimation described in Chapter 5. However, if the design satisfies the three conditions described by Vining *et al.* (2005), the coefficients (but not the standard errors) estimated by least squares will be the same as the coefficients estimated by the REML method. Therefore, when using this type of EESPRS design `proc mixed` should be used for hypothesis tests concerning the model coefficients for purposes of model simplification, but `proc rsreg` can be used for exploration of the fitted surface using canonical or ridge analysis.

10.11 Exercises

1. Create a central composite design for two factors using SAS ADX or data step commands.

   (a) Create the uniform precision CCD and store the design along with random numbers (simulated response) in a SAS file.

   (b) Create a fine grid of points over the unit square and augment your CCD design with this grid (leaving the simulated response variable as missing values).

   (c) Use `proc reg` to fit the general quadratic model and output the standard errors of the predicted value over the unit square.

   (d) Make a contour plot of the standard errors of the predicted values over the unit square.

   (e) Repeat (a) through (d) for a face-centered cube design (i.e., CCD with axial points at ± 1).

   (f) Based on the contour plots you made, which design do you prefer? State the reason.

2. Compare characteristics of standard response surface designs for four factors.

   (a) Create a uniform precision central composite design, a Box-Behnken design, a small composite design, and a hybrid design in SAS ADX.

   (b) Compare the variance dispersion graphs for each design, the number of levels and total number of runs.

   (c) Choose one of the designs you have created, export it to a SAS data set and use it to print a randomized data collection for the design.

3. Following up on the experiments described in Section 6.4, AlmeidaeSilva et al. (2003) ran a response surface design to find the optimum nutrient levels and fermentation time for producing biomass by fermentation of eucalyptus hemicellulosic hydrolyzate. The data is shown in the table on the next page. RB stands for the rice bran (g/l), AS is the ammonium sulfate (g/l), FT is fermentation time in hours, and Biomass is the mass of microorganisms produced. This was a face-centered cube design (fccd).

   (a) Fit the general quadratic model using `proc rsreg`. Check the model to see if it is adequate. Determine whether the model can be simplified by dropping any factor from the model.

   (b) Do the canonical analysis, ridge analysis or numerical optimization to determine the conditions that produce the maximum biomass within the cubical experimental region.

   (c) If the model can be simplified repeat (a) and (b) with the simpler model. Do the coordinates of the optimum change much? If the model has been simplified by eliminating one factor, make a contour plot of the response versus the two factors left in the model, holding the eliminated factor at its mid-level.

RB	AS	FT	Biomass
10	0	72	3.83
30	0	72	5.71
10	2	72	6.74
30	2	72	5.13
10	0	96	5.55
30	0	96	7.76
10	2	96	12.45
30	2	96	12.47
10	1	84	11.54
30	1	84	9.79
20	0	84	7.13
20	2	84	10.1
20	1	72	6.29
20	1	96	13.02
20	1	84	10.66
20	1	84	10.15
20	1	84	10.97

4. Consider performing a response surface experiment using the paper helicopters described in exercise 1 of Chapter 2 and Sections 3.2 and 3.3.

   (a) Create a response surface design of your choice by varying wing length between 3.5 inches and 7 inches and the width of the helicopters from 2.25 inches to 5.5 inches. Randomize your list.

   (b) Construct and test your helicopters by measuring their flight time with a stopwatch.

   (c) Fit the general quadratic model to your data, and test the adequacy of your model.

   (d) Using canonical analysis, ridge analysis or numerical optimization, determine the helicopter dimensions that you predict would result in the maximum flight time. Make a contour plot to visually verify your predicted optimum.

   (e) Actually construct a helicopter of the dimensions you predict to be optimal. Test this helicopter by dropping it repeatedly and recording the flight time. Are the flight times comparable to your predictions?

5. Consider conducting a response surface experiment with the wooden catapult described in exercise 2 of Chapter 3. The three factors that can be varied on the catapult are the start angle, stop angle and pivot height. Each factor has three possible levels. A full 3^3 factorial would allow estimation of all coefficients in the general quadratic model, but it would require 27 experiments with no replicates to check the model lack-of-fit.

   (a) Construct a face-centered cube (central composite) design for this problem.

EXERCISES

(b) Construct a Box-Behnken design for this problem.

(c) Since there are only 27 candidate design points, construct a 10-run design that is a D-optimal subset of the 27 candidates. Suggest one of the points that could be replicated three times to test lack of fit.

(d) Use SAS ADX to make variance dispersion graphs for each design. What design do you prefer? Why?

(e) Use a catapult to actually run the experiments you have planned.

(f) Fit the general quadratic model to the data, using proc rsreg and check the adequacy of the model.

(g) Using numerical optimization, contour plots or simply evaluating your prediction equation over the 27 possible settings, determine the settings (one of the 27 possible) that you predict would result in a distance closest to exactly 12 feet.

(h) Test the settings you determined in (g). Is the distance close to your predictions?

6. Consider the experimental region shown in Figure 10.11, where $-1 < x_1 < 1$, $-1 < x_2 < 1$, and $x_2 \geq -2x_1 - 2$, $x_2 \leq -2x_1 + 1$.

   (a) Construct a grid $\pm \frac{1}{2}$ of candidate points in the (x_1, x_2) design region.

   (b) Construct a D-optimal subset of the candidate points for fitting the general quadratic model.

7. Consider the nonlinear model $y = f(x_1, x_2) = \beta_1 e^{\beta_2 x_1} - \beta_3 e^{\beta_4 x_2}$.

   (a) Linearize the model by taking the partial derivatives $\frac{\partial f}{\partial \beta_i}$, $i = 1, 4$.

   (b) Construct a grid of candidate points in the experimental region range $1 \leq x_1 \leq 7$, $1 \leq x_2 \leq 7$.

   (c) Use proc optex to construct a D-optimal nonlinear design for this model using the initial estimates $\beta_1 = 10.0$, $\beta_2 = -0.5$, $\beta_3 = 5.0$, and $\beta_4 = -2.0$.

8. Consider the data in Table 10.1.

   (a) Fit the general quadratic model using proc rsreg as shown in Section 10.6.1.

   (b) Use canonical analysis or ridge analysis to determine the conditions that result in the maximum workability of the cement grout within the spherical region $-1.73 \leq x_i \leq 1.73$ for $i = 1$, to 3 in coded units.

9. Consider the model for the chemical reaction in Section 10.7.4.

   (a) Use the data shown on the next page that was obtained by running 20 experiments to fit model 10.13.

   (b) Make a contour plot of predicted concentration over the experimental region.

time	temperature	Conc.
2.5	380	26.62
2.5	287.5	36.00
2.5	395	38.83
2.5	405	40.99
5.0	380	43.20
5.0	387.5	50.60
5.0	395	51.12
5.0	405	27.11
10.0	380	57.71
10.0	387.5	51.71
10.0	395	46.34
10.0	405	7.13
2.5	420	12.82
5.0	420	1.61
10.0	420	0.02
20.0	380	54.32
20.0	387.5	30.47
20.0	395	9.24
20.0	405	0.38
25.0	425	0.00

(c) Use nonlinear programming to find the time and temperature (within the experimental limits) that maximize the predicted concentration of product. Does this result agree with your contour plot?

10. Create a blocked response surface design for four factors.

   (a) Use SAS ADX to create an orthogonally blocked central composite design. How many runs are in this design, and how many runs are in each block?

   (b) Use SAS ADX to create an orthogonally blocked Box-Behnken design. How many runs are in this design, and how many runs are in each block?

   (c) Supposing that only 6 runs can be made in each block, take the central composite design you created in (a) and use `proc optex` to block it into five blocks of six runs each.

   (d) Take the Box-Behnken design you created in (b), and add three center points, and then use `proc optex` to block the runs into five blocks of six runs each.

   (e) Import the designs you created in (c) and (d) into SAS ADX, create the variance dispersion graphs, and compare them to the variance dispersion graphs of the orthogonally blocked designs you created in (a) and (b). Do you lose much by reducing the block size?

11. Fit the general quadratic model to the data from the pastry dough experiment given in Table 10.8.

EXERCISES

(a) Use `proc rsreg` to fit the model including dummy variables for the block effects and check the adequacy of the model.

(b) Use canonical analysis or ridge analysis to determine the maximum response within the cubical experimental region.

12. A study was conducted to find the optimum ration of chloride and alkaline reserves (AR) in a product that was designed to add alkalinity to livestock rations. A small composite design was planned with the factors $x_1 = 1/\sqrt{chloride}$ and $x_2 = \ln(AR)$, but one combination of x_1 and x_2, had to be dropped from the study, and formulation problems with the other combinations resulted in an irregular design. Forty-eight cattle were blocked into 8 blocks based on initial weight and hip height. Three cattle had to be dropped from the study due to chronic health problems that were not related to the treatment. The goal was to find the combination of x_1 and x_2 that resulted in the highest average daily gain in weight (ADG) over the course of the study. The data from the study are shown on the next page.

    (a) Analyze the data with `proc rsreg` including dummy variables for the blocks, and check the assumptions of the fitted model.
    (b) Find the predicted maximum ADG.
    (c) Import the data to SAS ADX and make a variance dispersion graph. How far from rotatable is the irregular design?
    (d) Make a contour plot to visually confirm the coordinates of your predicted optimum.

13. Find the optimal moist cake for the data in Table 10.9.

    (a) Do the canonical analysis and make a contour plot of the moisture reading, y, using the least squares coefficients shown on the left side of Table 10.10.
    (b) Do the canonical analysis and make a contour plot of the moisture reading, y, using the REML estimates of the coefficients shown on the right side of Table 10.10.

14. Fit the general quadratic model to the ceramic pipe data in Table 10.12.

    (a) Fit the model with `proc mixed` including the random whole plot effects.
    (b) Fit the model in `proc rsreg` ignoring the random whole plot effects.
    (c) Verify that the regression coefficients are the same for both models, but that the standard errors of the coefficients are different.
    (d) What procedure would you use to fit the model in order to determine if the model can be simplified by dropping a factor?
    (e) What procedure would you use to do a canonical or ridge analysis to determine the strongest pipe?

15. Follow Table 10.11 and use the SAS data step to create an EESPRS split-plot central composite design with one whole-plot factor and three subplot factors.

16. Follow Table 10.11 and use the SAS data step to create an EESPRS split-plot Box-Behnken design with two whole-plot factors and two subplot factors.

Table 10.14 *Average Daily Gains for Cattle Experiment*

x_1	x_2	\multicolumn{8}{c}{Block}							
		1	2	3	4	5	6	7	8
6.723	0.1095	2.45	1.63	1.28	1.97	1.80	2.36	1.55	1.89
6.281	0.1273	2.08	2.86	3.37	2.37	2.59	2.16	4.14	2.12
5.788	0.1157	2.97	2.58	2.51	2.59	2.57	2.02	-	-
5.729	0.0669	1.92	2.63	2.08	3.26	2.64	2.17	1.35	4.21
6.411	0.0729	4.17	3.09	1.37	2.79	2.38	2.40	1.80	-
6.181	0.1004	3.86	2.39	2.93	2.41	2.59	3.23	2.19	3.04

CHAPTER 11

Mixture Experiments

11.1 Introduction

Many products, such as textile fiber blends, explosives, paints, polymers, and ceramics are made by mixing or blending two or more components or ingredients together. For example, a cotton-poly fabric is made by mixing cotton and polyester fibers together. The characteristics of a product that is composed of a mixture of components is usually a function of the proportion of each component in the mixture and not the total amount present.

If the proportion of the ith component is x_i, and there are k components in a mixture, then the proportions must satisfy the constraints

$$0.0 \le x_i \le 1.0, \text{ for each component, and } \sum_{i=1}^{k} x_i = 1.0. \qquad (11.1)$$

For example, in a three-component mixture, $0.0 \le x_1 \le 1.0$, $0.0 \le x_2 \le 1.0$, $0.0 \le x_3 \le 1.0$, and $x_1 + x_2 + x_3 = 1.0$.

If an experiment is conducted by varying the mixture components in an attempt to determine their effect on the product characteristics, the constraints prevent using standard factorial or response surface experimental design. If each component in the mixture can range from 0.0 to 100.0 percent of the total, a 2^3 factorial experiment would consist of all possible combinations of proportions 0.00 and 1.00 resulting in the corners of the cube (shown graphically in Figure 11.1).

However, the constraint $x_1+x_2+x_3 = 1.0$ reduces the three-dimensional experimental region to the two-dimensional shaded equilateral triangular plane shown in Figure 11.1. The coordinate system in this triangular experimental region can be represented as shown in Figure 11.2.

In this coordinate system, the proportion of component 3 can be read on the bottom or horizontal axis, the proportion of component 2 can be read on the left axis, and the proportion of component 1 can be read on the right axis. The proportion of component 1 is constant along horizontal lines that extend from the axis for component 2 on the left across to the axis for component 1 on the right. The proportion of component 2 is constant along diagonal lines that extend from the component 2 axis on the left down to the component 3 axis on the bottom, and the proportion of component 3 is constant along diagonal lines that extend up from the component 3 axis on the bottom to the component 1 axis on the right.

Along the line where the proportion of one component is constant, the

Figure 11.1 *Experimental Region for Factorial and Mixture Experiment*

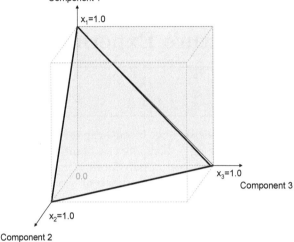

Figure 11.2 *Coordinate System for Three-Component Mixture*

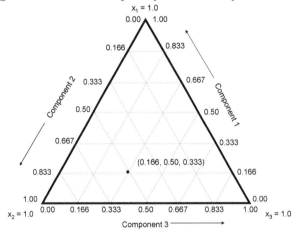

proportions of the other two components vary. For example, along the horizontal line where component 1 is constant at $x_1 = 0.166$, component 2 varies from $x_2 = 0.833$ at the left, where the line touches the component 2 axis, to $x_2 = 0.0$, where the line meets the component 3 axis on the right. The proportion of component 3 at any point along this line where $x_1 = 0.166$ is equal to $1 - x_1 - x_2$ and can be read by projecting from the point diagonally down to the left to the component 3 axis. For example, the point labeled just to the left of center of the region consists of a mixture of 16.6% of component 1, (by extending up along the diagonal line to the component 2 axis) 50% of component 2, and (by extending left downward to the component 3 axis)

MODELS AND DESIGNS FOR MIXTURE EXPERIMENTS

33.3% of component 3. The coordinates of this point can be determined by specifying the proportions of two of the three components, since the third component will always be determined as the sum of the other two subtracted from 1.0.

The experimental region for an experiment involving a four-component mixture is the three-dimensional equilateral tetrahedron shown in Figure 11.3. This figure, shows that constant proportions of component 1 are represented by planes parallel to the base where $x_1 = 0.0$. Likewise, constant proportions of the other components would be planes parallel to the sides where those proportions are equal to 0.00. The experimental region for experiments involving $k > 4$ mixture components cannot be represented graphically, but is always a $k - 1$ dimensional tetrahedron.

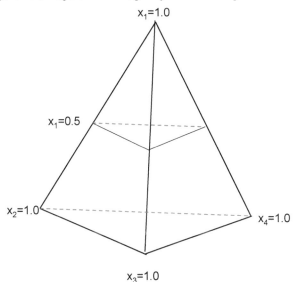

Figure 11.3 *Experimental Region for Four-Component Mixture*

11.2 Models and Designs for Mixture Experiments

As the experimental region in a mixture experiment is changed by the constraints, the mathematical model used to represent the data from a mixture experiment is also changed by the constraints. The linear model used for a three-factor factorial experiment can be represented as a linear regression model

$$y = \beta_0 + \beta_1 x_1 + \beta_2 x_2 + \beta_3 x_3 + \epsilon. \tag{11.2}$$

However, in a mixture experiment, the constraint $x_1 + x_2 + x_3 = 1.0$ makes one of the four coefficients in model 11.2 redundant. One way of modifying the model is to substitute $(1 - x_1 - x_2)$ for x_3, resulting in the so-called *slack variable* model

$$y = \beta_0^* + \beta_1^* x_1 + \beta_2^* x_2 + \epsilon. \tag{11.3}$$

In this model, the variable x_3 is the slack variable, and it is usually chosen to represent the most inert mixture component.

Many find the slack variable model objectionable because the coefficients β_1^* and β_2^* do not represent the effects of components x_1 and x_2 as they would in a factorial experiment, but rather they represent the effects of x_1 and x_2 confounded with the opposite of the effect of the slack variable x_3.

Scheffé (1958) described more suitable polynomial models for mixture experiments. In his form of the model, the coefficient for β_0 in Equation (11.2) is 1, and by substituting $x_1 + x_2 + x_3$ for 1, the model can be written in the Scheffé form as

$$y = \beta_1^* x_1 + \beta_2^* x_2 + \beta_3^* x_3 + \epsilon. \tag{11.4}$$

From this point on, the asterisks will be removed from the β_i^* as they were used to distinguish the coefficients in a mixture model from the coefficients in the general regression model.

Again the coefficients in the Scheffé form of the linear model do not represent the effects of the variables x_1, x_2, and x_3, as in a linear regression model, and a coefficient that is zero, or not significantly different from zero, does not mean that changing the corresponding mixture component will not affect product characteristics. Alternately, in the mixture model β_i represents the predicted response at the vertex of the experimental region where $x_i = 1.0$. This can be represented graphically for the three-component mixture in Figure 11.4 where the predicted response is a plane above the mixture experimental region.

Figure 11.4 *Interpretation of Coefficients in Scheffé Linear Model*

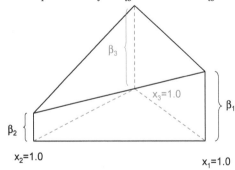

If the response surface over the mixture region is nonlinear, the general quadratic model in three variables,

MODELS AND DESIGNS FOR MIXTURE EXPERIMENTS

$$y = \beta_0 + \beta_1 x_1 + \beta_2 x_2 + \beta_3 x_3 + \beta_1 x_1^2 + \beta_2 x_2^2 + \beta_3 x_3^2 + \beta_{12} x_1 x_2$$
$$+ \beta_{13} x_1 x_3 + \beta_{23} x_2 x_3 + \epsilon, \tag{11.5}$$

usually used as an empirical response surface model for a nonlinear surface, is also different for mixture experiments. Multiplying β_0 by $x_1 + x_2 + x_3$ and substituting $x_1 \times (1 - x_2 - x_3)$ for x_1^2, etc., in model 11.5 results in the Scheffé quadratic model shown in Equation (11.6).

$$y = \beta_1 x_1 + \beta_2 x_2 + \beta_3 x_3 + \beta_{12} x_1 x_2 + \beta_{13} x_1 x_3 + \beta_{23} x_2 x_3 + \epsilon. \tag{11.6}$$

For k mixture components this model can be written as

$$y = \sum_{i=1}^{k} \beta_i x_i + \sum_{i<j}^{k} \sum^{k} \beta_{ij} x_i x_j + \epsilon. \tag{11.7}$$

The coefficients of the product terms $x_i x_j$ represent the quadratic curvature along an edge of the simplex experimental region, as illustrated for a three-component mixture in Figure 11.5.

Figure 11.5 *Interpretation of Coefficients of Product Terms in Scheffé Quadratic Model*

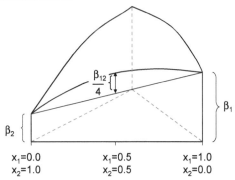

11.2.1 Experimental Designs To Fit Scheffé Models

Based on the interpretation of the coefficients in the linear and quadratic models for mixture experiments, Scheffé (1958) proposed the *simplex-lattice* designs for experiments with mixtures. These designs consist of all possible combinations of proportions, where the ith proportion takes on the values $x_i = 0, \frac{1}{m}, \frac{2}{m}, \ldots, 1$. A simplex-lattice design (SLD) in k components with $m+1$ equally spaced proportions for each component is designated as a simplex-lattice$\{k,m\}$ or SLD$\{k,m\}$ design.

Figure 11.6 shows the linear (SLD$\{3,1\}$) and the quadratic (SLD$\{3,2\}$) designs in three components. Only the pure components (i.e., $(x_1, x_2, x_3) = (1, 0, 0)$, $(x_1, x_2, x_3) = (0, 1, 0)$, and $(x_1, x_2, x_3) = (0, 0, 1)$) are required for a linear design and the coefficient β_i in model 11.4 can be estimated as average of all the response data at the pure component where $x_i = 1.0$. In the linear model, the effect of blending two or more components is assumed to be linear, and no intermediate points are necessary in the design. The 50/50 mixtures of each pair of components are required to estimate the coefficients β_{ij} of the quadratic blending effects in model 11.5, thus the mixtures $(x_1, x_2, x_3) = (\frac{1}{2}, \frac{1}{2}, 0)$, $(x_1, x_2, x_3) = (\frac{1}{2}, 0, \frac{1}{2})$, and $(x_1, x_2, x_3) = (0, \frac{1}{2}, \frac{1}{2})$ are required in addition to the pure components for a quadratic design.

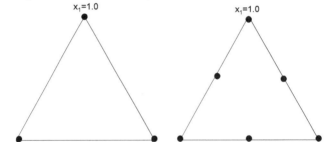

Figure 11.6 *Linear and Quadratic Simplex Lattice Designs*

In response surface experiments conducted with independent factors, as discussed in Chapter 10, the experimental region can be restricted so that the general quadratic model is usually a good approximation to the true nonlinear model relating the response to the factor levels in the restricted region. In a mixture experiment, the experimental region consists of the full $(k-1)$ dimensional simplex or tetrahedron and cannot be restricted. Therefore, higher order polynomial equations are sometimes necessary to approximate the true model over the entire simplex. A full cubic model in three-mixture components is shown in Equation (11.8).

$$\begin{aligned} y =& \beta_1 x_1 + \beta_2 x_2 + \beta_3 x_3 \\ & \beta_{12} x_1 x_2 + \beta_{13} x_1 x_3 + \beta_{23} x_2 x_3 \\ & \delta_{12} x_1 x_2 (x_1 - x_2) + \delta_{13} x_1 x_3 (x_1 - x_3) + \delta_{23} x_2 x_3 (x_2 - x_3) \\ & \beta_{123} x_1 x_2 x_3 \end{aligned} \tag{11.8}$$

MODELS AND DESIGNS FOR MIXTURE EXPERIMENTS 449

The coefficients δ_{ij} represent the cubic blending of binary mixtures along the edges of the simplex as shown in Figure 11.7.

Figure 11.7 *Cubic Blending of Binary Mixture of Components i and j*

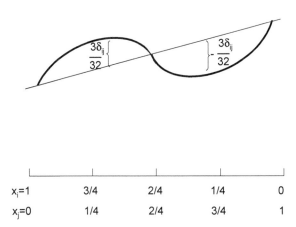

The β_{123} coefficient represents the ternary blending of components 1, 2, and 3 in the interior of the simplex.

To estimate the coefficients in the full cubic model requires an SLD{3,3} design consisting of the points

$$(x_1, x_2, x_3) = (1,0,0), (0,1,0), (0,0,1), (\frac{1}{3},\frac{2}{3},0), (\frac{2}{3},\frac{1}{3},0), (\frac{1}{3},0,\frac{2}{3})$$
$$(\frac{2}{3},0,\frac{1}{3}), (0,\frac{1}{3},\frac{2}{3}), (0,\frac{2}{3},\frac{1}{3},0), (\frac{1}{3},\frac{1}{3},\frac{1}{3}).$$

When the coefficients δ_{ij} are insignificant in the cubic Scheffé model, justifying a less complicated model, the simpler special cubic model can be used. The equation for the special cubic model in k mixture components is shown in Equation (11.9).

$$y = \sum_{i=1}^{k} \beta_i x_i + \sum_{i \le j}^{k} \sum^{k} \beta_{ij} x_i x_j + \sum_{i<j<l}^{k} \sum^{k} \sum^{k} \beta_{ijl} x_i x_j x_l + \epsilon. \qquad (11.9)$$

An alternate design, discussed by Cornell (2002), that allows estimation of all coefficients in the special cubic model is the *simplex centroid design* or SCD. A simplex centroid design in three-mixture components consists of the $\binom{3}{1}$ pure component blends (1,0,0), (0,1,0), and (0,0,1), the $\binom{3}{2}$ binary mixtures $(\frac{1}{2},\frac{1}{2},0),(\frac{1}{2},0,\frac{1}{2})$, and $(0,\frac{1}{2},\frac{1}{2})$, and the $\binom{3}{3}$ ternary mixture $(\frac{1}{3},\frac{1}{3},\frac{1}{3})$.

An SCD in k mixture components (SCD$\{k\}$) consists of the $\binom{k}{1}$ pure components, the $\binom{k}{2}$ permutations of the binary mixtures $(\frac{1}{2},\frac{1}{2},0,\ldots,0)$, the $\binom{k}{3}$ permutations of the ternary mixtures $(\frac{1}{3},\frac{1}{3},\frac{1}{3},0,\ldots,0)$, etc., up to the k-nary mixture $(\frac{1}{k},\frac{1}{k},\frac{1}{k},\ldots,\frac{1}{k})$.

The (SCD$\{k\}$) always includes the k-nary mixture, or centroid, which is interior to the design region, and it supports estimation of the coefficients in a polynomial model of the general form shown in Equation (11.10).

$$y = \sum_{i=1}^{k} \beta_i x_i + \sum_{i<j}^{k} \sum^{k} \beta_{ij} x_i x_j + \sum_{i<j<l}^{k} \sum^{k} \sum^{k} \beta_{ijl} x_i x_j x_l + \cdots +$$
$$\beta_{12\ldots k} x_1 x_2 x_3 \cdots x_k + \epsilon \qquad (11.10)$$

One criticism of quadratic simplex lattice designs in three components is that they do not contain any design points in the interior of the simplex design region where the accuracy of a fitted model can be tested. For this reason, these designs are often augmented by the overall centroid $(\frac{1}{3},\frac{1}{3},\frac{1}{3})$ and axial points of the form $(1-\Delta,\frac{\Delta}{2},\frac{\Delta}{2})$, $(\frac{\Delta}{2},1-\Delta,\frac{\Delta}{2})$, and $(\frac{\Delta}{2},\frac{\Delta}{2},1-\Delta)$ as shown in Figure 11.8.

11.2.2 Alternate Models for Mixture Experiments

Claringbold (1955), who was the first to use a simplex-type design for studying three-mixture components, described a two-step procedure for translating the origin and orthogonally rotating the axis of the component space to produce independent factors as shown graphically in Figure 11.9. The first step of the procedure is to translate the origin to the centroid of the simplex by subtracting the centroid value from each component and multiplying by the number of components. The second step is to post multiply the vector of translated coordinates by an orthogonal matrix $\mathbf{O}$ to rotate the axis. In three components, the rotated and translated axis labeled w_3 is perpendicular to the two-dimensional simplex experimental region, and all points in the experimental region will have $w_3 = 0$ in the translated and rotated factor space.

Figure 11.8 *Simplex Lattice{3,2} Augmented with Centroid and Axial Points*

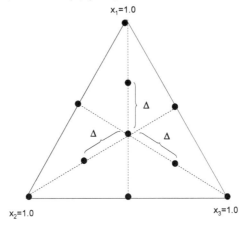

Figure 11.9 *Translation and Rotation of Axis To Form Independent Factors*

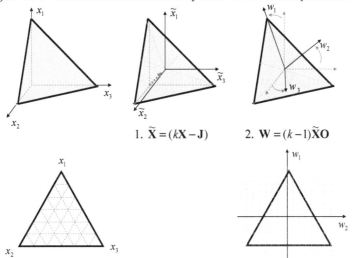

Cornell (2002) shows the translation and rotation from the k dimensional mixture component space to $k-1$ dimensional space for the independent factor space can be written as $k-1$ contrasts of the mixture components. For example, in three-mixture components (x_1, x_2, x_3),

$$w_1 = \sqrt{6}(2x_1 - x_2 - x_3) \tag{11.11}$$
$$w_2 = \sqrt{18}(x_2 - x_3). \tag{11.12}$$

The quadratic model can be written in the independent factors as shown in Equation (11.15), which is the same as model 10.2,

$$y = \beta_0 + \beta_1 w_1 + \beta_2 w_2 + \beta_{11} w_1^2 + \beta_{22} w_2^2 + \beta_{12} w_1 w_2 + \epsilon, \quad (11.13)$$

and standard regression programs and contour plotting programs like SAS proc gcontour can be used to fit the model and produce contour plots over the simplex experimental region. This fact is illustrated in an example in the appendix of this chapter.

Sometimes the response in a mixture experiment may increase or decrease rapidly as one or more of the mixture components approach their lower boundary. These are called edge effects. Edge effects can cause a lack of fit in first- and second-degree polynomial models. Rather than increasing the order of the polynomial model, Draper and John (1977) suggested including inverse terms in the model, as shown in Equation (11.14), to model edge effects.

$$y = \beta_1 x_1 + \beta_2 x_2 + \beta_3 x_3 + \beta_{-1} x_1^{-1} + \beta_{-2} x_2^{-1} + \beta_{-3} x_3^{-1} + \epsilon \quad (11.14)$$

Including inverse terms when there are edge effects can greatly improve the fit of the model as will be demonstrated in an exercise.

11.3 Creating Mixture Designs in SAS

Simplex lattice and simplex centroid designs can be easily created using the SAS macros included in the AUTOCALL library or with the SAS ADX automated system. The example SAS code below uses the macros to first create an SLD{3,2} in three components x_1, x_2, and x_3, and stores it in the file work.sld. The code then creates an SCD in four components and prints the resulting file work.scd.

```
%adxgen
%adxmix
%adxinit
%adxsld(sld,x1 x2 x3,2)
%adxgen
%adxmix
%adxinit
%adxscd(scd,x1 x2 x3 x4,4)
proc print data=scd; run;
```

The output at the top of the next page shows the simplex-centroid design.

CREATING MIXTURE DESIGNS IN SAS 453

Obs	x1	x2	x3	x4	dimen
1	1.00000	0.00000	0.00000	0.00000	0
2	0.00000	1.00000	0.00000	0.00000	0
3	0.00000	0.00000	1.00000	0.00000	0
4	0.00000	0.00000	0.00000	1.00000	0
5	0.50000	0.50000	0.00000	0.00000	1
6	0.50000	0.00000	0.50000	0.00000	1
7	0.50000	0.00000	0.00000	0.50000	1
8	0.00000	0.50000	0.50000	0.00000	1
9	0.00000	0.50000	0.00000	0.50000	1
10	0.00000	0.00000	0.50000	0.50000	1
11	0.33333	0.33333	0.33333	0.00000	2
12	0.33333	0.33333	0.00000	0.33333	2
13	0.33333	0.00000	0.33333	0.33333	2
14	0.00000	0.33333	0.33333	0.33333	2
15	0.25000	0.25000	0.25000	0.25000	3

The SAS automated design of experiments tool ADX also has many tools for the design and analysis of mixture experiments. Figure 11.10 shows the design choice template that is displayed when you choose **Select Design** after specifying three-mixture components. Once a design is selected (as shown in Figure 11.10), clicking the **Design Details** button gives you the option of looking at a description of the design, a listing of the runs in the design, or a two-dimensional graph showing the runs in the simplex design space. If there are more than three-mixture components, this graph is made at a slice with constant values for the components not shown on the graph.

Figure 11.10 *ADX Mixture Design Selection Template*

Factors	Runs	Maximum Model	Design Type
3	3	Linear	Simplex-Lattice
3	4	Linear	Simplex-Centroid
3	6	Quadratic	Simplex-Lattice
3	7	Special Cubic	Simplex-Centroid
* 3	10	Full Cubic	Simplex-Lattice
3	15	Quartic	Simplex-Lattice
3	21	Quartic	Simplex-Lattice

Figure 11.11 shows the graphical representation of the cubic simplex-lattice design that was selected in Figure 11.10.

Figure 11.11 *ADX Design Description Template*

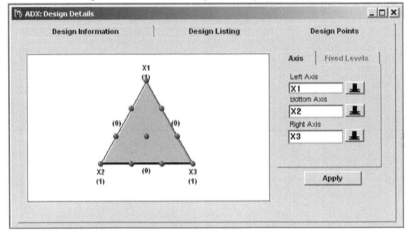

11.4 Analysis of Mixture Experiment

In attempting to find a pesticide formulation that would spread evenly on the leaves of cotton plants, researchers experimented by combining a synthetic pyrethroid insecticide with a mixture of three Niagara emulsifiers. They used an SLD{3,2} design, augmented by an overall centroid and six axial points. The experimental runs and the measured surface tension from each mixture is shown in Table 11.1. The formulations were made and tested in a completely random order.

Table 11.1 *Data Pesticide Formulation Experiment*

Run	x_1	x_2	x_3	surface tension (dyne/cm)
1	1.00000	0.00000	0.00000	48.7
2	0.80000	0.10000	0.10000	49.5
3	0.60000	0.20000	0.20000	50.2
4	0.50000	0.00000	0.50000	52.8
5	0.50000	0.50000	0.00000	49.3
6	0.33333	0.33333	0.33333	51.1
7	0.30000	0.20000	0.50000	52.7
8	0.30000	0.50000	0.20000	50.3
9	0.10000	0.10000	0.80000	60.7
10	0.10000	0.80000	0.10000	49.9
11	0.00000	0.00000	1.00000	64.9
12	0.00000	0.50000	0.50000	53.5
13	0.00000	1.00000	0.00000	50.6

ANALYSIS OF MIXTURE EXPERIMENT

The research goal was to determine how the proportions of the three Niagara emulsifiers affected the surface tension of the pesticide formulation. The lower the surface tension, the more likely the formulation would spread evenly on the cotton leaves.

The data analysis will first be illustrated with the ADX automated design of experiments tool. Since this specific design cannot be created easily in ADX, the SAS data step was used to read the data in Table 11.1 into the SAS data set work.form; from there, it was imported to the ADX system. Figure 11.12 shows the data in the ADX system.

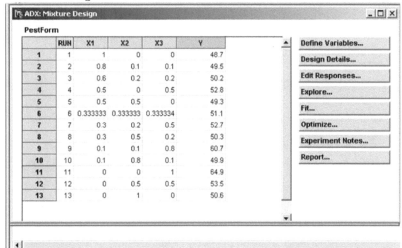

Figure 11.12 *Pesticide Formulation Data in ADX*

Clicking first on the Design Details button followed by the Design Points tab shows a graphical representation of the design points in the simplex region. The axial points can be seen in Figure 11.13.

The SLD{3,2} design augmented by the overall centroid supports fitting the special cubic model. A general strategy to follow when modeling the data from mixture experiments is to first fit the highest order model supported by the design, then try to simplify if higher order terms are insignificant. The variance of predicted values will be lower for simpler models with fewer terms.

The model can be selected in ADX by clicking the Model ▶ Change master model menus. ADX allows the user to choose the linear, quadratic, special cubic or full cubic models as defined earlier in Equations 11.4, 11.7, 11.8, and 11.9, respectively. Figure 11.14 below shows the result of fitting the Scheffé special cubic model. ADX fits the model by the method of least squares, and the same methods for checking the assumptions of the least squares fit (described in Chapter 2) can be accomplished within the ADX system. If there are replicated points, ADX automatically performs the lack of fit test as explained in Section 10.6.2. In the results shown in Figure 11.14, the β_{123}

Figure 11.13 *Design Points in ADX*

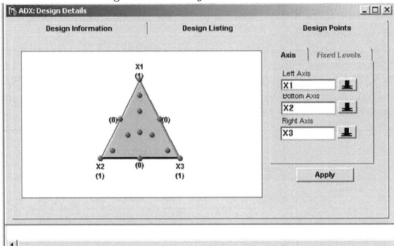

Figure 11.14 *ADX Fit of Scheffé Special Cubic Model*

Effect	Estimate	Std Error	t Ratio	P Value
X1	48.906	0.56456	86.627	<.0001
X2	50.395	0.56852	88.643	<.0001
X3	65.387	0.56852	115.01	<.0001
X1*X2	-0.91547	3.0422	-0.30092	0.7736
X1*X3	-16.364	3.0422	-5.379	0.0017
X2*X3	-17.144	3.0888	-5.5503	0.0014
X1*X2*X3	3.0993	18.035	0.17185	0.8692

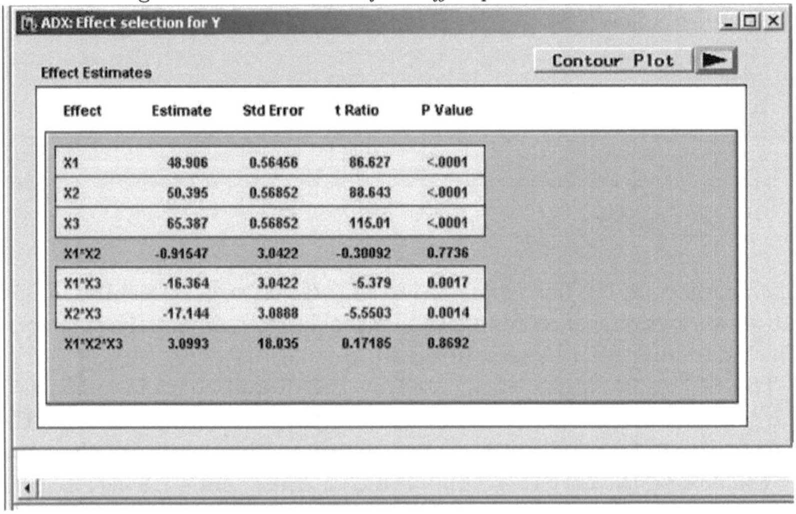

term is clearly insignificant indicating that there is no significant ternary blending effect. Figure 11.15 shows the output that results from changing the master model to the quadratic model and refitting the equation. Here, coefficients for the linear and quadratic coefficients are slightly different than they were in the special cubic model. The adjusted $R^2 = 0.9852$ for the quadratic model, while the adjusted $R^2 = 0.9828$ for the special cubic model. This together with the fact that there was no significant ternary blending, would justify using the simpler quadratic model to explore the mixture region to

ANALYSIS OF MIXTURE EXPERIMENT

identify mixtures that would satisfy the experimental goals. The negative coefficients for quadratic terms indicate that the fitted surface dips below a plane connecting the predicted response at the pure blends (the opposite of that shown in Figure 11.5).

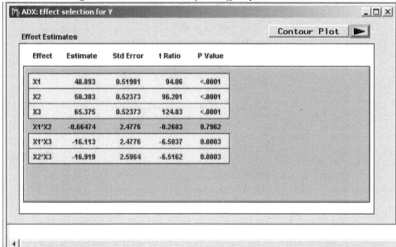

Figure 11.15 *ADX Fit of Scheffé Quadratic Model*

Once the model is fit, there are several ways of exploring the fitted surface. The first way to explore is to look at a contour plot over the simplex region. This, is accomplished by clicking the contour plot button in the upper right corner of the fitted model results. The contour plot can be made with either the master model that you specify, or by using only the terms in the model that ADX highlights as significant. The linear terms should always be included whether significant or not, since they represent the intercept. Therefore, using the master model for the contour plot may be the safest approach. Figure 11.16 shows the contour plot produced by ADX over the simplex region.

This figure also shows that mixtures with lower surface tension can be found on the left, along the axis where blends of emulsifiers x_1 and x_2 are found. The lowest surface tension is at the top vertex, where $x_1 = 1.0$. There, the surface tension is slightly less than 49 dyne/cm. Moving to the lower left where $x_2 = 1.0$, the surface tension increases slightly to just over 50 dyne/cm. However, if the proportion of emulsifier x_3 is increased, the surface tension begins to increase dramatically until it reaches the maximum where $x_3 = 1.0$. There the surface tension is over 62.5 dyne/cm.

By selecting the **Optimize** button, ADX offers three additional tools for exploring the fitted surface. One is the *contour optimizer* that allows the user to move crosshairs over the simplex region and read the proportions of each of the mixture components and the predicted response at any place in the experimental region. The **Contour Plot** tab found after selecting the optimize

Figure 11.16 *ADX Contour Plot of Quadratic Model Predictions*

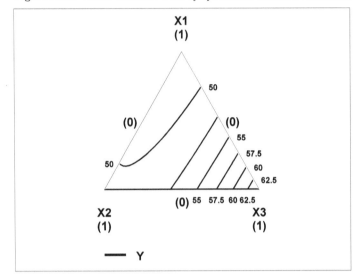

button, as well as the contour optimizer, allows the user to specify which mixture components to place on the left, right, and top vertices of the simplex region. Here the user can select three components for plotting, and to fix additional components when there are more than three mixture components in the problem. A third tab that appears after clicking the optimize button is the *prediction profiler*. This is a very useful tool that allows the user to see the response trace along the direction through the simplex introduced by Cox (1971) and illustrated in Figure 11.17. This tool plots the predicted value of the response and the 95% prediction intervals for the prediction using the directions shown in Figure 11.17. It is also very useful in visualizing the results of varying the mixture components when there are more than three components in the mixture, and the results cannot be displayed as a single contour plot.

Figure 11.18 shows the response trace for the three components in the pesticide formulation problem, and it can be seen that increasing the proportion of emulsifier x_2 has relatively little effect. Increasing the proportion of emulsifier x_1 has a slightly negative effect on surface tension, and increasing the proportion of emulsifier x_3 causes a large increase in surface tension. Therefore, just as can be seen on the contour plot, the conclusion reached by looking at the response traces would be to choose 100% of emulsifier x_1 or a blend of emulsifiers x_1 and x_2 to reach the lowest levels of surface tension.

In mixture experiments with several components, it is important to determine which components are most influential. This can simplify the problem and allow the experimenters to focus on optimization with respect to the most important mixture components. Since the coefficients in the models for

ANALYSIS OF MIXTURE EXPERIMENT 459

Figure 11.17 *Cox Directions Through Three-Dimensional Simplex*

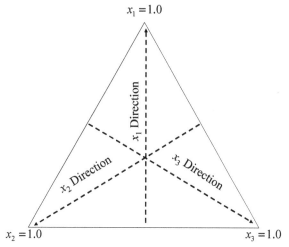

Figure 11.18 *ADX Response Trace Plots for Quadratic Model*

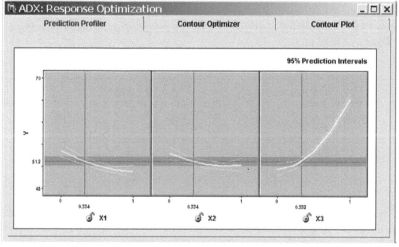

mixture experiments do not represent effects in the same way as they do for independent factors, tests of the significance of linear coefficients in the model are not a good way to determine the important components. However, the response trace is a good tool to use for this purpose. When a linear design is used as a screening experiment for several components, and the Scheffé linear model is fit to the data, the response traces will be straight lines and the least important components will have traces with relatively flat slopes, while the important components will have traces with steep slopes. Thus, the response trace plots can be used as a graphical tool to determine the important components similar to the way that the normal or half-normal plot of effects is used for screening experiments with independent factors. Cornell (2002) and Snee and Marquardt (1976) discuss other strategies for screening with mixture experiments.

The least squares fits performed by ADX for mixture experiments (illustrated in Figures 11.14 and 11.15) are designed specifically for fitting Scheffé models. Cornell (2002) explains that standard regression programs like SAS proc glm or SAS proc reg do not provide the correct analysis because when the no-constant option (model option /noint) is used, the fitted model sums of squares and total sums of squares are not corrected for the overall mean. Therefore, the R^2 value is inflated and gives the impression that the model fits better than it actually does. For example, Table 11.2 shows a comparison of the sums of squares and R^2 statistic produced by SAS ADX, proc glm, and proc reg when fitting the Scheffé quadratic model to the data in Table 11.1 and using the model /noint option with proc glm and proc reg.

Table 11.2 *Comparison of Different Programs for Fitting Scheffé Quadratic Model to Data in Table 11.1*

Program	Model df	Model SS	Error SS	Total SS	R^2
ADX Mixture	5	275.64	2.41	278.05	0.9913
proc glm	6	36285.60	2.41	36288	0.9999
proc reg	6	36285.60	2.41	36288	0.9999

One way the correct sums of squares and R^2 can be computed using proc reg or proc glm is to fit the model

$$y = \beta_0 + \beta_1 x_1 + \beta_2 x_2 + \beta_{12} x_1 x_2 + \beta_{13} x_1 x_3 + \beta_{23} x_2 x_3 + \epsilon. \qquad (11.15)$$

By including an intercept and removing x_3 from the model, the coefficients β_0, β_1, and β_2 take on a different meaning than in the Scheffé model shown in Equation (11.6). β_0 takes on the value of β_3 in Equation (11.6), and β_i represents the difference of $\beta_0 - \beta_i$. The code for fitting this model with proc glm and estimating β_1, β_2, and β_3 is shown in the listing on the next page.

```
data form;
input x1 x2 x3 y;
datalines;
1.0 0 0 48.7
.8 .1 .1 49.5
.6 .2 .2 50.2
.5 0 .5 52.8
.5 .5 0 49.3
.333333 .333333 .333334 51.1
.3 .2 .5 52.7
.3 .5 .2 50.3
.1 .1 .8 60.7
.1 .8 .1 49.9
0 0 1.0 64.9
0 .5 .5 53.5
0 1.0 0 50.6
proc glm data=form;
model y= x1 x2 x1*x2 x1*x3 x2*x3/ solution;
estimate 'beta1' intercept 1 x1 1;
estimate 'beta2' intercept 1 x2 1;
estimate 'beta3' intercept 1;
run;
```

When fitting the model 11.5 or 11.15 with `proc reg`, the mixture that produces the maximum or minimum response can be found numerically using `proc nlp` as shown in Section 10.7.4, but it is difficult to produce contour plots outside SAS ADX unless the transformation to orthogonal components is used, as described in the last section and illustrated in the appendix to this chapter.

11.5 Constrained Mixture Experiments

In some mixture experiments, it is impossible to test pure components. For example, fuel for solid rocket boosters is a mixture of binder, oxidizer, and fuel, and it must contain a percentage of each. Proportions can be varied but must remain within certain constraints in order for the propellant to work. Kurotori (1966) described an example of this situation where experiments with rocket propellant were performed. In that case, x_1: the binder, had to be at least 20% of the mixture; x_2: the oxidizer, could be no less that 40% of the mixture; and x_3: the fuel, had to comprise at least 20% of the mixture. This leads to three new constraints shown in Equation (11.16), and the experimental region is restricted to the shaded subset of the simplex shown in Figure 11.19.

$$x_1 \geq 0.20$$
$$x_2 \geq 0.40$$
$$x_3 \geq 0.20 \qquad (11.16)$$

When each mixture component only has a lower constraint as shown in Figure 11.19, the constrained region (if consistent) will always be a smaller simplex within the original simplex as shown in the figure. The component space

Figure 11.19 *Constrained Region for Rocket Propellant Experiment*

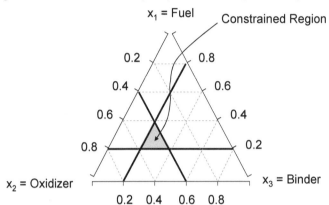

within this smaller simplex can be conveniently transformed into a pseudo-component space where the same simplex lattice and simplex centroid designs can be used for an experimental design. If the lower constraint for mixture component x_i is defined to be l_i, then the ith pseudo component is defined to be

$$x'_i = \frac{x_i - l_i}{1 - \sum_{i=1}^{k} l_i}, \qquad (11.17)$$

where k is the number of mixture components. A simplex lattice design can be constructed in the pseudo-components and the actual mixtures to be tested can then be obtained by solving Equation (11.17) for x_i as

$$x_i = l_i + \left(1 - \sum_{i=1}^{k} l_i\right) x'_i, \qquad (11.18)$$

For example, Table 11.3 shows an SLD{3,2} design in pseudo-components translated into a design with the actual components for the rocket propellent experiment using the lower bounds given in Equation (11.16). The experiments were run by making each mixture of fuel oxidizer and binder, then measuring the resulting elasticity. The general quadratic or special cubic model can be fit to either the pseudo or actual components.

11.5.1 Upper and Lower Constraints

In some special cases when there are only upper constraints or both upper and lower constraints, the constrained experimental region will result in a smaller inverted simplex within the simplex component space. However, in the more common situation, the constrained region will be an irregular hyperpolyhedron. In this case, McLean and Anderson (1966) recommend a design that consists of the extreme vertices (EVD) of the experimental region, possibly

Table 11.3 *Simplex Lattice Design in Pseudo and Actual Components*

Run	pseudo-component x'_1	x'_2	x'_3	x_1=Fuel	actual component x_2=Oxidizer	x_3=Binder	response elasticity
1	1	0	0	0.400	0.400	0.200	2350
2	0	1	0	0.200	0.600	0.200	2450
3	0	0	1	0.200	0.400	0.400	2650
4	$\frac{1}{2}$	$\frac{1}{2}$	0	0.300	0.500	0.200	2400
5	$\frac{1}{2}$	0	$\frac{1}{2}$	0.300	0.400	0.300	2750
6	0	$\frac{1}{2}$	$\frac{1}{2}$	0.200	0.500	0.300	2950
7	$\frac{1}{3}$	$\frac{1}{3}$	$\frac{1}{3}$	0.266	0.466	0.266	3000

augmented by edge and facet centroids. They propose an algorithm for finding the extreme vertices by forming a two-level factorial in $k-1$ mixture components (using the lower and upper constraints as levels), then setting the kth component equal to one minus the sum of the other $k-1$ components. Each point in this factorial, where the computed value of the kth component falls within its lower and upper constraints, is an extreme vertex. All the extreme vertices can be found by repeating this procedure using all possible subsets of $k-1$ mixture components for the two-level factorial.

Once the extreme vertices of the experimental region are found, the coordinates of the edge centroids can be found by first locating all vertices with a constant value for one component (i.e., two points on an edge) then averaging the other components. The centroids of two-dimensional facets can be found by first finding all vertices where two components are constant and then averaging the remaining $k-2$ components, etc.

Snee and Marquardt (1974) defined the XVERT algorithm for finding the extreme vertices, which is similar to McLean and Anderson's algorithm. When there are many components in the mixture problem, there may be many more extreme vertices in the experimental region than there are coefficients in the Scheffé linear or quadratic model. In that case Snee and Marquardt (1974) recommended choosing the A-optimal subset of the extreme vertices that minimizes the trace of the $(X'X)^{-1}$ matrix.

There are SAS macros, included in the AUTOCALL library, that use either McLean and Anderson's algorithm or Snee and Marquardt's XVERT algorithm to generate the vertices of a multi-constrained experimental region. To illustrate these macros, consider a problem studied by Barbuta and Lepadatu (2008). They investigated mechanical properties such as compressive strength, flexural strength, and adhesion stress of polymer concrete. Polymer concrete (PC) has many uses: precast components for buildings, bridge panels, repair of structural members, waterproofing and decorative overlay of pavements. Polymer concrete is formed by binding aggregates together with a resin that reacts with a hardener. The relatively high cost of PC led Barbuta and Lepadatu (2008) to study ways of reducing the dosage of polymer in the mix without diminishing the mechanical properties. The mixture components they studied were x_1 :Epoxy resin x_2: Silica Fume (SUF), x_3 :Aggregate Sort I and x_4: Aggregate Sort II. Constraints on the mixture components are shown in Equation (11.19) below.

$$0.124 \le x_1 \le 0.188$$
$$0.064 \le x_2 \le 0.128$$
$$0.374 \le x_3 \le 0.438$$
$$0.374 \le x_4 \le 0.438 \tag{11.19}$$

The SAS code below shows how to create the file **work.exvert** containing the extreme vertices using the McLean-Anderson algorithm.

```
%adxgen
%adxmix
%adxinit
%adxmamd(exvert, x1 .124-.188/ x2 .064-.128/ x3
    .374-.438/ x4 .374-.438)
proc print; run;
```

Running this code results in a list of four extreme vertices.

Obs	x1	x2	x3	x4
1	0.188	0.064	0.374	0.374
2	0.124	0.128	0.374	0.374
3	0.124	0.064	0.438	0.374
4	0.124	0.064	0.374	0.438

By changing the **%adxmamd** to **%adxxvert**, in the commands above, Snee and Marquardt's XVERT algorithm will be used instead of McLean and Anderson's algorithm to find the extreme vertices. The arguments for **%adxxvert** are the same as those for **%adxmamd**, except for one optional argument at the end that specifies the maximum order of centroids to be generated. The **%adxxvert** macro not only finds the extreme vertices using the XVERT algorithm but it also finds the centroids up to the order specified by the user, and, by default, it also finds the overall centroid. For example, the code at the top of the next page finds the four extreme vertices shown in the listing above, plus the six edge centroids and the overall centroid. To add two-dimensional facet centroids, change the last argument from 2 to 3.

CONSTRAINED MIXTURE EXPERIMENTS

```
%adxgen
%adxmix
%adxinit
%adxxvert(exvert, x1 .124-.188/ x2 .064-.128/ x3 .374-.438/
    x4 .374-.438,2)
proc print; run;
```

For many constrained mixture problems, the number of extreme vertices and edge centroids will be much greater than the number of coefficients in the Scheffé quadratic model. To reduce the number of mixtures and still allow fitting a quadratic model to the resulting data, a D-optimal subset can be selected using **proc optex** as shown in the example below. In this example, a D-optimal subset of 14 mixtures from the file **work.exvert** will be output to the file **work.dmax**.

```
proc optex data=exvert;
  model x1 x2 x3 x4 x1*x2 x1*x3 x1*x4 x2*x3 x2*x4 x3*x4/
      noint;
  generate n=14 ;
  output out=dmax; run;
proc print data=dmax; run;
```

If the constraint equations such as Equation (11.19) are inconsistent, there will be no feasible points within the constrained region. In that case, the SAS log will indicate the file **exvert.test** has zero observations.

The SAS ADX tool can also create a list of candidate design points for constrained mixture experiments composed of extreme vertices and centroids. To illustrate ADX, which can also graph the feasible region, consider the three-component mixture studied by Juan et al. (2006).

They studied consumer acceptance of polvoron composed of a mixture of x_1: sugar, x_2: peanut fines, and x_3: butter. Polvoron is a Philippine ethnic dessert or candy usually composed of milk powder, toasted flour, sugar, and butter. Ground nuts can be added to vary its flavor and texture properties. If acceptable to consumers, utilizing peanut fines (which are usually a discarded byproduct of the roasted peanut process) in polvoron could reduce waste and create an additional product line. The constraints on the mixture space are shown in Equation (11.20).

$$0.00 \leq x_1 \leq 0.80$$
$$0.10 \leq x_2 \leq 0.95$$
$$0.05 \leq x_3 \leq 0.50 \qquad (11.20)$$

Selecting three-mixture components and entering the constraints listed in Equation (11.20) into ADX, the constrained region is graphed as shown in Figure 11.20.

Table 11.4 shows data from the article. The response, overall liking, was the average ranking of 115 taste panelists who ranked the mixtures on a 9-point scale where 1 = dislike extremely, 5 = neither like nor dislike, and 9 = extremely like. Three additional mixtures in the interior of the simplex (that are not shown in the table) were tested to help discriminate between possible models for the data.

Once the mixture components are defined and the constraints entered to ADX, clicking the **Select Design** button causes ADX to construct a list of candidates which includes the five vertices, four edge centroids, and the overall

Figure 11.20 *Constrained Region for Peanut Fine Polvoron Experiment*

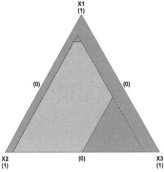

Table 11.4 *Data from Polvoron Mixture Experiment*

Run		x_1	x_2	x_3	overall liking
1	vertex	0.800	0.150	0.050	5.33
2	edge cent.	0.400	0.550	0.050	5.87
3	vertex	0.000	0.950	0.050	3.69
4	edge cent.	0.000	0.725	0.275	3.83
5	vertex	0.000	0.500	0.500	3.85
6	edge cent.	0.200	0.300	0.500	5.23
7	vertex	0.400	0.100	0.500	5.68
8	edge cent.	0.600	0.100	0.300	5.88
9	vertex	0.800	0.100	0.100	5.75
10	centroid	0.400	0.360	0.240	6.54
11	centroid	0.400	0.360	0.240	6.82
12	centroid	0.400	0.360	0.240	6.41

centroid shown in Table 11.4. One additional edge centroid at the top of the constrained region (shown in Figure 11.20) was found that is not shown in Table 11.4.

Once a list of candidates is found there are different options for creating the design. A first option is to select the status variable for each run in the list of candidate points and choose to include or exclude the point from the design. By doing this, a design with the exact mixture combinations shown in Table

CONSTRAINED MIXTURE EXPERIMENTS

11.4 can be constructed. A second option is to select a D-optimal or A-optimal subset of the candidate list. When using either of these options, the user is prompted to select the model for the data, since the X matrix depends on the model, and the number of runs included in the design. By selecting the special cubic model and asking for nine points to be included in the design, ADX produces the design points shown in Figure 11.21. Once a design is selected, individual points can be replicated by selecting the customize button on the main ADX template.

Figure 11.21 *Design Points for D-optimal Design for Special Cubic Model for Polvoron Experiment*

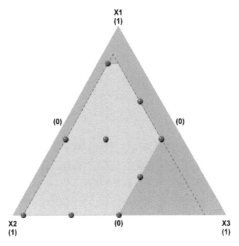

11.5.2 Analysis of Constrained Mixture Experiments

The models used to analyze constrained mixture experiments are the same as those used for unconstrained problems, unless one of the constraints is an equality (such as an ionic balance constraint). In that case the constraint must be incorporated into the model, resulting in a loss of dimension (see Schrevens and DeRijck, 2008, for an example).

Analysis of the data in Table 11.4 with standard Scheffé models will be illustrated in ADX. Figure 11.22 shows the results of fitting the special cubic model to the data in Table 11.4. Although two quadratic terms and the third order term are insignificant, the model fits the data well (shown in Figure 11.23). The reduced model that included only the terms x_1, x_2, x_3, and x_1x_2 had an adjusted R^2 of 0.754 compared to 0.9516 shown for the special cubic model in Figure 11.22, and it fit worse at three interior checkpoints that were made by the experimenters to check the model. These checkpoints are not shown in Table 11.4.

Figure 11.24 shows a contour plot over the constrained region produced by

Figure 11.22 *ADX Model Fit for Polvoron Data*

Effect	Estimate	Std Error	t Ratio	P Value
X1	4.4259	0.44827	9.8733	0.0002
X2	3.5181	0.30787	11.427	<.0001
X3	1.2367	1.615	0.76574	0.4784
X1*X2	6.9004	2.0179	3.4196	0.0188
X1*X3	8.9528	4.1427	2.1611	0.0831
X2*X3	5.3135	3.4988	1.5187	0.1893
X1*X2*X3	25.546	11.202	2.2804	0.0715

Figure 11.23 *Special Cubic Model Summary*

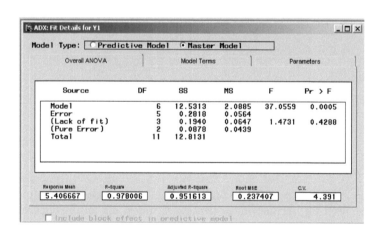

ADX, and Figure 11.25 shows the response trace plot for the three components. The conclusion the authors reached (that can be verified with Figures 11.24 and 11.25) was that mixtures containing 40 - 54% sugar (x_1), 22 - 36% peanut fines (x_2), and 24% butter (x_3) would have consumer acceptance ratings greater than 6.0.

The same model shown in Figures 11.22 and 11.23 can be fit using `proc glm` as shown in the commands below Figure 11.25.

Figure 11.24 *ADX Contour Plot in Restricted Region*

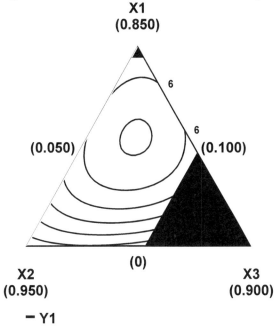

Figure 11.25 *ADX Response Trace Plot*

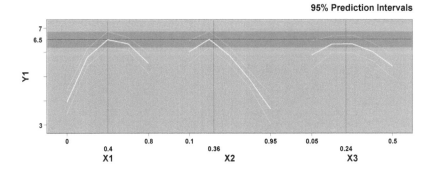

```
data polvoron;
input x1 x2 x3 y;
datalines;
.8 .15 .05 5.33
.4 .55 .05 5.87
    ...
proc print; run;
proc glm;
model y=x1 x2 x1*x2 x1*x3 x2*x3 x1*x2*x3/p solution;
estimate 'beta1' intercept 1 x1 1;
estimate 'beta2' intercept 1 x2 1;
estimate 'beta3' intercept 1;
```

By leaving x_3 out of the model and not using the /noint option, proc glm will calculate R^2 correctly for the Scheffé model, and by including the three estimate statements, proc glm will produce the estimates of β_1, β_2, and β_3 from the Scheffé quadratic model.

11.6 Blocking Mixture Experiments

Whenever experimental units for a mixture experiment (which may simply be the conditions that exist at the time a mixture is made and tested) are not homogeneous, a blocked design should be used. Levels of a blocking factor could include things such as the batches of mixture components or the times or different pieces of equipment used to test and measure a response. If runs can be blocked into groups that are large enough to include a complete replicate of the mixture experiment, a complete block design can be utilized. On the other hand, if the block size is smaller than a complete replicate of the mixture design, some kind of incomplete block design must be used.

In the discussion of incomplete block designs for 2^k factorials in Section 7.6 and of standard response surface experiments in Section 10.8, an important feature of the blocked design was to have the blocks orthogonal to the coded factor levels. In this way, the factor effects were unconfounded with the block effects or unaffected by the block differences. It is more difficult to achieve orthogonal incomplete block designs for mixture experiments because, except for the SLD{k,1}, the mixture designs themselves are not orthogonal.

If a mixture experiment in which N not necessarily unique mixture blends are to be arranged in t blocks, where the wth block contains n_w blends and $n_1 + n_2 + \cdots + n_t = N$, then Nigam (1970), Nigam (1977) and John (1984) suggested conditions on the moments that ensure orthogonal blocks. For the Scheffé second-degree model these conditions are

$$\sum_{u=1}^{n_w} x_{ui} = c_i \text{ for each block, } i = 1, 2, \ldots, k$$

$$\sum_{u=1}^{n_w} x_{ui}x_{uj} = c_{ij} \text{ for each block, } i, j = 1, 2, \ldots, k; \; i \neq j. \qquad (11.21)$$

This means that the sum of proportions for component i must be the same value, c_i, in each of the blocks, but it is not necessary that $c_1 = c_2, \ldots = c_k$; and the sum of the products of proportions for components i and j are the same within each block, but it is not necessary that $c_{ij} = c_{i'j'}$.

Cornell (2002) presented an example of an orthogonally blocked mixture design in three components (shown in Table 11.5). From this table, it can be verified that the conditions shown in Equation (11.21) hold, and this design can be visualized in Figure 11.26. The first block consists of all the mixtures in an SLD{3,2} design, and the second block consists of an SLD{3,1} augmented with three axial points. While the order of making and testing the mixtures for designs presented earlier in this chapter should be conducted in a completely

BLOCKING MIXTURE EXPERIMENTS

random order, the mixtures in Table 11.5 should be run one block at a time randomizing the order within the blocks.

Table 11.5 *Orthogonally Blocked Simplex Lattice Design Augmented by Axial Points*

	Block 1			Block 2		
run	x_1	x_2	x_3	x_1	x_2	x_3
1	1	0	0	1	0	0
2	0	1	0	0	1	0
3	0	0	1	0	0	1
4	$\frac{1}{2}$	$\frac{1}{2}$	0	$\frac{2}{3}$	$\frac{1}{6}$	$\frac{1}{6}$
5	$\frac{1}{2}$	0	$\frac{1}{2}$	$\frac{1}{6}$	$\frac{2}{3}$	$\frac{1}{6}$
6	0	$\frac{1}{2}$	$\frac{1}{2}$	$\frac{1}{6}$	$\frac{1}{6}$	$\frac{2}{3}$

Figure 11.26 *Diagram of Design in Table 11.5*

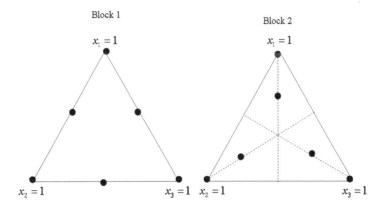

The SAS commands at the top of the next page can be used to analyze the blocked design in Table 11.5, and the fact that type I and type III sums of squares for blocks are the same shows that blocks are orthogonal to the terms in the Scheffé model.

```
data symsplx;
  input block x1 x2 x3 y;
  block=(block-1.5)/0.5;
  x12=x1*x2;  x13=x1*x3;  x23=x2*x3;
datalines;
   . . .
proc glm;
  model y2=block x1|x2|x3@2/noint solution;
run;
```

Orthogonal block designs can also be created for constrained mixture problems. For a three-component constrained mixture problem, where each component has the same upper and lower bounds (U, L) and the design region is hexagonal as shown in Figure 11.27, Cornell (2002) showed that an orthogonally blocked mixture design can be created using the two Latin squares shown in Table 11.6.

In this table, the Latin square treatment indicator A refers to the common lower constraint (L), B refers to the common upper constraint (U), and C the midpoint between A and B. Again, it can be verified that the conditions shown in Equation (11.21) hold for Table 11.6 when the upper and lower constraints are such that the constrained region is a hexagon. A centroid of the constrained region is added to each block so that all six coefficients in the Scheffé quadratic model can be estimated. Figure 11.27 shows that the extreme vertices of the constrained region are divided between the two blocks and a centroid is added to each block.

Table 11.6 *Orthogonally Blocked Constrained Mixture Using Latin Squares*

	Block 1				Block 2		
Blend	x_1	x_2	x_3	Blend	x_1	x_2	x_3
1	A^1	B	C	4	A	C	B
2	B^2	C	A	5	B	A	C
3	C^3	A	B	6	C	B	A
centroid	D^4	D	D	centroid	D	D	D

[1] A=Lower Constraint,
[2] B=Upper Constraint,
[3] C=(A+B)/2
[4] D=(A+B+C)/3

Other designs have been derived for orthogonally blocked constrained and unconstrained mixture experiments. Cornell (2002) summarizes many of these, but he also states that the number of practical designs that satisfy Equation (11.21) are rare.

Another approach to find blocked mixture designs that can be accomplished easily with SAS is to find a D_s optimal design using **proc optex** (this is sim-

BLOCKING MIXTURE EXPERIMENTS

Figure 11.27 *Orthogonally Blocked Constrained Mixture Design*

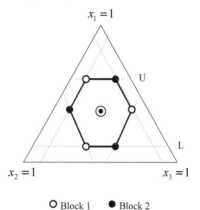

○ Block 1   ● Block 2

ilar to what was done for response surface designs in Section 10.8). Atkinson et al. (2007) show several examples of D-optimal blocked mixture experiments. Although D-optimal blocked designs are not orthogonally blocked, Goos and Donev (2006) show that they are actually more efficient in the sense of reducing variances of the estimated coefficients, and they can be created for a wide variety of situations. For example, the SAS code below shows how to create an unconstrained mixture experiment in three components blocked into two blocks of 4. In this example, a set of candidate mixtures is created using a simplex lattice{3,3} design or SLD{3,3} (created by the macro %adxsld) augmented by interior points created by the macro %adxfill that averages each pair of points.

```
%adxgen
%adxmix
%adxinit
%adxsld(scd,x1 x2 x3,3)
%adxfill(scd,x1 x2 x3)
proc print; run;
proc optex data=scd coding=orthcan seed=2347;
  model x1|x2|x3@2/noint;
  blocks structure=(2)4;
  generate n=8 niter=1000 keep=10;
  output out=bdesign blockname=block; run;
proc print data=bdesign; run;
```

The D-optimal design is not unique, but it always consists of points around the perimeter of the simplex, as shown in Figure 11.28, rather than including interior points, as did the orthogonally blocked design shown in Figure 11.26.

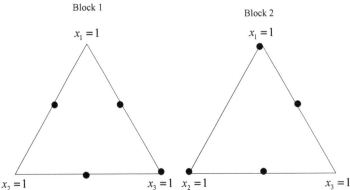

Figure 11.28 *D-optimal Blocked Mixture Experiment*

By changing the file of candidate mixtures, D-optimal blocked designs for constrained mixture experiments can also be created easily. To illustrate this, consider the example presented by Goos and Donev (2006). The experiment investigated the surface adhesive properties of polypropylene used in the Belgian car industry. The mixture components were x_1: talcum, x_2: mica, and x_3: polypropylene. The following constraints applied:

$$0.0 \leq x_1 \leq 0.1$$
$$0.0 \leq x_2 \leq 0.1$$
$$0.0 \leq x_1 + x_2 \leq 0.15$$

Ten tests could be run in two days, so the experiment had to be blocked in two sets of five.

The SAS code below creates a candidate file of extreme vertices, using the macro **%adxxvert**, augmented with interior points created by **%adxfill**. The next data step removes the vertices that violate the last constraint, and finally **proc optex** is used to create the design in two blocks of 5.

```
%adxgen
%adxmix
%adxxvert(ev,x1 0-.1/ x2 0-.1/x3);
%adxfill(ev,x1 x2 x3);
data ev; set ev;
if 0 <= x1+x2 <= 0.1501;
proc optex data=ev coding=orthcan seed=2347;
   model x1|x2|x3@2/noint;
   blocks structure=(2)5;
   generate n=10 niter=1000 keep=10;
   output out=bdesign blockname=block; run;
proc print data=bdesign; run;
```

A printout of the resulting file **work.bdesign** is shown on the next page.

MIXTURE EXPERIMENTS WITH PROCESS VARIABLES

Obs	BLOCK	x1	x2	x3
1	1	0.10	0.05	0.85
2	1	0.10	0.00	0.90
3	1	0.00	0.00	1.00
4	1	0.05	0.10	0.85
5	1	0.00	0.05	0.95
6	2	0.10	0.05	0.85
7	2	0.05	0.05	0.90
8	2	0.05	0.00	0.95
9	2	0.00	0.10	0.90
10	2	0.00	0.00	1.00

11.7 Mixture Experiments with Process Variables

In some mixture experiments, the qualities or characteristics of the product are influenced by process variables in addition to the proportions of the mixing components. For example, the strength of carbide ceramics used in advanced heat engines depends not only on the proportions of the mixture components but also on the sintering time and sintering temperature. In mixture experiments involving process variables, or MPV experiments, let z_l represent the coded level of the lth process variable. When there are three mixture components and one process variable z_1, the experimental region changes from a simplex like that shown in Figure 11.2 to a prism, as shown in Figure 11.29.

Figure 11.29 *Experimental Region with Three Mixture Components and One Process Variable*

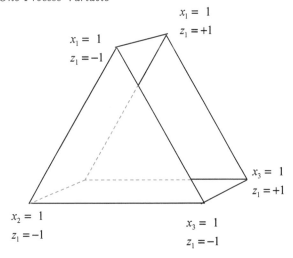

As an example of a mixture experiment with a process variable, or MPV, consider the situation studied by Chau and Kelly (1993). They studied the opacity of a printable coating material used for identification labels and tags.

The coating material was a mixture of two pigments, x_1 and x_2. and a polymeric binder (x_3). The opacity of the coating was not only influenced by the mixture of the three components, but also by the thickness of the coating (z). Constraints on the component proportions were

$$0.13 \leq x_1 \leq 0.45$$
$$0.21 \leq x_2 \leq 0.67$$
$$0.20 \leq x_3 \leq 0.34$$

If two levels (-1 and +1) of the coded process variable z are used, an appropriate experimental design would consist of the extreme vertices and edge centroids of the mixture space crossed with each level of the process variable. This list can be easily created with the SAS commands below. The macro %adxxvert creates the file work.ev that contains a list of the extreme vertices of the mixture region including the edge centroids and the overall centroid. The data step copies this file, deletes the overall centroid, and repeats each line at the low and high levels of the process variable z. The results are output to the file work.mp.

```
%adxgen
%adxmix
%adxinit
%adxxvert(ev,x1 .13-.45/x2 .21-.67/x3 .20-.34,2)
data mp; set ev; if dimen<2;
  z=-1; output;
  z= 1; output;
proc print data=mp; var x1 x2 x3 z; run;
```

The printout of work.mp appears below. The same design can be created using the SAS ADX system by specifying one process variable with two levels in addition to the three constrained mixture components.

Obs	x1	x2	x3	z
1	0.45	0.21	0.34	-1
2	0.45	0.21	0.34	1
3	0.45	0.35	0.20	-1
4	0.45	0.35	0.20	1
5	0.13	0.67	0.20	-1
6	0.13	0.67	0.20	1
7	0.13	0.53	0.34	-1
8	0.13	0.53	0.34	1
9	0.13	0.60	0.27	-1
10	0.13	0.60	0.27	1
11	0.45	0.28	0.27	-1
12	0.45	0.28	0.27	1
13	0.29	0.51	0.20	-1
14	0.29	0.51	0.20	1
15	0.29	0.37	0.34	-1
16	0.29	0.37	0.34	1

11.7.1 Models for Mixture Experiments With Process Variables

The model for a mixture experiment with process variables results from crossing the model for the mixture components and the model for the process variables. For fixed levels of the process variables, the model for the mixture components is

$$\eta_x = f(x), \tag{11.22}$$

and the model for the process variables is

$$\eta_z = g(z), \tag{11.23}$$

where $f(x)$ is the Scheffé linear, quadratic, or cubic mixture model, and $g(z)$ is the linear or general quadratic model for independent factors. The combined model is then found as the cross-product $\eta_{xz} = (f(x)) \times (g(z))$. For example, assuming a quadratic mixture model for the coating experiment where there were three mixture components and one two-level process variable,

$$f(x) = \beta_1 x_1 + \beta_2 x_2 + \beta_3 x_3 + \beta_{12} x_1 x_2 + \beta_{13} x_1 x_3 + \beta_{23} x_2 x_3,$$

and

$$g(z) = \alpha_0 + \alpha_1 z.$$

The combined model is

$$\begin{aligned}\eta_{xz} =& \beta_1(z)x_1 + \beta_2(z)x_2 + \beta_3(z)x_3 + \beta_{12}(z)x_1 x_2 + \beta_{13}(z)x_1 x_3 \\ &+ \beta_{23}(z)x_2 x_3,\end{aligned} \tag{11.24}$$

where $\beta_i(z) = \beta_i \times (\alpha_0 + \alpha_1 z)$.

Multiplying the terms we can write this model as:

$$\begin{aligned}\eta_{xz} =& \gamma_1^0 x_1 + \gamma_2^0 x_2 + \gamma_3^0 x_3 + \gamma_{12}^0 x_1 x_2 + \gamma_{13}^0(z)x_1 x_3 + \gamma_{23}^0(z)x_2 x_3 + \\ & \gamma_1^1 x_1 z + \gamma_2^1 x_2 z + \gamma_3^1 x_3 z + \gamma_{12}^1 x_1 x_2 z + \gamma_{13}^1 x_1 x_3 z + \gamma_{23}^0 x_2 x_3 z,\end{aligned} \tag{11.25}$$

where $\gamma_i^0 = \beta_i \times \alpha_0$ and $\gamma_i^1 = \beta_i \times \alpha_1$.

In this model, the first six terms represent the linear and quadratic blending of the mixture components since these terms involve only the mixture components. The last six terms in the model represent the changes in the linear and quadratic blending effects caused by changes in the process variable.

If the process variable has a constant effect on the response, which would cause the same change in all blends of the mixture components, there is no term in model 11.25 that can represent the constant effect of the process variable. Gorman and Cornell (1982) have suggested an alternate form of the model substituting $(1 - x_2 - x_3)$ for x_1 in the term $\gamma_1^1 x_1 z$, resulting in the model

$$\eta_{xz} = \gamma_1^0 x_1 + \gamma_2^0 x_2 + \gamma_3^0 x_3 + \gamma_{12}^0 x_1 x_2 + \gamma_{13}^0(z) x_1 x_3 + \gamma_{23}^0(z) x_2 x_3 +$$
$$\gamma_1^1 z + \gamma_2^2 x_2 z + \gamma_3^2 x_3 z + \gamma_{12}^1 x_1 x_2 z + \gamma_{13}^1 x_1 x_3 z + \gamma_{23}^0 x_2 x_3 z, \quad (11.26)$$

where $\gamma_i^2 = \gamma_i^1 - \gamma_1^1$ for $i = 2, 3$. If the process variable has a constant effect on the response, the last five terms in the model would be zero, and γ_1^1 will represent the constant effect of the process variable.

Another model that allows isolation of process variable effects can be obtained by first translating the mixture components (x_1, x_2, and x_3) to independent factors (w_1, w_2) as shown in Section 11.2.2, then writing the combined model as the cross-product of the model for the w's and the z's.

11.7.2 Multiple Process Variables and Mixture Components

Figure 11.30 shows the design for an MPV with three mixture components and two process variables. The 6-run SLD{3,2} for the mixture components is crossed with a 4-run 2^2 design in the process variables, resulting in a product array of $6 \times 4 = 24$ experiments. The runs in the design can be represented graphically as a 2^2 design with a simplex-lattice repeated at each corner, as shown on the left side of the figure, or as a simplex lattice with a 2^2 design repeated at each lattice point as shown on the right side of the figure.

Figure 11.30 *Experimental Region with Three Mixture Components and Two Process Variables*

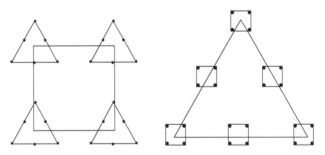

This design can be easily created in SAS by repeating the six runs of the simplex-lattice at each run of a 2^2 factorial created by `proc factex` as shown in the code at the top of the next page.

```
*Create Simplex-lattice design;
%adxgen
%adxmix
%adxinit
%adxsld(sld, x1 x2 x3,2)
*Repeat the sld for each run in a 2x2 factorial
    design;
proc factex;
  factors z1 z2;
  size design=4; model estimate=(z1|z2);
  output out=combdes pointrep=sld;
proc print; run;
```

However, as the number of mixture components and process variables increases, cross-product designs can become unwieldy. For example, if there are four mixture components and five process variables, the simplex-lattice{4,2} design has only 10 runs, but the 2^5 design in the process variables has 32 runs, and the cross-product design has 320 runs.

In most industrial experimentation, restrictions due to time and cost prohibit such large designs. Cornell and Gorman (1984) proposed running fractions of the 2^p design in the process variables. Linear designs in the mixture components can also be employed to reduce the total number of runs. For example, in the case of four mixture components and five process variables, if an 8-run resolution III fractional factorial in the process variables is substituted for the 2^5 design, and a 4-run SLD{4,1} is substituted for the SLD{4,2}, the total number of experiments could be reduced to $4 \times 8 = 32$ and would allow for fitting the model

$$\begin{aligned}\eta_{xz} =& \gamma_1^0 x_1 + \gamma_2^0 x_2 + \gamma_3^0 x_3 + \gamma_4^0 x_4 + \\ & \gamma_1^1 x_1 z_1 + \gamma_2^1 x_2 z_1 + \gamma_3^1 x_3 z_1 + \gamma_4^1 x_4 z_1 + \\ & \gamma_1^2 x_1 z_2 + \gamma_2^2 x_2 z_2 + \gamma_3^2 x_3 z_2 + \gamma_4^2 x_4 z_2 + \\ & \gamma_1^3 x_1 z_3 + \gamma_2^3 x_2 z_3 + \gamma_3^3 x_3 z_3 + \gamma_4^3 x_4 z_3 + \\ & \gamma_1^4 x_1 z_4 + \gamma_2^4 x_2 z_4 + \gamma_3^4 x_3 z_4 + \gamma_4^4 x_4 z_4 + \\ & \gamma_1^5 x_1 z_5 + \gamma_2^5 x_2 z_5 + \gamma_3^5 x_3 z_5 + \gamma_4^5 x_4 z_5\end{aligned}$$

But, this model only allows for linear blending and seeing how the predicted response at the pure components (as shown in Figure 11.4) is affected by the process variables. On the other hand, if the experimenter can propose a reasonable model to investigate, an economical design can be found using optimal design theory.

For example, the 40-term model in Equation (11.27) on the next page, that is a subset of the terms in an expanded version of Equation (11.26), was proposed by Kowalski et al. (2000). It includes the quadratic mixture model as well as interactions among the main effects of the process variables and interactions between the main effects of the process variables and the linear blending terms of the mixture components. In their article, Kowalski *et al.* (2000) considered a simple situation with three mixture components and two process variables, and they proposed two specific designs to allow fitting a subset of model 11.27. Their designs performed well in terms of D-efficiency for the specific situation with three mixture components and two process variables, but for the more general case, a D-optimal search will find a reasonable design.

$$\begin{aligned}\eta_{xz} =& \gamma_1^0 x_1 + \gamma_2^0 x_2 + \gamma_3^0 x_3 + \gamma_4^0 x_4 + \gamma_{12}^0 x_1 x_2 + \gamma_{13}^0 x_1 x_3 + \gamma_{14}^0 x_1 x_4 + \gamma_{23}^0 x_2 x_3 \\ &+ \gamma_{24}^0 x_2 x_4 + \gamma_{34}^0 x_3 x_4 + \gamma_{12}^1 z_1 z_2 + \gamma_{13}^1 z_1 z_3 + \gamma_{14}^1 z_1 z_4 + \gamma_{15}^1 z_1 z_5 + \gamma_{23}^1 z_2 z_3 \\ &+ \gamma_{24}^1 z_2 z_4 + \gamma_{25}^1 z_2 z_5 + \gamma_{34}^1 z_3 z_4 + \gamma_{35}^1 z_3 z_5 + \gamma_{45}^1 z_4 z_5 + \gamma_{11}^2 x_1 z_1 + \gamma_{12}^2 x_1 z_2 \\ &+ \gamma_{13}^2 x_1 z_3 + \gamma_{14}^2 x_1 z_4 + \gamma_{15}^2 x_1 z_5 + \gamma_{21}^2 x_2 z_1 + \gamma_{22}^2 x_2 z_2 + \gamma_{23}^2 x_2 z_3 + \gamma_{24}^2 x_2 z_4 \\ &+ \gamma_{25}^2 x_2 z_5 + \gamma_{31}^2 x_3 z_1 + \gamma_{32}^2 x_3 z_2 + \gamma_{33}^2 x_3 z_3 + \gamma_{34}^2 x_3 z_4 + \gamma_{35}^2 x_3 z_5 + \gamma_{41}^2 x_4 z_1 \\ &+ \gamma_{42}^2 x_4 z_2 + \gamma_{43}^2 x_4 z_3 + \gamma_{44}^2 x_4 z_4 + \gamma_{45}^2 x_5 z_5 \end{aligned} \quad (11.27)$$

At least 40 experiments must be performed to collect the data necessary for estimating the coefficients in model 11.27, and a 44-run D-optimal set can be easily obtained using **proc optex** as shown in the commands below.

```
*Create Simplex-lattice design;
%adxgen
%adxmix
%adxinit
%adxsld(sld, x1 x2 x3 x4,2)
*Repeat sld for each run in 2^5 factorial design;
proc factex;
   factors z1 z2 z3 z4 z5;
   size design=32;
   output out=combdes pointrep=sld;
run;
*Use Combined Design as candidates and find D-optimal design;
proc optex data=combdes coding=orthcan seed=2347;
   model x1|x2|x3|x4@2 z1*z2 z1*z3 z1*z4 z1*z5 z2*z3 z2*z4
         z2*z5 z3*z4 z3*z5 z4*z5 x1*z1 x1*z2 x1*z3 x1*z4
         x1*z5 x2*z1 x2*z2 x2*z3 x2*z4 x2*z5 x3*z1 x3*z2
         x3*z3 x3*z4 x3*z5 x4*z1 x4*z2 x4*z3 x4*z4 x4*z5
         /noint;
   generate n=44 niter=1000 keep=10;
   output out=dopt; run;
proc print data=dopt; run;
```

In these commands, the %adxsld macro is used to create the 10-run SLD{4,2} design, and **proc factex** is used to combine this design with the 2^5 design. This combined design, (work.combdes) is used as a set of candidates for **proc optex**, which finds a 44-run D-optimal subset.

MIXTURE EXPERIMENTS WITH PROCESS VARIABLES   481

11.7.3 An Example of a Mixture Experiment with Process Variables (MPV)

To illustrate the analysis of a mixture experiment with process variables, consider the problem discussed by Sahni et al. (2009). They studied a process to produce low-fat mayonnaise. The product was a mixture of three components x_1: stabilizer, x_2: starch 1, and x_3: starch 2. The response they were interested in was the viscosity of the final product that was influenced not only by the ingredients but also by two process variables: z_1: heat exchanger temperature and z_2: the flow rate through the system. The goal was to achieve a viscosity of 3657 at the lowest cost. The constraints on the mixture components are shown below

$$0.0 \le x_1 \le 0.0549$$
$$0.0 \le x_2 \le 0.9725$$
$$0.0 \le x_3 \le 0.9725,$$

and the constrained experimental region for the mixture components is shown as the shaded region in the simplex on the left side of Figure 11.31. This figure diagrammatically represents the subset of the authors' experiments that we will discuss. The dots on the constrained mixture region indicate the actual mixtures tested. Each of these seven mixtures was tested at five combinations

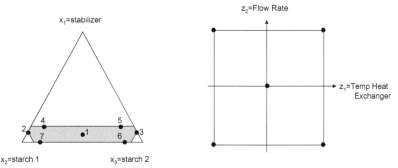

Figure 11.31 *Design for Mayonnaise Formulation Experiment*

of the two process variables as shown in the right side of Figure 11.31, resulting in a total of $7 \times 5 = 35$ experiments. A Scheffé quadratic model was used for the mixture components, and the model $\eta_z = \alpha_0 + \alpha_1 z_1 + \alpha_2 z_2 + \alpha_{12} z_1 z_2$ for the process variables. By crossing the two models, the resulting combined model is

$$\begin{aligned}\eta_{xz} =& \gamma_1^0 x_1 + \gamma_2^0 x_2 + \gamma_3^0 x_3 + \gamma_{12}^0 x_1 x_2 + \gamma_{13}^0 x_1 x_3 + \gamma_{23}^0 x_2 x_3 \\
& + \gamma_1^1 x_1 z_1 + \gamma_2^1 x_2 z_1 + \gamma_3^1 x_3 z_1 + \gamma_{12}^1 x_1 x_2 z_1 + \gamma_{13}^1 x_1 x_3 z_1 + \gamma_{23}^1 x_2 x_3 z_1 \\
& + \gamma_1^2 x_1 z_2 + \gamma_2^2 x_2 z_2 + \gamma_3^2 x_3 z_2 + \gamma_{12}^2 x_1 x_2 z_2 + \gamma_{13}^2 x_1 x_3 z_2 + \gamma_{23}^2 x_2 x_3 z_2 \\
& + \gamma_1^{12} x_1 z_1 z_2 + \gamma_2^{12} x_2 z_1 z_2 + \gamma_3^{12} x_3 z_1 z_2 + \gamma_{12}^{12} x_1 x_2 z_1 z_2 + \gamma_{13}^{12} x_1 x_3 z_1 z_2 \\
& + \gamma_{23}^{12} x_2 x_3 z_1 z_2.\end{aligned}$$

The viscosity measurements in Pa·s for the 56 experiments are shown in Table 11.7.

Table 11.7 *Viscosity Response Measurements (Pa·s) for Mixture-Process Variable Experiment with Mayonnaise*

	Mixture Components			Process Variables Factorial + Center				
#	x_1	x_2	x_3	(-1,1)	(1,1)	(-1,-1)	(1,-1)	(0,0)
1	0.0241	0.6018	0.3741	3010	3480	2780	4100	3840
2	0.0275	0.9725	0	8510	5670	7060	5210	6320
3	0.0275	0	0.9725	1600	2580	1660	2440	2210
4	0	0.6667	0.3333	4560	4350	3990	4130	5210
5	0	0.3333	0.6667	1930	3080	1810	3340	2600
6	0.0549	0.63	0.3151	1900	4740	2160	4330	2780
7	0.0549	0.3151	0.63	1780	3750	2000	3350	3140

The SAS commands to fit the combined model to the data in Table 11.7 are shown below.

```
data MPV;
input x1 x2 x3 z1 z2 y ;
datalines;
0.0241 0.6018 0.3741 -1 1 3010
0.0275 0.9725 0      -1 1 8510
0.0275 0      0.9725 -1 1 1600
0      0.6667 0.3333 -1 1 4560
   . .
proc glm data=MPV;
  model y=x1 x2 x3 x1*x2 x1*x3 x2*x3
          x1*z1 x2*z1 x3*z1 x1*x2*z1 x1*x3*z1 x2*x3*z1
          x1*z2 x2*z2 x3*z2 x1*x2*z2 x1*x3*z2 x2*x3*z2
          x1*z1*z2 x2*z1*z2 x3*z1*z2 x1*x2*z1*z2 x1*x3*z1*z2
          x2*x3*z1*z2/noint solution;
run;
```

Since the target viscosity of 3657 Pa·s is within the constrained mixture region for all combinations of the process variables, the processing cost can be reduced by first choosing to run at the lowest heat exchanger temperature ($z_1 = -1$

MIXTURE EXPERIMENTS WITH PROCESS VARIABLES

in coded units) and the fastest flow rate ($z_2 = +1$ in coded units). By setting $z_1 = -1$ and $z_2 = +1$, the fitted equation simplifies to

$$\text{Predicted viscosity} = 229186.40x_1 + 11487.06x_2 + 817.8356x_3 - 346975x_1x_2 \\ - 194919x_1x_3 - 12628.5x_2x_3.$$

Additional costs were incurred for the three raw materials and they could be summarized in the simple equation shown below (in units of Norwegian Kroner, NOK), where mixture component x_1: stabilizer is the most costly ingredient.

$$\text{Raw Material Cost} = 54.59x_1 + 5.69x_2 + 7.49x_3$$

To find the mixture combination that minimizes cost while meeting the target viscosity, nonlinear optimization similar to the example shown in Section 10.7.5 can be used. The SAS `proc nlp` commands to find the optimum are shown in the listing below. This example uses `proc nlp` to minimize cost of the raw materials subject to constraints. The bounds on the mixture components are given in the bounds statement, the constraint that the mixture components add to one is given as a linear constraint, and the constraint that the predicted viscosity is 3657 is given as a nonlinear constraint.

```
proc nlp;
min cost;
parms x1 x2 x3 = .0241 .6018 .3741;
bounds 0 <= x1 <= 0.0549,
       0 <= x2 <= 0.9725,
       0 <= x3 <= 0.9725;
lincon x1+x2+x3=1;
nlincon viscos=3657;
viscos=229186.4*x1+11487.06*x2+817.8356*x3-346975*x1*x2-194919
    *x1*x3-12628.5*x2*x3;
cost=54.59*x1+5.69*x2+7.49*x3;
run;
```

The resulting minimum cost for raw materials is 6.49 NOK at the combination of mixture components $x_1 = 0.0$, $x_2 = 0.558$, and $x_3 = 0.442$, where none of the more expensive stabilizer is used. This optimum can be visualized on the contour plot of predicted viscosity over the constrained mixture region shown in Figure 11.32.

Figure 11.32 *Contour Plot of Viscosity at Low Temperature, High Flow Rate*

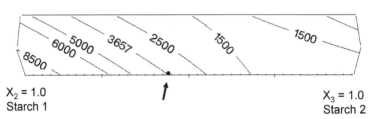

11.8 Mixture Experiments in Split-Plot Arrangements

When running mixture experiments with process variables, experiments are often large, and due to the combination of mixture proportions and process variables, it may be inconvenient to run all combinations in a random order. For example, the mayonnaise experiments were actually run by making large batches of each of the seven mixtures on separate days; then while running each batch through the processing equipment, the combinations of the heat exchanger temperature and flow rate were varied. The specific mixtures were randomized to seven days, and the order of the processing conditions were randomized within a day, but the entire sequence was not randomized. Due to this restricted randomization, the resulting experiment was actually a split-plot experiment like those described in Chapter 8 and Section 10.9. We call the designs for this type of an experiment a split-plot mixture process variable, or SPMPV, experiment.

The whole-plot experimental units were days, and the subplot experimental units were times within a day. The terms in the quadratic Scheffé mixture model were the whole-plot effects, since they were held constant within the whole plots, and the interactions between the mixture model terms and the process variables were the subplot factors. In split-plot experiments, the proper denominator for the F-test on whole-plot effects is the whole-plot mean square, while the proper denominator for the split-plot effects is the error mean square. In the mayonnaise experiments, the whole plot variance component, or day to day variability, was found to be negligible by the authors; therefore, it was dropped from the model resulting in the analysis presented in the last section.

To illustrate how to properly analyze a split-plot mixture process variable experiment when the whole plot variance is not negligible, consider another experiment presented by Kowalski *et al.* (2002). They modified an example taken from Cornell (1988). The experiment involved studying the mixtures of three plasticizers used in producing vinyl for automobile seat covers. The response was the thickness of extruded vinyl, which was affected by two process variables (z_1: extrusion speed and z_2: drying rate) in addition to the proportions of the three plasticizers (x_1, x_2, and x_3).

A simplex centroid design was used for the mixture components, and a 2^2

MIXTURE EXPERIMENTS IN SPLIT-PLOT ARRANGEMENTS

factorial plus three replicate center points were used for the process variables. In this example, the process variables were hard to change and were held constant in blocks, while a random subset of the mixtures in the simplex centroid design was formulated and processed. With this scheme, the process variables were the whole-plot effects, and the mixture components, which were varied within each set of constant process variable settings, were the subplot effects.

To combine the simplex-centroid in the mixture components, with the 2^2 factorial plus three replicate center points in the process variables would require a total of $7 \times 7 = 49$ runs. However, to fit a model of the form of Equation (11.28) (the general form of equation 11.27 proposed by Kowalski et al. (2000)) with only $k = 3$ mixture components and $n = 2$ process variables does not require that many experiments.

$$\eta_{zx} = \sum_{i=1}^{k} \beta_i x_i + \sum_{i<j}^{k} \beta_{ij} x_i x_j + \sum_{l<m}^{n} \alpha_{lm} z_l z_m + \sum_{i=1}^{k} \sum_{l=1}^{n} \gamma_{il} x_i z_l \quad (11.28)$$

The 28 experiments listed data in Table 11.8 are sufficient. This table shows the thickness measurements taken from a fractional split-plot design suggested by Kowalski et al. (2002) for fitting a model of the form of Equation 11.28 with three mixture components and three process variables. The table shows that the vertices of the simplex-centroid design plus one centroid were selected to be run in the whole plots where z_1: extrusion speed was low and z_2: drying time was high (and vice versa), while the binary mixtures plus the centroid were selected to be run in the whole plots where both process variables were either at their high or low settings. Four replicates of the mixture centroid were run in the three whole plots where both process variables were at their mid settings.

Table 11.8 *Vinyl Thickness Values for Example Experiment*

		Mixture Component Combinations						
z_1	z_2	(100)	(010)	(001)	($\frac{1}{2}\frac{1}{2}0$)	($\frac{1}{2}0\frac{1}{2}$)	($0\frac{1}{2}\frac{1}{2}$)	($\frac{1}{3}\frac{1}{3}\frac{1}{3}$)
-1	1	10	8	3				8
1	-1	10	5	9				9
1	1				5	4	7	10
-1	-1				7	8	4	7
0	0							8, 7
								7, 8
0	0							7, 8
								9, 9
0	0							12, 10
								9, 11

486   MIXTURE EXPERIMENTS

In order to get accurate estimates of the effects and their standard errors with data from a split-plot mixture process variable experiment, proc mixed can be used, which by default uses the REML method described in Section 10.9. For example, the SAS commands to fit the model to the data in Table 11.8 are shown below.

```
data SPMPV;
input WP     z1    z2    x1      x2      x3      y ;
datalines;
1      -1     1    1.00    0.00    0.00    10
1      -1     1    0.00    1.00    0.00     8
1      -1     1    0.00    0.00    1.00     3
1      -1     1    0.33    0.33    0.34     8
2       1    -1    1.00    0.00    0.00    10
2       1    -1    0.00    1.00    0.00     5
2       1    -1    0.00    0.00    1.00     9
2       1    -1    0.33    0.33    0.34     9
        ...
5       0     0    0.33    0.33    0.34     8
5       0     0    0.33    0.33    0.34     7
5       0     0    0.33    0.33    0.34     7
5       0     0    0.33    0.33    0.34     8
6       0     0    0.33    0.33    0.34     7
6       0     0    0.33    0.33    0.34     8
6       0     0    0.33    0.33    0.34     9
6       0     0    0.33    0.33    0.34     9
7       0     0    0.33    0.33    0.34    12
7       0     0    0.33    0.33    0.34    10
7       0     0    0.33    0.33    0.34     9
7       0     0    0.33    0.33    0.34    11
proc mixed;
  class wp;
  model y=z1*z2 x1 x2 x3 x1*x2 x1*x3 x2*x3 x1*z1 x2*z1 x3*z1
   x1*z2 x2*z2 x3*z2/noint solution ddfm=kenwardroger;
  random wp;
run;
```

A portion of the resulting output is shown on the next page. There, it can be seen that the variance of whole-plot experimental units is estimated as $\hat{\sigma}^2_{wp} = 1.578$ and the variance of subplot experimental is estimated as $\hat{\sigma}^2_{sp} = 1.9014$. If the standard least squares estimates are used, rather than the REML estimates, the subplot effects will be estimated inaccurately, and their standard errors will be underestimated, as was the case for split-plot response surface designs shown in Table 10.10.

The design shown in Table 11.8 is balanced, in that the whole simplex-centroid design in mixture components is repeated for both levels of one process variable when the other is ignored, and it has center points at $x_i = \frac{1}{3}$ and $z_j = 0$ for $i = 1, 3$ and $j = 1, 2$, which allow for testing the goodness of fit of model 11.28. However, Goos and Donev (2007) point out two shortcomings of the design shown in Table 11.8. The first shortcoming is that the parameters of interest in model 11.28 are not estimated precisely with this design because half of the runs in the design are center points used to estimate pure error and test lack of fit. The second shortcoming is that this design is not flexible for practical situations in which the number of runs possible within each whole

REVIEW OF IMPORTANT CONCEPTS

plot may be dictated by the experimental situation. They propose alternative tailor-made split-plot designs that can be constructed using the algorithms of Goos and Vandebroek (2003) or Jones and Goos (2007). The algorithm of Jones and Goos (2007) is available in the software package JMP, versions 6.0.2 and later.

```
                   Covariance Parameter
                        Estimates

                  Cov Parm        Estimate

                  WP              1.5780
                  Residual        1.9014

                Solution for Fixed Effects

                        Standard
Effect   Estimate        Error       DF    t Value    Pr > |t|

z1*z2    -1.1646        0.7816      3.51    -1.49      0.2199
x1        8.9194        1.2598     14.4      7.08     <.0001
x2        5.4194        1.2598     14.4      4.30      0.0007
x3        4.9242        1.2605     14.4      3.91      0.0015
x1*x2     3.3792        4.8962     13.8      0.69      0.5015
x1*x3     4.4435        4.8649     13.8      0.91      0.3767
x2*x3     9.4435        4.8649     13.8      1.94      0.0729
z1*x1    -2.0143        1.1435     11.8     -1.76      0.1041
z1*x2     0.7357        1.1435     11.8      0.64      0.5323
z1*x3     2.0154        1.1408     11.7      1.77      0.1033
z2*x1    -2.0082        1.1435     11.8     -1.76      0.1050
z2*x2     2.2418        1.1435     11.8      1.96      0.0741
z2*x3    -0.9782        1.1408     11.7     -0.86      0.4084
```

11.9 Review of Important Concepts

Mixture experiments are conducted when the response is a function of the proportion of each component in the design rather than the total amount. When the only restriction is that the component proportions add to one, the simplex lattice or simplex centroid designs shown in Figure 11.33 are appropriate. These designs can be used to fit the Scheffé linear, quadratic, special cubic, or cubic model. When there are constraints on the mixture proportions, an extreme vertices design, or EVD, can be used, which consists of all vertices of the constrained region possibly augmented by edge centroids. When there are many more extreme vertices and edge centroids than there are terms in the Scheffé model, a D-optimal subset can be selected using proc optex. These mixture designs can be created with SAS ADX or the %adx macros, and they should be run in completely random order. Scheffé models can be fit using either SAS ADX or using standard regression procedures like proc glm or proc reg.

When experimental units (normally trials for mixture experiments) are not homogeneous due to changes over time or in processing or measuring conditions, the experiments should be blocked rather than being run in a completely

Figure 11.33 *Design Selection Roadmap*

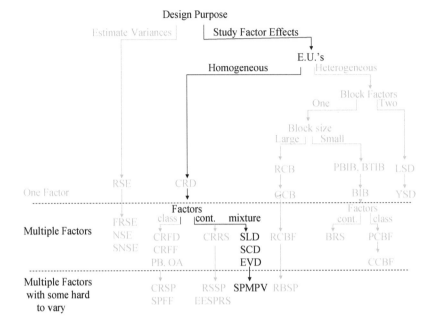

random order. There are two examples of orthogonally blocked mixture experiments in Section 11.6, but in general they are rare. D-efficient blocked mixture experiments can be found using `proc optex` as shown in Section 11.6.

When a mixture experiment includes process variables (MPV), a design and an appropriate model can be found by crossing the appropriate design used for the mixture components with the design used for the process variables. The appropriate models can be found in the same way. In many mixture experiments with process variables, the randomization is restricted, resulting in split-plot mixture process variable experiments (SPMPV). When there is non-negligible whole-plot variability, the REML method in `proc mixed` should be used to get accurate parameter estimates and hypothesis tests.

EXERCISES

11.10 Exercises

1. A mixture experiment was run in three components: x_1, x_2, and x_3.

   (a) If the Scheffé model that best represented the data was $y = 13.5x_1 + 3.2x_2 + 9.8x_3$, then what is the predicted response at the pure blend $x_2 = 1.0$? What is the predicted response at the 50-50 blend of x_1 and x_2?

   (b) If the Scheffé model that best represented the data was $y = 13.5X_1 + 3.2x_2 + 9.8x_3 - 2.9x_1x_2 + 8.7X_1x_3 + 0.9x_2x_3$, and the goal was to increase the response y, then does blending x_1 and x_2 have an synergistic effect or an antagonistic effect? Does blending x_1 and x_3 have a synergistic effect or an antagonistic effect? What is the predicted response at the mixture of equal proportions for all three components?

2. Belloto et al. (1985) studied the relation between $y=$ Soluability of phenobarbital and mixture components x_1: ethanol, x_2: propylene gycol and x_3: water.

   (a) Use the %adx macros to list the experiments required to fit a linear Scheffé model.

   (b) Use the %adx macros to list the experiments required to fit a Scheffé quadratic model.

   (c) Use the %adx macros to list the experiments required to fit a special cubic model.

   (d) Use the %adx macros to list the experiments required to fit a full cubic model.

3. Cornell (2002) describes an experiment to make a fruit punch composed of three types of juice: x_1: watermellon, x_2: pineapple, and x_3: orange. The general acceptance of the fruit punch was to be determined by a taste panel that would rate them on a 1 to 9 scale where 1=dislike extremely, 5=neither like nor dislike and 9=like extremely. If the proportions of the three juices are restricted by cost according to the inequalities below

$$0.20 \leq x_1 \leq 0.80$$
$$0.10 \leq x_2 \leq 0.40$$
$$0.10 \leq x_3 \leq 0.50$$

   (a) Graph the restricted experimental region within the simplex.
   (b) Create a design appropriate for fitting the Scheffé linear model.
   (c) Create a design appropriate for fitting the Scheffé quadratic model.
   (d) Would there be any value in including an overall centroid or other interior points in the design?

4. In agricultural field tests, two or more herbicides are often mixed together in so-called tank mixes in order to find a mixture that is more effective than individual herbicides in controlling a multitude of pest weeds. In a specific test, various mixtures of x_1: a herbicide formulated to control broad leaf weeds, x_2: a herbicide formulated to control grass seedlings, and x_3: a general-purpose herbicide were tested. The data from the tests are shown in the table below. The response is the proportion of weeds controlled.

Table 11.9 *Data from Herbicide Tank Mix Experiment*

Run	Mixture Component			% Weed Control
	x_1	x_2	x_3	y
1	1	0	0	73
2	0	1	0	68
3	0	0	1	80
4	$\frac{1}{2}$	$\frac{1}{2}$	0	77
5	$\frac{1}{2}$	0	$\frac{1}{2}$	86
6	0	$\frac{1}{2}$	$\frac{1}{2}$	75
7	$\frac{1}{3}$	$\frac{1}{3}$	$\frac{1}{3}$	92
8	$\frac{1}{3}$	$\frac{1}{3}$	$\frac{1}{3}$	93
9	$\frac{1}{3}$	$\frac{1}{3}$	$\frac{1}{3}$	88

(a) Create the design shown in the table above using SAS ADX.

(b) Input the data and fit the Scheffé special cubic model.

(c) Make a contour plot of your fitted model and identify a mixture that would result in the highest % weed control.

(d) Transform the mixture components to independent factors and fit the general quadratic model in the independent factors using proc glm or proc reg.

(e) Make a contour plot in the independent factor using proc gcontour as shown in the appendix of this chapter.

5. Consider the problem discussed by Anik and Sukumar (1981). They studied the solubility of a butoconazole nitrate imidazole antifungal agent in a mixture of x_1: polyethylene glycol 400, x_2: glycerin, x_3: polysorbate 60, x_4: water, and x_5: poloxamer 407. Constraints on the mixture components are shown below.

EXERCISES

$$0.10 \leq x_1 \leq 0.40$$
$$0.10 \leq x_2 \leq 0.40$$
$$0.00 \leq x_3 \leq 0.08$$
$$0.30 \leq x_4 \leq 0.70$$
$$x_5 = 0.10$$

(a) Use the %adx macros to find the extreme vertices of the experimental region, and verify that the ones listed in Table 11.10 are correct.
(b) Given the data from Anik and Sukumar (1981) experiments shown in Table 11.10, fit the Scheffé quadratic model to the four mixture components that are not constant.

Table 11.10 *Design and Response Data for Solubility Experiments*

run	x_1	x_2	x_3	x_4	x_5	Solubility, mg/ml
vertices						
1	0.1	0.1	0.0	0.70	0.10	3.0
2	0.1	0.1	0.08	0.62	0.10	7.3
3	0.15	0.4	0.0	0.35	0.10	4.9
4	0.11	0.4	0.08	0.31	0.10	8.4
5	0.4	0.15	0.0	0.35	0.10	8.6
6	0.4	0.11	0.08	0.31	0.10	12.7
centroids						
7	0.1	0.1	0.04	0.66	0.10	5.1
8	0.4	0.13	0.04	0.33	0.10	10.8
9	0.13	0.4	0.04	0.33	0.10	6.6
10	0.216	0.216	0.0	0.468	0.10	4.4
11	0.203	0.203	0.08	0.414	0.10	7.9
12	0.255	0.255	0.08	0.31	0.10	9.4
13	0.275	0.275	0.0	0.35	0.10	5.8
overall centroid						
14	0.21	0.21	0.04	0.44	0.10	5.6

6. Consider the mixture experiment described by Anderson and McLean (1974). They presented a problem where the formula for a flare was obtained by mixing four chemicals x_1: magnesium, x_2: sodium nitrate, x_3: strontium nitrate, and x_4: binder. Constraints on the components were

$$0.40 \le x_1 \le 0.60$$
$$0.10 \le x_2 \le 0.50$$
$$0.10 \le x_3 \le 0.50$$
$$0.03 \le x_4 \le 0.08$$

(a) Given the data from the experiments in Table 11.11, where the response is y=illumination, fit the Scheffé quadratic model to the data.
(b) Fit the model in Equation (11.21) to the data.
(c) Which model appears to fit best? Why? What does this imply?

Table 11.11 *Design and Response Data for Flare Experiments*

run	x_1	x_2	x_3	x_4	y
1	0.4	0.1	0.47	0.03	75
2	0.4	0.1	0.42	0.08	480
3	0.6	0.1	0.27	0.03	195
4	0.6	0.1	0.22	0.08	300
5	0.4	0.47	0.1	0.03	145
6	0.4	0.42	0.1	0.08	230
7	0.6	0.27	0.1	0.03	220
8	0.6	0.22	0.1	0.08	350
9	0.5	0.1	0.345	0.055	220
10	0.5	0.345	0.1	0.055	260
11	0.4	0.2725	0.2725	0.055	190
12	0.6	0.1725	0.1725	0.055	310
13	0.5	0.235	0.235	0.03	260
14	0.5	0.21	0.21	0.08	410
15	0.5	0.2225	0.2225	0.055	425

EXERCISES

7. Shumate and Montgomery (1996) developed a TiW plasma etch process for semiconductor manufacturing. Studying a mixture of three gasses x_1: SR_6, x_2: He and x_3: N_2. The partial pressures of the three gasses were forced to add to a constant value of total pressure to form the mixture constraint. The constraints in micrometers of pressure were

$$100 \leq SF_6 \leq 160$$
$$100 \leq He \leq 160$$
$$SF_6 + He + N_2 = 650$$

Expressing the components as mixture proportions, these constraints are

$$0.153846 \leq x_1 \leq 0.246154$$
$$0.153846 \leq x_2 \leq 0.246154$$

The authors created a design composed of the extreme vertices of the constrained region plus the overall centroid. Six responses were measured for each mixture and the table below shows the responses, and their specifications.

Response	Name	Specification
y_1	Delta critical dimension (CD) positive PR	< 0.25 μm
y_2	Positive PR etch uniformity	$< 2.5\%$
y_3	Delta CD negative PR	< 0.25 μm
y_4	Negative PR etch uniformity	$< 2.5\%$
y_5	Selectivity (TiW: positive PR)	> 0.85
y_6	Selectivity (TiW: negative PR)	> 0.85

(a) Create the design below in SAS ADX and enter data for the six responses that are shown in Table 11.12.

Table 11.12 Data from Plasma Etch Experiment

x_1	x_2	x_3	y_1	y_2	y_3	y_4	y_5	y_6
0.153846154	0.153846154	0.692307692	0.26	0.90	0.24	0.93	0.91	0.94
0.153846154	0.153846154	0.692307692	0.30	0.94	0.22	0.98	0.88	1.00
0.246153846	0.246153846	0.507692308	0.33	1.82	0.14	1.49	0.99	1.04
0.246153846	0.246153846	0.507692308	0.36	1.84	0.14	1.64	1.02	1.04
0.246153846	0.153846154	0.600000000	0.62	7.14	0.22	7.35	0.84	0.84
0.246153846	0.153846154	0.600000000	0.68	7.10	0.17	7.27	0.87	0.87
0.153846154	0.246153846	0.600000000	0.23	2.62	0.25	2.16	0.77	0.92
0.153846154	0.246153846	0.600000000	0.23	2.73	0.31	1.88	0.76	0.88
0.153846154	0.200000000	0.646153846	0.27	3.01	0.15	2.70	0.91	0.98
0.200000000	0.153846154	0.646153846	0.39	4.50	0.17	4.25	0.87	0.92
0.246153846	0.200000000	0.553846154	0.30	4.79	0.20	4.50	0.99	0.96
0.200000000	0.246153846	0.553846154	0.31	2.88	0.19	2.36	0.91	1.01
0.200000000	0.200000000	0.600000000	0.34	1.22	0.13	1.54	0.92	0.99

(b) Fit the Scheffé special cubic model for each response, check the adequacy of the fits, and reduce the model to the quadratic for responses where the ternary mixture coefficient is not significant.

(c) Make contour plots of the fitted surface for each response.

(d) By examining your contour plots, or using the contour optimizer in ADX, find a region in the constrained space where the fitted models predict that all response specifications can be met.

8. If all three components in a mixture design have the same lower and upper constraints (L=0.20, U=0.80), use Table 11.6 to create an orthogonally blocked mixture design, for fitting the Scheffé quadratic model, with two blocks of four mixtures in each block.

EXERCISES 495

9. The numbers in parentheses below are the responses for the design shown in Figure 11.26.

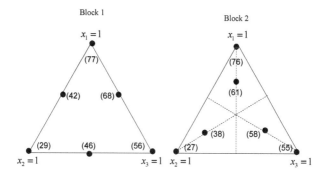

(a) Use the SAS code below Figure 11.26 to fit the Scheffé quadratic model to the data above.
(b) Verify that the design blocks orthogonally by checking to see that the type I and type III sums of squares for blocks are the same.

10. The Table 11.13 below shows the results of Chau and Kelly (1993) mixture process variable experiments (MPV) with printable coating material, which were described in Section 11.7.

Table 11.13 *Mixture Experiments with Printable Coating Material*

Run	x_1	x_2	x_3	z	Opacity
1	0.13	0.53	0.34	-1	0.698
2	0.13	0.53	0.34	-1	0.711
3	0.13	0.53	0.34	1	0.912
4	0.13	0.53	0.34	1	0.930
5	0.13	0.60	0.27	-1	0.700
6	0.13	0.67	0.20	-1	0.710
7	0.13	0.67	0.20	-1	0.680
8	0.13	0.67	0.20	1	0.908
9	0.13	0.67	0.20	1	0.901
10	0.29	0.37	0.34	-1	0.772
11	0.29	0.51	0.20	-1	0.772
12	0.45	0.21	0.34	-1	0.823
13	0.45	0.21	0.34	-1	0.798
14	0.45	0.21	0.34	1	0.992
15	0.45	0.28	0.27	-1	0.818
16	0.45	0.35	0.20	-1	0.802
17	0.45	0.35	0.20	1	0.976
18	0.45	0.35	0.20	1	0.940

(a) Create this design and merge the response values.
(b) Fit a simplified model of the form of Equation (11.28) with only one process variable z.
(c) Does the process variable thickness affect the opacity?
(d) Make a contour plot over constrained mixture space holding the process variable thickness constant at is midpoint (i.e., coded value $z=0$).
(e) Can you identify a region where predicted opacity > 0.86?

11. Steiner et al. (2007) describe a mixture-process variable (MPV) experiment conducted by students to find an optimal homemade bubble solution for use in a battery operated bubble-blowing toy. The response was the average number of bubbles produced by the bubble-blowing toy. The homemade bubble solution was made as a mixture of x_1: dish soap, x_2: water, and x_3: glycerin. The following constraints were placed on the mixture components, based on recipes found in children's books and cost,

$$0.04 \leq x_1 \leq 0.35$$
$$0.60 \leq x_2 \leq 0.98$$
$$0.00 \leq x_3 \leq 0.15.$$

Two process variables were also included in the study, which were z_1: brand of soap (-1=Joy, +1=Ivory) and z_2: type of water (-1=spring water, +1=tap water).

(a) Create a design for this experiment by crossing a design appropriate for fitting a Scheffé quadratic model in the mixture components and a 2^2 factorial in the process variables.
(b) Write the model that can be fit from the design you created.
(c) Given the actual data from Steiner et al. (2007) experiments shown in Table 11.14, that were run in a completely random order, fit the model you propose in (b) to the data.
(d) Is the process water significant? Can it be safely removed from the model?
(e) Make a contour plot over the constrained mixture space for each level of the process variable soap type.
(f) Make a recommendation for the optimal bubble solution.

12. Cornell (1988) presented the original example of mixture experiment for producing vinyl for automobile seat covers. The data is shown in Table 11.15, and the variable names are the same as those given in Section 11.8.

(a) Given that this is a cross between an extreme vertices design (plus the overall centroid appropriate for fitting a Scheffé linear model in the mixture components) and a 2^2 factorial in the process variables, what is the appropriate model for this data?

EXERCISES

Table 11.14 Data from Mixture Process Experiments with Bubble Mixture

| | | | z_1: | -1 | -1 | 1 | 1 |
| | | | z_2: | -1 | 1 | -1 | 1 |
x_1	x_2	x_3					
0.350	0.600	0.05		23.8	14.2	4.2	3.6
0.250	0.600	0.15		12.6	13.6	2.6	3.4
0.350	0.650	0.00		7.8	10.6	2.4	1.0
0.250	0.650	0.10		12.0	12.0	5.8	2.6
0.250	0.750	0.05		12.8	15.4	4.0	4.4
0.150	0.700	0.15		10.8	9.2	3.2	3.4
0.145	0.775	0.08		8.8	11.2	4.4	3.0
0.200	0.800	0.00		10.2	13.6	3.8	2.2
0.040	0.810	0.15		2.4	4.0	3.0	4.2
0.120	0.850	0.03		12.8	8.0	1.6	3.4
0.040	0.880	0.08		2.0	4.6	0.6	0.6
0.050	0.950	0.00		4.0	2.8	3.0	3.0

(b) Fit the model by least squares using `proc glm`.

(c) In this experiment, the runs were not made in a completely random order, but rather randomized within whole plots and run as a split-plot experiment. Refit the model using `proc mixed` including a random term for whole plots like the example shown in Section 11.8. Is there any difference in the parameter estimates you get in (b) and (c)? What does this imply?

Table 11.15 Data for Mixture Experiments for Producing Vinyl

| | | | | z_1: | 1 | -1 | -1 | 1 |
| | | | | z_2: | -1 | 1 | -1 | 1 |
Whole plot	x_1	x_2	x_3					
1	0.85	0	0.15		8	12	7	12
2	0.72	0	0.28		6	9	7	10
3	0.6	0	0.15		10	13	9	14
4	0.47	0.25	0.28		4	6	5	6
5	0.66	0.125	0.215		11	15	9	13
6	0.85	0	0.15		7	10	8	11
7	0.72	0	0.28		5	8	6	9
8	0.6	0	0.15		11	12	10	12
9	0.47	0.25	0.28		5	3	4	5
10	0.66	0.125	0.215		10	11	7	9

11.11 Appendix–Example of Fitting Independent Factors

The SAS code below first illustrates how to transform mixture components to independent factors using the translation and rotation of axis described in Figure 11.9 and Equations (11.11) and (11.12). The data used is the pesticide formulation data from Table 11.1. Next, a grid of independent factors is generated over the simplex experimental region.

```
data form;
input x1 x2 x3 y;
x1x2=x1*x2;  x1x3=x1*x3;  x2x3=x2*x3;
datalines;
1. 0 0 0 48.7
.8 .1 .1 49.5
.6 .2 .2 50.2
.5 0 .5 52.8
.5 .5 0 49.3
.333333 .333333 .333334 51.1
.3 .2 .5 52.7
.3 .5 .2 50.3
.1 .1 .8 60.7
.1 .8 .1 49.9
0 0 1.0 64.9
0 .5 .5 53.5
0 1.0 0 50.6
*Creates independent factors w1 and w2;
data orth; set form;
w1=sqrt(6)*(2*x1-x2-x3);
w2=sqrt(18)*(x2-x3);
* Creates Grid for Predicted Values;
data grid;
do;
 y=.; flag=1;
  do w1=-2.5 to 5.0 by .05;
   do w2=-4.25 to 4.25 by .05;
   output;
   end;
  end;
end;
*Trims Grid to Simplex Region;
data grid; set grid;
if w1<-2.44944 then delete;
if w1<4.89898+((4.89898+2.44944)/4.24264)*w2;
if w1<4.89898-((4.89898+2.44944)/4.24264)*w2;
```

The next section of code illustrates how to use **proc rsreg** to create a grid of predicted values and make a contour plot over the simplex region with **proc gcontour** similar to the way it was done in Section 10.7.1.

APPENDIX–EXAMPLE OF FITTING INDEPENDENT FACTORS 499

```
* Combines data with Grid ;
data grid; set grid orth;
*Uses proc rsreg to fit quadratic model in w1 and w1
     and
 file of predicted values;
proc rsreg data=grid out=predict noprint;
model y=w1 w2/predict;
id flag;
run;
*Reverses direction of w2 axis to match ADX results;
data plot;
   set predict;
w2=-w2;
*Uses proc  gcontour to make the contour plot;
goptions reset=all
         htext=1.3
         ftext=swiss;
proc gcontour data=plot(where=(flag=1));
     plot w1*w2=y/nlevels=6 nolegend autolabel
     levels=50 to 62.5 by 2.5
     vaxis=axis1 haxis=axis2;
  axis1 label=(a=90 h=2.0) value=(h=2.0);
     axis2 label=(h=2.0) value=(h=2.0) ;
  symbol1 height=2.0 color=black;
  symbol2 height=2.0 color=black;
  symbol3 height=2.0 color=black;
  symbol4 height=2.0 color=black;
  symbol5 height=2.0 color=black;
  symbol5 height=2.0 color=black;
  symbol7 height=2.0 color=black;
  run;
```

Figure 11.34 below shows the resulting contour plot. The upper boundaries of the simplex region were added manually using the edit graph menu in SAS.

Figure 11.34 *Contour Plot of Pesticide Formulation Data*

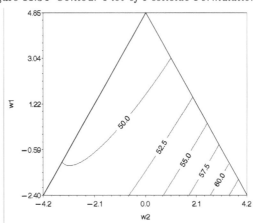

CHAPTER 12

Robust Parameter Design Experiments

12.1 Introduction

In this chapter no new designs are presented, but rather a slightly different application of previous designs is demonstrated. In Chapter 5, sampling experiments were presented for the purpose of characterizing the variability in the response. In the other chapters, the main purpose has been to establish cause and effect relationships between the factors and the average response, so that in the future the average response at selected levels of the factors can be predicted, or the levels of the factors can be chosen to produce a desired value of the response. In this chapter the purpose is to determine what levels of the factors should be chosen to simultaneously produce a desired value of the response and at the same minimize the variability of the response.

After World War II, as Japan was attempting to re-establish telephone communications throughout the country, many problems were encountered because of the poor quality switching systems that had been manufactured in Japan. American advisors such as W.E. Deming visited Japan and taught the principles of quality control that were most useful to Americans during the war production effort. During this industrial recovery period, a Japanese engineer from the Electrical Communications Laboratory, Genichi Taguchi, came to devote his life's work to the application of designed experiments in industry (see Taguchi, 1987). Quality control has been traditionally thought of as an activity that is performed in the manufacturing process, but Taguchi took the ideas of experimental design that he learned at the Statistical Research Laboratory in India and at the Bell Telephone laboratories in the United States and directed them to quality improvement activities upstream in the product design stage. This was a more cost effective way that is now referred to as off-line quality control. In off-line quality control, experimental design techniques are utilized to find nominal settings of system parameters that will make the products robust to noise encountered in the manufacturing process and in the use environment.

To show appreciation for what the American advisors had done for the Japanese after WWII, Taguchi introduced his ideas for optimizing the design of products and production processes in a cost effective way to the United States in 1980. He not only presented his methods but also sought appraisal and comments (see Taguchi, 1987). Since then, Taguchi's contributions have been recognized as major advances in the history of engineering and manufacturing industries. However, some of the statistical design and analysis methods recommended by Taguchi have been found to be somewhat inefficient, and suggestions have been made for simpler, more efficient alternatives to implement some of Taguchi's ideas. In this chapter, Taguchi's engineering philosophy is presented and illustrated, when necessary, by more modern statistical methods.

12.2 Noise-Sources of Functional Variation

Taguchi has defined quality as the loss a product causes society after it is shipped (see Taguchi, 1986). We might think of this more as a definition of the lack of quality. By loss, Taguchi specifically means (1) loss caused by variability in function or (2) loss due to harmful side effects. For example, losses to a customer could occur due to subpar performance, early failure, or product deterioration, while losses to third parties (or society in general) could be caused by things such as emitted pollutants or safety concerns. Variation in product function is caused by what is commonly called noise. Taguchi has defined three distinct classes of noise that cause functional variation in products and manufacturing processes. These classes are (1) inner or deterioration noise, (2) outer or environmental noise, and (3) variational noise or piece-to-piece variation caused in manufacturing.

Inner noise or deterioration noise refers to changes in product or components that occur as the product ages or wears out preventing it from achieving its target function. Outer or environmental noise refers to external factors in the use environment such as temperature, humidity, vibration or human treatment that affect product performance. Finally, variational noise refers to manufacturing imperfections. As an example, Phadke (1989) described the three sources of noise in a refrigerator temperature control. The inner noise would consist of things such as the leakage of refrigerant and mechanical wear of compressor parts. The outer noise would consist of things such as the number of times the door is opened and closed, the amount of food kept, the ambient temperature, and the voltage supply variation. The variational noise would consist of the tightness of the door closure, quality of the seals on the refrigerant, amount of refrigerant charged, and the fit and finish of moving parts in the compressor.

Taguchi pointed out that only variational noise can be reduced with traditional quality control activities in manufacturing, but that in the product design stage it may be possible to find counter-measures for all three sources of noise using experimental design techniques. This is done through a procedure called robust parameter design.

The performance of products and manufacturing processes are affected by factors that are controlled by the designers of these products and processes (called *control factors*) and by other hard-to-control factors such as environmental conditions, raw material quality and aging (often called *noise factors* or *environmental factors*). In conducting a robust parameter design experiment, the levels of the noise factors are varied systematically to represent

NOISE-SOURCES OF FUNCTIONAL VARIATION

their variation in normal conditions. The levels of the control factors are varied over levels which are thought to include the optimal levels. The purpose of the parameter design experiment is to discover interactions between the control factors and the noise factors. Once discovered, these interactions can be exploited to reduce variation in the product or process function caused by hard-to-control variation in the noise factors.

The model for a robust design experiment can be represented symbolically as

$$y = X\beta + Z\gamma + \epsilon, \tag{12.1}$$

where y is the response or vector of measures of the product or process function. X represents the design matrix for the control factors and β represents the control factor effects or regression coefficients. Z represents the design matrix for the noise factors and the interactions between the noise factors and control factors. Obviously variation in the levels of the noise factors will be transmitted to variation in the response y. If there are interactions between the control factors and the noise factors, it may be possible to reduce variation in the response caused by changes in the levels of the noise factors through choice of the levels of the control factors.

The procedure for conducting a parameter design is shown below:

1. *Select control factors that are easily controlled.*
2. *Determine the sources of noise, and choose the most important noise sources to be represented as factors in the design. This sometimes requires some ingenuity to define factors that will represent the inner, outer, and variational noise you are trying to mitigate.*
3. *Select an appropriate experimental design.*
4. *Perform the experiment.*
5. *Analyze the data and detect significant main effects and interactions.*
6. *Select levels of the control factors to reduce variation caused by the noise factors and meet the target performance.*

The aspect that is different about parameter design experiments is the emphasis on including noise factors in the design, and the ability to determine control factor settings that not only ensure that the response will achieve the target level but will also reduce variation in the response caused by the sources of noise.

Two experimental strategies have been used in parameter design experiments. One is to use a cartesian product design which is composed of all combinations of levels of the control factors (in a control factor array) with all combinations of levels of the noise factors (in a noise factor array). The other approach is to combine both control and noise factors into one single array design. Two separate analysis strategies have been used. One strategy is called the loss model approach or location-dispersion modeling. This strategy can be used when the experiment is conducted as a product array. A second analysis strategy that has been used is called the response modeling strategy, and is the same as the approach described in previous chapters. This strategy is the only analysis strategy that can be used when the experiment is conducted as a single array, but it can also be used when the experiment is conducted as a product array. The next four sections will present examples of both product and single array parameter design experiments along with both strategies for analysis.

12.3 Product Array Parameter Design Experiments

To begin illustrating product array parameter design experiments we will consider a classic example that has been used by Taguchi (1987). This experiment was conducted in the summer of 1953 at Ina Seito Tile and Porcelain Company in Japan. Figure 12.1 is a flow diagram of the production process of wall tiles. The raw materials are clays and various stones.

Figure 12.1 *Ina Tile Process*

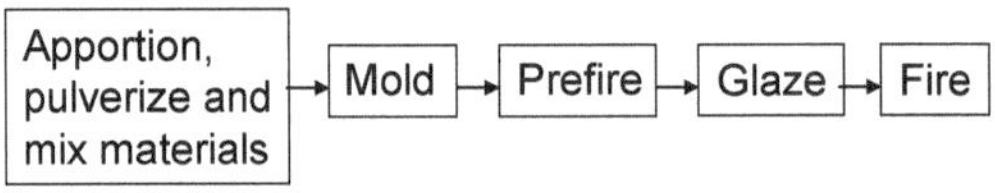

Apportion, pulverize and mix materials → Mold → Prefire → Glaze → Fire

After pulverizing the stones, they are mixed with the clays, finely ground in a ball mill and molded in a press. Next, they are loaded on a kiln car and pre-fired in a tunnel kiln. After a glaze is applied to the surface, the tiles are again loaded on the kiln car and the main firing is performed to produce the final product. When considering a cross-section of the tunnel kiln shown in Figure 12.2, there is a steep temperature gradient with higher temperatures near the outer wall closer to the burner. Many of the colored tiles processed in this region were defective due to improper shape and appearance.

Figure 12.2 *Ina Tile Kiln*
Tunnel Kiln

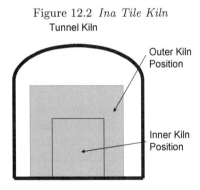

Since there weren't resources to replace the kiln in 1953, a less costly option was to experiment with the clay mixture to see if varying the components in the mixture could to some extent reduce the effect of the temperature gradient

PRODUCT ARRAY PARAMETER DESIGN EXPERIMENTS

in the kiln. This is an example of using a parameter design experiment to reduce variational noise in manufacturing. The control factors in this example are the composition elements of the raw materials (clay), and the noise factor is the position of a tile in the tunnel kiln. Table 12.1 shows the control factors that were used in the experiment and their levels. The noise factor had two levels: the low level stands for tiles fired in an inner kiln position and the high level stands for tiles fired in the outer kiln position closer to the burner.

Table 12.1 *Control Factors and Levels for Tile Experiment*

Factor	Levels −	+
A: kind of agalmatolite	existing	new
B: fineness of additive	courser	finer
C: content of lime	5%	1%
D: content of waste return	4%	0%
E: raw material charging quantity	1200kg	1300kg
F: content of agalmatolite	43%	53%
G: content of feldspar	0%	5%

The control factor array or experimental design was an eight run 2^{7-4} fractional factorial, and the noise factor array was a 2^1 design. The product array then consisted of all possible combinations of the runs in the control factor array and the noise factor array that resulted in $2^{7-4} \times 2^1 = 16$ runs. The design and resulting responses are shown in Table 12.2.

Table 12.2 *Product Design and Response for Tile Experiment*

A	B	C	D	E	F	G	H −	+
−	−	−	+	+	+	−	16	47
+	−	−	−	−	+	+	17	41
−	+	−	−	+	−	+	12	20
+	+	−	+	−	−	−	6	28
−	−	+	+	−	−	+	21	74
+	−	+	−	+	−	−	68	100
−	+	+	−	−	+	−	42	80
+	+	+	+	+	+	+	26	100

generators $D = AB, E = AC, F = BC, G = ABC$
note: responses on line 5 have been modified
(Bisgaard and Sutherland, 2004)

The response was the observed number of defectives per 100 tiles. Since the clay mixture is made and molded first, then tiles from the same clay mixture are fired in both the inner and outer regions of the kiln at the same time. It should be recognized that this particular product array design is a split-plot design that is quite similar to the experiment used to formulate a boxed cake mix (presented in exercise 9 of Chapter 8). Bisgaard and Sutherland (2004) noticed unusual observations on line five in the original Ina tile data. These observations have been modified in Table 12.2 to simplify the discussion of the analysis in the next section.

As a second example of a product array, consider an experiment described by Taguchi and Byrne (1984) to optimize the installation procedure for an elastomeric connector used in the automotive industry. The desire was to maximize the separation pull-off force and minimize the variability in this force. There were four control factors and three noise factors. These factors are shown in Table 12.3 and illustrated in Figure 12.3.

Table 12.3 *Control and Noise Factors for Connector Experiment*

	Control Factors:		levels	
A:	Interference	Low	Medium	High
B:	Connector wall thickness	Thin	Medium	Thick
C:	Insertion depth	Shallow	Medium	Deep
D:	Percentage of adhesive	Low	Medium	High
	Noise Factors:		levels	
E:	Conditioning time	24 hours	120 hours	
F:	Conditioning temperature	72°F	150°F	
G:	Conditioning relative humidity	25%	75%	

The interference (or allowance between the force fit elastometric connector and the metal tube it is connected to), the wall thickness of the connector, insertion depth of the connector onto the metal tube, and percent adhesive that is in the solution the connector is dipped into before insertion on the metal tube are specified by the designer and are therefore designated as control factors. The conditioning or set-up time, the temperature during the set-up time, and the relative humidity during the set-up time may affect the pull-off force but are usually difficult to control during routine production conditions. Therefore these three factors are designated as noise factors. During the parameter design experiments, extra care was taken to control these factors.

The complete layout of the control factor array, noise factor array and resulting data is shown in Figure 12.4. There, the control and noise factor arrays are shown in symbolic levels (1, 2, etc.) and actual levels. The control factor array was a 3^{4-2} fractional factorial. The noise factor array is a full 2^3 factorial where the symbolic levels are 1, and 2 rather than $-$ and $+$. The run numbers for the two designs are shown in the far left and top rows, respectively. The randomization of runs in this experiment was not restricted, and each of the $3^{4-2} \times 2^3 = 72$ runs could have been run in a completely random order.

PRODUCT ARRAY PARAMETER DESIGN EXPERIMENTS   507

Figure 12.3 *Illustration of Control Factors for Connector Experiment*

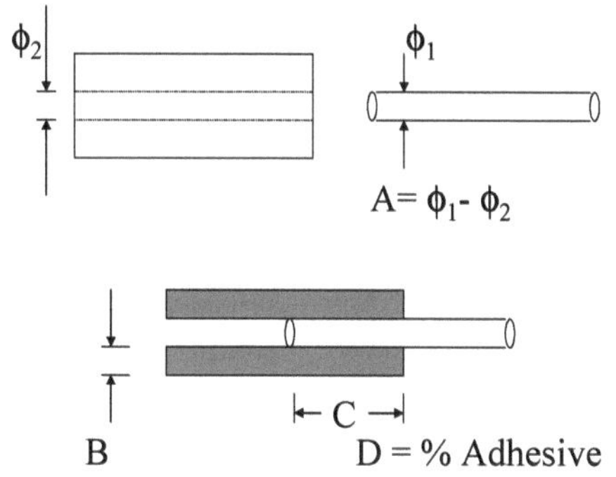

$$A = \phi_1 - \phi_2$$

B $\leftarrow C \rightarrow$
     D = % Adhesive

Therefore, this was not a split-plot experiment like the tile example. The `pointrep` option of `proc factex` can be used to create a cartesian product design similar to the way it was shown in Section 8.2.1.

```
*Create noise factor design with factex;
proc factex;
  factors E F G;
  output out=s1 E nvals=(24 120) F nvals=(72 150) G nvals
    =(25 75);
*Add the control factor design;
  factors A   B C D/nlev=3;
  model estimate=(A B C D);
  size design=9;
  output out=s2 pointrep=s1 A cvals=('Low' 'Medium' 'High')
  B cvals=('Thin' 'Medium' 'Thick')
  C cvals=('Shallow' 'Medium' 'Deep')
  D cvals=('Low' 'Medium' 'High');
proc print; run;
```

One of the comments that has been made about product array parameter design experiments popularized by Taguchi is that they require more experiments than are necessary. One of the main purposes of parameter design experiments is to detect interactions between control factors and noise factors. By understanding the interactions it may be possible to select levels of the control factors that will minimize the effect of the noise factors. By utilizing a cartesian product array design, all interactions between the factors in the control array and the factors in the noise array will be estimable. However, it may be possible to do this with fewer runs by utilizing resolution IV or resolution V fractional factorial designs where all interactions between control and noise factors are clear. Examples of this will be shown in Section 12.5.

508 ROBUST PARAMETER DESIGN EXPERIMENTS

Figure 12.4 *Complete Layout of Control Factor Array and Noise Factor Array*

Design matrix

Experimental conditions								
No.	A	B	C	D	Inteference (A)	Wall thickness (B)	Insertion depth (C)	Percentage adhesive (D)
1	1	1	1	1	Low	Thin	Shallow	Low
2	1	2	2	2	Low	Medium	Medium	Medium
3	1	3	3	3	Low	Thick	Deep	High
4	2	1	2	3	Medium	Thin	Medium	High
5	2	2	3	1	Medium	Medium	Deep	Low
6	2	3	1	2	Medium	Thick	Shallow	Medium
7	3	1	3	2	High	Thin	Deep	Medium
8	3	2	1	3	High	Medium	Shallow	High
9	3	3	2	1	High	Thick	Medium	Low

Noise matrix

No.	8	7	6	5	4	3	2	1	
	2	2	2	2	1	1	2	1	E
	2	2	1	1	2	2	1	1	F
	1	1	2	2	2	2	1	1	E×F
	2	1	2	1	2	1	2	1	G
	1	2	1	2	1	1	2	1	E×G
	1	2	1	1	1	2	2	1	F×G
	2	1	1	2	1	2	2	1	e
Conditioning time (E)	120h	120h	120h	120h	24h	24h	24h	24h	
Conditioning temperature (F)	150F	150F	72F	72F	150F	150F	72F	72F	
Conditioning r.h. (G)	75%	25%	75%	25%	75%	25%	75%	25%	
1	19.1	20.0	19.6	19.6	19.9	16.9	9.5	15.6	
2	21.9	24.2	19.8	19.7	19.6	19.4	16.2	15.0	
3	20.4	23.3	18.2	22.6	15.6	19.1	16.7	16.3	
4	24.7	23.2	18.9	21.0	18.6	18.9	17.4	18.3	
5	25.3	27.5	21.4	25.6	25.1	19.4	18.6	19.7	
6	24.7	22.5	19.6	14.7	19.8	20.0	16.3	16.2	
7	21.6	24.3	18.6	16.8	23.6	18.4	19.1	16.4	
8	24.4	23.2	19.6	17.8	16.8	15.1	15.6	14.2	
9	28.6	22.6	22.7	23.1	17.3	19.3	19.9	16.1	

PRODUCT ARRAY PARAMETER DESIGN EXPERIMENTS

One situation where product arrays are actually more efficient is when the experiment is a split-plot. For example, in the Ina tile experiment, the noise factor array consisted of simply measuring the response (number of defective tiles in a sample of 100) once for tiles produced in the inner kiln position and again for tiles produced in the outer kiln position. Therefore no additional experimental effort was required. Only eight batches of clay were mixed and molded, but more information was produced by measuring multiple responses for each run of the control factor array. Split-plot designs are sometimes described as super efficient since the sup-plot factors and interactions between sub-plot and whole-plot factors can be tested with the (usually) smaller sub-plot error term.

A third example of a product array parameter design experiment involves the design of a product. The product is the simple circuit for a temperature controller that uses resistance thermometer R_T. The circuit diagram that is shown in Figure 12.5 was described by Taguchi and Phadke (1985). The

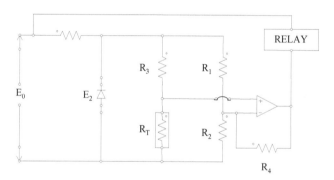

Figure 12.5 *Temperature Controller Circuit*

response, or measure of product performance, is the resistance value of R_T at which the relay turns on. Through circuit relations, the response is given by the rational function

$$R_T = \frac{R_3 R_2 (E_2 R_4 + E_0 R_1)}{R_1 (E_2 R_2 + E_2 R_4 - E_0 R_2)}. \tag{12.2}$$

The adjustment factor is R_3, which is known by the circuit relations. The control factors that a designer can specify are the nominal ratings of the resistors and diodes $A = R_1$, $B = R_2/R_1$, $C = R_4/R_1$, $D = E_0$ and $F = E_2$. The noise factors are deviations from the nominal values of the control factors that would be expected due to manufacturing imperfections and deterioration while the product is in use. The objective was to find the nominal settings of the control factors that minimize the variation in R_T since the desired value of R_T can always be obtained by adjusting R_3. This is an example of the type of problems encountered in the electronics industry in Japan which were successfully solved using parameter design experiments. In problems similar to this where there is no clear functional relation like Equation (12.2), Taguchi recommends the values of the response be determined experimentally.

Since the response is a known function of the factors or independent variables, statisticians evaluating the methods used by Taguchi have shown that this particular kind of problem can be solved more efficiently using computer optimization as described by Box and Fung (1986) and illustrated by Lawson and Madrigal (1994). However, computer optimization was not possible in the early 1960s and the solution to this problem will be illustrated using a product array parameter design experiment to give a historical perspective.

The control factors and levels used in the control factor array are shown in Table 12.4.

Table 12.4 *Factors and Levels for Circuit Parameter Design*

Control Factors	Levels		
	-1	0	1
$A: R_1$	2.67	4.0	6.0
$B: R_2/R_1$	1.33	2.0	3.0
$C: R_4/R_1$	5.33	8.0	16.0
$D: E_0$	8.00	10.0	12.0
$F: E_2$	4.80	6.0	7.2
Noise Factors	97.96% of Nominal	Nominal	102.04% of Nominal

The experimental design for the control factors is the 18-run orthogonal array shown in Table 12.5. The coded factor levels on the left side of the table were created for this example using the %mktEx macro that was described in Section 6.7 as shown in the commands below

```
%mktex(2 3**7, n=18, levels=i)
proc print data=design; run;
```

The %mktex macro creates the design in a non-randomized order in the SAS file named design and a randomized list in the SAS file named randomized. For this experiment there is no experimental error since the response consists of evaluating a function, so there is no need for a randomized list. The default levels for the factors produced by %mktex are 1, 2, ..., k where k is the number of levels of the factor. The levels=i option creates centered integers the levels (i.e., −1 and 1 for two-level factors, −1, 1 and 0 for the three-level factors, etc.)

The design that is created by %mktex is the $OA(18)$ design that was popularized by Taguchi for use in parameter design experiments. Wang and Wu (1995) found this design to have properties similar to the Plackett-Burman Designs, in that main effects plus linear×linear two-factor interactions can be

Table 12.5 *OA18 Design for Control Factors*

		Coded Factor Levels							Actual Factor Levels			
X_1	X_2	X_3	X_4	X_5	X_6	X_7	X_8	A	B	C	D	F
-1	-1	-1	-1	-1	-1	-1	-1	2.67	1.33	5.33	8	4.8
-1	-1	-1	1	0	0	1	0	2.67	1.33	16	10	6
-1	-1	0	1	1	-1	0	1	2.67	2	16	12	4.8
-1	1	-1	0	0	1	-1	1	6	1.33	8	10	7.2
-1	1	1	1	1	1	1	-1	6	3	16	12	7.2
-1	1	1	0	-1	-1	0	0	6	3	8	8	4.8
-1	0	1	-1	-1	0	1	1	4	3	5.33	8	6
-1	0	0	-1	1	1	-1	0	4	2	5.33	12	7.2
-1	0	0	0	0	0	0	-1	4	2	8	10	6
1	-1	1	-1	0	1	0	0	2.67	3	5.33	10	7.2
1	-1	1	0	1	0	-1	-1	2.67	3	8	12	6
1	-1	0	0	-1	1	1	1	2.67	2	8	8	7.2
1	1	-1	-1	1	0	0	1	6	1.33	5.33	12	6
1	1	0	-1	0	-1	1	-1	6	2	5.33	10	4.8
1	1	0	1	-1	0	-1	0	6	2	16	8	6
1	0	-1	1	-1	1	0	-1	4	1.33	16	8	7.2
1	0	-1	0	1	-1	1	0	4	1.33	8	12	4.8
1	0	1	1	0	-1	-1	1	4	3	16	10	4.8

fit using subset selection regression analysis even though the design is resolution III. Since there were only five three-level factors in this problem, the -1's, 0's and 1's in columns X_2 through X_6 were replaced by the actual factor levels in Table 12.4 to form the design shown in the right half of Table 12.5, and X_1, X_7, and X_8 were unassigned or not used in this problem.

The experimental design in coded factor levels for the noise factors is the same $OA(18)$ used for the control factors. The effect of changing levels of the noise factors is a perturbation of the levels of the control factors. For example, when the coded level of the noise factor is -1, the nominal settings of the control factors are modified by multiplying them by 0.9796. When the coded level of the noise factor is 1, the nominal settings of the control factors are modified by multiplying them by 1.0204. When the noise factor is at the 0, or mid-level, the nominal value of the control factor is left unchanged. This is done to simulate the noise environment where the actual values of the control factors may differ from the specified nominal values by a tolerance of $\pm 2.04\%$ due to manufacturing imperfections and deterioration over time in use. For a particular run in the control factor design, a set of modified control factor levels are calculated for each run in the noise factor array and the response, R_T, is calculated using Equation (12.2).

In product array parameter design experiments the noise factor array is usually shown horizontally across the top similar to Figure 12.4. For this example the product array is too large (18×18) to represent as a single table. In Table

12.6 the coded factor levels of the noise factor array are shown transposed in the first five columns. The next five columns of the table illustrate the calculations of the modified control factor levels for the first run of the control factor array where $A = 2.67$, $B = 1.33$, $C = 5.33$, $D = 8.0$ and $F = 4.8$. The last column of Table 12.6 shows the response, R_T, calculated with Equation (12.2). This would be repeated for each row of the control factor array to obtain all the simulated response data for the product array design. Although the values shown in Table 12.6 were calculated with a spreadsheet, it would have been possible to calculate them manually as Taguchi might have in 1960.

Table 12.6 *Transposed Noise Factor Array and Modified Control Factor Levels and Response for the First Run of Control Factor Array*

X_2	X_3	X_4	X_5	X_6	A	B	C	D	F	y
-1	-1	-1	-1	-1	2.62	1.30	5.22	7.84	4.70	103.09
-1	-1	1	0	0	2.62	1.30	5.44	8.00	4.80	101.28
-1	0	1	1	-1	2.62	1.33	5.44	8.16	4.70	106.98
1	-1	0	0	1	2.72	1.30	5.33	8.00	4.90	100.70
1	1	1	1	1	2.72	1.36	5.44	8.16	4.90	106.34
1	1	0	-1	-1	2.72	1.36	5.33	7.84	4.70	107.29
0	1	-1	-1	0	2.67	1.36	5.22	7.84	4.80	106.61
0	0	-1	1	1	2.67	1.33	5.22	8.16	4.90	105.67
0	0	0	0	0	2.67	1.33	5.33	8.00	4.80	104.71
-1	1	-1	0	1	2.62	1.36	5.22	8.00	4.90	106.64
-1	1	0	1	0	2.62	1.36	5.33	8.16	4.80	108.94
-1	0	0	-1	1	2.62	1.33	5.33	7.84	4.90	101.69
1	-1	-1	1	0	2.72	1.30	5.22	8.16	4.80	104.66
1	0	-1	0	-1	2.72	1.33	5.22	8.00	4.70	107.35
1	0	1	-1	0	2.72	1.33	5.44	7.84	4.80	102.29
0	-1	1	-1	1	2.67	1.30	5.44	7.84	4.90	98.46
0	-1	0	1	-1	2.67	1.30	5.33	8.16	4.70	105.31
0	1	1	0	-1	2.67	1.36	5.44	8.00	4.70	107.98

12.4 Analysis of Product Array Experiments

12.4.1 Loss Model or Location-Dispersion Modeling

One method of analyzing the data from a product array parameter design experiment is to summarize the responses across the combinations of factor levels in the noise array for each run of the control factor array and then analyze the summary statistics as a function of the control factor settings. Usually two summary statistics are calculated, one being the mean and the other being a measure of the dispersion or variability. Taguchi employed three different performance measures he called SN ratios as a measure of dispersion. Finding the control factor levels to maximize Tagughi's SN_L performance

ANALYSIS OF PRODUCT ARRAY EXPERIMENTS

measure was designed to make the response as large as possible, and finding the control factor levels to maximize his SN_S was designed to make the response as small as possible. However, because these performance measures confuse the mean response level and variability about the mean, Box (1988) has shown they are inefficient. Wu and Hamada (2000) instead recommend a two-step procedure shown below:

1. *Select the levels of the control factors (that affect the variation in the response) to minimize the variation.*
2. *Select the levels of the control factors (that affect the mean response but not the variation) to bring the response to the desired target.*

In the first step the control factors that affect $ln(s^2)$, where the sample variance s^2 is calculated across the levels of the noise factor array for each run in the control factor array, are identified and their factor levels are chosen to minimize $ln(s^2)$. Other factors may affect the average response but not $ln(s^2)$. These control factors are called *adjustment factors*. The second step involves identifying the levels of the adjustment factors that bring the response to the desired target level. When the objective is to minimize or maximize the response with minimum variability, the order of the two steps is reversed.

To illustrate this procedure, consider the data in the Ina tile experiment. Since the responses were binomial counts, the $\sin^{-1}\sqrt{y/n}$ transformation described in Section 2.5.2 was used. The SAS commands to read the data and create the transformed values are.

```
data tile;
input A B C D E F G y1 y2 ;
y1=arsin(sqrt(y1/100)); y2=arsin(sqrt(y2/100));
ybar=(y1+y2)/2; lns2=log(((y1-y2)**2)/2);
datalines;
-1 -1 -1 1 1 1 -1 16 47
...
```

Table 12.7 shows the control factor settings, the transformed response, and the two summary statistics $\bar{y}$ and $ln(s^2)$.

Table 12.7 *Control Factor Array and Summary Statistics*

A	B	C	D	E	F	G	H		$\bar{y}$	$\ln(s^2)$
							−	+		
−	−	−	+	+	+	−	0.412	0.755	0.583	-2.828
+	−	−	−	−	+	+	0.425	0.695	0.560	-3.312
−	+	−	−	+	−	+	0.354	0.464	0.409	-5.109
+	+	−	+	−	−	−	0.247	0.558	0.402	-3.035
−	−	+	+	−	−	+	0.476	1.036	0.756	-1.854
+	−	+	−	+	−	−	0.970	1.570	1.270	-1.711
−	+	+	−	−	+	−	0.705	1.107	0.906	-2.515
+	+	+	+	+	+	+	0.535	1.571	1.053	-0.623

The commands to fit the model to the average response are shown below.

```
proc glm;
    model ybar   = A B C D E F G/solution;
    ods output ParameterEstimates=sol;
run;
```

The resulting output is shown below and Figure 12.6 shows a half-normal plot of the regression coefficients.

Parameter	Estimate	Standard Error	t Value	Pr > |t|
Intercept	0.7424626418	.	.	.
A	0.0789317559	.	.	.
B	-.0498971676	.	.	.
C	0.2538069161	.	.	.
D	-.0437639741	.	.	.
E	0.0863475789	.	.	.
F	0.0331447819	.	.	.
G	-.0480989659	.	.	.

Here it can be seen that control factor C, the content of lime, appears to be the only one to have a significant effect on the mean response. Since the effect is positive and the low level was defined in Table 12.1 as 5%, the number of defective tiles can be reduced by mixing 5% of lime in the clay mixture rather than 1% as was previously the practice. Since the objective is to minimize the number of defective tiles, the first step in the procedure shown above would be to choose the 5% level of lime in the clay mix.

The second step starts by identifying the control factors that affect the log variance. The output below the figure on the next page shows the results of fitting the same model using the summary statistic $\ln(s^2)$ as the response. Here, it can again be seen that control factor C has the largest effect.

ANALYSIS OF PRODUCT ARRAY EXPERIMENTS

Figure 12.6 *Half-Normal Plot of Effects on Average Response- Ina Tile Experiment*

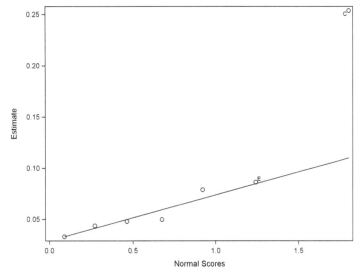

```
                                  Standard
Parameter            Estimate     Error       t Value    Pr > |t|

Intercept         -2.623421027       .           .           .
A                  0.453263556       .           .           .
B                 -0.197151337       .           .           .
C                  0.947747578       .           .           .
D                  0.538505766       .           .           .
E                  0.055644066       .           .           .
F                  0.303715062       .           .           .
G                 -0.101246512       .           .           .
```

The half-normal plot of the regression coefficient in Figure 12.7 shows that it is the only control factor having a significant effect on variability. Luckily in this case the effect is positive (like it was for the mean response) and therefore the low or 5% level of lime minimizes the variability in the percentage of defective tiles.

If the goal were to minimize the response with minimum variability and one or more control factors have significant effects of different directions on the mean and log variance, then a compromise much be reached. In this case that was not necessary, and the Ina tile company was able to greatly improve the quality of their finished tiles without upgrading their kiln. Simply increasing the amount of lime in the clay mixture, which was an inexpensive change to make, had the desired effect. This example is one of the first examples that Taguchi used when explaining his ideas on using experimental designs to find cost effective ways to improve products and production processes.

Figure 12.7 *Half-Normal Plot of Effects on Log Variance of Response- Ina Tile Experiment*

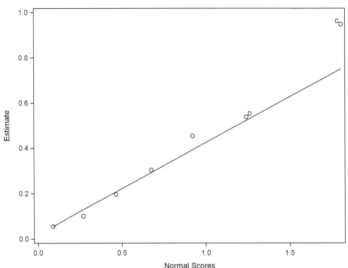

A second example of location-dispersion modeling of a product array parameter design experiment will be illustrated with the data from the temperature controller circuit design experiment. The purpose of the parameter design was to discover control factor levels, or the design specifications for the nominal ratings of resistors and diodes, that would minimize the variation in R_T caused by small deviations from the nominal settings due to manufacturing imperfections and deterioration while in use. The experiments were actually done by simulation since there is a rational function (Equation (12.2)) relating the control factor settings and the response. In this example, it is only necessary to calculate one summary statistic (the log variance) across the levels of the noise factor array since the mean value of R_T can be set to any desired value by adjusting R_3. Table 12.8 shows the control factor array in actual levels along with the log variance of the simulated responses calculated over the noise factor array. For example, the variance calculated from the simulated responses in the first row of the control factor array (shown as a column in Table 12.6) is Var(103.09, 101.28, ...,107.98) = 8.421, and the $ln(s^2) = 2.131$.

In this table, it can be seen that changes in control factors cause large differences in the variance of R_T. The minimum variance in R_T is exp(0.468) = 1.597 for run number 16, while the maximum variance is exp(4.892) = 133.219 for run 14, and the average variance being 14.33. Although there is a rational function relating the value of R_T to the settings on the control factors, there is not a function relating the log of the variance of R_T to the settings of the control factors. To solve this problem with computer optimization an approximation formula relating the variance of R_T can be developed and numerically

ANALYSIS OF PRODUCT ARRAY EXPERIMENTS

minimized as a function of the control factor settings (Box and Fung (1986), Lawson and Madrigal (1994). To solve the problem using a parameter design experiment, a function relating the log of the variance of R_T to the settings of the control factors is found by fitting a model to the data in Table 12.8.

Table 12.8 Actual Factor levels and Log Variance for Circuit Design Experiment

A	B	C	D	F	$ln(s^2)$
2.67	1.33	5.33	8	4.8	2.131
2.67	1.33	16	10	6	0.778
2.67	2	16	12	4.8	2.537
6	1.33	8	10	7.2	1.046
6	3	16	12	7.2	2.839
6	3	8	8	4.8	4.192
4	3	5.33	8	6	4.032
4	2	5.33	12	7.2	3.478
4	2	8	10	6	2.629
2.67	3	5.33	10	7.2	4.180
2.67	3	8	12	6	5.069
2.67	2	8	8	7.2	1.656
6	1.33	5.33	12	6	2.705
6	2	5.33	10	4.8	4.892
6	2	16	8	6	1.500
4	1.33	16	8	7.2	0.468
4	1.33	8	12	4.8	2.546
4	3	16	10	4.8	3.447

Although the $OA(18)$ control factor array is resolution III, Wang and Wu (1995) have shown that it has similar hidden projection properties as the Plackett-Burman Designs. Therefore, it is possible to identify significant interactions using a regression subset procedure like the example shown for a Plackett-Burman Design in Section 6.6. The SAS data step commands below read in the data from Table 12.8, create coded factor levels X_A - X_D, squares of the coded factor levels, and linear×linear interaction terms. The square terms were created to fit quadratic terms to the three-level factors. Next, `proc reg` is used to perform an all subsets regression.

The input commands to do this are shown on the next page. In the output shown on the following page, it can be seen that R^2 increases as the number of variables in the model increases, but that when the number of variables exceeds six, the rate of increase in R^2 diminishes.

Increasing the number of variables from six to seven only increases R^2 for the best model from 0.9600 to 0.9711, and the seventh term is probably not necessary. The best model for six variables includes the terms X_B, X_C, X_D, X_F, $X_A \times X_F$ and $X_B \times X_D$. This appears to be a reasonable model, since all interactions in the model involve at least one main effect in the model (effect heredity).

```
*data for temp control design;
data cont;
input A B C D F lnvar;
XA=(A-4)/2; XB=(B-2); XC=(C-8)/8; XD=(D-10)/2; XF=(F-6)/1.2;
datalines;
2.67 1.33 5.33 8 4.8 1.163
 . . .
data mod; set cont;
  XA2=XA*XA;
  XB2=XB*XB;
  XC2=XC*XC;
  XD2=XD*XD;
  XF2=XF*XF;
  XAB=XA*XB;
  XAC=XA*XC;
  XAD=XA*XD;
  XAF=XA*XF;
  XBC=XB*XC;
  XBD=XB*XD;
  XBF=XB*XF;
  XCD=XC*XD;
  XCF=XC*XF;
  XDF=XD*XF;
proc reg data=mod;
  model lnvar=XA XB XC XD XF XA2 XB2 XC2 XD2 XF2 XAB XAC XAD
              XAF XBC XBD XBF XCD XCF XDF/
  selection=rsquare best=4 stop=8; run;
```

The SAS commands for fitting the best six variable model are shown below.

```
proc reg data=mod;
  model lnvar=XB XC XD XF XAF XBD XC2/p;
  output out=s p=yhat r=resid;
run;
```

The results are below the all subsets results on the next page. All the model terms appear to be significant, and residual plots showed that the least squares assumptions appear justified. Now that a model is found relating the control factor settings to the log variance of R_T, the next step is to determine the factor settings that minimize the log variance. The negative coefficient for the linear effect of factor C indicates that the low level should be chosen for this ratio of nominal resistor ratings in order to minimize the $ln(s^2)$. For factors B, D and F the interpretation is not so straightforward since there are significant interactions with these factors.

```
Number in
  Model     R-Square    Variables in Model

    1        0.5106     XB
    1        0.2436     XC
    1        0.1936     XAF
    1        0.1848     XC2
-----------------------------------------------------
    2        0.7541     XB XC
    2        0.6953     XB XC2
    2        0.6068     XB XF
    2        0.5842     XB XAC
-----------------------------------------------------
    3        0.8504     XB XC XF
    3        0.8245     XB XC XD
    3        0.7924     XB XC XAC
    3        0.7916     XB XF XC2
-----------------------------------------------------
    4        0.9207     XB XC XD XF
    4        0.8679     XB XC XF XBD
    4        0.8660     XB XC XD XAF
    4        0.8641     XB XC XF XAB
-----------------------------------------------------
    5        0.9425     XB XC XD XF XAF
    5        0.9396     XB XC XD XF XCD
    5        0.9313     XB XC XD XF XC2
    5        0.9295     XB XC XD XF XBC
-----------------------------------------------------
    6        0.9600     XB XC XD XF XAF XBD
    6        0.9547     XB XC XD XF XC2 XAF
    6        0.9515     XB XC XD XF XAF XCF
    6        0.9509     XB XC XD XF XBD XCD
-----------------------------------------------------
    7        0.9711     XB XC XD XF XC2 XAF XBD
    7        0.9676     XA XB XC XD XF XAF XBD
    7        0.9645     XB XC XD XF XAF XBD XCD
    7        0.9638     XB XC XD XF XC2 XAF XCF
-----------------------------------------------------
    8        0.9799     XA XB XC XD XF XAF XBC XBD
    8        0.9797     XB XC XD XF XC2 XAD XAF XBD
    8        0.9786     XA XB XC XD XF XC2 XAF XBD
    8        0.9774     XB XC XD XF XC2 XAF XBD XBF
```

```
                         Parameter Estimates

                    Parameter   Standard
    Variable    DF  Estimate    Error      t Value   Pr > |t|

    Intercept   1    2.89028    0.08842     32.69    <.0001
    XB          1    1.27702    0.12303     10.38    <.0001
    XC          1   -1.10764    0.15933     -6.95    <.0001
    XD          1    0.50917    0.10585      4.81    0.0005
    XF          1   -0.53056    0.10539     -5.03    0.0004
    XAF         1   -0.52822    0.17849     -2.96    0.0130
    XBD         1    0.37328    0.17012      2.19    0.0506
```

The prediction equation for the fitted model can be written as

$$ln(s^2) = 2.890 + 1.277\left(\frac{B-2}{1}\right) - 1.108\left(\frac{C-8}{8}\right)$$
$$+ .509\left(\frac{D-10}{2}\right) - .531\left(\frac{F-6}{1.2}\right)$$
$$- .528\left(\frac{A-4}{2}\right)\left(\frac{F-6}{1.2}\right) + .373\left(\frac{B-2}{1}\right)\left(\frac{D-10}{2}\right).$$
(12.3)

Using this equation with factors A, C and F held constant at their mid-levels, the predicted $ln(s^2)$ shown in Table 12.9 were calculated. Here it can be seen that the combination of factors B and D that are predicted to have the lowest $ln(s^2)$ is $B = 1.33$ (or the nominal rating of resistor R_2 should be specified as $1.33 \times R_1$) and $D = 8.0$ (or the nominal rating of diode E_0 should be specified as 8.0).

Table 12.9 Predicted $ln(s^2)$ as Function of B and D

		$B=R_2/R_1$	
		1.33	3.0
	8.0	1.776	3.285
$D=E_0$			
	12.0	2.294	5.050

A similar table of predicted values for the four combinations of low and high levels for factors A and F reveals that factor A should be set at its high level (or the nominal rating of R_1 should be specified as 6.0) and factor F should be set at its high level (or the nominal rating for diode E_2 should be specified as 7.2). Therefore, the optimum combination of factor levels for minimizing the log variation in R_T is (A=6.0, B=1.33, C=16, D=8.0, and F=7.2). The factor settings for the 16th run in the control factor design shown in Table 12.8 are practically the same, the only difference being A=4.0 instead of A=6.0, and it had the smallest log variance (0.468) of any of the experimental conditions. Testing the predicted optimal conditions in Equation (12.2) across all the values in the noise factor array resulted in exactly the same predicted mean and log variance response, since the level for A can be factored from the numerator and denominator of Equation (12.2) and canceled.

Although not the optimal way of solving this problem, the parameter design experiment would enable a circuit designer to find settings for the nominal ratings of the resistors and diodes that will reduce the variance in R_T tenfold over the average control factor settings in Table 12.8, and make the circuit perform much more consistently in the presence of variation from the nominal ratings caused by manufacturing imperfections and deterioration in use.

12.4.2 Response Modeling

The other method of analyzing data from a product array parameter design experiment is to analyze the response directly as a function of the control and noise factor settings rather than analyze a summary statistic calculated over the levels of the noise factor array. This is the approach that has been used for analysis of all the experimental data discussed prior to this chapter. For parameter design experiments special attention will be given to discovering interactions between control and noise factors and to interpreting these interactions in order to find control factor settings that minimize the effect of noise factors.

As an example consider a re-analysis of the data from the Ina tile experiment. The experiment was a split plot and the whole-plot factors or control factors were A through G shown in Table 12.1 that defined the clay mixture. The whole-plot design or control factor array was a 2^{7-3} saturated resolution III fractional factorial. The noise factor array or sub-plot design consisted of one factor, H:position of tile in the kiln, with two levels that specified where the tiles were sampled. The split-plot cartesian product design guarantees that all interactions between the control factors and noise factor will be estimable. The model for this split-plot design can be written in the form

$$y_{ijklmnop} = \mu + \alpha_i + \beta_j + \gamma_k + \delta_l + \xi_m + \phi_n + \eta_o + w_{ijklmno}$$
$$+ \pi_p + \alpha\pi_{ip} + \beta\pi_{jp} + \gamma\pi_{kp} + \delta\pi_{lp} + \xi\pi_{mp} + \phi\pi_{np} + \eta\pi_{op} + \epsilon_{ijklmnop}, \quad (12.4)$$

where α-η represent the control factor effects, π represents the noise factor effect, $w_{ijklmno}$ is the whole-plot error term, and $\epsilon_{ijklmnop}$ is the sub-plot error term. Since there were no replicates in this experiment, $w_{ijklmno}$ and $\epsilon_{ijklmnop}$ cannot be estimated, and the significance of effects must be judged with normal plots like the examples in Section 8.4 and Section 8.5.2. The SAS commands to fit Equation (12.4) to the data are shown below and the resulting parameter estimates are shown on the next page.

```
Data strng;
  set tile;
  H=-1;  y=y1; output;
  H=1;   y=y2; output;
  keep A B C D E F G H y;
run;
proc glm data=strng;
   model y=A B C D E F G H A*H B*H C*H D*H E*H F*H G*H/solution;
   ods output ParameterEstimates=sol;
run;
```

A half-normal plot of the whole-plot or control factor effects is shown in Figure 12.8. Here it can be seen that Factor C, the content of lime, appears to

Parameter	Estimate	Standard Error	t Value	Pr > |t|
Intercept	0.7424626418	.	.	.
A	0.0789317559	.	.	.
B	-.0498971676	.	.	.
C	0.2538069161	.	.	.
D	-.0437639741	.	.	.
E	0.0863475789	.	.	.
F	0.0331447819	.	.	.
G	-.0480989659	.	.	.
H	0.2270371580	.	.	.
A*H	0.0500925488	.	.	.
B*H	0.0051952379	.	.	.
C*H	0.0978100068	.	.	.
D*H	0.0541393761	.	.	.
E*H	0.0343077204	.	.	.
F*H	0.0294129474	.	.	.
G*H	0.0198677644	.	.	.

have a significant effect on the $\sin^{-1}$ transformation of the percent defective tiles.

Figure 12.8 *Half-Normal Plot of Whole-plot Effects - Ina Tile Experiment*

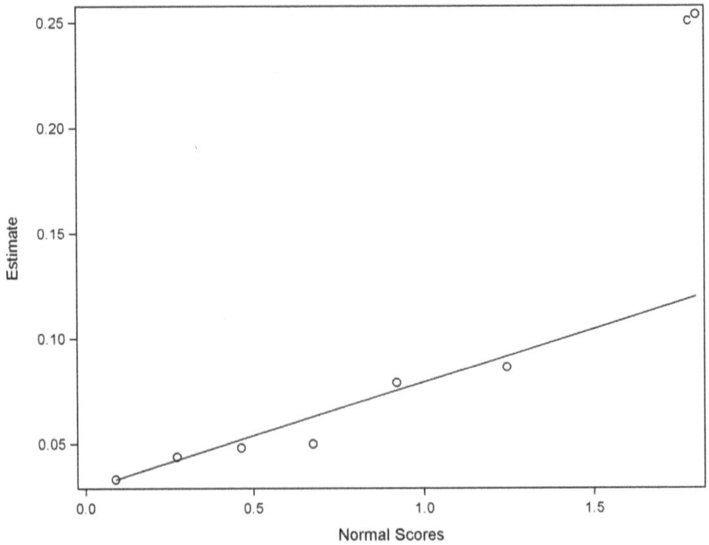

Since the effect is positive and the low level for factor C was defined in Table 12.1 as 5%, the number of defective tiles can be reduced by mixing 5% of lime in the clay. This is the same conclusion that was reached earlier by analyzing the average response across the levels of the noise factor H, position of tile in the kiln, in Section 12.4.1.

Figure 12.9 shows a half-normal plot of the sub-plot effects which include the noise factors and interactions between the control and noise factors. Here

Figure 12.9 *Half-Normal Plot of Sub-Plot Effects - Ina Tile Experiment*

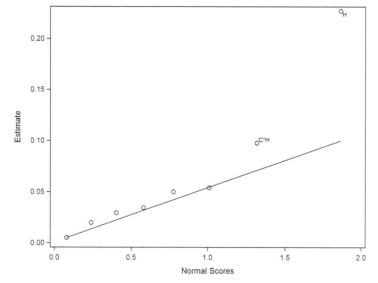

it can be seen that both the noise factor H and the interaction between control factor C and the noise factor H appear to be significant. The effect of the noise factor H in conjunction with the control factor C can best be visualized by examining the interaction plot shown in Figure 12.10. In this plot, shown on the next page, it can be seen that in general there is a higher proportion of defective tiles in the outer region of the kiln which is closer to the burner. However, if 5% lime is added to the clay mixture, rather than 1%, the effect of kiln position on the proportion of defective tiles (i.e., the slope of the line in Figure 12.10) is reduced. Therefore, it can be said that not only is the average proportion of defective tiles reduced by using 5% lime, but also the variability in proportion of defective tiles is reduced.

This is the same conclusion that was reached in Section 12.4.1 by analyzing the log variance of the response across the levels of the noise factor. However, when there are multiple noise factors, the response modeling approach can often provide much more information than can be obtained by simply modeling the mean and log variance as a function of the control factors alone. The next example illustrates this.

When a product array parameter design experiment is performed as a split plot experiment, Box and Jones (1992) have recommended, when possible, to make the noise factors the whole-plot factors and the control factors the subplot factors (exactly opposite of Table 12.2). This is because split-plot experiments have more precision for detecting sub-plot effects and interactions between whole-plot and subplot effects. If the experiment is conducted this way, the control factor effects and the control-by-noise factor effects can be

Figure 12.10 *Interaction between Percent Lime and Kiln Position - Ina Tile Experiment*

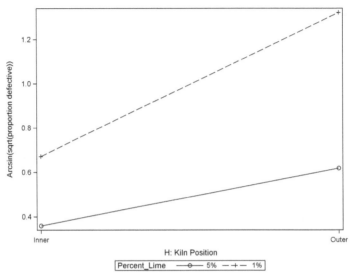

tested with the smaller subplot error term. In experiments where noise factors represent environmental variables like ambient temperature and humidity, this arrangement makes more sense. All combinations of levels of the control factors could be run in random order within blocks where the environmental variables are constant. Resolution III designs could be utilized for the noise factor whole-plot array since estimating interactions between noise factors is usually not of interest.

As an example of the response modeling approach with multiple noise factors, consider analyzing the data from the experiment to optimize the installation procedure for an elastometric connector used in the automotive industry described by Taguchi and Byrne (1984) and shown in Figure 12.4. The design was a $3^{4-2} \times 2^3$ resulting in a total of 72 experiments with no replicates. The runs were made in a completely random order, and this was not a split-plot design. From the 3^{4-2} control factor array only the main effects A, B, C, and D are estimable. Each of these main effects have two degrees of freedom. From the noise factor array, the main effects (E, F, and G), all two-factor interactions (EF, EG, FG), and the three-factor interaction EFG are estimable. Since the design is a cartesian product design, all interactions between estimable effects from the control factor array and estimable effects from the noise factor array are estimable, i.e., AE, AF, AG, AEF, AEG, AFG, $AEFG$, BE, ..., $DEFG$. However, if a model involving all of these effects is fit to the response (pull-off force) there will be zero degrees of freedom for error and no way to test for significance.

A normal probability plot of the estimated effects or regression coefficients can be used to get an initial idea about which effects may be significant. However, it is not as straightforward as the normal plots of regression coefficients described previously in Chapters 3 and 6. When all the factors have two levels as the examples shown in Chapters 3 and 6, each regression coefficient represents one single degree of freedom. When there are factors with more than two levels, like the control factors in the elastometric connector problem, the main effects and interactions have more than one degree of freedom so the first step in making a normal plot is to represent each main effect by an orthogonal set of single-degree-of-freedom contrasts. This can be accomplished by using the orthogonal polynomial contrasts described in Section 2.8.1. Since the control factor main effects have three levels the commands below illustrate the creation of linear and quadratic contrasts for main effect A, creation of the coded factor level, X_E, for main effect E, and the creation of the linear$\times X_E$ and quadratic$\times X_E$ parts of the $A \times E$ interaction. The linear and quadratic contrasts for the other control factors, coded factor levels for the other noise factors and the remainder of the interactions were created similarly.

```
data prod;
input A B C D E F G pf;
  if A=1 then Al=-.707107;
  if A=2 then Al=0;
  if A=3 then Al=.707107;
  if A=2 then Aq=-.816497; else Aq=.4082483;
  XE=(E-1.5)/.5;
  AlXE=Al*XE;
  AqXE=Aq*XE;
. . .
datalines;
1 1 1 1 1 1 1 15.6
```

When all the factors have two levels, the X matrix for the regression is made up of orthogonal columns of -1's and 1's, and the $X'X$ matrix is diagonal with n on the diagonal where n is the total number of runs in the unreplicated design. The variance of the regression coefficient $\hat{\beta}_i$ is $\sigma^2 \times c_{ii}$ where c_{ii} is the ith diagonal of $X'X^{-1}$. Even though σ^2 cannot be estimated in an unreplicated 2^k or 2^{k-p} design, all the regression coefficients have the same theoretical variance since diagonal elements c_{ii} are all equal to $1/n$. Because they all have the same variance, they can be plotted on a normal plot.

However, when some of the regression coefficients represent orthogonal polynomial contrasts of factors with more than two levels and other regression coefficients represent coded two-level factors, their variances are not equal. For example, the sum of squares of the linear contrast for a three-level factor will be equal to $r \times ((-.707107)^2 + 0^2 + .707107^2)$, where r is the number of times each level of the factor is repeated in the design, the sum of squares for the quadratic contrast for a three-level factor will be $r \times (.4082483^2 + (-.816497)^2 + .4082483^2)$, and the sum of squares for a coded two-level factor will be $q \times ((-1)^2 + 1^2)$, where q is the number of times each level of the two-level factor is repeated in the design.

Since the sum of squares $((-.707107)^2 + 0^2 + .707107^2) = (.4082483^2 + .(-.816497)^2 + .4082483^2) = 1$, and there are 24 replicates of each level of the three-level factors in the 72 runs of the design, the diagonal elements of $X'X$ for each contrast of the three-level factors is 24. Since there are 36 replicates of each level of the two-level factors in the 72 runs, the diagonal element of the $X'X$ for each two-level factor and interaction between two-level factors is 72. All of the regression coefficients in the 72-run design can be standardized to have the same variance by dividing each contrast by the square root of the diagonal element of the unstandardized $X'X$ matrix. For example, in the commands below this is illustrated for factors A, B and E and F. This was done for all the coded factor levels and contrasts in the design resulting in an $X'X$ that is equal to a 72×72 identity matrix. Recall that no interactions were created between three-level control factors because they are not estimable from the 3^{4-1} control factor array.

```
data prodstd; set prod;
Al=Al/sqrt(24);
Aq=Aq/sqrt(24);
Bl=Bl/sqrt(24);
Bq=Bq/sqrt(24);
E=XE/sqrt(72);
F=XF/sqrt(72);
EF=XEF/sqrt(72);
AlE=AlXE/sqrt(24);
AqE=AqXE/sqrt(24);
AlF=AlXF/sqrt(24);
AqF=AqXF/sqrt(24);
AlEF=AlXEF/sqrt(24);
AqEF=AqXEF/sqrt(24);
BlE=BlXE/sqrt(24);
BqE=BqXE/sqrt(24);
BlF=BlXF/sqrt(24);
BqF=BqXF/sqrt(24);
BlEF=BlXEF/sqrt(24);
BqEF=BqXEF/sqrt(24);
. . .
keep pf Al Aq Bl Bq Cl Cq Dl Dq E F G EF EG FG EFG
AlE AqE AlF AqF AlG AqG AlEF AqEF AlEG AqEG AlFG AqFG AlEFG AqEFG
BlE BqE BlF BqF BlG BqG BlEF BqEF BlEG BqEG BlFG BqFG BlEFG BqEFG
ClE CqE ClF CqF ClG CqG ClEF CqEF ClEG CqEG ClFG CqFG ClEFG CqEFG
DlE DqE DlF DqF DlG DqG DlEF DqEF DlEG DqEG DlFG DqFG DlEFG DqEFG
;
```

The estimated regression coefficients were determined by proc glm and output to a file for plotting similar to the examples in Chapters 3 and 6 using the commands shown at the top of the next page.

The normal plot of the 71 regression coefficients (excluding the intercept) was constructed by modifying the code from Section 3.7.5 and the result is shown in Figure 12.11. Examining the plot and the list of coefficients printed

```
proc glm data=prodstd;
  model pf=Al Aq Bl Bq Cl Cq Dl Dq E F G EF EG FG EFG
          AlE AqE AlF AqF AlG AqG AlEF AqEF AlEG AqEG
          AlFG AqFG AlEFG AqEFG BlE BqE BlF BqF BlG
          BqG BlEF BqEF BlEG BqEG BlFG BqFG BlEFG BqEFG
          ClE CqE ClF CqF ClG CqG ClEF CqEF ClEG CqEG
          ClFG CqFG ClEFG CqEFG DlE DqE DlF DqF DlG
          DqG DlEF DqEF DlEG DqEG DlFG DqFG DlEFG
          DqEFG/solution;
*this outputs the parameter estimates to a file;
  ods output ParameterEstimates=sol;
```

Figure 12.11 *Normal Plot of Standardized Contrasts from Connector Experiment*

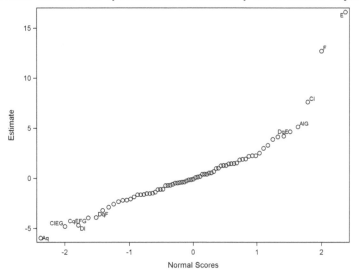

by `proc glm`, the largest coefficients were found to be E, F, Cl, AlG, DqE, DqF $CqEFG$, Dl, $ClEG$, Aq and $AlEF$. These single degree of freedom contrasts appear to be significant, and the others, that fall along the straight line in the normal plot, would appear to represent experimental error. A more formal test of the significance of effects can be made by running a model that includes all main effects plus the interactions that include the single degree of freedom interaction contrasts identified on the normal plot (i.e., interaction DE includes both DlE and DqE that was identified on the plot). The error mean square from this model will estimate the variance of experimental error and can be used as the denominator of the F-tests for terms in the model. This model was fit using the commands shown at the top of the next page with the results directly following.

The results show that control factor main effects A, C, and D, noise factor main effects E, F, and interactions between control and noise factors AG, and DE are significant at the 5% significance level.

Since there were no significant interactions with factor C, the insertion

```
proc glm data=prod;
  class A B C D E F G;
  model pf= A B C D E F G A*G A*E*F B*E*G C*E*G C*E*G*F D*E D*F;
run;
```

Source	DF	Type III SS	Mean Square	F Value	Pr > F
A	2	50.5769444	25.2884722	7.98	0.0022
B	2	13.3836111	6.6918056	2.11	0.1429
C	2	68.5936111	34.2968056	10.82	0.0004
D	2	23.6736111	11.8368056	3.74	0.0387
E	1	275.7334722	275.7334722	87.02	<.0001
F	1	161.7001389	161.7001389	51.03	<.0001
G	1	1.0512500	1.0512500	0.33	0.5700
A*G	2	26.5675000	13.2837500	4.19	0.0274
A*E*F	7	28.4043056	4.0577579	1.28	0.3014
B*E*G	7	22.4531944	3.2075992	1.01	0.4475
C*E*G	6	35.5463889	5.9243981	1.87	0.1278
C*E*F*G	10	24.6069444	2.4606944	0.78	0.6501
D*E	2	21.7452778	10.8726389	3.43	0.0489
D*F	2	15.4502778	7.7251389	2.44	0.1087

depth, the lsmeans shown below indicate that medium or deep level should be chosen to maximize the pull-off force.

Least Squares Means	
C	pf LSMEAN
1	18.3625000
2	20.2666667
3	20.5666667

To determine where control main effects A and D should be set, we must examine the interaction plots. Examining the interaction between control factor A, interference, and noise factor G, conditioning relative humidity, shown in Figure 12.12, it can be seen that choosing the mid-level of interference can simultaneously minimize the effect of uncontrollable changes in relative humidity and maximize the pull-off force.

Finally, by examining the interaction between between D, percentage of adhesive, and E, conditioning time, shown in Figure 12.13, it can be seen that the effect of uncontrolled variation in conditioning time can be reduced by choosing the medium level percent adhesive.

In summary, the results of the response modeling analysis indicate that all three noise factors had significant effects on the pull-off force, but by choosing the mid level of A: interference, minimizes the effect of noise factor G: conditioning relative humidity; choosing the medium D: percent adhesive, reduces

SINGLE ARRAY PARAMETER DESIGN EXPERIMENTS

Figure 12.12 *Interaction between A=Interference and G=Conditioning Relative Humidity*

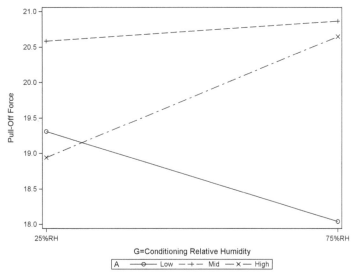

the effect of E: conditioning time, and choosing the medium to deep level of C: insertion depth, will maximize the pull-off force.

Choosing the levels of factors A and D to reduce the effects of G and E makes the pull-off force more consistent. Control factor B: connector wall thickness, does not have a significant effect on pull-off force and can be set at the level to minimize cost. Even though noise factor F: the conditioning temperature, affects the pull-off force there was no easily manipulated control factor that could counteract its effect. Therefore, to make the pull-off force more consistently high, efforts should be made in the manufacturing process to control the conditioning temperature as much as possible. The effects of conditioning humidity and time are less critical since control factor levels have been found that reduce their effects.

Had the loss model or location-dispersion modeling approach been used (left for an exercise) with this data, the optimal levels of the control factors could still be identified, but there is no way to determine which noise factors are important and which noise factor effects are reduced or minimized by choice of the control factor levels. Thus the response modeling approach results in more information than the location-dispersion modeling approach when there is more than one noise factor.

12.5 Single Array Parameter Design Experiments

In a single-array parameter design experiment, both control and noise factors are included as factors in a single experimental design, and the analysis is

Figure 12.13 *Interaction between D=Percent Adhesive and E=Conditioning Time*

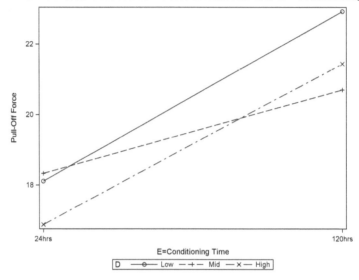

completed using the response modeling approach. Product design experiments usually provide more information than single-array experiments. However, when the levels of the noise factors are actual settings rather than sampling locations or simulations like those illustrated in the Ina tile example and the circuit design and each experiment consists of making an actual assembly like the elastometric connector example, the product array design often requires more effort than is necessary.

As an example, consider the following experiment described by Lawson and Erjavec (2001) that is patterned after an experiment that was conducted in a motorscooter manufacturer in Taiwan*. The purpose for the experiment was to select the nominal values for certain parameters in an electric starting motor. The desired function of the starting motor was to deliver a torque of 0.25kgs or more with high probability. The four parameters of the starting motor that are under the control of the designer are B: the inside diameter of the starting motor casing, C: the spring constant, D: the R.L.C., and the E: the type of weld (spot or line), used in the assembly process. These were the control factors. Three levels were chosen for factors B, C and D and there were just two alternatives for factor E.

In order to simulate noise caused by manufacturing imperfections and deterioration while in use, a composite noise factor, A, was defined. At the low level of this noise factor, starting motors were assembled with loose rusty springs and motor casings were used with a rough inside surface finish. At the high level of the noise factor, starting motors were assembled with tight springs

---

* The experiment was conducted with consultation help from G. Taguchi and Y. Wang.

SINGLE ARRAY PARAMETER DESIGN EXPERIMENTS

without rust and smooth finished casings. These two extremes were thought to represent the worst and best case that might occur in the use environment. The factors and levels for the experiment are shown in Table 12.10.

Table 12.10 *Factors and Levels for Motorscooter Starting Motor Experiment*

Factor	levels		
	1	2	3
A: simulated noise	worst case	best case	
B: inside diameter	small	medium	large
C: spring constant	weak	medium	strong
D: R.L.C.	low	medium	high
E: weld type	spot	line	

To use a product design for this problem, a control array would be chosen for the control factors, and then it would be repeated for both levels of the noise factor. Since there are three three-level control factors and one two-level control factor, a call to the %mktruns(2 3**3) macro in SAS reveals that the smallest orthogonal array for one two-level and three three-level factors is an $OA(18)$. Thus, the product array would require $18 \times 2 = 36$ experiments. Since each experiment actually required assembling a prototype starting motor and then putting it on the test stand to measure the torque generated in the locked position, it would be desirable to reduce the total number of experiments as long as it would be possible to still estimate the control factor by noise factor interactions.

Since the $OA(18)$ has hidden projection properties described by Wang and Wu (1995), it can allow estimation of many two-factor interactions. A call to the %mktex(2 3**3 2, interact=X1*X2 X1*X3 X1*X4 X1*X5, n=18) macro in SAS shows that an 18-run main effect plan can be found with D-efficiency of 98.53% that would allow estimation of all main effects and all interactions between the control and the noise factors. Thus, combining the noise factor and the control factors in this single array would reduce the total number of prototype starting motors that must be assembled and tested from 36 to 18.

Table 12.11 shows the single-array design and the measured response for each of the 18 prototype starting motors. The runs in this design could be discharged in a completely random order.

The SAS commands to fit a model that includes all main effects and control-factor by noise-factor interactions is shown below Table 12.11 on the next page. The resulting ANOVA table, shown below the commands, indicates that the noise factor, A, control factors B: the inside diameter and C: the spring constant, along with the interaction between the noise factor A and B, were all significant.

Table 12.11 *Single Array for Starting Motor Design*

Run	A	B	C	D	E	torque
1	1	1	1	1	1	0.225353
2	1	1	2	2	2	0.257185
3	1	1	3	3	2	0.276032
4	1	2	1	1	2	0.224485
5	1	2	2	2	2	0.242203
6	1	2	3	3	1	0.258791
7	1	3	1	2	1	0.208673
8	1	3	2	3	2	0.231972
9	1	3	3	1	2	0.257377
10	2	1	1	3	2	0.241026
11	2	1	2	1	1	0.259373
12	2	1	3	2	2	0.284073
13	2	2	1	2	2	0.266845
14	2	2	2	3	1	0.278359
15	2	2	3	1	2	0.307101
16	2	3	1	3	2	0.248256
17	2	3	2	1	2	0.270575
18	2	3	3	2	1	0.293259

```
Data Smotor;
input Run A  B C D E torque;
datalines;
1   1   1   1   1   1    0.225353
    . . .
proc glm;
  class A B C D E;
  model torque=A B C D E A*B A*C A*D A*E;
run;
```

Source	DF	Type III SS	Mean Square	F Value	Pr > F
A	1	0.00342407	0.00342407	467.76	0.0021
B	2	0.00038172	0.00019086	26.07	0.0369
C	2	0.00320145	0.00160072	218.68	0.0046
D	2	0.00002682	0.00001341	1.83	0.3531
E	1	0.00004227	0.00004227	5.77	0.1382
A*B	2	0.00100676	0.00050338	68.77	0.0143
A*C	2	0.00000312	0.00000156	0.21	0.8241
A*D	2	0.00000518	0.00000259	0.35	0.7388
A*E	1	0.00002577	0.00002577	3.52	0.2015

SINGLE ARRAY PARAMETER DESIGN EXPERIMENTS

The least squares means (shown below) for factor C: the spring constant show that the torque generated by the starting motor increases nearly linearly as the spring constant increases. Therefore, the strong level of the spring constant should be used to maximize the torque.

```
                    torque
    C               LSMEAN

    1            0.23679975
    2            0.25424225
    3            0.27848238
```

The interaction between A and B should be studied in order to determine which level of B, inside diameter, that should be used. The output below shows the least squares means for the different combinations of levels of factors A and B and Figure 12.14 shows the interaction plot.

```
                         torque
    A    B               LSMEAN

    1    1            0.25149225
    1    2            0.24046192
    1    3            0.23130958
    2    1            0.26132267
    2    2            0.28393367
    2    3            0.27052867
```

In the plot (shown in Figure 12.14) it can be seen that on the average the torque is higher when the noise factor is set at the best case than it is when the noise factor is set at the worst case. However, when the inside diameter is small, the difference in the torque generated at the best and worst cases for the noise factor is minimized. Although the average torque appears to be slightly higher when B: the inside diameter is at the medium level, the range in differences (0.28393367-0.24046192=0.04347175) caused by the noise factor is over four times larger than it is when B is set at its small level (0.26132267-0.25149225=0.00983042). Therefore, in order to maximize the probability that the starting motor will deliver a torque of 0.25kgs or more, the small inside diameter should be used.

As a second example of a single-array parameter design, consider an experiment described by Montgomery (1990) to improve an injection molding process for producing plastic parts. Prior to the experiment there was excessive shrinkage of parts after molding that was causing downstream assembly problems. Four factors that could easily be controlled in the process were A: mold temperature; B: screw speed; C: holding time ; and D: gate size. These were the control factors. Process variables that could affect shrinkage but that

Figure 12.14 *Interaction between Casing Inside Diameter and Simulated Noise Factor*

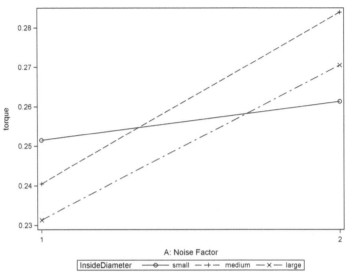

were normally difficult to control are E: cycle time; F: moisture content; and G: holding pressure. These were identified as noise factors. The operational personnel and quality engineer proposed to use three levels (low, mid and high) for each factor. The design initially considered was a product-array design using a 3^{4-1} design for the control factors and a 3^3 for the noise factors. This would have required $27^2 = 729$ runs, and since each run required actually making parts and measuring the percent shrinkage, an alternative plan requiring less experimentation was desired.

A single array experiment utilizing a 2^{7-3} fractional factorial with four added center points to check for curvature in the response only required 20 experiments. This was a resolution IV design with generators $E = ABC$, $F = BCD$, and $G = ACD$, and each two-factor interaction is confounded with two other two-factor interactions. In general this design would appear to be unsuitable for a parameter design experiment since the interactions between control and noise factors are not clear of other two-factor interactions. However, after completing this design, if any of the confounded strings of two-factor interactions appear to be significant, additional runs consisting of a foldover fraction reversing signs on some of the factors (as described in Section 6.5.1) could be used to break the confounding. If no two-factor interactions appear significant, then no additional experiments will be required and the single array experiment will require less than 1/30 the runs of the product array design originally considered.

Table 12.12 shows the factor settings and the response for the experiment in standard order.

Table 12.12 *Factors and Response for Injection Molding Experiment*

Run	A	B	C	D	E	F	G	shrinkage
1	−	−	−	−	−	−	−	6
2	+	−	−	−	+	−	+	10
3	−	+	−	−	+	+	−	32
4	+	+	−	−	−	+	+	60
5	−	−	+	−	+	+	+	4
6	+	−	+	−	−	+	−	15
7	−	+	+	−	−	−	+	26
8	+	+	+	−	+	−	−	60
9	−	−	−	+	−	+	+	8
10	+	−	−	+	+	+	−	12
11	−	+	−	+	+	−	+	34
12	+	+	−	+	−	−	−	60
13	−	−	+	+	+	−	−	16
14	+	−	+	+	−	−	+	5
15	−	+	+	+	−	+	−	37
16	+	+	+	+	+	+	+	52
17	0	0	0	0	0	0	0	25
18	0	0	0	0	0	0	0	29
19	0	0	0	0	0	0	0	24
20	0	0	0	0	0	0	0	27

An analysis including the center points showed no significant curvature and Figure 12.15 shows a normal plot of the regression coefficients from the saturated model.

Here it can be seen that factor B, the screw speed; factor A, the mold temperature; and the confounded string of two factor interactions $AB + CE + FG$ appear to be significant. The effect-heredity principle would lead one to the conclusion that the confounded string of interactions represents the interaction between mold temperature and screw speed. None of the noise factors E, F, or G seem to have any effect on percent shrinkage. Figure 12.16 shows the interaction between factors A and B.

This plot shows that when B, screw speed, is at its low level, shrinkage is minimized, and that changing A, mold temperature, has a minimal effect when screw speed is low. Therefore, choosing the low screw speed will reduce shrinkage to an average of about 10%, and the mold size can be adjusted so that the average part size will be on target. However, the unexplained variability about the average due to experimental error (shown in the box plot in Figure 12.17) is still too large, with values ranging from 4 to 16, and will cause problems in assembly. None of the noise factors used in the experiment can explain this variability, but by making box plots of the percent shrinkage (with screw speed at the low level) versus each control factor separately, one

Figure 12.15 *Normal Plot of Regression Coefficients From Injection Molding Experiment*

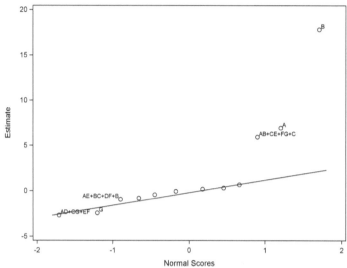

Figure 12.16 *Interaction Between Mold Temperature and Screw Speed*

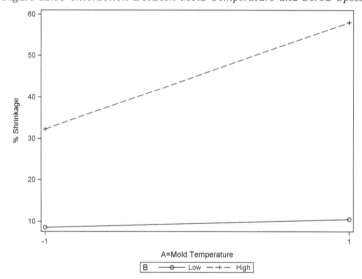

potential dispersion effect appears as illustrated in Figure 12.18. A dispersion effect is a control factor that affects the variability in the response. In Figure 12.18 it can be seen that at the low level of C: holding time, there appears to be less variability in percent shrinkage than there is when holding time is at

Figure 12.17 *Box Plot of % Shrinkage at Low Screw Speed*

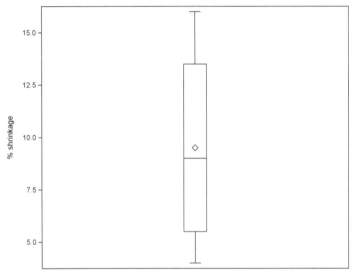

the high level. Therefore the conclusion is to fix both B: screw speed, and C: holding time, at their low levels. This should result in shrinkage of about 9% plus or minus 3% as shown in the left side of Figure 12.18. In this example

Figure 12.18 *Box Plots of % Shrinkage at Low Screw Speed at Each Level of Holding Time*

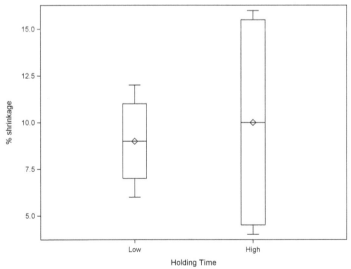

there were no significant noise factors, and no significant interactions between the control and noise factors. Therefore, variability in percent shrinkage could not be accomplished by reducing or nullifying the effect of the suspected noise factors used in the experiment. However, there are obviously other unknown or lurking noise factors that are causing variability in the response. By identifying a dispersion effect and choosing the level which minimizes the variability, one actually reduces the effects of the unknown noise factors resulting in a more consistent response.

Box and Meyer (1986b) described a more formal method of detecting dispersion effects. They propose to first fit a model to the response (for this example A, B, and $A \times B$) and calculate the residuals from the model. Next, they calculate the dispersion statistic for each factor as the natural log of the ratio of the sample variances of the residuals at the high and low levels with the formula

$$F_i^* = \ln \frac{S^2(i^+)}{S^2(i^-)}, \qquad (12.5)$$

where $S^2(i^+)$ is the sample variance of the residuals at the high level of factor i and $S^2(i^-)$ is the sample variance of the residuals at the low level of factor i. They show that these statistics are approximately normally distributed with mean zero and standard deviation 1.

12.6 Joint Modeling of Mean and Dispersion Effects

12.6.1 Joint Modeling

One of the assumptions that justifies fitting a linear model by least squares is constancy of the variance of the experimental error across all levels of the factors in the model. If there is a dispersion effect, this assumption will obviously be violated. Therefore, the accuracy of the coefficients in the model for the average response will be reduced because too much weight will be given to minimizing errors in prediction in regions where the variance is larger. Nelder and Lee (1998) and Lee and Nelder (2003) have suggested an iterative weighted least squares approach to estimation. This consists of:

1. Fitting a model

$$\hat{\beta} = (X'X)^{-1}X'y \qquad (12.6)$$

   which includes all terms that affect the mean response, and calculating and storing the residuals from this model.

2. Fitting a model

$$\hat{\gamma} = (X'X)^{-1}X'z \qquad (12.7)$$

   which includes all dispersion effects is fit, using $z_i = \ln r_i^2$ as the response, where r_i are the residuals from the model 12.6. The predicted values from this model, $\hat{z}_i$ are calculated and stored.

JOINT MODELING OF MEAN AND DISPERSION EFFECTS

3. Finally fitting a model using the method of weighted least squares,

$$\hat{\beta} = (X'WX)^{-1}X'Wy. \quad (12.8)$$

This model includes all terms that affect the mean response, and $W = \text{diag}\left(1/\sqrt{\exp \hat{z}_i}\right)$, where $\hat{z}_i = \ln(\hat{r}_i^2)$ are the predicted values from model 12.7. After the residuals are calculated and stored from this model, return to step 2.

This process is continued until convergence.

An alternate way to do the same thing is to use proc mixed, as described in Section 5.8, to estimate the effects and variances of the residuals simultaneously using the method of maximum likelihood or REML.

To illustrate this procedure, consider the following example described by Shoemaker et al. (1991) that involves a process for growing a uniform layer of silicon on top of a silicon wafer. This is one of the earliest steps in manufacturing integrated circuits. It is important that the added layer be as uniform as possible because electrical devices will be formed within these layers in later process steps. The layers are grown in a reactor that holds 14 wafers simultaneously on a susceptor. The reactor is heated and gasses are introduced at the top. The susceptor rotates to mix the gasses and silicon is deposited on the upper surface of the wafers. When the deposited layer approaches the target thickness, the heat in the reactor is reduced to stop the deposition process.

There were eight control factors in the study shown in Table 12.13. The response was the thickness of the silicon layer deposited. Four replicate measurements of thickness were taken from wafers in different positions of the reactor to represent the noise array.

Table 12.13 *Control Factors and Levels for Silicon Layer Growth Experiment*

Factor	levels	
	−	+
A: rotation method	continuous	oscillating
B: wafer code	668G4	678D4
C: deposition temperature	1,210	1,220
D: deposition time	High	Low
E: arsenic flow rate	57%	59%
F: HCI etch temp.	1,180	1,215
G: HCI flow rate	10%	14%
H: nozzle position	2	6

The control factor array was a resolution IV 2^{8-4} fractional factorial with generators $D = ABC$, $F = ABE$, $G = ACE$ and $H = BCE$. It is shown in Table 12.14 along with the summary statistics calculated across the noise array.

Figure 12.19 is a normal plot of the regression coefficients from the saturated model relating $\bar{y}$ to the factors, and Figure 12.20 is a normal plot of the

540 ROBUST PARAMETER DESIGN EXPERIMENTS

Table 12.14 *Mean and Variance for Each Row in Control Array*

A	B	C	D	E	F	G	H	$\bar{y}$	s_i^2
−	−	−	−	−	−	−	−	14.804	0.00171
+	−	−	+	−	+	+	−	13.979	0.10942
−	−	−	−	+	+	+	+	14.880	0.00372
−	−	−	+	+	−	−	+	14.120	0.10615
−	+	−	+	−	+	−	+	14.158	0.00060
+	+	−	−	−	−	+	+	14.766	0.19199
−	+	−	+	+	−	+	−	13.837	0.00612
+	+	−	−	+	+	−	−	14.846	0.22907
−	−	+	+	−	−	+	+	14.034	0.00305
+	−	+	−	−	+	−	+	14.792	0.38227
−	−	+	+	+	+	−	−	13.891	0.00096
+	−	+	−	+	−	+	−	14.348	0.21191
−	+	+	−	−	+	+	−	14.728	0.00144
+	+	+	+	−	−	−	−	13.738	0.26182
−	+	+	−	+	−	−	+	14.884	0.01923
+	+	+	+	+	+	+	+	13.973	0.07569

regression coefficients from the saturated model relating $\ln(s_i^2)$ to the factors. In Figure 12.19 it can be seen that D: deposition time and H: nozzle position appear to have significant effects on the average thickness. In Figure 12.20 it can be seen that A: rotation method appears to have a positive effect on $\ln(s_i^2)$. This is a dispersion effect and should be set to its low level (continuous) to minimize variability of the silicon layer thickness.

The levels of the adjustment factors (D: deposition time, and H: nozzle position) can be chosen to bring the average thickness to the desired target. However, since A: rotation method affects $\ln(s_i^2)$, the variance of the residuals will not be constant and the assumptions required for the least squares estimates of the D and H effects are violated. Therefore, inaccuracy will result if the least squares coefficients for D and H are used in predicting the levels of D and H needed to bring the average thickness to the desired target.

In the commands below, use of the `repeated/group=A;` statement causes `proc mixed` to simultaneously compute weighted least squares estimates of the

```
proc mixed data=epitaxstr;
  class A;
  model y=D H/ solution;
  repeated/group=A;
run;
```

adjustment factor effects and REML estimates of the variance of the residuals within each level of the dispersion factor. A portion of the output is shown following Figure 12.20.

JOINT MODELING OF MEAN AND DISPERSION EFFECTS 541

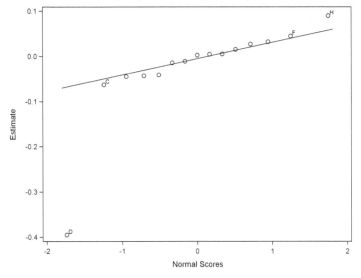

Figure 12.19 *Normal Plot of Regression Coefficients for Mean Silicon Thickness*

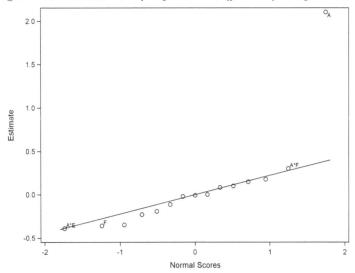

Figure 12.20 *Normal Plot of Regression Coefficients for Log Variance*

Covariance Parameter Estimates

Cov Parm	Group	Estimate
Residual	A -1	0.006409
Residual	A 1	0.1772

Null Model Likelihood Ratio Test

DF	Chi-Square	Pr > ChiSq
1	59.24	<.0001

Solution for Fixed Effects

Effect	Estimate	Standard Error	DF	t Value	Pr > |t|
Intercept	14.3993	0.01390	61	1035.69	<.0001
D	-0.4202	0.01390	61	-30.22	<.0001
H	0.08713	0.01390	61	6.27	<.0001

In the output above the weighted least squares estimates of the adjustment factors are shown, and it can be seen that the $\hat{\sigma}^2_{(A=-)} = 0.006409$ and $\hat{\sigma}^2_{(A=+)} = 0.1772$. From these coefficients, the coded level of D: deposition time, and H: nozzle position, can be determined by solving the equation:

$$\text{desired thickness} = 14.3993 - 0.4202 X_D + 0.08713 X_H \qquad (12.9)$$

for X_D and X_H.

12.6.2 Designing a Noise Array

When the noise-factor array consists of replicate samples over time or space like the tile experiment or the silicon layer growth experiment, and the analysis is to be completed using the location-dispersion modeling approach, a fractional factorial or orthogonal array design may not be the most efficient way to design the noise array. When the analysis is completed using the location dispersion modeling, the purpose of the noise array is not to efficiently estimate the effects of the noise factors, but rather to allow accurate estimation of the variance in the response across the noise space. Therefore, rather than using an orthogonal array or fractional factorial that distributes points in the noise space to the extremes, a plan that provides representative points from the noise space would be better. Wang et al. (1995) introduced the concept of a quantizer to select optimally representative points from the noise space. They tabulated their representative points for practical use. Table 12.15, shown on the next page, is a part of their table of representative points.

Table 12.15 *Optimal Representative Points in $[0,1]^k$ Assuming Uniform Distribution*

Number of runs n	\multicolumn{6}{c}{Number of Noise Factors k}					
	1	2	3	4	5	6
3	0.1667					
	0.5000					
	0.8333					
4	0.1250	0.3750	0.8750	0.6250		
	0.3750	0.8750	0.6250	0.1250		
	0.6250	0.1250	0.3750	0.8750		
	0.8750	0.6250	0.1250	0.3750		
5	0.1000	0.3000	0.7000	0.5000		
	0.3000	0.7000	0.5000	0.1000		
	0.5000	0.1000	0.3000	0.7000		
	0.7000	0.5000	0.1000	0.3000		
	0.9000	0.9000	0.9000	0.9000		
6	0.0833	0.4167	0.2500	0.9167	0.5833	0.7500
	0.2500	0.9167	0.5833	0.7500	0.0833	0.4167
	0.4167	0.2500	0.9167	0.5833	0.7500	0.0833
	0.5833	0.7500	0.0833	0.4167	0.2500	0.9167
	0.7500	0.0833	0.4167	0.2500	0.9167	0.5833
	0.9167	0.5833	0.7500	0.0833	0.4167	0.2500
7	0.0714	0.3571	0.2143	0.7857	0.5000	0.6429
	0.2143	0.7857	0.5000	0.6429	0.0714	0.3571
	0.3571	0.2143	0.7857	0.5000	0.6429	0.0714
	0.5000	0.6429	0.0714	0.3571	0.2143	0.7857
	0.6429	0.0714	0.3571	0.2143	0.7857	0.5000
	0.7857	0.5000	0.6429	0.0714	0.3571	0.2143
	0.9286	0.9286	0.9286	0.9286	0.9286	0.9286
8	0.0625	0.4375	0.8125	0.1875	0.9375	0.5625
	0.1875	0.9375	0.5625	0.4375	0.8125	0.0625
	0.3125	0.3125	0.3125	0.6875	0.6875	0.6875
	0.4375	0.8125	0.0625	0.9375	0.5625	0.1875
	0.5625	0.1875	0.9375	0.0625	0.4375	0.8125
	0.6875	0.6875	0.6875	0.3125	0.3125	0.3125
	0.8125	0.0625	0.4375	0.5625	0.1875	0.9375
	0.9375	0.5625	0.1875	0.8125	0.0625	0.4375

As one example of the use of these tables, Wang, Fang and Lin used them to construct an improved noise array for an experiment that was described by Phadke (1989). The experiment involved a process to deposit polysilicon on silicon wafers. In this process wafers are placed on two quartz carriers that were positioned in a hot-wall-reduced pressure reactor. Silane and nitrogen gasses were introduced at one end of the reactor and pumped out the other. As the silane gas pyrolyzes a polysilicon layer is deposited on the wafers. The quartz carriers each held twenty-five wafers and occupied half the reactor so that a total of fifty wafers were deposited along the flow of gasses from the inlet to outlet. Gas flow patterns caused different concentrations along the length of the reactor. The noise array consisted of three measurements made along the two quartz carriers at locations 3, 23, and 48. The optimal spacing of samples is shown in Table 12.15 with $n=3$ and $k=1$ (0.1667, 0.5000, 0.8333). Therefore, a better noise array could be constructed using the table. It would consist of sampling locations $8=50\times0.1667$, $25=50\times0.5000$ and $42=50\times0.8333$.

As a second example of using the tables, consider again the design of the temperature controller circuit described in Section 12.3. In that example, a $OA(18)$ was used for the noise factor array to simulate the noise environment where actual values of the control factors, could deviate from the specified nominal values by $\pm2.04\%$. To get a more accurate estimate of the variance of R_T for any combination of levels of the control factors Wang, Fang and Lin's table can be used. For example, the left side of Table 12.16 shows the coded factor levels for five noise factors and $n=8$ runs taken from Table 12.15. The right side of Table 12.16 shows the perturbations of the actual control factor levels across the noise-factor array when all control factors are set at their low levels (i.e., $A=2.67$, $B=1.33$, $C=5.33$, $D=8.00$, $F=4.80$).

Table 12.16 *Representative Noise Samples for Temperature Controller*

Coded levels on $[0,1]^5$					Actual Factor Levels				
1	2	3	4	5	A	B	C	D	F
.0625	.4375	.8125	.1875	.9375	2.62	1.33	5.40	7.90	4.89
.1875	.9375	.5625	.4375	.8125	2.64	1.35	5.34	7.98	4.86
.3125	.3125	.3125	.6875	.6875	2.65	1.32	5.29	8.06	4.84
.4375	.8125	.0625	.9375	.5625	2.66	1.35	5.23	8.14	4.81
.5625	.1875	.9375	.0625	.4375	2.68	1.31	5.43	7.86	4.79
.6875	.6875	.6875	.3125	.3125	2.69	1.34	5.37	7.94	4.76
.8125	.0625	.4375	.5625	.1875	2.70	1.31	5.32	8.02	4.74
.9375	.5625	.1875	.8125	.0625	2.72	1.33	5.26	8.10	4.71

The perturbed level for factor A in the first line of the table was created as $2.62 = 2.67 + ((2\times.0625)-1.0)\times 0.0204 \times 2.67$. Here the term $(2\times.0625)-1.0$ converts the coded level from the $[0,1]$ scale to the $[-1,1]$ scale. Multiplying by 0.0204×2.67 changes the coded level on the $[-1,1]$ scale to a perturbation of the specified nominal level in the $\pm2.04\%$ of nominal range. Similar calculations were made to convert the other coded factor levels to perturbations of the specified nominal values for factors B through F on runs one through eight. Notice that by using Wang, Fang and Lin's table to create the noise array, perturbed values for the control factors will be created within the tolerance range of deviations from the specified nominals, while when using the orthogonal array perturbed values are pushed to the extremes of the tolerance range. A more accurate representation of the true variance of R_T can be made when the noise array more uniformly spaces sampling points over the tolerance range.

The response R_T could now be calculated for each combination of levels of factors A through F using Equation (12.2), as was shown in Table 12.6 for the $OA(18)$ noise array, and the variance of these values could be calculated like the log variance statistics shown in Table 12.8. The method of using the noise factor array from Wang, Fang and Lin's tables to simulate the variance of R_T could be used whether a parameter design experiment were used to find the optimal nominal values of the control factors or the more accurate method of computer optimization.

Wang, Fang and Lin's tables for representative points in noise arrays include up to twelve noise factors and twenty-five runs. They also include a table for representative samples over a disk, ball or spherical surface. However, for cases where these tables are unavailable, another option for designing the noise array that would be better than a fractional factorial or orthogonal array would be to use a Latin Hypercube design or the space filling `criterion=u` option of `proc optex`.

12.7 Review of Important Concepts

Taguchi's ideas of using experimental designs to improve products and production processes has been hailed as one of the major advances in the twentieth century in engineering and manufacturing. Taguchi showed that variation from ideal product and process function is caused by noise. Utilizing experimental designs that contain control factors (which are easy for designers or operators to manipulate) and noise factors (that represent difficult to control sources of noise), he showed that settings of the control factors can often be found to nullify or reduce the effects of the noise factors. This chapter has illustrated how product-array designs or single-array designs can be used to accomplish this. Product array designs provide more information but require more effort if changing levels of noise factors actually involve physical changes. If the noise factors simply represent sampling over space, the product array designs may not involve much additional work.

Analysis of data can be accomplished using location-dispersion modeling, where summary statistics (mean and log variance) are calculated across the levels of the noise array and analyzed as the responses for the control factor array. Control factors that affect the mean, but not the log variance of the response, are called adjustment factors. Their levels can be chosen to bring the response to a maximum, minimum or target value. Control factors that affect the log variance should be set at a level that will minimize the variance in the response. A two-step procedure is usually used to identify the optimal levels of the control factors and adjustment factors.

The other method of analysis is to use the response model which is the same as described in the other chapters of this book. When using the response modeling approach, identifying interactions between control and noise factors becomes very important. The levels of control factors that interact with noise factors can sometimes be selected to minimize the effect of the noise factor. When using the response modeling approach, it is easier to determine exactly what noise factors cause functional variation, and what control factor settings can reduce their effects. When no significant noise factors can be found, but experimental error is too high and there is too much functional variation at optimal control factor settings, residuals should be plotted versus the levels of each control factor to identify any dispersion effects. Dispersion effects are factors that have an effect on the variation of the response. Choosing the level of the dispersion effect to minimize residual variation will reduce the effect of unknown or lurking noise factors.

When levels of the noise factors represent replicate samples over the noise space, the product array may actually be more efficient than a single-array design. When analyzing product arrays using location-dispersion modeling, Wang, Fang, and Lin's tables, or uniformly spaced points generated by `proc optex`, will give better estimates of the variance in the response over the noise array.

Sometimes product arrays are actually split-plot experiments, and when modeled using the response modeling, they should be analyzed as a split-plot experiment. Box and Jones (1992) have recommended, when possible, to conduct product array parameter design experiments with the noise factors as the whole-plot factors and the control factors as the subplot factors. This way the control factor effects and the control-by-noise factor effects can be tested with the smaller subplot error term.

12.8 Exercises

1. Consider the commercial product test of erasers conducted by Sachiyo and Kosaka in 1971 and discussed by Taguchi (1987). The purpose of the test was to determine the quality of erasure when lines of the same length drawn on different types of paper were rubbed with different erasers. For the experiment lines of the same length were drawn on each type of paper by different pencils. The lines were rubbed three times back and forth by each eraser and then judged for quality. A response of $0 =$ top means that no pencil trace was left on the paper. A response of $1 =$ middle means only a faint pencil trace was left on the paper, and response of $2 =$ bottom means it did not erase very well.

   Factors and levels for the experiment are shown in Table 12.17. The hope was to identify the brand and material that produced the best quality erasure consistently regardless of the type of paper or the hardness of the pencil lead used.

   Table 12.17 *Factors and Levels for Commercial Product Test of Erasers*

Factor	1	Levels 2	3	4	5
Control Factors:					
A: Manufacturer	Rabbit	Staedtler			
B: Material	Rubber	Plastic			
Noise Factors:					
C: Paper	Note Paper	Tracing Paper			
D: Lead Hardness	4H	2H	HB	2B	4B

   The control-factor array was a 2^2 factorial. One cell is missing since one manufacturer did not make rubber erasers, and one cell is duplicated since the other manufacturer made two types of plastic erasers. The noise factor array was a 2×5 full factorial. The product array design and the judged responses are shown in Table 12.18.

   Table 12.18 *Product Array Design and Response For Commercial Product Test*

A	B	D	C=1 1	2	3	4	5	C=2 1	2	3	4	5
1	1		1	0.5	1	1	1	0	0.5	0.5	2	1.5
1	2		0.5	1	1	1	1	0	0	0	0.5	1.5
2	2		0	0	0	0	0.5	0	0	0	0.5	0
1	2		0	0.5	1	1	0	0	0	0.5	1.5	1.5

   (a) Analyze the data using location-dispersion modeling.
   (b) Use the two-step procedure in Section 12.4.1 to identify the brand and material that produce the best quality erasure.

(c) Repeat the analysis using the response modeling approach.

(d) Are there significant interactions between control and noise factors? Can they be exploited to reduce the effect of the noise factors?

(e) Is there an adjustment factor?

2. Reconsider the cake mix experiment described in exercise 9 of Chapter 8. The control factors that can be selected by the manufacturer are F: the amount of flour; S: the amount of shortening; and E: the amount of egg powder in the mix. The noise factors are T: baking temperature, and t: baking time.

   (a) Did you detect any control-by-noise interactions in the split-plot analysis conducted in Chapter 8? If so, can these interactions be exploited to reduce the effects of variation in baking time or baking temperature?

   (b) Reanalyze the data using the location-dispersion modeling approach. Do you reach the same conclusion about the optimal choice of levels of F, S and E?

   (c) Which method of analysis gives you more information?

3. Lawson (1990) describes an experiment in a chemical process with the aim of producing less byproduct tars. The control factors were A, reaction temperature; B, catalyst concentration; and C, excess of reagent 1. The noise factors were E, purity of the recycled solvent; and D, purity of reagent 1 that comes from a supplier. A product-array design was used for the experiment that is shown in Table 12.19. The control-factor array is a Box-Behnken design that will allow estimation of all terms in a full quadratic model involving factors A, B and C. The noise factor array is a 2^2 factorial.

   (a) What interactions are estimable in this design?

   (b) Analyze the data using the response modeling method. Are there any interactions between control and noise factors that would allow you to reduce or nullify the effect of uncontrollable variability in the purity of the solvent stream and purity of reagent 1?

   (c) Reanalyze the data using the location-dispersion modeling method. Do you reach the same conclusions about optimal settings of the control factors? Is any information lost in analyzing the data in this way?

   (d) Explain how the response modeling analysis would change if this experiment had been conducted as a split-plot design, where the noise factors were the whole-plot factors and the control factors were the sub-plot factors? Is there any advantage to running a product array split-plot design with the noise factors as the whole-plot factors, rather than having the control factors as the whole-plot factors as in the Ina tile example presented in Section 12.3?

EXERCISES 549

Table 12.19 Product Array Design For Chemical Process Experiment

			E −	+	−	+
			D −	−	+	+
A	B	C				
−	−	0	37.29	57.81	42.87	47.07
+	−	0	4.35	24.89	8.23	14.69
−	+	0	9.51	13.21	10.10	11.19
+	+	0	9.15	13.39	10.30	11.23
−	0	−	20.24	27.71	22.28	24.32
+	0	−	4.48	11.40	5.44	8.23
−	0	+	18.40	30.65	20.24	24.45
+	0	+	2.29	14.94	4.30	8.49
0	−	−	22.42	42.68	21.64	30.30
0	+	−	10.08	13.56	9.85	11.38
0	−	+	13.19	50.60	18.84	30.97
0	+	+	7.44	15.21	9.78	11.82
0	0	0	12.29	19.62	13.14	14.54
0	0	0	11.49	20.60	12.06	13.49
0	0	0	12.20	20.15	14.06	13.38

4. Because the elastometric connector experiments described in Section 12.3 required physical experimentation, Song and Lawson (1988) suggested using a single array design to save on the number of experiments required. Table 12.20 shows a resolution IV 2^{7-2} fractional factorial design and the resulting pull-off force. The factor names and levels are the same as those shown in Table 12.3 of Section 12.3. The generators for the design were $F = ABC$ and $G = ABD$.

(a) What is the defining relation for this design, and how many of the 12 control-by-noise factor interactions can be estimated clear of other main effects or two-factor interactions?

(b) Using the SAS ADX, can you find a 32-run resolution IV design that has more control-by-noise factor interactions clear? If so, what are the generators?

(c) Calculate a set of 31 saturated effects for this design, and make a normal plot to determine which effects, interactions and confounded strings of two-factor interactions appear to be significant. Is there a clear interpretation of any confounded strings of interactions?

(d) Make interaction and contour plots of any two-factor interactions between control factors and noise factors and use them to choose the level of control factors that will minimize the effect of noise factors.

(e) Are there any adjustment factors? If so, what levels should be chosen to maximize the pull-off force?

Table 12.20 *Single Array Experiment for Elastometric Connector*

A	B	C	D	E	F	G	pull-off force
Low	High	High	Low	120	150	75	22.0
High	High	High	High	120	150	75	27.6
High	High	High	Low	120	72	25	22.1
High	High	Low	High	120	72	75	20.2
Low	Low	Low	High	120	72	75	18.9
High	High	High	High	24	150	25	13.8
Low	High	High	High	24	72	75	15.2
Low	Low	Low	Low	120	150	25	23.1
High	Low	High	High	24	72	75	16.1
High	Low	High	High	120	72	25	20.1
Low	Low	High	High	120	150	75	21.9
High	Low	Low	Low	24	72	25	17.1
High	Low	High	Low	24	150	25	18.1
Low	High	Low	High	24	150	75	9.6
Low	Low	High	Low	24	72	75	18.3
Low	High	Low	Low	120	72	75	19.1
High	Low	High	Low	120	150	75	27.0
Low	Low	Low	Low	24	150	75	17.3
Low	Low	High	High	24	150	25	17.7
Low	Low	Low	High	24	72	25	14.7
High	Low	Low	Low	120	72	75	22.2
Low	High	High	Low	24	150	25	21.1
Low	Low	High	Low	120	72	25	24.3
Low	High	Low	Low	24	72	25	13.9
High	Low	Low	High	24	150	75	19.4
High	Low	Low	High	120	150	25	22.7
High	High	Low	High	24	72	25	14.6
High	High	Low	Low	120	150	25	23.3
High	High	High	Low	24	72	75	20.4
Low	High	Low	High	120	150	25	22.6
Low	High	High	High	120	72	25	21.0
High	High	Low	Low	24	150	75	17.5

EXERCISES

5. Taguchi (1987) described an experiment to find the source of clutch slipping in automobiles. Prototype clutches were assembled using some new components and some components taken from disassembled clutches that had complaints. Each prototype was tested and the response, y, was obtained by dividing the slipping time by the number of revolutions of the engine and then multiplying by 10,000. The control and noise factors and their levels are shown in Table 12.21.

Table 12.21 *Factors and Levels for Auto Clutch Slipping*

Control Factors:	Levels −	Levels +
B:Spring Allignment		5 levels
A:Force	current	+50kg
D:Clutch Plate Running Distance	≤10,000km	≥30,000km
F:Cover Running Distance	≤10,000km	≥30,000km

Noise Factors:	Levels −	Levels +
C:Clutch Plate	ordinary product	object of complaint
E:Cover	ordinary product	object of complaint

A 20-run orthogonal array (shown in Table 12.22) was used to study both control and noise factors in a single array.

(a) Read the data into a SAS data set and construct −1 and +1 coded factor levels for factors A, D, F, C and 4 orthogonal polynomial contrasts for factor B. Create columns for the interactions between control and noise factors.

(b) Find a model for the data using an all-subsets regression, as shown for analysis of a Plackett-Burman design in Section 6.6 and for the analysis of the log variance for the temperature controller circuit in Section 12.4.

(c) From your model, identify any significant adjustment factors and control-by-noise interactions, and determine the levels of the control factors that will minimize clutch slippage with minimum variation.

(d) Are all control-by-noise factor interactions estimable from this design? If not, construct an orthogonal-array design or orthogonal main-effect plan that allows estimation of all control by noise interactions by using the `interact=` option of the `%mktex` macro as shown in Section 6.7.

6. An experiment originally performed by the National Railway Corporation of Japan (Taguchi and Wu, 1980) was re-analyzed by Box and Meyer (1986b). The control factors in the design were A, kind of welding rods;

Table 12.22 Orthogonal Array Experiment for Clutch Slipping

B	A	D	F	E	C	y
1	1	1	1	1	1	1.0
1	1	1	1	1	2	1.4
1	2	2	2	2	1	1.1
1	2	2	2	2	2	1.0
2	1	1	2	2	1	1.2
2	1	1	2	2	2	1.6
2	2	2	1	1	1	0.9
2	2	2	1	1	2	1.2
3	1	2	1	2	1	1.6
3	1	2	1	2	2	1.8
3	2	1	2	1	1	1.2
3	2	1	2	1	2	1.0
4	1	2	2	1	1	1.3
4	1	2	2	1	2	1.3
4	2	1	1	2	1	1.0
4	2	1	1	2	2	1.6
5	1	2	2	1	1	1.3
5	1	2	2	1	2	1.1
5	2	1	1	2	1	1.2
5	2	1	1	2	2	1.1

B, period of drying; C, welded materials; D, thickness; E, angle; F, opening; G, current; H, welding method; and J, preheating. Taguchi and Wu also considered the interactions AC, AG, AH, and GH. The design and response $y=$ tensile strength of welds is shown in Table 12.23 where e_1 and e_2 are unassigned columns that represent confounded interactions.

(a) Calculate effects for each of the fifteen columns in Table 12.23, and make a normal plot to identify factors that affect the tensile strength.

(b) Fit a model to the data including only the effects that appear significant on the normal plot, and calculate the residuals from the fitted model.

(c) Calculate the sample variance of residuals from the last model for the $-$ and $+$ level of each column in Table 12.23, and calculate the dispersion statistic (given by Equation (12.5)) for each column. Make a normal plot of the dispersion statistics and identify any potential dispersion effects.

(d) Modify the SAS macro in the appendix to simultaneously estimate the location-dispersion effect using the method of iterative weighted least squares.

(e) What factor levels do you recommend to maximize the tensile strength with minimum variation?

EXERCISES

Table 12.23 *Design and Tensile Strength for Weld Experiment*

D	H	G	A	F	GH	AC	E	AH	AG	J	B	C	e_1	e_2	y
−	−	−	−	+	+	−	+	+	+	−	−	+	+	−	43.7
+	−	−	−	−	+	+	−	+	+	+	−	−	−	+	40.2
−	+	−	−	+	−	+	+	−	+	−	+	−	−	+	42.4
+	+	−	−	−	−	−	−	+	+	+	+	+	+	−	44.7
−	−	+	−	−	−	+	+	+	−	+	+	−	+	−	42.4
+	−	+	−	+	−	−	−	+	−	−	+	+	−	+	45.9
−	+	+	−	−	+	−	+	−	−	+	−	+	−	+	42.2
+	+	+	−	+	+	+	−	−	−	−	−	−	+	−	40.6
−	−	−	+	+	+	−	−	−	+	+	−	+	+	42.4
+	−	−	+	−	+	+	+	−	−	+	+	−	−	45.5
−	+	−	+	+	−	+	−	+	−	+	−	+	−	−	43.6
+	+	−	+	−	−	−	+	+	−	−	−	−	+	+	40.6
−	−	+	+	−	−	+	−	−	+	−	−	+	+	+	44.0
+	−	+	+	+	−	−	+	−	+	+	−	−	−	−	40.2
−	+	+	+	−	+	−	−	+	+	−	+	−	−	−	42.5
+	+	+	+	+	+	+	+	+	+	+	+	+	+	+	46.5

7. In the design of a thin film redistribution layer (cited by J. Lorenzen, IBM Kingston and discussed by Lawson and Madrigal (1994)), the circuit impedance (Z) is a function of three design factors: the A, insulator thickness; B, linewidth; and C, line height as shown in Figure 12.21. From engineering first principles, it can be shown that the impedance is given by Equation (12.10), where ϵ is the dielectric constant of the insulator and is assumed to be constant at 3.10.

$$Z = f(A, B, C) = \frac{87.0}{\sqrt{\epsilon + 1.41}} \ln\left(\frac{5.98A}{0.8B + C}\right) \qquad (12.10)$$

Figure 12.21 *Thin Film Distribution Layer*

The nominal or mean values of A, B, and C can be specified by the design engineer, but in actual circuits these characteristics will vary from their nominal values due to manufacturing imperfections and wear during use. The table above shows the feasible range and tolerance limits for these variables.

554   ROBUST PARAMETER DESIGN EXPERIMENTS

Control Factor	Feasible Range (μm)	Low-Cost Tolerance Range (μm)
A: Insulator thickness	20-30	± 1.0
B: Linewidth	12.5-17.5	± 0.67
C: Line height	4-6	± 0.33

(a) Construct a control-factor array using a 2^3 factorial design.

(b) Construct a noise-factor array (in deviations from the nominal settings in the control-factor array) using the table of Wang, Fang and Lin.

(c) Evaluate Equation (12.10) for each combination of values in the product array, and calculate the mean and log variance impedance across the noise-factor array for each of the eight points in the control-factor array.

(d) Fit a model to the mean and log variance and identify the control factor settings that will make the impedance equal to 85Ω with minimum variability.

(e) Approximate the variance of Z, using the formula $\sigma_Z^2 \approx \left(\frac{\partial f}{\partial A}\right)^2 \sigma_A^2 + \left(\frac{\partial f}{\partial B}\right)^2 \sigma_B^2 + \left(\frac{\partial f}{\partial C}\right)^2 \sigma_C^2$ (where σ_A^2=tolerance range/6, etc.), and use nonlinear programming (proc nlp as described in Chapter 10) to minimize σ_Z^2 as a function of A, B, and C subject to the constraint that $Z = 85\Omega$. Do the results differ substantially from the answer you got in (d)?

8. Consider the data for the product array design for the elastometric connector shown in Figure 12.4.

   (a) Calculate the mean pull-off force and log variance of the pull-off force across the noise array for each run in the control-factor array.

   (b) Analyze the data using the location-dispersion modeling method.

   (c) Do you find the same optimal levels of the control factors as identified in Section 12.4.2?

   (d) Is any information lost when analyzing this data using location-dispersion modeling?

9. Reconsider the data from the injection molding experiment in Table 12.12.

   (a) Fit the model $y = \beta_0 + \beta_A X_A + \beta_B X_B + \beta_{A \times B} X_A X_B$ in the adjustment factors using the method of least squares.

   (b) Simultaneously estimate the variance of the residuals at each level of the dispersion factor, C, and compute the weighted least squares estimates of the coefficients in the model in (a) using proc mixed as shown in Section 12.6.1.

CHAPTER 13

Experimental Strategies for Increasing Knowledge

13.1 Introduction

Experimental designs are the basis for collecting data to discover cause and effect relationships. A number of different experimental designs have been presented in this book. The experimental design most appropriate for a particular situation will depend upon the circumstances. Throughout this book, the design selection roadmap (Figure 1.2) has been used to indicate when specific experimental designs should be used based on the number of factors, homogeneity of the experimental units, and the ease of randomizing treatments to experimental units. However, another way of viewing experimental designs is as a collection of tools that are useful for increasing the state of knowledge about a particular phenomenon.

The appropriate design (or tool) depends on how much knowledge is available, and what additional information the experimenter would like to discover. The different experimental designs and corresponding methods of data analysis presented in this book can be used in a sequence to advance knowledge, often referred to as sequential experimentation. This chapter presents an outline of how experimental designs can be used in sequential experimentation. No detailed examples are presented in this chapter, but reference is made to specific examples that have been presented in previous chapters.

13.2 Sequential Experimentation

Figure 13.1 illustrates the state of knowledge about a phenomenon as a continuous line that goes from zero knowledge on the left to a high state of knowledge on the right. The purpose for experimenting is to advance the state of knowledge. Various stages along the line, descriptive names for the stages, purposes for experimenting at each stage, and appropriate experimental designs for each stage are shown below the knowledge line in the figure.

When there is a very low state of knowledge, the experimenter may know that the response is variable but he or she may not know what factors cause variability in the response. When this is the case, determining the sources of variability may give an experimenter clues to potential factors. Random sampling experiments, nested, and staggered nested designs described in Chapter 5 are useful for determining sources of variability.

When the experimenter has a list of potential factors that could influence

EXPERIMENTAL STRATEGIES FOR INCREASING KNOWLEDGE

the response, screening designs such as fractional factorials, Plackett-Burman designs, and orthogonal array designs are useful for separating the list of potential factors into those that (1) influence the response and (2) those that do not. When there are many factors to be tested in few runs, the experimenter should not expect to do much more than identify the factors that are important. This stage of experimentation is called screening, and the so-called screening designs are described in Chapters 6, 7 and 8. When experimental units are homogeneous and there is no restriction on randomization, the designs in Chapter 6 are appropriate. Blocked designs are discussed in Chapter 7 and split-plot designs in Chapter 8. The design roadmap (Figure 1.2) helps in determining which of these screening designs should be used.

Figure 13.1 *The State of Knowledge Line*

State of Knowledge	0%				100%
Stage	Preliminary Exploration	Screening Factors	Effect Estimation	Optimization	Mechanistic Modeling
Purpose	Determine Sources of Variability	Identify Important Factors	Estimate Main Effects, Interactions and Optimum Conditions Tested	Fit Empirical Models and Interpolate to Find Optimum in Bounded Region	Estimate Parameters of Theoretical Model and Extrapolate Predictions Outside Region of Experiments
Useful Designs	RSE NSE SNSE	CRFF PB OA SPFF	CRFD RCBF CCBF PCBF CRSP RBSP	CRRS D-optimal Designs BRS RSSP EESPRS	D-optimal Designs

In most problems, the number of factors that have a practical and statistically significant effect on the response will reasonably small. Once they have been identified, the next stage of experimentation is called estimation. Here, the purpose is to estimate the factor effects and interactions. Experimental designs useful at this stage are full factorial designs, two-level factorial designs, and resolution V fractional factorials. These are described in Chapters 3, 4, 6, 7 and 8. When the main effects and interactions have been estimated, it is also possible to identify the conditions (among those tested) that result in a maximum or minimum predicted response.

Once the most desirable levels of the qualitative factors have been chosen, experimenters can further advance the state of knowledge, during the estimation stage, by finding an empirical equation relating the response to the levels of the quantitative factors. Using this equation, the settings of the quantitative factors that result in an optimum response within the region studied can be identified. This is called the optimization stage. The experimental designs

SEQUENTIAL EXPERIMENTATION

useful for this stage are the response surface designs and mixture experiment designs discussed in Chapters 10 and 11.

Box and Wilson (1951), who were the first to propose response surface methods, also described an intermediate stage of experimentation between the estimation of effects and optimization. They proposed to estimate first the direction of maximum improvement from the 2^k design run in the estimation stage, and next conduct a series of one-at-a-time experiments outside the original experimental region along this path as illustrated in Figure 13.2. When the response from individual experiments continues to improve, additional experiments are run along the path. When the response from experiments along the path stops improving, a second order or response surface experiment is then conducted to identify the optimum conditions. This whole process is called the method of steepest ascent. Myers and Montgomery (2002) show the details of how to determine the path of steepest ascent. In many cases the method of steepest ascent is impractical because (1) the experimental error makes it very difficult to determine if individual experiments are improving or not, and (2) sometimes it is impossible to go outside the range of the original experiments.

Figure 13.2 *The Method of Steepest Ascent*

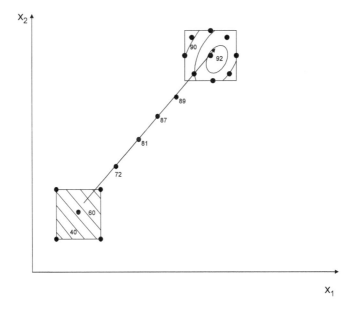

When it is not possible to experiment outside the region used in the estimation stage, the transition from the estimation stage to the optimization stage can be accomplished by running a standard response surface design in blocks. This is illustrated in Figure 13.3. The first block includes a 2^{k-p} fractional factorial design plus center points involving the quantitative factors.

558 EXPERIMENTAL STRATEGIES FOR INCREASING KNOWLEDGE

After collecting data from the first block, the significance of factor main effects can be determined. If there are insignificant main effects, they can be ignored and the design collapses to a factorial plus center points as shown in the bottom leg of the next stage of experimentation in Figure 13.3. If all main effects are significant, a second block of experiments is performed to create a full factorial plus center points as shown in the top leg of Figure 13.3. In the analysis of data at the second stage, a test for curvature should be performed as described in Section 10.6.1. If there is significant curvature, a third block of experiments involving central-composite-design-axial points and additional center points should be run to complete a response surface design.

Figure 13.3 *Blocked Response Surface*

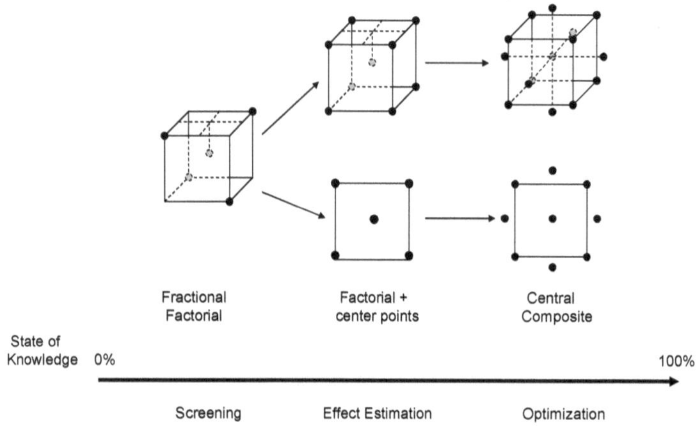

The orthogonally blocked central composite designs, which are produced automatically by SAS ADX, can be used as illustrated in Figure 13.3 if they have at least three blocks. For example, there is a three-factor central composite design that blocks orthogonally into three blocks. The first block is the half-fraction plus center points. The second block is the mirror image half-fraction plus additional center points, and the third block is the axial points plus center points. There is a similar four-factor central composite design in three blocks. For five- and six-factor designs there are central composite designs that block into five blocks. The first four blocks consist of $\frac{1}{4}$ fractions ($\frac{1}{8}$ fractions for the six-factor design) and it may be possible to eliminate some of these blocks if not all factors are significant.

An example of using a sequence of fractional factorial, factorial, and central composite design to increase knowledge in the way shown in Figure 13.3 was illustrated in the example of producing an amino acid rich biomass from eucalyptus wood given in Section 6.4 and continued in exercise 3 of Chapter 10.

After the data is collected in the optimization stage of experimentation, general quadratic or cubic (in the case of some mixture experiments) models

ONE STEP SCREENING AND OPTIMIZATION

are fit to the resulting data, and various tools are used to explore the fitted surface and identify optimum operating conditions. Since the polynomial models used in the optimization stage are just approximations to the true relationship between the quantitative factors and the response, predictions should never be made with these models outside the region of experimentation.

When more knowledge is available in the form of a theoretical model, like the compartment model described in Section 10.5 or the chemical degradation model determined from first-order kinetics in Section 10.7.4., there is no need to conduct a traditional response surface experiment. When a theoretical model is available, the parameters of this model can be estimated from data collected using custom experimental designs that are based on the form of the model and prior estimates of the parameters. Once data is collected and a theoretical model is fit, predictions or extrapolations can be made outside the experimental region.

Utilizing all the stages of experimentation shown in Figure 13.1, the state of knowledge can be advanced through a sequence of experiments as illustrated in Figure 13.4. A plan of action to gain knowledge in this way is called a sequential experimentation strategy.

Figure 13.4 *Sequential Experimentation*

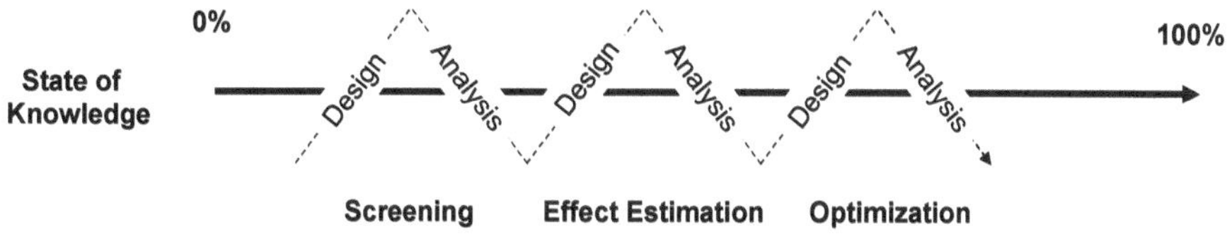

13.3 One Step Screening and Optimization

In some cases when experiments are conducted in a manufacturing facility, a pilot plant (that can only be scheduled for a limited time), in marketing research studies or engineering prototype testing (where there is a fixed deadline), the sequential experimentation strategy may not be possible. After beginning experiments, some experimental plans must be modified when an experimenter finds that factors cannot be varied independently, the chosen range for one or more factors is infeasible, or a previously unknown but highly influential factor is discovered. However, when the time allotted for experimentation is tight, rarely is it possible to start with a screening experiment and proceed with additional designs and analysis to arrive at the desired state of knowledge.

For the reasons above, the sequential experimentation strategy is sometimes abandoned in favor of a single design or set of experiments. Even so, the designs used for screening and optimization are mutually exclusive. If an experimenter

560 EXPERIMENTAL STRATEGIES FOR INCREASING KNOWLEDGE

can only afford to run one set of experiments, he is forced to choose between one objective or the other. If the experimenter chooses a classical response surface design, he must select a subset of the factors to work with, and he risks the possibility of missing other important factors. On the other hand, if the experimenter chooses a screening experiment to avoid missing important factors, important interactions and quadratic effects will be missed.

It would be desirable in some situations to have experimental design plans that allow for (1) screening a reasonably large number of factors to identify the important ones, and (2) refitting a model to the data in only the important factors, so that important interactions or optimum operating conditions can be determined after one set of experiments. Recent literature has shown this is possible using certain orthogonal array main effect plans.

Lawson (2003) illustrates how screening followed by estimation of interactions can be accomplished using Plackett-Burman Designs. The example in Section 6.6 illustrates how this can be accomplished. Cheng and Wu (2001) proposed a strategy for using non-regular fractions of 3^k designs for simultaneous screening and response surface exploration. Lawson (2003) presents 18-, 27- and 36-run examples of these non-regular fractions and an example using Chen and Wu's strategy with an 18-run design. Many additional non-regular fractions of 3^k designs are stored in a vast catalog by SAS and can be accessed using the %mkt macros described in Section 6.7. For example, the SAS commands below search for an orthogonal array design with seven factors at three levels, and then create an 18-run design.

```
*Search for orthogonal arrays in catalog of designs;
%mktruns(3**7)
*Find an 18 run design for 7 factors with 3 levels;
%mktex(3**7,n=18)
*Print a randomized list of the runs;
proc print data=randomized; run;
```

The analysis of data from non-regular fractions of 3^k designs, like this, is similar to the analysis of Plackett-Burman Designs as illustrated in Section 6.6.

13.4 Evolutionary Operation

The idea of using optimization experiments, or any experimentation for that matter, is often met with resistance in an operating manufacturing facility. In manufacturing, the key process variables thought to influence production are usually tightly controlled in order to ensure the quality of the manufactured product. Tampering or experimenting with the levels of key process variables could result in the production of off-grade or scrap product. However, the levels at which the process variables are controlled in a manufacturing process, may not be optimal. These levels may have been determined in pilot plant or laboratory experiments where conditions may differ in many ways from the actual manufacturing facility. In other cases, where the levels of key process

EVOLUTIONARY OPERATION

variables were actually determined through preliminary experiments in the manufacturing facility, they may no longer be optimal due to drift over time in raw materials or environmental conditions. Therefore, although acceptable product is being produced with current process variable settings, there may be benefits in experimenting to obtain improved settings or to counteract drift over time.

Box (1957) proposed a method for finding improved settings of key process variables without upsetting production. He believed that a manufacturing process should not only produce product, but also information that can be used to improve the process over time. Data is normally collected in manufacturing facilities regarding (1) the settings of key process variables during each production run (to ensure operating procedures are being followed) and (2) measured characteristics of product (to monitor quality). While recording this informaton, Box suggested making small perturbations in key process variable settings from time to time according to a statistical plan like a 2^2 or 2^3 experiment. The planned changes in the key process variables would be small enough that the product output would not be degraded, yet large enough that potential process improvements could be recognized after a number of cycles through the planned settings. He called this method *Evolutionary Operation* or EVOP for short.

When using EVOP, the center point in a 2^2 or 2^3 design is assigned to the current operating conditions (in terms of what is believed to be the key process variables). Figure 13.5 illustrates the plan for two process variables. The five different process settings would each be run during one block of manufacturing. Running all five of these conditions completes one cycle of the EVOP.

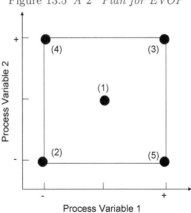

Figure 13.5 *A 2^2 Plan for EVOP*

If the average measure of product characteristic is y_{ijk} during the kth EVOP cycle at process variable settings i and j, then after two or more cycles, the main effect of process variable 1 ($\bar{y}_{+..} - \bar{y}_{-..}$), the main effect of process variable 2 ($\bar{y}_{.+.} - \bar{y}_{.-.}$), their interaction effect (($\bar{y}_{++.} + \bar{y}_{--.}) - (\bar{y}_{+-.} + \bar{y}_{-+.})$) and the curvature effect ($\bar{y}_{00.} - \bar{y}_{...}$) can be calculated. The standard error of the main effect and interaction estimates is $2s_p/\sqrt{4 \times r}$, and the standard error of the curvature effect is $2s_p \times \sqrt{1/r + 1/4r}$, where s_p is the pooled standard deviation over the five treatment combinations and r is the number of cycles of EVOP performed. The statistical significance of the effects can be determined by dividing the effect by its appropriate standard error and comparing to the reference t-distribution with $5 \times (r-1)$ degrees of freedom. Since r is in the denominator of the standard errors, they will decrease as the number of cycles of EVOP increases.

If after five or more cycles of EVOP there is no practical or statistically significant effects, then either the process performance is optimal and stable over the experimental region, or improved operating conditions may exist outside the current experimental region. If it is believed that further improvement may be possible outside the range of the current EVOP, the ranges on the key process variables should be increased and another phase of EVOP should begin. If the effects of the key process variables (or their interaction effect) are statistically significant after two or more cycles of EVOP, a new phase of EVOP should begin where the center point of the new phase is selected to be the process condition that produced the optimum result in the last phase. If the curvature effect is statistically significant after several cycles of EVOP, and the measured product characteristics are better at the center point, the current operating conditions are optimal and EVOP should be discontinued.

Given the objective and sufficient training, EVOP should be carried out by the operators in the manufacturing facility without intervention from engineering or production management. In their book on EVOP, Box and Draper (1969) provide detailed worksheets which facilitate the EVOP calculations in a manufacturing environment. These can also be carried out simply in a modern spreadsheet program. Box and Draper suggest displaying a bulletin board in plain sight where the current EVOP status can be displayed. This will motivate the operators conducting the EVOP and inform them of the latest results.

13.5 Concluding Remarks

For each type of experimental design presented in this book, there are examples of SAS commands to create the design. Earlier chapters emphasized how to randomize lists of experiments to avoid bias from unknown factors and ensure valid analyses of data. In later chapters, the need for randomization was assumed, but the specific details were not emphasized since they are the same as shown in Chapter 2. In Chapter 4, the purpose for blocking experimental designs was described and the methods to create blocked designs in SAS were illustrated. The importance of blocking was emphasized throughout the remainder of the book, and situations where blocking should be used with each type of design were discussed and illustrated in the appropriate chapters.

For each type of experimental design presented, an appropriate model for the data analysis was given and a description of the way to fit that model using SAS software. In Chapter 8, restricted randomization or split-plot designs were discussed and it was shown how to construct designs and analyze the data when randomization is restricted. Again the situations where this occurs and the means to handle them were emphasized in the remaining chapters of the book with all the types of designs presented. In addition, this book emphasizes the interpretation and presentation of results which is amply described and demonstrated in every chapter. The exercises at the end of each chapter provide practice for design creation and analysis and interpretation of data. They were selected to reinforce all the concepts presented in the book.

In the author's experience, the experimental design topics presented in the book cover most of the situations encountered in practice. With the power of the software in SAS Stat, SAS QC, SAS OR, SAS Graph, the automated SAS ADX system and the %adx and %mkt macros, a researcher has a powerful set of tools that simplify most research design and data analysis problems. By emphasizing where to use each of the different types of experimental designs, illustrating the creation of designs and analysis of data with SAS, and presenting many examples of interpretation and presentation of results, this book provides an updated guide to researchers when compared to the earlier books by Fisher (1935), Cochran and Cox (1950), and Kempthorne (1952).

After completing the study of this book, you should be able to (1) choose an experimental design that is appropriate for a research problem; (2) construct the design (including performing proper randomization and determining the proper number of replicates); (3) execute the plan (or advise a colleague to do it); (4) determine a model appropriate for the data; (5) fit the model to the data; (6) interpret the results; and (7) present them in a meaningful way to answer the research question at hand. The use of SAS software simplifies many of these tasks, demonstrated by the examples throughout the book.

Bibliography

AlmeidaeSilva, J. B., Lima, U. A., S., M. E., Taqueda, and Guaragna, F. G. (1998). Use of fractional factorial design for selection of nutrients for culturing *Paecilomyces variotii* in eucalyptus hemicellulosic hydrolysate. *Brazilian Journal of Chemical Engineering*, **15**, 273–279.

AlmeidaeSilva, J. B., Lima, U. A., Taqueda, M. E. S., and Guaragna, F. G. (2003). Use of response surface methodology for selection of nutrient levels for culturing *Paecilomyces variotii* in eucalyptus hemicellulosic hydrolyzate. *Biosource Technology*, **87**, 45–50.

Anderson, M. (2003). Messing with medieval missle machines. *Stateaser*, **Sept.** URL www.statease.com.

Anderson, V. L. and McLean, R. A. (1974). *Design of Experiments: A Realistic Approach*. Marcel Dekker, Inc., New York.

Anik, S. T. and Sukumar, L. (1981). Extreme vertexes design in formulation development: Solubility of butoconazole nitrate in a multi-component system. *Journal of Pharmaceutical Sciences*, **70**, 897–900.

Apple, D. (2006). Dish detergent experiment. Term Paper Stat 431, BYU Dept. of Statistics.

Atkinson, A. C., Donev, A. N., and Tobias, R. D. (2007). *Optimum Experimental Designs with SAS*. Oxford University Press, Oxford.

Bainbridge, T. R. (1965). Staggered nested designs for estimating variance components. *Industrial Quality Control*, **22**, 12–20.

Barbuta, M. and Lepadatu, D. (2008). Mechanical characteristics investigation of polymer concrete using mixture design of experiments and response surface method. *Journal of Applied Sciences*, **8**, 2242–2249.

Bastable, A. (2006). Tee it high or low. *Golf Magazine*, **June**, 173–179.

Bates, D. M. and Watts, D. G. (2007). *Nonlinear Regression and Its Applications*. John Wiley & Sons, Hoboken N.J.

Bechhofer, R. E. and Tamhane, A. C. (1981). Incomplete block designs for comparing treatments with a control: General theory. *Technometrics*, **23**, 45–57.

Belloto, J., Dean, R. J., Moustafa, M. A., Molokhia, A. M., Gouda, M. W., and Sokoloski, T. D. (1985). Statistical techniques applied to solubility predictions and pharmaceutical formulations: An approach to problem solving using mixture response surface methodology. *International Journal of Pharmaceutics*, **23**, 195–207.

Bennett, C. A. and Franklin, N. L. (1954). *Statistical Analysis in Chemistry and the Chemical Industry*. John Wiley & Sons, New York.

Bingham, D. R. and Sitter, H. R. (1999). Minimum aberration two-level fractional factorial split-plot designs. *Technometrics*, **41**, 62–70.

Bingham, D. R. and Sitter, H. R. (2001). Design issues in fractional factorial split-plot experiments. *Journal of Quality Technology*, **33**, 2–15.

Bisgaard, S. (1999). Quality quandries-proposals: A mechanism for achieving better experiments. *Quality Engineering*, **11**, 645–649.

Bisgaard, S. (2000). The design and analysis of $2^{k-p} \times 2^{q-r}$ split plot experiments. *Quality Engineering*, **8**, 705–708.

Bisgaard, S. and Kulahci, M. (2001). Quality quandries: Robust product design saving trials with split-plot confounding. *Quality Engineering*, **13**, 525–530.

Bisgaard, S. and Sutherland, M. (2004). Quality quandries-split plot experiments: Taguchi's Ina tile experiment reanalyzed. *Quality Engineering*, **16**, 157–164.

Bisgaard, S., Fuller, H. T., and Barrios, E. (1996). Quality quandries: Two-level factorials run as split plot experiments. *Journal of Quality Technology*, **32**, 39–55.

Bose, R. C., Clatworthy, W. H., and Shrikhande, S. S. (1954). Tables of partially balanced designs with two associate classes. Technical report, North Carolina Agricultural Experiment Station. Technical Bulletin 107.

Box, G. E. P. (1957). Evolutionary operation: A method for increasing industrial productivity. *Applied Statistics*, **6**, 81–101.

Box, G. E. P. (1988). Signal-to-noise ratios, performance criteria, and transformations. *Technometrics*, **30**, 1–17.

Box, G. E. P. (1991). George's column: Finding bad values in factorial designs. *Quality Engineering*, **3**, 249–254.

Box, G. E. P. and Behnken, D. W. (1960). Some new three level designs for the study of quantitative variables. *Technometrics*, **2**, 455–460.

Box, G. E. P. and Cox, D. R. (1964). An analysis of transformations. *Journal of the Royal Statistical Society, Series B*, **26**, 15–20.

Box, G. E. P. and Draper, N. R. (1969). *Evolutionary Operation*. John Wiley & Sons, New York, first edition.

Box, G. E. P. and Fung, C. A. (1986). Minimizing transmitted variation by parameter design. Technical Report 8, Center for Quality and Productivity Improvement, University of Wisconsin-Madison.

Box, G. E. P. and Hunter, J. S. (1957). Multifactor experimental designs for exploring response surfaces. *Annals of Mathematical Statistics*, **28**, 195–241.

BIBLIOGRAPHY

Box, G. E. P. and Hunter, J. S. (1961). The 2^{k-p} fractional factorial designs. *Technometrics*, **3**, 311–351 and 449–458.

Box, G. E. P. and Jones, S. (1992). Split-plot designs for robust product experimentation. *Journal of Applied Statistics*, **19**, 3–26.

Box, G. E. P. and Meyer, R. D. (1986a). An analysis for unreplicated fractional factorials. *Technometrics*, **28**, 11–18.

Box, G. E. P. and Meyer, R. D. (1986b). Dispersion effects from fractional designs. *Technometrics*, **28**, 19–27.

Box, G. E. P. and Meyer, R. D. (1993). Finding the active factors in fractionated screening experiments. *Journal of quality technology*, **25**, 94–105.

Box, G. E. P. and Wilson, K. B. (1951). On experimental attainment of optimum conditions. *Journal of the Royal Statistical Society B*, **13**, 1–45.

Box, G. E. P., Hunter, W. G., and Hunter, J. S. (1978). *Statistics for Experimenters*. John Wiley & Sons, New York, first edition.

Burdick, R. and Graybill, F. (1992). *Confidence Intervals on Variance Components*. Marcel Dekker, Inc., New York.

Butler, N. A. (2006). Optimal blocking of two-level factorial designs. *Biometrika*, **93**, 289–302.

Casella, G. (2008). *Statistical Design*. Springer, New York.

Chau, K. W. and Kelly, W. R. (1993). Formulating printable coatings via d-optimality. *Journal of Coatings Technology*, **65**, 71–78.

Chen, J., Sun, D. X., and Wu, C. F. J. (1993). A catalogue of two-level and three-level fractional factorial designs with small runs. *International Statistical Review*, **61**, 131–145.

Cheng, S. and Wu, C. F. J. (2001). Factor screening and response surface exploration. *Statistica Sinica*, **11**, 553–604.

Chi, E. M. (1994). M-estimation in cross-over trials. *Biometrics*, **50**, 486–493.

Chipman, H., Hamada, M., and Wu, C. F. J. (1997). A Bayesian variable-selection approach for analyzing designed experiments with complex aliasing. *Technometrics*, **39**, 372–381.

Chipman, J. (2006). Sprinting experiment. Term Paper Stat 431, BYU Dept. of Statistics.

Claringbold, P. J. (1955). Use of the simplex design in the study of the joint action of related hormones. *Biometrics*, **11**, 174–185.

Cochran, W. G. and Cox, G. M. (1950). *Experimental Designs*. John Wiley & Sons, New York, first edition.

Cook, R. D. and Nachtsheim, C. J. (1989). Computer-aided blocking of factorial and response surface designs. *Technometrics*, **31**, 339–346.

Cornell, J. A. (1988). Analyzing data from mixture experiments containing process variables: A split-plot approach. *Journal of Quality Technology*, **20**, 2–23.

Cornell, J. A. (2002). *Experiments with Mixtures: Designs Models and Analysis of Mixture Data*. John Wiley & Sons, New York, third edition.

Cornell, J. A. and Gorman, J. W. (1984). Fractional design plans for process variables in mixture experiments. *Journal of Quality Technology*, **16**, 20–38.

Cox, D. R. (1971). A note on polynomial response functions for mixtures. *Biometrika*, **58**, 155–159.

Curtis, B., Warburton, R., Gannon, J., Gloss-Soler, S., Goel, A., Miyamoto, I., and Panzl, D. (1982). The application of fractional factorial techniques to the design of software experiments. *Software Engineering Notes*, **7**, 21–30.

Czitrom, V. and Spagon, P. D. (1997). *Statistical Case Studies for Industrial Process Improvement- Chapter 8*. Siam, Philadelphia.

Daniel, C. (1959). Use of half-normal plots in interpreting factorial two-level experiments. *Technometrics*, **1**, 311–341.

Daniel, C. (1960). Locating outliers in factorial experiments. *Technometrics*, **2**, 149–156.

Daniel, C. (1976). *Applications of Statistics to Industrial Experimentation*. John Wiley & Sons, New York, first edition.

Das, M. N. (1960). Fractional replicates as asymmetrical factorial designs. *Journal of the Indian Society of Agricultural Statistics*, **12**, 159–174.

Davies, O. L. (1949). *Statistical Methods in Research and Production*. Oliver and Boyd, Edinburg, first edition.

Deeb, O., Youssef, K. M., and Hemmateenejad, B. (2008). QSAR of novel hydroxyphenylureas as antioxidant agents. *QSAR Combinatorial Science*, **27**, 417–424.

Derringer, G. and Suich, R. (1980). Simultaneous optimization of several response variables. *Journal of Quality Technology*, **12**, 214–219.

Derringer, G. C. (1974). An empirical model for viscosity of filled and plasticized elastomer compounds. *Journal of Applied Polymer Science*, **18**, 1083–1101.

Diggle, P. J., Liang, K., and Zeger, S. L. (1994). *The Analysis of Longitudinal Data*. Oxford University Press, Oxford.

Dossett, T., Lefler, R., and Moore, P. (2007). Apple browning experiment. Term Paper Stat 431, BYU Dept. of Statistics.

Draper, N. R. (1963). Ridge analysis of response surfaces. *Technometrics*, **5**, 469–479.

Draper, N. R. (1985). Small composite designs. *Technometrics*, **27**, 173–180.

Draper, N. R. and John, R. C. S. (1977). A mixtures model with inverse terms. *Technometrics*, **19**, 37–46.

Draper, N. R. and Lin, D. K. J. (1990). Small response surface designs. *Technometrics*, **32**, 187–194.

Dunnett, C. W. (1955). A multiple comparison procedure for comparing several treatments with a control. *Journal of the American Statistical Association*, **509**, 1096–1121.

Dykstra, O. (1971). The augmentation of experimental data to maximize $|X^T X|$. *Technometrics*, **13**, 682–688.

Easterling, R. G. (2004). Teaching experimental design. *The American Statistician*, **58**, 244–252.

Erhardt, E. B. (2007). Designing a better paper helicopter using response surface methodology. *Stats*, **48**, 14–21.

Fannin, T. E., Marcus, M. D., Anderson, D. A., and Bergman, H. L. (1981). Use of a fractional factorial design to evaluate interactions of environmental factors affecting biodegradation rates. *Applied and Environmental Microbiology*, **42**, 936–943.

Faraway, J. (2004). *Linear Models with R*. Chapman & Hall/CRC, Boca Raton, FL.

Festing, M. F. (2003). Principles: The need for better experimental design. *Trends Pharmacol Sci*, **24**, 341–345.

Fisher, R. A. (1926). The arrangement of field experiments. *Journal of the Ministry of Agriculture of Great Britain*, **33**, 503–513.

Fisher, R. A. (1935). *The Design of Experiments*. Oliver and Boyd, Edinburgh, first edition.

Fisher, R. A. (1940). An examination of possible different solutions of a problem in incomplete blocks. *Annals of Eugenics*, **9**, 353–400.

Franklin, M. F. (1985). Selecting defining contrasts and confounded effects in p^{n-m} factorial experiments. *Technometrics*, **27**, 165–172.

Freeman, P. (1989). The performance of the two-stage analysis of two treatment two period crossover trials. *Statistics in Medicine*, **8**, 1421–1432.

Freund, R. J. and Littell, R. C. (1981). *SAS for Linear Models: A Guide to ANOVA and GLM Procedures*. SAS Institute, Cary, NC, first edition.

Fries, A. and Hunter, W. G. (1980). Minimum aberation 2^{k-p} designs. *Technometrics*, **22**, 601–608.

Gilmour, S. G. and Goos, P. (2006). Analysis of data from nonorthogonal multi-stratum designs. Technical report, Faculty of Applied Economics Universiteit Antwerpen Belgium. Tech. Rep. 2006/05.

Gilmour, S. G. and Trinca, L. A. (2000). Some practical advice on polynomial regression analysis from blocked response surface designs. *Comunications in Statistics: Theory and Methods*, **29**, 2157–2180.

Giovannitti-Jensen, A. and Myers, R. H. (1989). Graphical assessment of the prediction capability of response surface designs. *Technometrics*, **31**, 159–171.

Goodnight, J. H. (1980). Tests of hypothesis in fixed effects linear models. *Communication in Statistics - Theory and Methods*, **A9**, 167–180.

Goos, P. (2002). *The Optimal Design of Blocked and Split Plot Experiments*. Springer, New York.

Goos, P. and Donev, A. N. (2006). The d-optimal design of blocked experiments with mixture components. *Journal of Quality Technology*, **38**, 319–332.

Goos, P. and Donev, A. N. (2007). Tailor-made split plot designs for mixture and process variables. *Journal of Quality Technology*, **39**, 326–339.

Goos, P. and Vandebroek, M. (2001). Optimal split-plot designs. *Journal of Quality Technology*, **33**, 436–450.

Goos, P. and Vandebroek, M. (2003). D-optimal split plot designs with given numbers and sizes of whole plots. *Technometrics*, **45**, 235–245.

Gorman, J. W. and Cornell, J. A. (1982). A note on model reduction for experiments with both mixture components and process variables. *Technometrics*, **24**, 243–247.

Greenhouse, S. W. and Geisser, S. (1959). On methods in the analysis of profile data. *Psychometrika*, **24**, 95–112.

Grizzle, J. E. (1965). The two-period change-over design and its use in clinical trials. *Biometrics*, **21**, 467–480.

Hamada, M. and Wu, C. F. J. (1992). Analysis of designed experiments with complex aliasing. *Journal of Quality Technology*, **24**, 130–137.

Hare, L. B. (1988). In the soup: A case study to identify contributors to filling variability. *Journal of Quality Technology*, **20**, 36–43.

Hartley, H. O. (1959). Smallest composite design for quadratic response surfaces. *Biometrics*, **15**, 611–624.

Hartley, H. O. (1967). Expectations, variances and covariances of ANOVA mean squares by synthesis. *Biometrics*, **34**, 105–114.

Heffner, T. G., Drawbaugh, R. B., and Zigmond, M. J. (1974). Amphetamine and operant behavior in rats: Relationship between drug effect and control response rate. *Journal of Comparative and Physiological Psychology*, **86**, 1031–1043.

Henderson, L. O., Hannon, W., Smith, S. J., and Cooper, G. R. (1987). An international collaborative study on standardization of apolipoproteins a-i and b. part ii. evaluation of contributions of antisers to among-laboratory variance components. *Clinical Chemistry*, **33**, 2250–2256.

Hinkelmann, K. and Kempthorne, O. (1994). *Design and Analysis of Experiments Volume I*. John Wiley & Sons, New York.

Hoerl, A. E. (1959). Optimum solution of many variables equations. *Chemical Engineering Progress*, **55**, 69–78.

Horiuchi, Y., Imai, K., and Taniguchi, N. (2005). Designing and analyzing randomized experiments. In *Proceedings of the Annual Meeting of the American Political Science Association*. American Political Science Association.

Hsu, J. C. (1984). Constrained simultaneous confidence intervals for multiple comparisons with the best. *Annals of Statistics*, **12**, 1136–1144.

Huang, P., Chen, D., and Voelkel, J. (1998). Minimum aberration two-level split-plot designs. *Technometrics*, **40**, 314–326.

Hunter, G. B., Hodi, F. S., and Eager, T. W. (1982). High-cycle fatigue of weld repaired cast t-6a1-4v. *Metallurgical Transactions*, **13A**, 1589–1594.

Hunter, J. S. (1989). Let's all beware the latin square. *Quality Engineering*, **1**, 453–465.

Huynh, H. and Feldt, L. S. (1970). Conditions under which mean square ratios in repeated measurements designs have exact f-distributions. *Journal of the American Statistical Association*, **65**, 1582–1589.

Huynh, H. and Feldt, L. S. (1976). Estimation of the box correction for degrees of freedom from sample data in the randomized block and split-plot designs. *Journal of Educational Statistics*, **1**, 69–82.

Immer, F. R., Hayes, H. K., and Powers, L. R. (1934). Statistical determination of barley varietal adaptation. *Journal of the American Society of Agronomy*, **25**, 403–419.

Jarrett, R. G. and Hall, W. B. (1978). Generalized cyclic incomplete block designs. *Biometrika*, **65**, 397–401.

John, P. W. M. (1984). Experiments with mixtures involving process variables. Technical report, Center for Statistical Sciences, University of Texas, Austin. Technical Report #8.

Jones, B. (1980). The crossover experiment for clinical trials. *Biometrics*, **36**, 69–79.

Jones, B. and Goos, P. (2007). A candidate-set-free algorithm for generating d-optimal split plot designs. *Journal of the Royal Statistical Society, Series C: Applied Statistics*, **56**, 347–364.

Jones, B. and Kenward, M. G. (2003). *Design and Analysis of Crossover Trials*. Chapman & Hall/CRC, Boca Raton, FL, second edition.

Juan, E. M. S., Edra, E. V., Sales, J. M., Lustre, A. O., and Resurreccion, A. V. A. (2006). Utilization of peanut fines in the optimization of peanut polvoron using mixture respose surface methodology. *International Journal of Food Science and Technology*, **41**, 768–774.

Kamohara, S., Kawakami, M., Kosaka, E., Ito, F., Okuyama, K., and Ohji, Y. (2001). Robust design of 0.18 μm ASIC MOSFETs using Taguchi method with TCAD. In *Statistical Methodology, IEEE International Workshop*, pages 21–24, Kyoto, Japan. IEEE.

Kamon, A., Samet, B., Bouaziz, J., and Châabouni, M. (1999). Application of a rotatable orthogonal central composite design to the optimization of the formulation and utilization of a useful plasticizer for cement. *EDP Sciences*, **27**, 91–96.

Kempthorne, O. (1952). *The Design and Analysis of Experiments*. John Wiley & Sons, New York, first edition.

Kenett, R. S. and Steinberg, D. M. (1987). Some experiences teaching factorial design in introductory statistics courses. *Journal of Applied Statistics*, **14**, 219–227.

Kerry, S. M. and Bland, J. M. (1998). Analysis of a trial randomized in clusters. *British Journal of Medicine*, **316**, 54–54.

Keuls, M. (1952). The use of the studentized range in connection with analysis of variance. *Euphytica*, **1**, 112–122.

Kiefer, J. (1958). On the nonrandomized optimality and randomized nonoptimality of symmetrical designs. *Annals of Mathematical Statistics*, **29**, 675–699.

Kowalski, S. M., Cornell, J. A., and Vining, G. G. (2000). A new model and class of designs for mixture experiments with process variables. *Communications in Statistics-Theory and Methods*, **29**, 2255–2280.

Kowalski, S. M., Cornell, J. A., and Vining, G. G. (2002). Split-plot designs and estimation methods for mixture experiments with process variables. *Technometrics*, **44**, 72–79.

Kshirsager, A. M. (1958). A note on incomplete block designs. *Annals of Mathematical Statistics*, **29**, 907–910.

Kuehl, R. O. (2000). *Design of Experiments Statistical Principles of Research Design and Analysis*. Duxbury, Pacific Grove, CA, second edition.

Kuhfeld, W. F. (2009). Marketing research methods in SAS - report mr2009. Technical report, SAS Institute, Cary, NC. URL http://support.sas.com/resources/papers/tnote/tnote_marketresearch.html.

Kuhfeld, W. F. and Tobias, R. D. (2005). Large factorial designs for product engineering and marketing research applications. *Technometrics*, **47**, 132–141.

Kulahci, M., Ramirez, J. G., and Tobias, R. (2006). Split plot fractional designs: Is minimum aberration enough? *Journal of Quality Technology*, **38**, 56–64.

Kurotori, I. S. (1966). Experiments with mixtures of components having lower bounds. *Industrial Quality Control*, **22**, 592–596.

Kutner, M. H., Nachtsheim, C. J., and Neter, J. (2004). *Applied Linear Regression Models*. McGraw-Hill Irwin, Boston, fourth edition.

Lawson, J. (1982). Applications of robust regression in designed industrial experiments. *Journal of Quality Technology*, **14**, 19–33.

Lawson, J. (1990). Improve a chemical process through use of a designed experiment. *Quality Engineering*, **3**, 215–235.

Lawson, J. (2002). Regression analysis of experiments with complex confounding patterns guided by the alias matrix. *Computational Statistics and Data Analysis*, **39**, 227–241.

Lawson, J. (2008a). SAS macros for analysis of unreplicated 2^k and 2^{k-p} designs with a possible outlier. *Journal of Statistical Software*, **25**(13), 1–17.

Lawson, J. and Erjavec, J. (2001). *Modern Statistics for Engineering and Quality Improvement*. Duxbury, Pacific Grove, CA, first edition.

Lawson, J., Grimshaw, S., and Burt, J. (1998). A quantitative method for identifying active contrasts in unreplicated factorial experiments based on the half normal plot. *Computational Statistics and Data Analysis*, **26**, 425–436.

Lawson, J. S. (2003). One step screening and process optimization experiments. *The American Statistician*, **57**, 15–20.

Lawson, J. S. (2008b). Bayesian interval estimates of variance components used in quality improvement studies. *Quality Engineering*, **20**, 334–345.

Lawson, J. S. and Gatlin, J. (2006). Finding bad values in factorials - revisited. *Quality Engineering*, **18**, 491–501.

Lawson, J. S. and Madrigal, J. L. (1994). Robust design through optimization techniques. *Quality Engineering*, **6**, 593–608.

Lawson, J. S., Schaalje, G. B., and Collings, B. J. (2009). Blocking mixed level factorials with SAS. *Journal of Statistical Software*, **32**, 1–19.

le Riche, W. H. and Csima, A. (1964). A clinical evaluation of four hypnotic agents, using a Latin square design. *Canadian Medical Association Journal*, **91**, 435–438.

Lee, Y. and Nelder, J. A. (2003). Robust design via generalized linear models. *Journal of Quality Technology*, **35**, 2–12.

Leitnaker, M. G. and Cooper, A. (2005). Using statistical thinking and designed experiments to understand process operation. *Quality Engineering*, **17**, 279–289.

Lenth, R. V. (1989). Quick and easy analysis of un-replicated factorials. *Technometrics*, **31**, 469–473.

Letsinger, J. D., Myers, R. H., and Letner, M. (1996). Response surface methods for bi-randomization structures. *Journal of Quality Technology*, **28**, 381–397.

Lew, M. (2007). Good statistical practice in pharmacology problem 2. *British Journal of Pharmacology*, **152**, 299–303.

Lim, D. H. and Wolfe, D. A. (1997). Nonparametric tests for comparing umbrella pattern treatment effects with a control in a randomized block design. *Biometrics*, **53**, 410–418.

Lin, D. K. J. and Draper, N. R. (1992). Projection properties of Plackett-Burman designs. *Technometrics*, **34**, 423–428.

Lin, L. I. and Stephenson, W. R. (1998). Validating an assay of viral contamination. In R. Peck, L. D. Haugh, and A. Goodman, editors, *Statistical Case Studies: A Collaboration Between Academe and Industry*, pages 43–48. SIAM, Philadelphia.

Lucas, H. L. (1957). Extra-period Latin square change-over designs. *Journal of Dairy Science*, **40**, 225–239.

Mason, R. L., Gunst, R. F., and Hess, J. L. (1989). *Statistical Design and Analysis of Experiments: with Applications to Engineering and Science*. John Wiley & Sons, New York.

Mauchly, J. W. (1940). Significance test for sphericity of a normal n-variate distribution. *Annals of Mathematical Statistics*, **11**, 204–209.

McCullagh, P. and Nelder, J. A. (1989). *Generalized Linear Models*. Chapman & Hall, London, second edition.

McLean, R. A. and Anderson, V. L. (1966). Extreme vertices designs of mixture experiments. *Technometrics*, **8**, 447–454.

Meier, P. (1972). The biggest health experiment ever: The 1954 field trial of the salk polio vaccine. In J. M. Tanur, F. Mosteller, W. H. Kruskal, R. F. Link, and R. Pieters, editors, *Statistics: A Guide to the Unknown*, pages 2–13. Holden Day, San Francisco.

Melo, I. R., Pimentel, M. F., Lopes, C. E., G., and Calazans, M. T. (2007). Application of fractional factorial design to levan production by *Zymomonas mobilis*. *Brazilian Journal of Microbiology*, **38**, 45–51.

Montgomery, D. C. (1990). Using fractional factorial designs for robust process development. *Quality Engineering*, **3**, 193–205.

Montgomery, D. C. and Runger, G. C. (1996). Foldovers of 2^{k-p} resolution iv experimental designs. *Journal of Quality Technology*, **28**, 446–450.

Moskowitz, H. (1988). *Applied Sensory Analysis of Foods*. CRC Press, Boca Raton, FL.

Myers, R. H. and Montgomery, D. C. (2002). *Response Surface Methodology: Process and Product Optimization Using Designed Experiments*. John Wiley & Sons, New York, second edition.

Myers, R. H., Vining, G. G., Giovannitti-Jensen, A., and Myers, S. L. (1992). Variance dispersion properties of second order response surface designs. *Journal of Quality Technology*, **24**, 1–11.

Nelder, J. A. and Lee, Y. (1998). Letter to the editor. *Technometrics*, **40**, 168–175.

Newman, D. (1939). The distribution of the range in samples from a normal population, expressed in terms of an independent estimate of standard deviation. *Biometrika*, **31**, 20–20.

BIBLIOGRAPHY

Nigam, A. K. (1970). Block designs for mixture experiments. *Annals of Mathematical Statistics*, **41**, 1861–1869.

Nigam, A. K. (1977). Corrections to blocking conditions for mixture experiments. *Annals of Mathematical Statistics*, **47**, 1294–1295.

Phadke, M. S. (1989). *Quality Engineering Using Robust Design*. Prentice Hall, New York.

Plackett, R. L. and Burman, J. P. (1946). The design of optimum multifactorial experiments. *Biometrika*, **33**, 305–325.

Porter, W. P. and Busch, R. L. (1978). Fractional factorial analysis of growth and weaning success in *Peromyscus maniculatus. Science*, **202**, 907–910.

Prairie, R. R. and Anderson, R. L. (1962). Optimal designs to estimate variance components and to reduce product variability for nested classifications. *North Carolina State College, Raleigh, Inst. of Stat Mimeo Series*, **313**, 80–81.

Prince, E. (2007). Real-time PCR optimization for *S. pyogenes* detection assay - a fractional factorial approach. Term Paper Stat 431, BYU Dept. of Statistics.

Ramakrishna, D. M., Viraraghavan, T., and Jin, Y. (2006). Iron oxide coated sand for arsenic removal: Investigation of coating parameters using a factorial design approach. *Practice Periodical of Hazardous, Toxic, and Radioactive Waste Management*, **10**, 198–206.

Ramirez, J. G. and Tobias, R. (2007). Split and conquer! using SAS/QC to design quality into complex manufacturing processes. In *Global Forum Paper 190-2007*. SAS Institute. URL:www2.sas.com/proceedings/forum/190-2007.pdf.

Rao, P. S. and Rao, R. S. (1997). *Variance Component Estimation*. Chapman & Hall, London, first edition.

Rogness, N. and Richardson, M. (2003). Using chewing gum to illustrate intermediate statistical concepts. *Proceedings of the 2003 Joint Statistical Meetings - Section on Statistical Education*, pages 3545–3547.

Roquemore, K. G. (1976). Hybrid designs for quadratic response surfaces. *Technometrics*, **18**, 419–423.

Sahni, N. S., Piepel, G. F., and Naes, T. (2009). Product and process improvement using mixture-process variable methods and robust optimization techniques. *Journal of Quality Technology*, **41**, 181–197.

SAS Institute (2004a). *Gantt Procedure SAS/QC*. Cary, NC.

SAS Institute (2004b). *Ishakawa Procedure SAS/QC*. Cary, NC.

Scheffé, H. (1958). Experiments with mixtures. *Journal of the Royal Statistical Society, Series B*, **20**, 344–360.

Scheffé, H. (1959). *The Analysis of Variance*. John Wiley & Sons, New York, first edition.

Schrevens, E. and DeRijck, G. (2008). Defining nutrient solutions as constrained double mixture amount systems. In *Proceedings of the Section on Physical and Engineeing Statistics*. American Statistical Association.

Searle, S. R., Casella, G., and McCulloch, C. E. (1992). *Variance Components*. John Wiley & Sons, New York, first edition.

Selwyn, M. R. and Hall, N. R. (1984). On Bayesian methods for bioequivalence. *Biometrics*, **40**, 1103–1108.

Senn, S. (2002). *Cross-over Trials in Clinical Research*. John Wiley & Sons, New York, second edition.

Sheehe, P. R. and Bross, I. D. J. (1961). Latin squares to balance immediate residuals and other effects. *Biometrics*, **17**, 405–414.

Shoemaker, A. C., Tsui, K. L., and Wu, C. F. J. (1991). Economical experimentation methods for robust design. *Technometrics*, **33**, 415–427.

Shumate, D. A. and Montgomery, D. C. (1996). Development of a TiW plasma etch process using a mixture experiment and response surface optimization. *IEEE Transactions on Semiconductor Manufacturing*, **9**, 335–343.

Smith, J. R. and Beverly, J. M. (1981). The use and analysis of staggered nested designs. *Journal of Quality Technology*, **13**, 166–173.

Snee, R. D. (1983). Graphical analysis of process variation studies. *Journal of Quality Technology*, **15**, 76–88.

Snee, R. D. and Marquardt, D. W. (1974). Extreme vertices designs for linear mixture models. *Technometrics*, **16**, 399–408.

Snee, R. D. and Marquardt, D. W. (1976). Screening concepts and designs for experiments with mixtures. *Technometrics*, **18**, 19–29.

Song, J. and Lawson, J. (1988). Use of 2^{k-p} designs in parameter design. *Quality and Reliability Engineering International*, **4**, 151–158.

Sower, V. E., Savoie, M. J., and Renick, S. (1999). *An Introduction to Quality Management and Engineering*. Prentice Hall, Upper Saddle River, NJ.

Steiner, S. H., Hamada, M., White, B. J. G., Kutsyy, V., Mosesova, S., and Salloum, G. (2007). A bubble mixture experiment project for use in an advanced design of experiments class. *Journal of Statistical Education*, **15**, 1–18.

Street, D. J. and Burgess, L. (2007). *The Construction of Optimal Stated Choice Experiments*. John Wiley & Sons, Hoboken, N.J.

Taguchi, G. (1986). *Introduction to Quality Engineering, Designing Quality into Products and Processes*. Asian Productivity Organization, Tokyo.

Taguchi, G. (1987). *System of Experimental Design: Engineering Methods to Optimize Quality and Minimize Costs*. UNIPUB/Kraus International Publications, White Plains, New York.

Taguchi, G. and Phadke, M. S. (1985). Quality engineering through design optimization. *The International Quality Forum*, **11**, 27–48.

BIBLIOGRAPHY

Taguchi, G. and Wu, Y. (1980). *Introduction to Off-Line Quality Control.* Central Japan Quality Control Association, Nagoya Japan.

Taguchi, S. and Byrne, D. (1984). The concept of parameter design. *Auto Division Newsletter of ASQC*, pages 5–7.

Tukey, J. W. (1949a). Comparing individual means in the analysis of variance. *Biometrics*, **5**, 99–114.

Tukey, J. W. (1949b). One degree of freedom for non-additivity. *Biometrics*, **5**, 232–242.

unknown (1937). Rothamstead experimental station report.

Usage note: Stat-40 (1997). Usage note: Stat-40. University of Texas at Austin Statistical Services, 45-9372. Copywright 1995-1997 ACITS.

Varma, A. O. and Chilton, N. W. (1974). Crossover designs involving two treatments. *Journal of Periodontal Research*, **14**, 160–170.

Vining, G. G., Kowalski, S. M., and Montgomery, D. C. (2005). Response surface designs within a split-plot structure. *Journal of Quality Technology*, **37**, 115–129.

Wagner, J. G. (1967). Use of computers in pharmacokinetics. *Journal of Clinical Pharmacology and Therapeutics*, **8**, 201–220.

Wallenstein, S. and Fisher, A. C. (1977). The analysis of the two-period repeated measures crossover design with application to clinical trials. *Biometrics*, **33**, 261–269.

Wang, B., Wang, X., and Gong, L. (2009). The construction of a Williams design and randomization in cross-over clinical trials using SAS. *Journal of Statistical Software*, **29**(Code Snippet 1), 1–10.

Wang, J. C. and Wu, C. F. J. (1995). A hidden projection property of Plackett-Burman and related designs. *Statistica Sinica*, **5**, 235–250.

Wang, Q., Shi, G., and Chan-Halbrendt, C. (2004). Market potential for fine furniture manufactured from low-grade hardwood: Evidence from a conjoint analysis in the northeastern United States. *Forest Products Journal*, **54**, 19–25.

Wang, Y., Lin, D. K. J., and Fang, K. (1995). Designing outer array points. *Journal of Quality Technology*, **27**, 226–241.

Westlake, W. J. (1965). Composite designs based on irregular fractions of factorials. *Biometrics*, **21**, 324–336.

Wheeler, R. E. (1974). Portable power. *Technometrics*, **16**, 193–201.

Wilk, M. B., Gnanadesikan, R., and Huyett, M. J. (1962). Probability plots for the gamma distribution. *Technometrics*, **4**, 1–20.

Williams, E. J. (1949). Experimental designs balanced for the estimation of residual effects of treatments. *Australian Journal of Scientific Research*, **2**, 149–168.

Woodward, W. F. (1970). *A Comparison of Base Running Methods in Baseball*. Master's thesis, Florida State University.

Wu, C. F. J. and Hamada, M. (2000). *Experiments-Planning, Analysis, and Parameter Design Optimization*. John Wiley & Sons, New York.

Yang-Gyu, K. and Woolley, D. J. (2006). Effect of plant growth regulators and spear bud scales on growth of asparagus officinalis. *Scientia Horticulturae*, **108**, 238–242.

Yates, F. (1937). *The Design and Analysis of Factorial Experiments*. Hafner, Harpenden Herts, England. Bulletin 35, Imperial Bureau of Soil Science.

Index

ssE, 23, 24
ssT, 24
$ssTotal$, 24

effect sparsity, 241

A-Optimal, 231, 463
aberration criteria, 212, 213, 215, 247
adjustment factor, 513
alias structure, 202, 205, 238, 245
approximation formula for calculating number of runs for a two-level factorial, 93
asymmetric factorial, 283
asymptotic confidence interval on the variance components, 158

background variable, 3
balanced incomplete block design, 255, 256
bias, 4
BIB, 255, 256
block defining contrasts, 267, 274, 277
blocking, 115
Box-Cox Power transformations, 31
BTIB, 263
BTIB design, 263

canonical analysis, 413
cartesian product design, 324
CCBF, 268
CCBFF, 273
cell means model, 19
cell standard deviations, 34
coded factor levels, 90
completely confounded blocked factorial, 268
completely confounded blocked fractional factorial, 273
completely randomized design, 15

completely randomized factorial design, 58
completely randomized fractional factorial design, 204
confidence intervals on variance components, 154
confounded, 4, 200, 206, 209, 212, 213, 216, 222, 228, 234, 245
contour plot, 410
control factors, 502
CRD, 15
CRFD, 58
CRFF, 204
crossed factor, 170
CRRS, 385
CRSP, 303

D-Optimal, 230, 236, 257, 384, 395
D-optimal, 287, 465
define objectives, 8
dependent variable, 3
determine the number of replicates, 67
dispersion effect, 536
distribution-based transformations, 34
Dunnett's procedure, 47
duplicates, 4, 16, 23, 26

EBLUPs, 192
edge effects, 452
effect, 3
effect estimation, 556
effect heredity, 323, 517
effect heredity principle, 220, 271, 535
effect sparsity, 209, 222, 246
effect sparsity principle, 201, 267
effects model, 20
empirical best linear unbiased predictors, 192
EMS, 169
error control, 5, 15, 115

estimable function, 23
estimation equivalent split-plot response surface design, 427
evolutionary operation, 561
expected mean squares, 152, 164, 169, 182, 185
experiment, 3
experimental design, 4
experimental error, 4, 23
experimental unit, 3, 8, 15

factor, 3, 15
factorial random sampling experiment, 161
FFSP, 324
fixed factor, 148
fixed factors, 179
fractional factorial split-plot design, 324
FRSE, 161

G-Optimal, 384
Gage R&R study, 161
gage repeatability, 162
gage reproducibility, 162
gamma probability plots, 187
GCB, 128
general complete block design, 116
generalized complete block design, 128
generalized cyclic incomplete block designs, 263
generalized interactions, 267
generalized inverse, 61
generalized likelihood ratio test, 25
generalized linear model, 35
growth curve, 373

half-normal plot of effects, 99
heterogeneity of variance, 30
hidden replication, 54, 301
hierarchical ordering principle, 201, 203, 209, 212, 221, 222, 246, 267

incomplete block designs, 255
independent variable, 3, 8
interaction, 55
interaction graph, 65
interaction plot, 57, 65, 186

interactions, 53, 301
inverse terms in mixture model, 452

Latin square, 132
least squares estimator, 21, 181
least squares means, 73
LSD, 132
lurking variable, 3

mathematical model, 19
maximum likelihood estimators, 155
MCB procedure, 47
mechanistic modeling, 559
method of moments estimators, 151, 153
method of steepest ascent, 557
mixed level factorial, 283
mixed model, 183
mixture model in independent factors, 450
model for a completely randomized factorial, 60
model for an incomplete block design, 259
model for BIB, 259
model for CRD, 19
model for CRFD, 60
model for FRSE, 163
model for GCB, 128
model for LSD, 132
model for NSE, 171
model for PBIB, 259
model for RCB design, 119
model for RCBF, 126
model for RSE, 149

nested factor, 170
nested sampling design, 171
Newman-Keuls method, 45
noise factors, 502
non-centrality parameter, 38, 39, 62
noncentral F-distribution, 38
nonlinear programming, 416
normal probability plot of residuals, 29, 60
normal probability plot of the effects, 98
NSE, 171

INDEX

observational study, 1
observational unit, 3, 26
optimization, 557
order of estimability, 278
orthogonal polynomial, 42
orthogonality, 88, 90, 200, 232, 236

pairwise comparisons, 43
partial confounding, 290
partially balanced incomplete block design, 255, 261
PBIB, 255, 261
pilot tests, 8
planning experiments, 7
plot of the residuals versus the cell means, 27
practical difference in cell means, 39
practical significance, 38, 257
predicted cell means, 73
proc factex, 281
proc genmod, 35
proc glm estimate command, 24
proc glmpower, 40, 70
proc power, 40, 351
pseudo factors, 283

quadratic trend line, 122

R. A. Fisher, 2, 5, 15
random factor, 148, 179
random sampling experiment, 149
randomization, 15
randomized complete block design, 116
randomized complete block factorial, 125
RBSP, 312
RCB, 116
RCBF, 125
RE, 120
reference line, 99
regression equation, 90
relative efficiency, 120
REML, 155
replicate, 3
replication, 15
residual versus predicted plot, 60
resolution, 212, 213, 215, 222, 224, 232, 239, 247, 279

response, 3, 8, 15
response surface methods, 558
response trace, 458
ridge analysis, 414
rotatable design, 384
Rothamsted Experimental Station, 2
RSE, 149
run, 3

sample survey, 1
SAS ADX, 95, 278
SAS Power and Sample Size Application, 40, 351
saturated set of orthogonal contrasts, 43
Scheffé model for mixture experiments, 446
screening, 556
sequential experimentation, 555
simplex centroid, 450
simplex lattice, 447
size of a practical effect, 94
slack variable, 446
SNSE, 173
special cubic model, 449
split-plot confounding, 325
split-plot design with an RCB design in the whole plots, 312
split-plot design with CRD in whole plots, 302, 303
split-plot experiment, 301
staggered nested sampling experiment, 173
standard order, 87
sub-experiment, 283
sub-sample, 3, 16, 23, 26
sub-unit, 3
super efficient, 303, 509

three-way interaction, 83
transformation, 30
treatment design, 6, 53
treatment factor, 3
Tukey's HSD method, 43, 130, 294
Tukey's single degree of freedom test for interaction, 78
two block factors, 132
two-level factorials, 86
type I, 73, 167

type III, 73, 167

unequal number of replicates, 167
unequal number of replicates per cell, 72
uniform precision, 384
unreplicated designs, 95
unweighted mean squares, 168

variance components, 147

W. E. Deming, 1
weighted least squares, 35

Yates' algorithm, 87
Yates' order, 87
Youden square design, 265
YSD, 265